HEAVY METALS
EFFECTS, REMOVAL AND PROTECTION

Kruger Brentt
Publishers

HEAVY METALS
EFFECTS, REMOVAL AND PROTECTION

Jennifer Powers
Stephanie Kampf

Kruger Brentt
Publishers
2025

Kruger Brentt Publishers UK. LTD.
Company Number 9728962
Regd. Office: 68 St Margarets Road, Edgware, Middlesex HA8 9UU

ISBN: 978-1-78715-321-9

For information on all our publications visit our website at http://krugerbrentt.com/

PREFACE

Heavy metals are ubiquitous environmental pollutants that pose significant risks to human health and the ecosystem due to their persistence, bioaccumulation, and toxic effects. From industrial activities to agricultural runoff, heavy metals enter the environment through various pathways and can accumulate in soil, water, and air, leading to adverse impacts on human health, wildlife, and ecosystems. Understanding the effects of heavy metals, as well as developing effective strategies for their removal and protection, is essential for mitigating their adverse effects and safeguarding public health and the environment.

This handbook, titled "**Heavy Metals: Effects, Removal and Protection**" is a comprehensive exploration of the sources, effects, remediation techniques, and regulatory measures related to heavy metals pollution. Drawing upon the expertise of environmental scientists, engineers, policymakers, and health professionals, this volume offers theoretical insights, practical guidance, and case studies for understanding and addressing the challenges posed by heavy metals contamination.

From the toxicological effects of heavy metals on human health and ecosystems to the latest advancements in remediation technologies and regulatory frameworks, the "Heavy Metals: Effects, Removal, and Protection" covers a wide range of topics essential for mastering the field of heavy metals pollution management. Through clear explanations, scientific evidence, and practical examples, readers will gain a deep understanding of the mechanisms of heavy metal toxicity, as well as the strategies and technologies available for mitigating their impacts on the environment and human health.

Moreover, this handbook explores the emerging trends and best practices in heavy metals remediation and protection, including innovative treatment technologies, phytoremediation approaches, and risk assessment methodologies. By examining case studies and regulatory guidelines from around the world, we highlight the importance of interdisciplinary collaboration and policy interventions in addressing the complex challenges posed by heavy metals pollution.

As we confront the urgent need to protect our environment and public health from the impacts of heavy metals contamination, it is our hope that the "Heavy Metals: Effects, Removal and Protection" handbook will serve as a valuable resource for researchers, policymakers, practitioners, and students in the field of environmental science and engineering. Whether you are studying heavy metals pollution, designing remediation strategies, or advocating for regulatory measures, this handbook invites you to explore the multifaceted dimensions of heavy metals contamination and contribute to building a cleaner, safer, and more sustainable future.

Jennifer Powers
Stephanie Kampf

CONTENTS

1 Chapter

INTRODUCING HEAVY METALS

1. INTRODUCTION: AN OVERVIEW

Heavy metals are naturally occurring elements that are distinguished by their very large atomic mass and high density. While usually present in relatively small amounts, they are distributed throughout the Earth's crust. Typically, a minimum density of 5 g cm^{-3} is employed to classify a metal as heavy and distinguish it from other metals that are considered "light." Alternative, more inclusive definitions of "heavy metals" stipulate an atomic mass more than 23 or an atomic number beyond 20. However, these definitions are prone to significant errors and can cause confusion. Both alternative definitions encompass nonmetals as well. By applying the criterion of atomic mass, the upper limit for the number of elements categorized as "heavy metals" soars to 99 out of the total 118 fundamental components of our cosmos. When examining the periodic chart of elements, we discover that heavy metals, specifically according to their density, make up the majority of columns 3-16 in periods 4 to 6. These columns include the transition metals, post-transition metals, and lanthanides [1]. Copper, selenium, and zinc are crucial trace elements that play important roles in different biological processes and drive human metabolism as a whole [2]. Cobalt, a heavy metal, serves as the core atom in the vitamin B12 complex and plays a crucial role in the reductive branch of the propionic acid fermentation pathway [3]. Without this specific heavy metal combination, the distinct flavor of Emmentaler cheese would be absent. Several heavy metals, such as iron, zinc, tin, lead, copper, and tungsten, have significant technical importance. In recent times, many heavy metals have been utilized as the core atom in artificially engineered "bioinorganic" catalysts for certain chemical reactions [4]. Furthermore, inside this group, we encounter valuable and esteemed elements like as gold, silver, iridium, rhodium, and platinum [5]. However, certain elements such as mercury, cadmium, arsenic, chromium, thallium, lead, and others are commonly associated with the negative aspects of chemistry. These elements have harmful effects even at low concentrations [6]. Within this particular setting, certain heavy metals have acquired a questionable level of notoriety due to their utilization as primary components in the commission of significant criminal acts [7]. Heavy metals were physically delivered from the heavens by asteroid collisions. Heavy metals are generally present in the earth's crust in relatively low quantities, ranging from low parts per billion (ppb) levels for noble metals to as high as 5% for iron. These heavy metals are primarily found in chemical compounds such as carbonates, sulfates, oxides, or silicates, although they may exist in their metallic form. The process of weathering and erosion caused the leaching and mitigation of substances into the soil, rivers, and groundwater. Approximately 4-5 billion years ago, during a time when Earth's mantle was in a liquid state, dense metallic elements descended into the heart of the planet and coalesced to create Earth's core. This core, which is primarily composed of iron and nickel, exists in its current form today [1].

2. THE IMPACT OF HEAVY METALS ON THE ECOLOGICAL AND BIOSPHERIC SYSTEMS

It is crucial to highlight that there exist some trace elements within the heavy metal group that play a vital role in numerous biological processes. These elements are primarily located in period 4 of the periodic table of elements. As strictly aerobic organisms, our survival would be impossible without cytochromes, which have allowed aerobic living forms to breathe from their inception [8]. Iron is a crucial component of our respiratory system, serving as the core, oxygen-binding atom in the blood pigment heme. Copper has a comparable function in the transportation of electrons and oxygen, particularly as the core atom in hemocyanin found in mollusks and arthropods [9]. Zinc is essential as a component of zinc finger enzymes [10]. Selenium is classified as an antioxidant and plays a role in hormone production [11]. Cobalt has been identified as playing a crucial role in the production of complex molecules and several stages of cellular metabolism. It serves as a pivotal atom in vitamin B12, which is essential for cell division, blood formation, the nervous system, and the manufacture of propionic acid (as mentioned before) [12]. In addition, vanadium and manganese have a crucial role in the regulation and functioning of several enzymes [13]. Furthermore, chromium, arsenic, and nickel, which are commonly considered harmful, also have metabolic activities. Arsenic, a recently discovered element, has been found to naturally occur in herring caviar. It has been observed to replace phosphorus in a type of lipid termed phosphatidylcholine-like lipids, which are known as arsenolipids [14]. Additionally, it is important to acknowledge the involvement of molybdenum in some redox reactions [15], as well as the importance of cadmium in the metabolic processes of specific microalgae belonging to the Diatomophyceae class [16]. Furthermore, the scientific confirmation of the role of tungsten, a heavy metal, in the metabolism of prokaryotes is established [17]. In order to convey the number of heavy metals in our bodies, which we refer to as the "heavy metal load," we can estimate that only approximately 0.01% of our body mass comes from heavy metals. The top four heavy metals in terms of quantity are iron (approximately 5 g in a person weighing 70 kg), zinc (2 g), lead, and copper (0.1-0.2 g each). The remaining heavy metals can be considered insignificant in terms of mass. [18] In addition to the previously described processes of leaching and the reduction of heavy metals through erosion and weathering, these elements are primarily mobilized by human activities such as physical extraction and smelting, as well as chemical release from ores and subsequent processing for various purposes. Additional mechanisms that contribute to the release of heavy metals into the environment include their usage in agriculture, industry, households, automobiles, medicine, electronics, and other technologies. This leads to the widespread presence of heavy metals in both water and land ecosystems. Presently, there is a growing concern worldwide about the potential detrimental health effects and long-lasting harmful affects on ecosystems. Several heavy metals, such as hexavalent chromium, arsenic, cobalt, nickel, antimony, vanadium, and mercury, have been reported or are at least thought to have carcinogenic properties. Arsenic and vanadium are also believed to be mutagenic, while arsenic is known to be teratogenic. Nickel is known to be allergic, and silver, copper, zinc, and selenium are considered to be endocrine-disrupting. Some substances, such as thallium, can cause changes in the brain and behavior, especially in children. They can also damage the central nervous system (mercury, lead, thallium, manganese, and tin), bone marrow, and cause osteoporosis (cadmium). These substances can be toxic to the liver and kidneys (cadmium, mercury, heptavalent manganese), disrupt heart rhythm (thallium), or have a negative impact on the immune system (lead) [19].

3. ANALYZING THE PRESENCE OF HEAVY METALS

Traditionally, the measurement of heavy metals is done using established methods like wet chemical techniques (gravimetric, titrimetric, colorimetric, etc.), coupled plasma/atomic emission spectrometry

(ICP/AES), inductively coupled plasma with mass spectrometric detection (ICP/MS), or atomic absorption spectroscopy (AAS) [20, 21, 22]. In addition, a wide range of ion selective electrodes are commonly used for the purpose of determining heavy metals [22, 23]. Currently, there is ongoing development of optical [24, 25, 26], chemical [27, 28], and biological [29, 30, 31, 32, 33] sensory systems that are innovative, robust, sensitive, selective, affordable, and rapid. The progress in analytical chemistry is currently closely linked to nanotechnology [26, 28, 34]. In addition, researchers have developed lab-on-paper sensors for the detection of heavy metals, such as mercury, silver, copper, cadmium, lead, chromium, and nickel. This was proved in a study [29] where the quantification of these metals was successfully achieved. This sensor, which utilizes an immobilized enzyme, is an exemplar of biosensors, which integrate the scientific disciplines of biotechnology and microelectronics. These "biosensors" comprise of a fixed biological component combined with a transducer [30]. Genetically encoded fluorescence sensors, a new technology, can be utilized to monitor heavy metals within biological cells. These sensors have already been evaluated for detecting heavy metals such as zinc, copper, lead, cadmium, mercury, and arsenic [35].

4. REDUCING THE PRESENCE OF HEAVY METALS

In response to mounting public demand to mitigate existing environmental dangers and provide a sustainable future for future generations, remediation procedures for heavy metals are being continuously enhanced to minimize their adverse effects. Conventional physical, thermal, chelating, and other chemical methods sometimes suffer from significant drawbacks, including expensive costs, considerable labor requirements, and disruptive alterations to soil characteristics and microbiota [36]. Historically, the conventional method for addressing soil contamination caused by heavy metals involves excavating the contaminated soil and then disposing of it in landfills. Indeed, this kind of disposal only delays the environmental issue by transferring it from one place to another. Additionally, it creates risks associated with transporting unstable soil and the release of toxic metals at the final disposal site. Alkaline lime precipitation is well recognized as a superior and very effective traditional method for treating extensively polluted effluents contaminated with heavy metals. Lime precipitation is a highly successful method for treating wastewater that contains metal concentrations more than 1 g L^{-1}. Nevertheless, the leftover heavy metal-alkali-sludge stills require final disposal [37]. Contemporary methods for addressing heavy metal pollution include utilizing new adsorbents like nano-carriers, employing ion exchange techniques, employing enhanced membrane filtering techniques for removal, using electrodialysis, or employing photocatalysis. Out of all the new physicochemical techniques, the methods that involve adsorption and membrane filtration are the most extensively studied and routinely used to treat contaminated wastewater [37, 38]. Recent research has investigated the use of inorganic substances such as kaolinite and montmorillonite, as well as organic materials like agricultural waste or bio-char, as absorbents for the recovery of heavy metals. Within this particular framework, the utilization of carbon-, metal-, or metal oxide-based nanoparticles as adsorbents is advantageous due to their large surface areas that are highly receptive to metal adsorption and their rapid reactivity. In order to optimize the recovery procedures, it is necessary to understand the mechanisms by which nanomaterials interact with heavy metals and with additional wastewater constituents that have metal-binding groups [44]. Photocatalysis harnesses photons from the UV-near vis region of the electromagnetic spectrum of light. When utilized intelligently, it can simultaneously break down harmful organic contaminants and recover metals in a single step. This method utilizes photocatalytic semiconductors, such as TiO_2, which, when exposed to UV light, produce extremely reactive electrons that subsequently lower the concentration of heavy metal ions in polluted wastewater. Photocatalysis was effectively used to reduce the very dangerous hexavalent chromium to its far less toxic trivalent form, resulting in a reduction of toxicity by around 500 times. For precious noble metals such as gold, this technique not only reduces environmental pollution but also extracts valuable minerals for future use (referred to as "photorecovery") [45]. Phytoremediation is a developing technology that addresses the limitations of the approaches mentioned above. Phytoremediation involves the collaborative action of plants and various soil microbes. These microbes transform heavy metals into a form that may be easily absorbed by plants, ultimately reducing the levels of toxins in impacted areas. Phytoremediation relies on the capacity of certain plants to effectively absorb, transport, and store harmful substances that have chemical characteristics similar to those necessary for plant growth [46]. Phytoremediation, a relatively new technique, is known for its efficiency, cost-effectiveness, and ecological friendliness. It relies solely on sunlight as its energy source and has gained widespread public support [47]. Currently, there is ongoing evaluation of novel robust plants known as "heavy metal hyperaccumulators" in the field of heavy metals. These plants are being studied for their ability to remove undesired heavy metals through phytoremediation, as well as their potential to accumulate valuable heavy metals for future utilization through phytomining. Hyperaccumulators are plants that have the ability to absorb harmful metal ions at concentrations of thousands of parts per million (ppm). In the ideal situation, the toxic metals are carried from the plant's rhizosphere to the shoots in the plant's outer region. These shoots, which now contain high levels of the desired pollutants, can be easily collected and burned to produce energy. If it is economically viable, the metal can be recycled from the remaining ash [48]. Phytoremediation can be employed to treat contaminated water bodies by utilizing aquatic plants. Diverse plant species have effectively absorbed heavy metals, including arsenic, cadmium, chromium, lead, and even radionuclides, from soil that is contaminated. Phytoextraction is a type of phytoremediation that can be used to reduce heavy metal contamination in soil. It takes advantage of plants' ability to absorb and store these metals, which are important for their growth, such as iron, copper, manganese, molybdenum, and nickel. Furthermore, certain heavy metals like cadmium, chromium, silver, lead, cobalt, selenium, or mercury, which have comparable chemical properties, can also be extracted from plants, although their biological function is still unknown or not yet verified [49]. The fern species *Pteris vittata L.* shown outstanding efficacy in phytoremediation of arsenic, as evidenced by the reported data. The plant exhibited a remarkable absorption capability, accumulating more than 4 grams of heavy metal per kilogram of plant material. Lead can be accumulated by various plants, particularly different species of mustards (*Brassica* ssp.), with levels ranging from 50 to 100 mg per gram of plant dry mass [46].

REFERENCES

1. Duffus JH. Heavy metals' – A meaningless term? International Union of Pure and Applied Chemistry (IUPAC). 2002;74:793-807. DOI: 10.1351/pac2002740507
2. Mertz W. The essential trace elements. Science. 1981;213(4514):1332-1338
3. 3.Stowers CC, Cox BM, Rodriguez BA. Development of an industrializable fermentation process for propionic acid production. Journal of Industrial Microbiology & Biotechnology. 2014;41(5):837-852
4. Terfassa B, Schachner JA, Traar P, Belaj F, Zanetti NCM. Oxorhenium(V) complexes with naphtholate-oxazoline ligands in the catalytic epoxidation of olefins. Polyhedron. 2014;75:141-145
5. Rao CRM, Reddi GS. Platinum group metals (PGM); occurrence, use and recent trends in their determination. TrAC Trends in Analytical Chemistry. 2000;19(9):565-586
6. Duruibe JO, Ogwuegbu MOC, Egwurugwu JN. Heavy metal pollution and human biotoxic effects. International Journal of physical sciences. 2007;2(5):112-118

7. Kaiser J. Toxicologists shed new light on old poisons. Science. 1998;279(5358):1850-1851
8. Groves JT, Haushalter RC, Nakamura M, Nemo TE, Evans BJ. High-valent iron-porphyrin complexes related to peroxidase and cytochrome P-450. Journal of the American Chemical Society. 1981;103(10):2884-2886
9. Redfield AC, Coolidge T, Shotts MA. The respiratory proteins of the blood I. The copper content and the minimal molecular weight of the Hemocyanin of Limulus polyphemus. Journal of Biological Chemistry. 1928;76(1):185-195
10. Maret W. Zinc biochemistry: From a single zinc enzyme to a key element of life. Advances in Nutrition. 2013;4(1):82-91
11. Stadtman TC. Selenium biochemistry. Annual Review of Biochemistry. 1990;59(1):111-127
12. Hudson TS, Subramanian S, Allen RJ. Determination of pantothenic acid, biotin, and vitamin B12 in nutritional products. Journal-Association of Official Analytical Chemists. 1984;67(5):994-998
13. Egami F. Origin and early evolution of transition element enzymes. The Journal of Biochemistry. 1975;77(6):1165-1169
14. Viczek SA, Jensen KB, Francesconi KA. Arsenic-containing phosphatidylcholines: A new group of arsenolipids discovered in herring caviar. Angewandte Chemie International Edition. 2016;55(17):5259-5262
15. Mendel RR. Molybdenum: Biological activity and metabolism. Dalton Transactions. 2005;21:3404-3409
16. Lane TW, Saito MA, George GN, Pickering IJ, Prince RC, Morel FM. Biochemistry: A cadmium enzyme from a marine diatom. Nature. 2005;435(7038):42
17. Andreesen JR, Makdessi K. Tungsten, the surprisingly positively acting heavy metal element for prokaryotes. Annals of the New York Academy of Sciences. 2008;1125(1):215-229
18. Iyengar GV. Reevaluation of the trace element content in reference man. Radiation Physics and Chemistry. 1998;51(4-6):545-560
19. Tchounwou PB, Yedjou CG, Patlolla AK, Sutton DJ. Heavy metal toxicity and the environment. In: Molecular, Clinical and Environmental Toxicology. Basel: Springer; 2012. pp. 133-164
20. Butcher DJ. Advances in inductively coupled plasma optical emission spectrometry for environmental analysis. Instrumentation Science and Technology. 2010;38(6):458-469
21. Feldmann J, Salaün P, Lombi E. Critical review perspective: Elemental speciation analysis methods in environmental chemistry Moving towards methodological integration. Environmental Chemistry. 2009;6(4):275-289
22. Saar RA, Weber JH. Comparison of spectrofluorometry and ion-selective electrode potentiometry for determination of complexes between fulvic acid and heavy-metal ions. Analytical Chemistry. 1980;52(13):2095-2100
23. Wang J. Stripping analysis at bismuth electrodes: A review. Electroanalysis. 2005;17(15-16):1341-1346
24. Oehme I, Wolfbeis OS. Optical sensors for determination of heavy metal ions. Microchimica Acta. 1997;126(3-4):177-192

25. Rasheed T, Bilal M, Nabeel F, Iqbal HM, Li C, Zhou Y. Fluorescent sensor based models for the detection of environmentally-related toxic heavy metals. Science of the Total Environment. 2018;615:476-48577

26. Ullah N, Mansha M, Khan I, Qurashi A. Nanomaterial-based optical chemical sensors for the detection of heavy metals in water: Recent advances and challenges. TrAC Trends in Analytical Chemistry. 2018;100:155-166

27. Vlasov Y, Legin A, Rudnitskaya A. Cross-sensitivity evaluation of chemical sensors for electronic tongue: Determination of heavy metal ions. Sensors and Actuators B: Chemical. 1997;44(1-3):532-537

28. Salimi F, Kiani M, Karami C, Taher MA. Colorimetric sensor of detection of Cr(III) and Fe(II) ions in aqueous solutions using gold nanoparticles modified with methylene blue. Optik-International Journal for Light and Electron Optics. 2018;158:813-825

29. Hossain SZ, Brennan JD. β-Galactosidase-based colorimetric paper sensor for determination of heavy metals. Analytical Chemistry. 2011;83(22):8772-8778

30. Verma N, Singh M. Biosensors for heavy metals. Biometals. 2005;18(2):121-129

31. Ghica ME, Carvalho RC, Amine A, Brett CM. Glucose oxidase enzyme inhibition sensors for heavy metals at carbon film electrodes modified with cobalt or copper hexacyanoferrate. Sensors and Actuators B: Chemical. 2013;178:270-278

32. Ghica ME, Brett CM. Glucose oxidase inhibition in poly(neutral red) mediated enzyme biosensors for heavy metal determination. Microchimica Acta. 2008;163(3-4):185-193

33. Zhylyak GA, Dzyadevich SV, Korpan YI, Soldatkin AP, El'Skaya AV. Application of urease conductometric biosensor for heavy-metal ion determination. Sensors and Actuators B: Chemical. 1995;24(1-3):145-148

34. Aragay G, Merkoçi A. Nanomaterials application in electrochemical detection of heavy metals. Electrochimica Acta. 2012;84:49-61

35. Finney LA, O'halloran TV. Transition metal speciation in the cell: Insights from the chemistry of metal ion receptors. Science. 2003;300(5621):931-936

36. Fu F, Wang Q. Removal of heavy metal ions from wastewaters: A review. Journal of Environmental Management. 2011;92(3):407-418

37. Barakat MA. New trends in removing heavy metals from industrial wastewater. Arabian Journal of Chemistry. 2011;4(4):361-377

38. Premkumar MP, Thiruvengadaravi KV, Kumar PS, Nandagopal J, Sivanesan S. Eco-friendly treatment strategies for wastewater containing dyes and heavy metals. In: Environmental Contaminants. Singapore: Springer; 2018. pp. 317-360

39. Bhattacharyya KG, Gupta SS. Adsorption of a few heavy metals on natural and modified kaolinite and montmorillonite: A review. Advances in Colloid and Interface Science. 2008;140:114-131

40. Sud D, Mahajan G, Kaur MP. Agricultural waste material as potential adsorbent for sequestering heavy metal ions from aqueous solutions – A review. Bioresource Technology. 2008;99:6017-6027

41. Cimino G, Passerini A, Toscano G. Removal of toxic cations and Cr(VI) from aqueous solution by hazelnut shell. Water Research. 2000;34(11):2955-2962

42. Inyang MI, Gao B, Yao Y, Xue Y, Zimmerman A, Mosa A, Pullammanappallil P, Ok YS, Cao X. A review of biochar as a low-cost adsorbent for aqueous heavy metal removal. Critical Reviews in Environmental Science and Technology. 2016;46(4):406-433

43. Poo KM, Son EB, Chang JS, Ren X, Choi YJ, Chae KJ. Biochars derived from wasted marine macro-algae (Saccharina japonica and Sargassum fusiforme) and their potential for heavy metal removal in aqueous solution. Journal of Environmental Management. 2018;206:364-372

44. Tang WW, Zeng GM, Gong JL, Liang J, Xu P, Zhang C, Huang BB. Impact of humic/fulvic acid on the removal of heavy metals from aqueous solutions using nanomaterials: A review. Science of the Total Environment. 2014;468:1014-1027

45. Ku Y, Jung IL. Photocatalytic reduction of Cr(VI) in aqueous solutions by UV irradiation with the presence of titanium dioxide. Water Research. 2001;35(1):135-142

46. Ali H, Khan E, Sajad MA. Phytoremediation of heavy metals – Concepts and applications. Chemosphere. 2013;91(7):869-881

47. Ghosh M, Singh SP. A review on phytoremediation of heavy metals and utilization of it's by products. Asian Journal on Energy and Environment. 2005;6(4):18

48. Tangahu BV, Sheikh Abdullah SR, Basri H, Idris M, Anuar N, Mukhlisin M. A review on heavy metals (As, Pb, and Hg) uptake by plants through phytoremediation. International Journal of Chemical Engineering. 2011;2011:31. Article ID 939161

49. Sarwar N, Imran M, Shaheen MR, Ishaque W, Asif M, Kamran MA, Matloob A, Rehim A, Hussain S. Phytoremediation strategies for soils contaminated with heavy metals: Modifications and future perspectives. Chemosphere. 2017;171:710-721

50. Francesconi K, Pornsawan V, Sridokchan W, Goessler W. Arsenic species in an arsenic hyperaccumulating fern, Pityrogramma calomelanos: A potential phytoremediator of arsenic-contaminated soils. Science of the Total Environment. 2002;284:27-35

51. Fayiga AO, Ma LQ, Cao X, Rathinasabapathi B. Effects of heavy metals on growth and arsenic accumulation in the arsenic hyperaccumulator Pteris vittata L. Environmental Pollution. 2004;132(2):289-296

Chapter 2 SYNTHESIS AND CHARACTERIZATION OF CADMIUM IODATE

1. INTRODUCTION: AN OVERVIEW

Cadmium is an elemental substance denoted by the symbol Cd and atomic number 48. This metal, which has a soft and bluish-white appearance, shares similar chemical properties with two other stable metals in group 12, namely zinc and mercury. Similar to zinc, it exhibits an oxidation state of +2 in the majority of its compounds. Additionally, like mercury, it possesses a lower melting point compared to the transition metals found in groups 3 through 11. Cadmium and other elements in group 12 are typically not classified as transition metals because they lack partially filled d or f electron shells in their elemental or common oxidation states. The mean concentration of cadmium in the Earth's crust ranges from 0.1 to 0.5 parts per million (ppm). In 1817, Stromeyer and Hermann, who were both in Germany, independently found it as a contaminant in zinc carbonate.Although cadmium is a perilous heavy metal that is harmful to nearly all living organisms, it is utilized in batteries, electroplating, nuclear reactors, and semiconductors. Rechargeable nickel-cadmium batteries accounted for the majority of cadmium usage in 2009, with a proportion of 86%. The nominal cell potential of nickel-cadmium cells is 1.2 V. The cell has a positively charged nickel hydroxide electrode and a negatively charged cadmium electrode plate, which are separated by an alkaline electrolyte (potassium hydroxide) [1]. Cadmium electroplating, which accounts for 6% of global output, is employed in the aircraft industry to mitigate the corrosion of steel components [2]. It is also utilized in the control rods of nuclear reactors, where it acts as a highly efficient "neutron poison" to regulate the neutron flux in nuclear fission [2]. Cadmium rods, when placed in the core of a nuclear reactor, serve to absorb neutrons and prevent them from causing further fission reactions. This effectively regulates the level of reactivity in the reactor. The pressurized water reactor developed by Westinghouse Electric Company utilizes an alloy composed of 80% silver, 15% indium, and 5% cadmium [2]. Cadmium oxide was employed in the phosphors of black and white television screens, as well as in the blue and green phosphors of color television cathode ray tubes [3]. Cadmium sulfide (CdS) is utilized as a photoconductive layer on the surface of photocopier drums [4]. Cadmium salts of various types are utilized in the production of paint pigments, with cadmium sulfide (CdS) being the most prevalent choice for a yellow pigment. Cadmium selenide, sometimes known as cadmium red, is a red pigment. Cadmium is highly valued by painters for its ability to produce vibrant and long-lasting yellows, oranges, and reds. However, before being mixed with oils, binders, or other paint formulations, these colors are intentionally reduced in intensity during production. To mitigate potential toxicity, it is advisable for users to apply a barrier cream on their hands to limit the absorption of these pigments through the skin [5]. However, it is worth noting that the reported amount of cadmium taken into the body through the skin is less than 1% [6]. Cadmium was employed as stabilizers for heat, light, and weathering in PVC [2, 7]. Presently, cadmium stabilizers have been entirely substituted with barium-zinc, calcium-zinc, and organo-tin stabilizers. Cadmium is utilized in several types of solder and bearing alloys due to its low coefficient of friction and resistance to fatigue [2]. It can also be found in certain alloys with low melting points, such as Wood's metal [8]. Second Harmonic Generation (SHG) is a nonlinear optical phenomenon when two photons of identical frequency interact with a nonlinear material and effectively merge to produce new photons with double the energy, frequency, and half the wavelength of the original photons. Second harmonic generation, being a nonlinear optical process of even order, can only occur in mediums that lack inversion symmetry [9]. It is an instance that falls within the category of sum frequency generation. Half-harmonic generation is a phenomenon that occurs as a reverse process of spontaneous parametric down-conversion. In this process, one photon produces a pair of photons with half the energy. It is similar to second harmonic production but happens with a lower probability. [10] The second-order nonlinear susceptibility of a medium generates second harmonic generation (SHG), which can convert either a little or a significant portion of the excitation wave, depending on certain interference conditions occurring inside the medium. Frequency doubling, also known as frequency multiplication, is a technique used in radio transmission. It was invented in the early 20th century and is commonly employed with frequencies in the megahertz range. It is an instance that falls within the category of frequency multiplication. Metal iodates play a significant role in the study of nonlinear optical (NLO) materials. The IO3 unit is a preferred anionic group for non-linear optical (NLO) activity due to the existence of stereochemically active lone-pair electrons on IV ions. This group exhibits a significant microscopic second-order NLO susceptibility. The arrangement of IO_3 units in crystal structures that are either polar or non-centrosymmetric (NCS) can lead to the formation of materials that exhibit exceptional second harmonic generation (SHG) capabilities [11]. Recently, several efficient methods have been devised for creating new non-linear optical (NLO) iodates. These methods involve including a combination of chloride (Cl^-), fluoride (F^-), and hydroxide (OH^-) ligands. These research endeavors resulted in the discovery of numerous metal iodates that possess exceptional second harmonic generation (SHG) capabilities, such as $Bi(IO_3)F_2$ [11] and $Zn(IO_3)OH$ [12]. Cadmium iodate ($Cd(IO_3)_2$) is a compound formed by the combination of cadmium with the iodate ligand. Its crystal structure is noteworthy. In 1838, Rammelsberg produced anhydrous cadmium iodate by vigorously combining concentrated solutions of sodium iodate and cadmium nitrate through heating [13]. In 1940, Oelke and Wagner conducted a study on the solubility of this chemical in water [14] and its thermal behavior [15]. The first diffraction pattern was then shown in 1968 [16]. The structure of anhydrous cadmium iodate was initially reported in 1978 as a sole orthorhombic phase P212121 [17, 18], which was later referred to as δ-$Cd(IO_3)_2$. However, in 1980, Liang et al. [19] provided powder diffraction findings that do not align with this particular chemical [20]. In 2005, research discovered a variation in the structure of anhydrous $Cd(IO_3)_2$ iodate. The $CdCl_2$-IO_3 system has demonstrated the presence of at least five different polymorphic phases for cadmium, namely α-$Cd(IO_3)_2$, α-$Cd(IO_3)_2$, α-$Cd(IO_3)_2$, α-$Cd(IO_3)_2$, and α-$Cd(IO_3)_2$. These phases were obtained under specific synthesis conditions [13, 21, 22]. Additionally, in 2017, Hebboul discovered a new polymorph called α-$Cd(IO_3)_2$ [23]. The chemical system under consideration is highly intricate and exhibits a significant degree of polymorphism, which poses challenges in developing optical crystals. The complexity of the system has made it difficult to obtain large crystals. Specifically, the polymorphic phases α-$Cd(IO_3)_2$, α-$Cd(IO_3)_2$, and α-$Cd(IO_3)_2$ are all non-centrosymmetric. It is worth noting that α-$Cd(IO_3)_2$ does exhibit a signal of second harmonic generation (SHG), although it is lower compared to that of LiIO3. Ravi Kumar et al. adjusted the growth conditions of the α-Cd $(IO_3)_2$ crystal using the slow-cooling approach in an aqueous solution. The structure was verified using single-crystal X-ray diffraction (XRD) research. The compound $Cd(IO_3)_2$ exhibits transparency in the wavelength range of 350 to 2000 nm, indicating that it can be used for many applications in the visible and near-infrared (NIR) regions. The optical band gap is determined to be 3.85 eV based on the analysis of Tauc's Fig.. As an inorganic substance, the crystal that has been created exhibits a modest efficiency in second harmonic generation (SHG), along with good resistance to changes in temperature and strong mechanical properties. $Cd(IO_3)_2$ shows potential for use in applications involving nonlinear optics (NLO) and the production of photonic devices. In 2008, the first-ever report on the optical band gap, mechanical, dielectric, and electrical characteristics of $Cd(IO_3)_2$ was published [22]. This chapter will examine the literature on synthetic techniques and powder X-ray diffraction patterns for several polymorphs of $Cd(IO_3)_2$ and the monohydrate form of cadmium iodate. Furthermore, the bibliography focuses on two cadmium mixed ligand compounds: cadmium chloro-iodate ($CdIO_3Cl$) and cadmium hydroxy-iodate ($CdIO_3OH$).

2. SYNTHESIS METHODS AND POWDER SPECTRA FOR VARIOUS $CD(IO_3)_2$ POLYMORPHS AND MONOHYDRATE CADMIUM IODATE

At 20°C and constant concentration of cadmium, it is possible to precipitate in sequence; this fact is used to assign a name—the first polymorph that precipitates is called α-$Cd(IO_3)_2$, the following α-$Cd(IO_3)_2$, then α-$Cd(IO_3)_2$, and finally α-$Cd(IO_3)_2$—heating these four polymorphs or working in nitric acid at 30°C can lead to α-$Cd(IO_3)_2$ [20]. The kinetic study of the $CdCl_2$-HIO_3 [13] system in aqueous solution is extremely complicated and very sensitive to the procedure. A thorough study of this system by systematically varying the operating conditions was carried out. Table 1 summarizes the different results obtained according to the four studied items:

- Variable cadmium concentration at fixed temperature and time
- Variable reaction time at fixed cadmium concentration and at T=20°C
- Variable reaction time with fixed cadmium concentration and at T= 60°C
- The synthesis in a concentrated nitric acid medium

Table 1. Summary of various results obtained according to synthesis conditions [20].

(1) *Variable cadmium concentration at fixed temperatures and time* **(*T* = 20°C, *t* = 4 hours)**				
Concentration	C_1 = 94.510^{-3} mol/l	C_2 = 56.810^{-3} mol/l	C_3 = 35.510^{-3} mol/l	C_4 ≤ 25.810^{-3} mol/l
Polymorphs present in the precipitate	α~β~γ	β~δ > α~γ	γ > β ~α	γ > > α
(2) *Variable reaction time with fixed cadmium concentration* **(*T* = 20°C, *CCd* = 94.510^{-3} mol/l)**				
Reaction time	t = 10 min	t = 1 h	t = 4 h	t = 7 h
Polymorphs present in the precipitate	A	α > > β	α~β~γ	α ≥ γ > > δ
(3) *Variable reaction time with fixed cadmium concentration* **(*T* = 60°C, *CCd* = 94.510^{-3} mol/l)**				
Reaction time	t = 10 min	t = 1 h	t = 4 h	t = 7 h
Polymorphs present in the precipitate	α ≥ γ	γ > α ≥ δ	γ > > δ> > α	γ > > δ> > α

From acid solutions Abrahams and Nassau have shown that the transition metal iodate d or f could be obtained in the anhydrous state from nitric acid solutions [26, 27, 28, 29, 30, 31, 32]. Nitric acid is the medium used by Bach for the polymorph crystallogenesis of the δ-$Cd(IO_3)_2$ [18].

The fifth polymorph ε can be prepared under the same previous conditions but in the 30% nitric acid medium [13, 20].

The comparison of all results shown in the literature shows that the XRD diagram of the polymorph ε has been already published in 1980 without being identified [19].

2.1. Preparations and characterizations of pure phases

2.1.1. The polymorph α-$Cd(IO_3)_2$

This product precipitates in the form of a very fine white powder from a solution of cadmium chloride and the highly concentrated iodic acid, C1 = 94.510–3 mol/l; all very slow evaporation efforts in order to obtain single crystals for structural study proved vain. The powder diagram of this variety is shown in Fig. 1 indicating that the spectrum of polymorph α is characterized by the main peak d = 3.5784 Å [13].

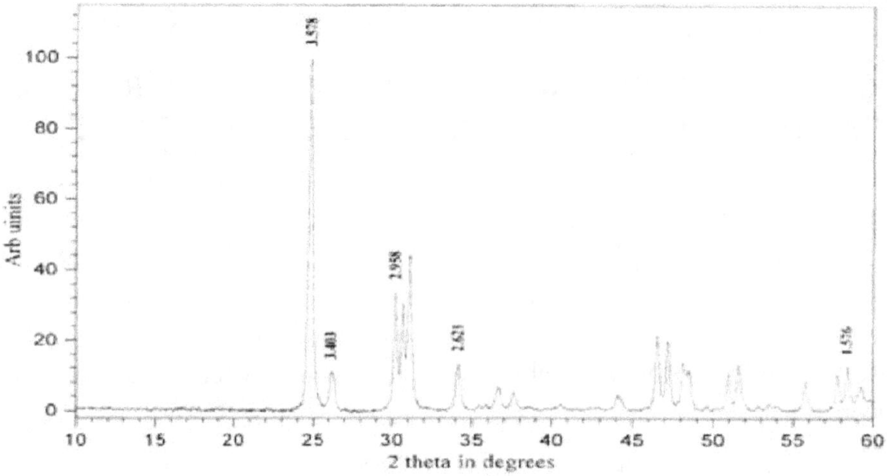

Fig. 1. Diffraction diagram of α-$Cd(IO_3)_2$ [20].

2.1.2. The polymorph β-$Cd(IO_3)_2$

The pure form of polymorph β can be created by heating the product resulting from the reaction of 0.26 g of cadmium chloride with 0.5 g of lithium iodate in 70 ml of water at a temperature of 250°C for a few days. Fig. 2 displays the powder diagram of this variety, revealing that the spectrum of the polymorphic β is distinguished by a prominent peak at d = 3.296 Å [13].

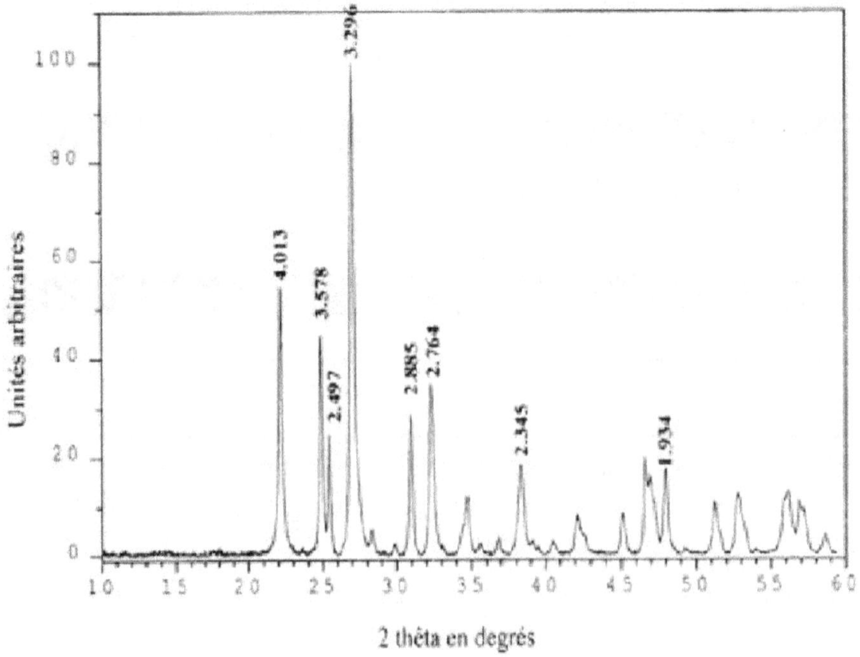

Fig. 2. Diffraction diagram of β-$Cd(IO_3)_2$ [20].

2.1.3. The polymorph γ-$Cd(IO_3)_2$

This polymorph can be obtained in its pure form as a well-crystallized powder through two distinct methods. The first method involves evaporating a diluted aqueous solution of cadmium chloride and iodic acid (with a concentration of CCd < 25 10-3 mol/l) at a temperature of 60°C. The second method involves evaporating a diluted aqueous solution of cadmium chloride with 0.26 g of potassium iodate and 0.607 g of potassium iodate in 100 ml of water at a temperature of 70°C. The attempts to form crystals of this chemical have been unsuccessful. It is important to mention that this compound maintains stability up to a temperature of 380°C [13, 20, 25]. Additionally, it crystallizes in a non-centro-symmetric space group, as confirmed by the positive optical activity test in second harmonic generation (Kurtz and Perry test [33]), which is visually less pronounced compared to $LiIO_3$ [13, 20, 24, 25]. The X-ray diffraction pattern depicted in Fig. 3 exhibits a flawlessly crystalline product that is absent from the JCPDS database. It is important to mention that within the 2θ = 10°–60° range of its powder spectrum, there are 44 peaks with modest intensity in comparison to the primary peak with a value of d = 3.22 Å.

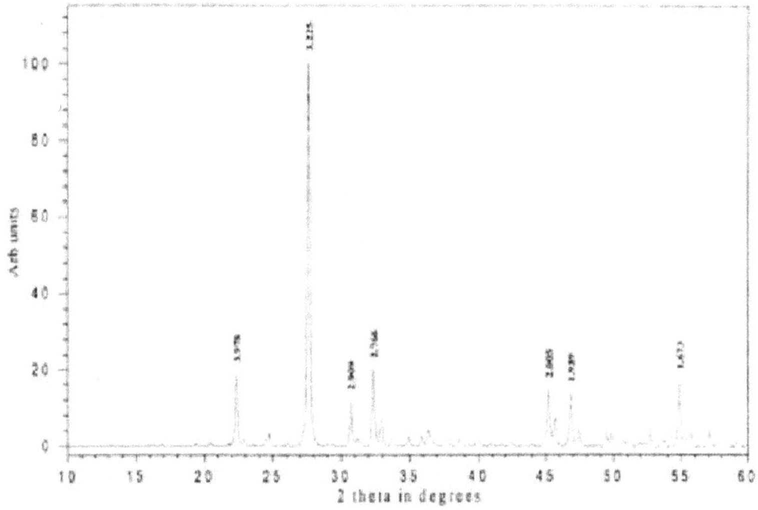

Fig. 3. Diffraction diagram of γ-$Cd(IO_3)_2$ [20].

2.1.4. The polymorph δ-$Cd(IO_3)_2$

By dissolving 0.2 grams of γ-$Cd(IO_3)_2$ in 30 milliliters of 60% HNO_3 and allowing the solution to slowly evaporate at 70°C, prismatic transparent crystals with a slight yellow color are formed [13, 20]. A single crystal measuring 11 × 10 × 2 mm^3, as seen in Fig. 4, was acquired [22] from a concentrated solution of $Cd(IO_3)_2$ white powder in a 60% nitric acid medium (HNO_3) at a temperature (T) of 60°C. The entire process of crystal formation took approximately 20-25 days. The white powder utilized is the resultant substance from the chemical reaction between cadmium chloride and iodic acid (1/2) in an aqueous solution.

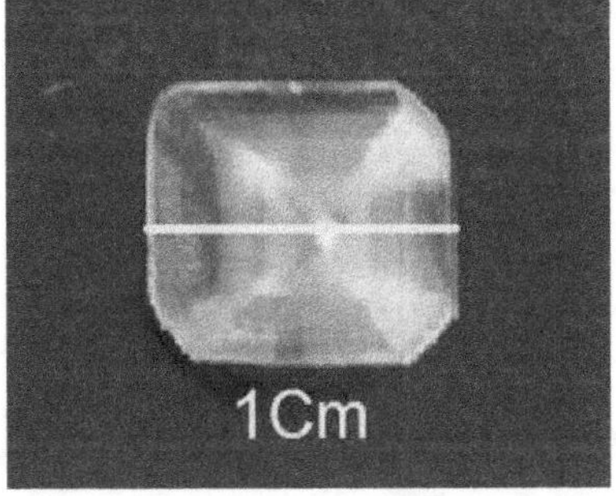

Fig. 4. δ-$Cd(IO_3)_2$ morphology [22].

Unlike other polymorphs, the spectrum of δ-$Cd(IO_3)_2$ shown in Fig. 5 of 2θ = 10° to 2θ = 60° is characterized by several peaks because it contains more than 60 peaks, six among which are intense (d = 3.93 Å, d = 3.66 Å, d = 3.44 Å, d = 3.32 Å, d = 2.49 Å, and d = 1.77 Å).

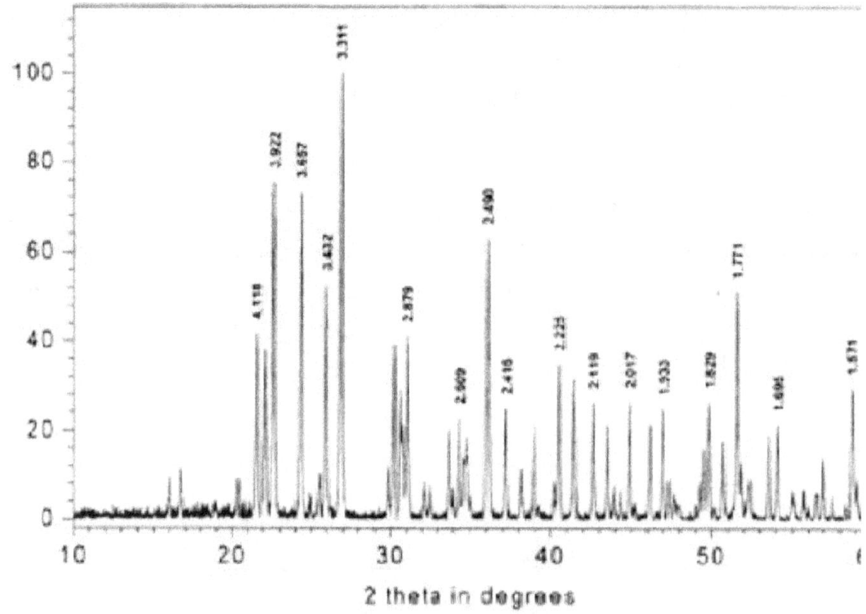

Fig. 5. Diffraction diagram of δ-$Cd(IO_3)_2$ [20].

2.1.5. The polymorph ε-$Cd(IO_3)_2$

This particular polymorph is the ultimate result of the thermal assessment conducted on all the previously known polymorphs, prior to their disintegration [20]. The γ-$Cd(IO_3)_2$ polymorph can be obtained in solution through the process of gradual evaporation at a temperature of 70°C. This involves using a solution comprising 0.2 g of γ-$Cd(IO_3)_2$ dissolved in 30 ml of 30% HNO_3 [13]. The crystalline structure of this polymorph, which crystallizes in the centro-symmetric group Pca21, is determined using clear colorless needles measuring approximately 0.2 mm, as illustrated in Fig. 6. This polymorph is inactive in GSH but remains thermally stable up to 550°C. The powder diagram is presented in Fig. 7, with the major peak having a characteristic value of d = 3.26 Å.

Fig. 6. SEM view on the polymorph ε-$Cd(IO_3)_2$ [20].

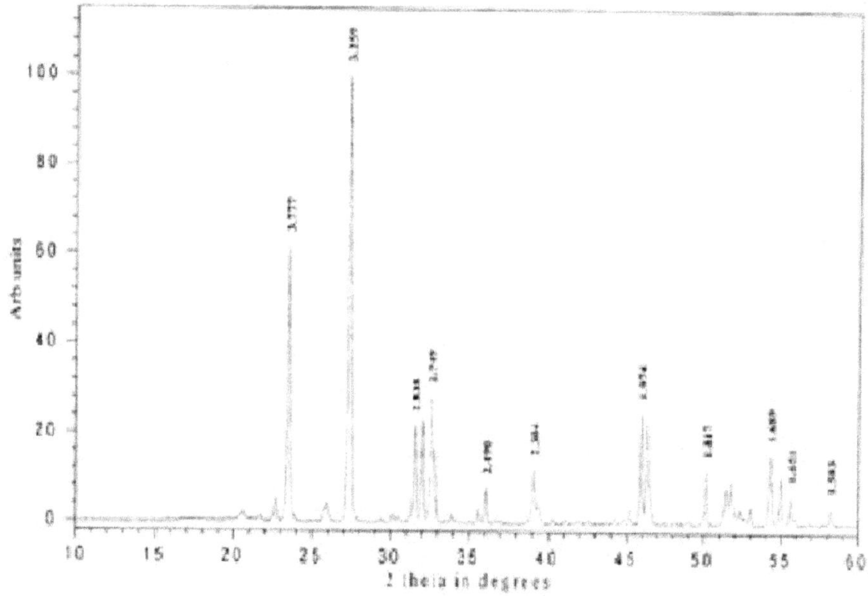

Fig. 7. Diffraction diagram of ε-$Cd(IO_3)_2$ [20].

2.1.6. The polymorph ζ-$Cd(IO_3)_2$

A solution was prepared by dissolving 2 millimoles of anhydrous cadmium chloride and 1 millimole of potassium iodate in 20 milliliters of deionized water. The solution was gradually evaporated at ambient temperature. After a period of 2 days, translucent and colorless prismatic crystals of $Cd(IO_3)_2$H2O are separated from the solution using filtration. The solution was allowed to evaporate gradually at room temperature once more; after a period of 4 days, ζ-$Cd(IO_3)_2$ precipitates as a whitish crystalline powder. The samples underwent filtration, were rinsed with deionized water, and then air-dried at room temperature, resulting in a yield of 30%. The second approach involves synthesizing the polymorph γ-$Cd(IO_3)_2$ in a single step. The compound forms as a colorless crystalline powder when a low concentrated aqueous solution containing 2 mmol of $CdCl_2$ and 1 mmol of KIO_3 is heated to a temperature of 60°C. The ζ-$Cd(IO_3)_2$ compound was synthesized by subjecting γ-$Cd(IO_3)_2$ to a temperature of 450°C for a duration of 2 hours [23]. The X-ray powder diagram of this newly synthesized chemical, obtained using the first approach, is depicted in Fig. 8. Analysis reveals that it exhibits a distinct polymorph, distinct from the α, β, γ, δ, and ε polymorphs. Notably, the picture is characterized by a prominent peak at a distance of 3.14 Å. We utilized the powder diagram for peak indexing, as shown in Table 2. Within the range of 15°–60°, there are 18 peaks with a relative intensity exceeding 0.58%. Out of these, 16 peaks can be identified and classified using the monoclinic cell provided below: The values of a, b, c, and β are 8.4672 Å, 11.4365 Å, 14.4432 Å, and 91.12° respectively. The agreement factor R is 0.057 for all 16 peaks. The only two peaks that deviate from the expected pattern in this cell are located at 2θ = 20.796° and 2θ = 31.474°, with relative intensities of I = 1.649 and 5.4%, respectively. The ζ-$Cd(IO_3)_2$ polymorph demonstrates significant second harmonic generation (SHG) activity, indicating that it crystallizes in an acentric space group [23].

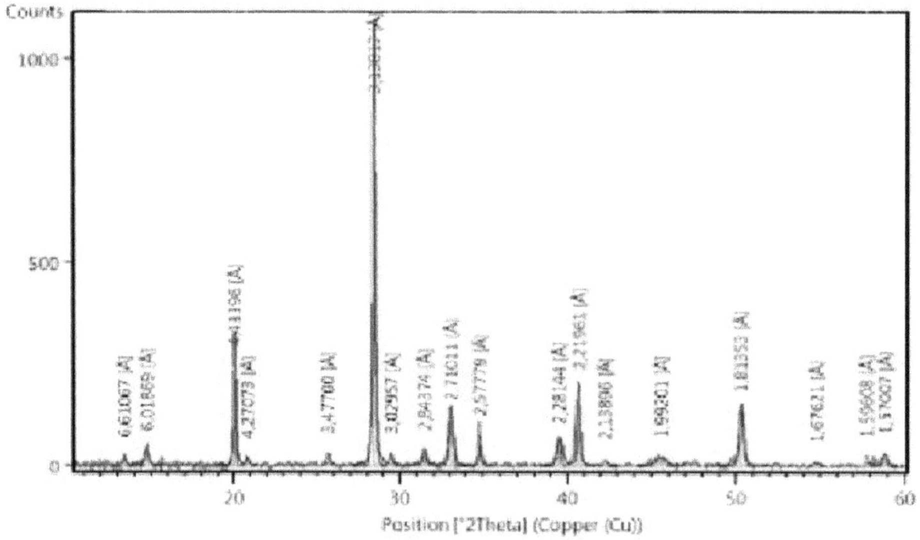

Fig. 8. X-ray powder diagram of ζ-$Cd(IO_3)_2$ polymorph [23].

Table 2. Indexed reflections, standardized intensities, and 2θ values (°) of ζ-$Cd(IO_3)_2$ polymorph used for cell determination [23].

h	k	l	Int	2θ(obs)	2θ(cal)	diff
0	1	3	25.69	20.0196	20.0145	0.0051
*	*	*	1.649	20.7960	******	******
1	3	0	2.22	25.6282	25.6287	-0.0005
-1	3	2	100	28.4564	28.4086	0.0478
2	1	3	1.36	29.4866	29.4287	0.0579
*	*	*	5.40	31.4740	******	*****
1	4	0	10.87	33.0587	33.0709	-0.0122
-3	1	2	7.519	34.8143	34.8034	0.0109
-1	1	6	6.959	39.4989	39.5209	-0.0220
3	2	3	20.99	40.6123	40.5692	0.0431
1	2	6	1.013	42.2563	42.3072	-0.0509
3	0	5	3.915	45.3698	45.3740	-0.0042
-1	4	5	2.662	45.7851	45.7821	0.0030
4	3	1	1.739	49.7340	49.7907	-0.0567
3	0	6	15.11	50.3116	50.3107	0.0009
-1	7	1	1.832	57.8014	57.8082	-0.0068
2	2	8	0.583	58.2564	58.2407	0.0157
-4	3	5	3.258	58.8234	58.8067	0.0167
* Peak wich do not fit with this cell.						

2.2. The Synthesis Technique and Powder Spectrum for $Cd(IO_3)_2H2O$, a Monohydrate Iodate of Cadmium

The slow evaporation of a very dilute solution of $CdCl_2$ (0.26 g) and 0.607 g of KIO_3 in 100 ml of distilled water gives, after a few days, clear transparent crystals of prismatic shape and millimetric size of

$Cd(IO_3)_2H_2O$. This compound is characterized by its theoretical powder diagram retraced by the Poudrix software [34] (Fig. 9). This compound crystallizes in the triclinic space group P1¯ a = 7.119(2), b = 7.952(2), c = 6.646(2) Å, α = 102.17(2)°, β = 114.13(2)°, and γ = 66.78(4)° [20].

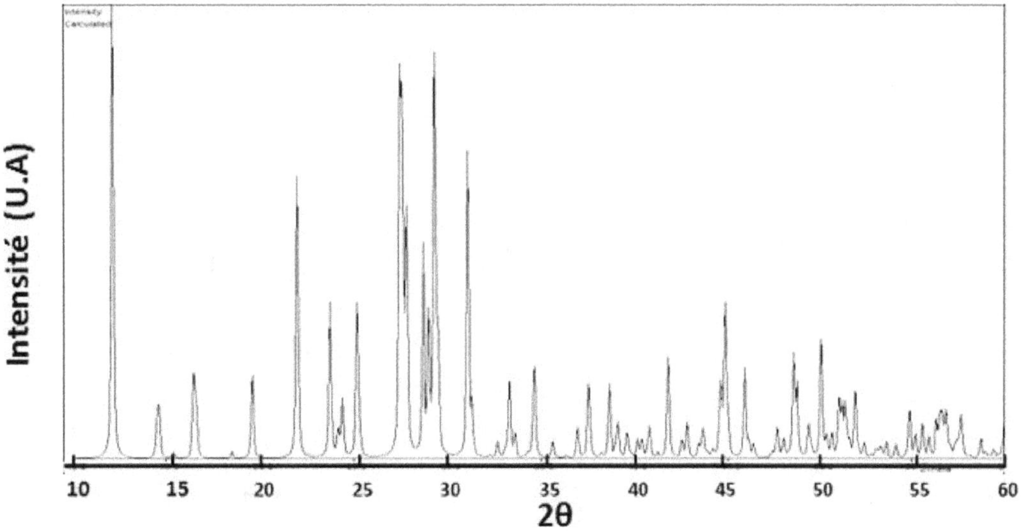

Fig. 9. Theoretical powder diagram of $Cd(IO_3)_2H_2O$.

3. THE SYNTHESIS PROCESS AND POWDER SPECTRA ARE SHOWN FOR TWO MIXED CADMIUM-BASED LIGAND COMPOUNDS: CADMIUM CHLORO-IODATE ($CDIO_3CL$) AND CADMIUM HYDROXIDE-IODATE ($CDIO_3OH$).

In 2014, the cadmium chloro-iodate compound $CdIO_3Cl$ in the form of a prism is synthesized by Yang and Mao. The mixture of 1mmol of $CdCl_2$ and 0.5 mmol of I_2O_5 and 1ml of water sealed in an autoclave aquipped with teflon liner (23 ml) at 200°C for 4 days and than cooled to 30°C at 6°C/h. The crystal structure has been determined by single-crystal X-ray diffraction methods; the unit cell is orthorhombic with a = 7.293(6) Å, b = 16.04(2) Å, c = 7.207(5) Å, V = 843(1) Å3, and Z = 8. The space group is CmCa [35].

The cadmium hydroxy-iodate CdIO3OH is synthesized in the form of colorless platelike in the same way as $CdIO_3Cl$ except that they have used the starting reagents: 0.25 mmol of $Cd(CH_3COO)22H_2O$, 2.5 mmol I_2O_5, and 0.625 mmol of K_2CO_3 in a volume of 3 ml of distilled water [35]. The powder diagram for both compounds is shown in Fig. 10. The crystal structure has been determined by single-crystal X-ray diffraction methods; the unit cell is orthorhombic with a = 11.5245(0) Å, b = 6.7985 (7) Å, c = 4.7303 (4) Å, V = 304.31(1) Å3, and Z = 4. The space group is Pnma [35].

In 2016, Hebboul [36] published two new production methods for obtaining the $Cd(IO_3)Cl$ compound. The first method, known as Procedure 1, involves a synthesis by double decomposition. This method entails dissolving KIO_3 in nitric acid 16 N, as well as $CdCl_2$ in a 2:1 molar ratio. The resulting reaction mixture is then conducted into an ambient environment. Within the initial minute, the solution becomes visually unclear. After a period of 24 hours, this opacity may be observed under an optical microscope, revealing the development of minuscule needle-shaped structures. However, after 7 days, we saw a notable growth in the size of the crystals, which took on the appearance of spherulite needles (as depicted in Fig. 11).

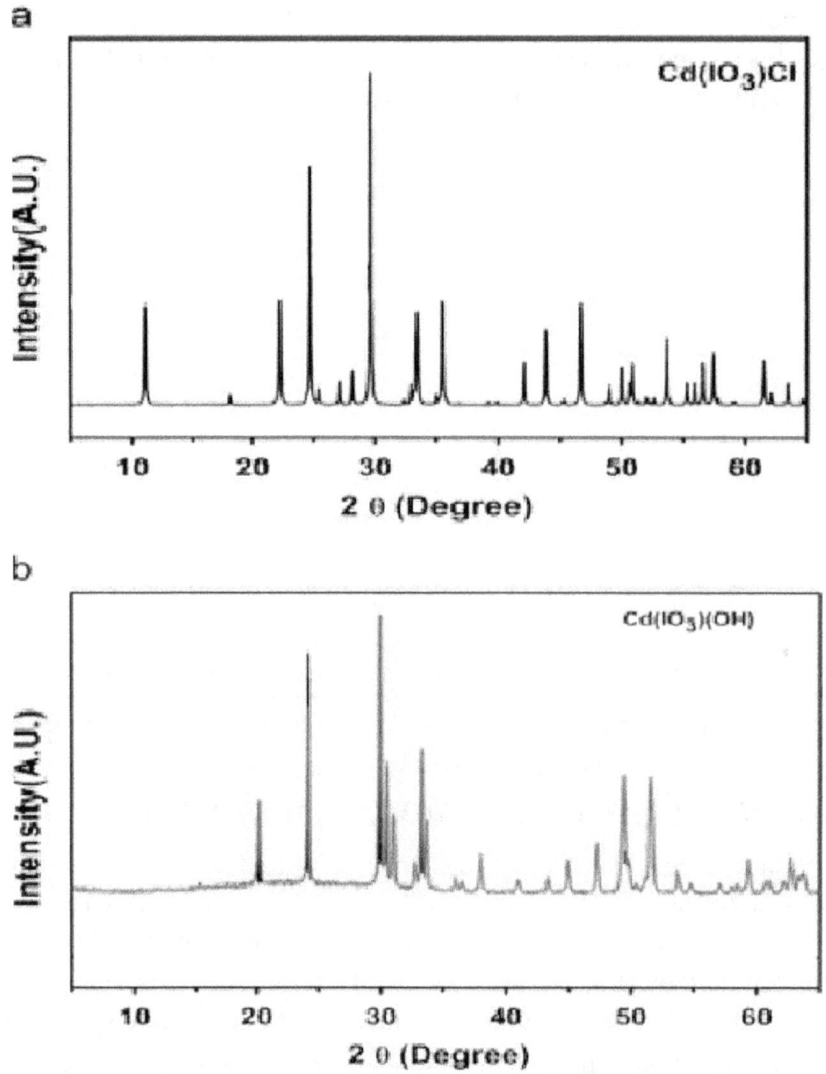

Fig. 10. Mixed ligand compound powder diagram [35]. a-Cd(IO3)Cl and b-Cd(IO3)OH.

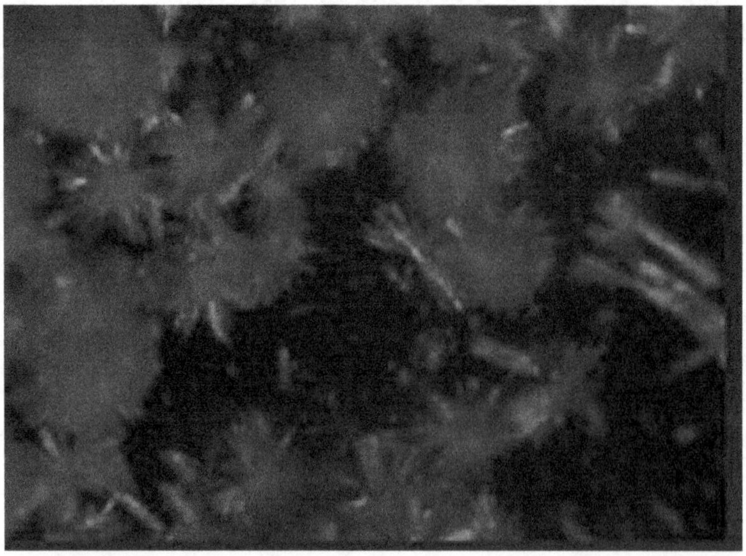

Fig. 11. A view taken by the optical microscope after 7 days [37].

The timing of crystal formation in the solution is highly dependent on the concentration of nitric acid. Therefore, a dilute solution of nitric acid necessitates nearly complete evaporation of the solvent, but a more concentrated solution, such as 7 M, permits crystallization once half of the solvent has evaporated. The crystals undergo filtration, followed by rinsing with distilled water, and are subsequently dried at a temperature of 60 °C in a drying oven. The reaction has a molar yield of approximately 90%. Procedure 2 involves synthesizing γ-$Cd(IO_3)_2$ by substituting one of the two groups [IO_3^-] with a chloride [Cl^-]. 0.148 grams of iodic acid (HIO_3) and 0.48 grams of cadmium chloride ($CdCl_2$) are dissolved in 20 milliliters of distilled water. The resulting combination is then exposed to the surrounding environment. After a few days, we see the creation of two distinct phases: rod-shaped crystals and white powder. The resulting mixture is subjected to a temperature of 400°C in a muffle furnace for a duration of 2 hours. Ultimately, the target molecule is acquired as a powdered substance, containing small amounts of remaining cadmium chloride. These traces can be eliminated by rinsing with distilled water.

3.1. $CdIO_3Cl$ compound characteristics

The SEM image of Fig. 12a shows the crystalline appearance of $CdIO_3Cl$ compound obtained by method 1; the crystal has needle morphology. Fig. 12b shows the morphology of the mixture produced by method 2; a powder of γ-$Cd(IO_3)_2$ is stuck to the $CdCl_2H_2O$ rods

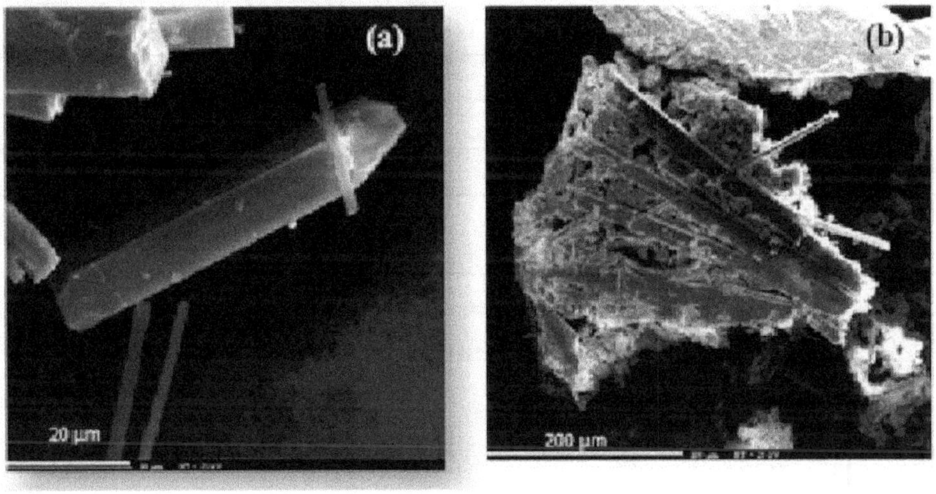

Fig. 12. Morphology of the produced samples [36]. a-$CdIO_3Cl$ and b-$CdCl_2H_2O$ et γ-$Cd(IO_3)_2$.

A monocrystalline needle is obtained by Procedure 1 of dimensions .09 × 0.03 × 0.03 mm3 and was chosen and then mounted on a goniometric head. The conditions of the intensity collection and the refinement parameters of the structure are given in Table 3. The $CdIO_3Cl$ crystallizes in the orthorhombic system, CmCa space group (No. 64) with the following mesh parameters: a = 7.270(0) Å, b = 15.995 (0) Å, c = 7.1980 (0) Å, V = 837.009 Å3, and Z = 8.

Table 3. Crystallographic data and structural refinement of $CdIO_3Cl$ [36].

Formula	$CdIO_3Cl$
Shape color	Colorless needle
Crystal size (mm^3)	0.09 × 0.03 × 0.03
Molecular weight ($g.mol^{-1}$)	322.75
Crystalline system	Orthorhombic
Temperature (K)	293

Formula	$CdIO_3Cl$
λ (Ag) (Å)	0.56087
Space group (n°)	CmCa(64)
a (Å) b (Å) c (Å)	7.270(0) 15.995(0) 7.1980(0)
V (Å3)	837.01
Z	8
Dx (g.cm^{-3})	5.122
Absorption factor (mm^{-1})	6.897
F(000)	1136
Scan area in/θ (°)	3.00 – 21.38
Indication limits (h k l)	--9 ≤ 9; -20 ≤ 20;-9 ≤ 9;
Measured reflections	16,006
Independent reflections	522
Number of refined parameters	33
Rint	0.0457
R1	0.0385
ωR2	0.0888
Refinement quality (S) _ρmax/_ρmin (e.Å^{-3})	1.247 1.406/-1.565

The powder spectrum of $CdIO_3Cl$ (Fig. 13) was recorded and compared to the spectrum calculated with the Poudrix program [34]. The similarity of the diagrams shows that the phase obtained is pure.

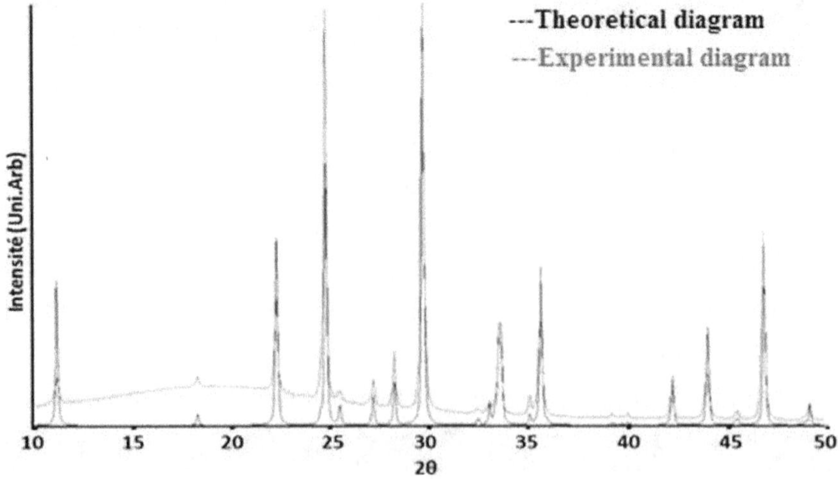

Fig. 13. (XRD) Experimental and calculated powder spectra of the $CdIO_3Cl$ compound [36].

The experimental diagrams (Procedure 2) are refined using the High Score Plus program. The identification of the observed phases is conducted by comparing the obtained experimental diagram with the reference file. The analysis provided information about the types of prepared phases, as shown in Fig. 14. Comparing the results with the JCPDS files, it was determined that one of the phases is hydrate cadmium chloride $CdCl_2H_2O$ No.00-027-0073. The other phase is not listed in the JCPDS files, but comparing the powder spectra published on the polymorphs of cadmium iodate [13] confirms that it is γ-$Cd(IO_3)_2$. Following the heat treatment, the analysis of the powder diagram before and after washing (Fig. 15a) and (Fig. 15b) reveals the successful creation of the $CdIO_3Cl$ phase.

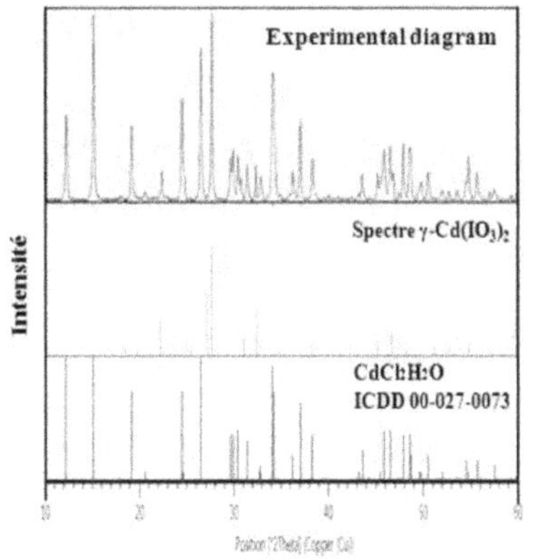

Fig. 14. DRX spectrum of the produced mixture [36].

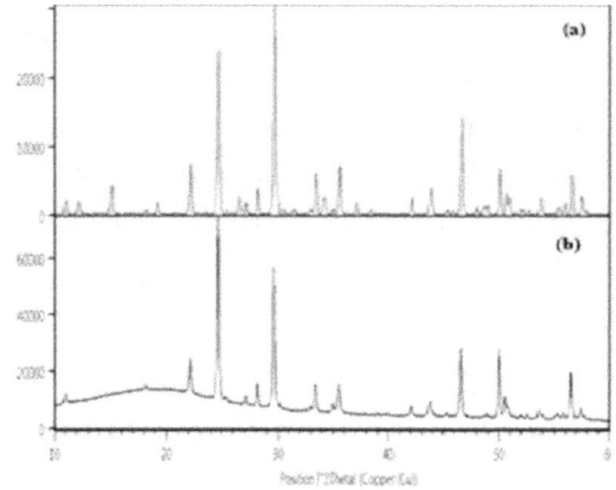

Fig. 15. $CdIO_3Cl$ DRX spectrum [36]. a-Before washing and b-After washing.

4. CONCISE SUMMARY

Summarizing the links between the structures or chemical compositions of materials and their NLO qualities remains a significant difficulty. Compounds with identical chemical compositions can display several phases with distinct structures. This chapter presents the conditions for obtaining six different forms of anhydrous cadmium iodate in a pure state. We also provide the powder diffractograms of these forms, three of which exhibit second harmonic generation activity (SHG). The research was conducted between 2001 and 2005 by the Sciences Fondamentales Laboratory at the University of Laghouat, in collaboration with the Crystallography Laboratory at CNRS Grenoble, France. Additionally, crystallogenesis work by Ravi Kumar et al. in 2008 and further research conducted in 2017 by LPCM at the University of Laghouat is described. Specifically, we outline the conditions for obtaining a triclinic single crystal of $Cd(IO_3)_2$H2O. There are three chemical methods available for preparing chloro-cadmium iodate $CdIO_3Cl$. The produced material underwent characterization using X-ray diffraction (XRD) and scanning electron microscopy (SEM). The crystal structure has been ascertained using single-crystal X-ray diffraction techniques. Cadmium iodate hydroxide (CdIO3OH) is produced as colorless platelike crystals. The crystal structure has been ascertained using single-crystal X-ray diffraction techniques.

REFERENCES

1. Krishnamurthy N. Engineering Chemistry (2nd ed). New York: PHI Learning Private Limited; July 2, 2013. pp. 82-83. ISBN 978-81-203-3666-7
2. Scoullos MJ, Vonkeman GH, Iain T, Zen M. Mercury, Cadmium, Lead: Handbook for Sustainable Heavy Metals Policy and Regulation. Springer; 2001. ISBN 978-1-4020-0224-3
3. Lee C-H; Hsi CS. Recycling of scrap cathode ray tubes. Environmental Science & Technology. 2002;36(1):69-75. Bibcode:2002EnST...36...69L. DOI:10.1021/es010517q. PMID 11811492
4. Miller LS, Mullin JB. Crystalline Cadmium Sulfide. Electronic materials: From silicon to organics. Springer; 1991. p. 273. ISBN 978-0-306-43655-0

5. Buxbaum G, Pfaff G. Cadmium Pigments: Industrial inorganic pigments. Wiley-VCH; 2005. pp. 121-123. ISBN 978-3-527-30363-2

6. Case Studies in Environmental Medicine (CSEM) Cadmium Toxicity. Agency for Toxic Substances and Disease Registry. Archived from the original on 6 June 2011. Retrieved 30 May 2011

7. Jennings TC. Cadmium Environmental Concerns. PVC handbook. Hanser Verlag; 2005. p. 149. ISBN 978-1-56990-379-7

8. Brady GS, Brady GS, Clauser HR, Vaccari JA. Materials handbook: An encyclopedia for managers, technical professionals, purchasing and production managers, technicians, and supervisors. McGraw-Hill Professional; 2002. p. 425. ISBN 978-0-07-136076-0

9. Boyd R. The nonlinear optical susceptibility. Nonlinear Optics (3rd ed); 2007. p. 2. DOI:10.1016B978-0-12-369470-6.00001-0

10. Mandel L, Wolf E. Some Quantum Effects in Nonlinear Optics. Optical Coherence and Quantum Optics; 1995. p. 1079. DOI: 10.1017/CBO9781139644105.023

11. Mao FF, Hu CL, Yang D, Yang BP, Mao JG. Angewandte Chemie International Edition. 2017;56:2151-2155. DOI:/10.1002/anie.201611770

12. Lee DW, Kim SB, Ok KM. Dalton Transactions. 2012;41:8348-8353. DOI: 10.1039/C2DT30947E

13. Bentria B, Benbertal D, Hebboul Z, Bagieu BM, Mosset A. Zeitschrift für Anorganische und Allgemeine Chemie. 2005;631(5):894-901

14. Oelke WC, et Ch. Wagner, pro. Iowa Academy of Science. 1940;46:187

15. Varhelyi C, Kekedy E. Studia Universitatis Babes-Bolyai. Serbian Chemical Society. 1962;1:11

16. Pacesova L, Bohanesova E. Collection of Czechoslovak Chemical Communications. 1968;33:1

17. Svensson C, Abrahams SC, Bernstein JL. Journal of Solid State Chemistry. 1981;36:195-204

18. Bach H, Kuppers H. Acta Crystallographica. 1978;B34:263

19. Liang JK,Yu YD, Ding SL. Acta PhisicaSinica. 1980;29(2):252

20. Bentria B. Ingénierie cristalline pour l'optique non linéaire quadratique - préparation et caractérisation de nouveaux iodates métalliques [Doctoral Thesis]. University of Batna Algeria; 2005

21. Rammelsberg CF. Annalen der Physik. 1838;55:588

22. Ravi-Kumar SM, Melikechi N, Selvakumar S, Sagayaraj P. Journal of Crystal Growth. 2009;311:337-341

23. Hebboul Z, Benbertal D. Synthesis and characterization of new anhydrous cadmium iodate zeta polymorph ζ-$Cd(IO_3)_2$. Journal of Materials and Environmental Science. 2018;7(9):565-569. DOI:org/10.26872/jmes.2018.9.2.62

24. Hebboul Z. Eaboration et caractérisation de nouveaux iodates bimétalliques pour l'optique non linéaire [Ingeniorate Thesis]. Algeria: University of Laghouat; 2001

25. Hebboul Z. Elaboration et caractérisation de nouveaux iodates métalliques pour l'optique non linéaire [Magister Thesis]. Algérie: University of Laghouat; 2004.

26. Nassau K, Shiever JW, PrescottBE. Journal of Solid State Chemistry. 1973;7:186

27. Abrahams SC Sherwood RC, Brenstein JL, Nassau K. Journal of Solid State Chemistry. 1973;7:205

28. Nassau K, Cooper AS, Shiever JW, Prescott BE. Journal of Solid State Chemistry. 1973;8:260

29. Abrahams SC, Sherwod RC, Brenstein JL, Nassau K. Journal of Solid State Chemistry. 1973;8:274

30. Nassau K, Shiever JW, Prescott BE, Cooper AS. Journal of Solid State Chemistry. 1974;11:314

31. Nassau K, Shiever JW, Prescott BE. Journal of Solid State Chemistry. 1975;14:122

32. Abrahams SC, Brenstein JL, Nassau K. Journal of Solid State Chemistry. 1976;16:173

33. Kurtz SA, Perry TT. Journal of Applied Physics. 1968;39:3798

34. Laugier J, Bochau B: Poudrix, a suite of programme for interprétation of X-ray experiments. ENSP / LMGP, BP 46, 38042 sain martin D(Héres, France. WWW: http://www.Inpg.fr/LMGp and http://www.ccp14.ac.Uk/tutorial/Impg/

35. Yang BP, Mao JG. Journal of Solid State Chemistry. 2014;219:185-190

36. Hebboul Z, Taouti MB, Benbertal D. Synthesis and characterization of CdIO3Cl compound. Journal of Materials and Environmental Science. 2016;7(5):1607-1613

37. Hebboul Z. Elaboration et caractérisation de nouveaux matériaux pour l'optique non linéaire. [Doctoral Thesis]. Algérie: University of Laghouat; 2016

3 Chapter

UTILIZATION OF BENTONITE CLAY AND INORGANIC COAGULANTS FOR HEAVY METAL REMOVAL

1. INTRODUCTION: AN OVERVIEW

Heavy metals are typically characterized as elements that possess a density more than 5 g/cm3 [1, 2]. In contrast, the majority of heavy metals possess a density that is either equal to or less than 3 g/cm^3. As a result, they have been reclassified as hazardous metals. Geologically occurring substances, known as heavy metals, undergo physicochemical reactions to create valuable natural resources including coal, gold, and copper, among others. These substances possess a trait of being semi- or non-degradable and have the tendency to build up in the food chain, posing a risk to human health [3, 4]. In addition to their harmful effects, they also have a crucial function in both household and industrial operations. Consuming water that is contaminated with even trace amounts of heavy metals can have harmful effects on both human and aquatic organisms [4]. Various research efforts have been conducted to explore novel methods of eliminating heavy metals from industrial wastewater [5]. Certain metals, such as uranium, possess both poisonous and radioactive properties. Uranium is typically found within mineral formations alongside other metals including gold, radium, selenium, and thorium. Arsenic, cadmium, chromium, lead, and mercury are considered priority metals due to their significant toxicity to public health [6]. Although they possess hazardous properties, their utilization is inevitable due to their indispensability in economic activities, particularly in the production of metal-containing compounds [7, 8, 9, 10]. In addition to typical natural causes, heavy metals are also delivered into the environment by natural phenomena such as weathering and volcanic eruptions [7, 11, 12, 13, 14]. The majority of these elements are transition elements, with some being metalloids such as arsenic. These elements have been found to demonstrate toxicity at low levels, affecting soil, flora, and rivers [6, 10, 11]. Environmental degradation is a significant worldwide concern resulting from the emissions produced by power utilities, mining, and chemical companies. The latter refers to the lack or poor filtration systems, which are primarily responsible for the elevated levels of lead and chromium in various water bodies [15]. However, certain heavy metals such as lead (Pb2+) and chromium (Cr^{3+}) are hazardous and carcinogenic in their respective oxidation states [16]. Wastes or effluent from coal and gold mining (Fig. 1) contain a high concentration of heavy metals, making them highly contaminated. Alternatively, they serve as the origin of pyrite (FeS_2), a mineral that is created through geological, chemical, or microbiological processes by the interaction of iron and sulfur under specific heat circumstances [17].

Fig. 1. Geochemical surface feature along the river concentrated with metals (www.earth.illinois.edu).

In addition to their economic worth as mineral resources, heavy metals also play a crucial role in the production of clay. Conversely, clay minerals make up a significant portion of coal and are also utilized as catalysts in the process of coal combustion (oxidation). The extensive transformation of rocks into clay minerals results in the creation of clay deposits that are relatively pure. These deposits have various economic uses, including the production of cooking pots, bricks, porcelain, drainage pipes, floor and wall tiles, tobacco pipes, oil drilling, cat litter, heat resistant tiles, lime mortars for construction, and building materials and equipment, among others [18, 19]. Certain substances act as fluxing agents and contain troublesome chemicals that cause slagging and clogging in power utility boilers. Certain pollutants are released into the atmosphere as a result of industrial activities, leading to a significant environmental disaster. The primary cause of environmental degradation in dumping sites is the disposal of chemicals and barren ore deposits contaminated with heavy metals.

2. ENVIRONMENTAL CONSEQUENCES OF WASTE MATERIALS

This evaluation does not encompass the utilization of heavy metals, such as cyanide and mercury, in industrial procedures. The presence of heavy metals is predominantly evident in tailing dumps. Tailings contribute to environmental damage, irrespective of the quantities of harmful substances they contain. They include high concentrations of heavy metals, primarily in the form of gangue minerals such as silicates, oxides, hydroxides, carbonates, and sulfides. Certain heavy metals present in the tailings are in the form of iron-sulfide minerals, such as pyrite, pyrrhotite, chalcopyrite, arsenopyrite (FeAsS), sphalerite (ZnS), and galena (PbS). Other heavy metals in the tailings consist of sulfur-containing minerals (Pb, Zn, Cd, Se, As) and compounds that are highly toxic. In addition, sulfide-containing waste materials can undergo oxidation during the process of weathering, leading to the production of acid mine drainage (AMD) and the release of various metals and metalloids such as arsenic, cadmium, cobalt, chromium, copper, iron, mercury, manganese, nickel, lead, and zinc [9]. The dangerous metals disperse throughout the ecosystem through extensive leaching, leading to significant health issues as a result of their non-degradable nature [20]. The process of heavy metal leaking into the soil poses a threat to the natural population of bacteria, resulting in the breakdown of bacterial species responsible for nutrient cycling. This has a detrimental impact on the ecosystem [21]. In such circumstances, they must create methods of survival in the face of such unfavorable conditions, namely by developing and implementing mechanisms for detoxification (such as biotransformation, biosorption, and bioaccumulation) [22]. Uranium, the element with the highest level of radioactivity, possesses a molecular weight of 238.03 mol/g and poses a significant threat to human health due to its chemical toxicity, radiation, and other by-products derived from uranium. It is quite detrimental and can result in kidney, pulmonary, and neurological impairments [23]. The element exists in six radioactive isotopes (232–238) which are classed based on their mother element and contribute to the instability of the highly reactive nuclei. The radioactivity of a substance becomes evident when its nucleus releases tiny particles during a process called radioactive decay. This includes the emission of α and β particles as well as γ-rays. Fig. 2 depicts the dispersion of radioactive uranium. In many countries, vast expanses of land are covered by millions of tons of tailings. One such example is South Africa (SA), which generated more than 43,500 tons of gold and 73,000 tons of uranium between 1953 and 1995. The basin encompasses a total area of 1600 square kilometers, which includes 400 square kilometers of tailings dams. These dams include approximately 6 billion metric tons of pyrite tailings and 430,000 metric tons (600,000 metric tons) of low-grade uranium. The gold mining operations led to the pollution of approximately 6000 km^2 of soil [24].

Fig. 2. Spillages and remaining of dumps of uranium in West Rand (SA) [24].

The most harmful effects of uranium on humans and the environment include (1) defects and deaths in aquatic organisms, (2) disruption of aquatic physiological and biochemical processes, (3) inhibition of aquatic activities due to the presence of iron, zinc, copper, manganese, and lead, (4) disturbance of biodiversity in aquatic ecosystems, (5) reduction of dissolved oxygen levels in water bodies, (6) alteration of nutrient availability leading to vegetation loss, (7) redirection of root growth, and (8) accumulation of uranium in aerial plant organs and human bodies causing illnesses. Lead (Pb) and chromium (Cr) are highly poisonous metals that pose a significant threat to plants and animals. These metals are difficult to cure because they are not easily broken down by natural processes and have a tendency to accumulate in living cells, leading to further harm. [25, 26] Chromium has the ability to exist in several oxidation states, including Cr^{2+}, Cr^{3+}, and Cr^{6+}. These forms are highly poisonous, reactive, and soluble in solutions with different pH levels [27]. However, the toxicity of these forms is lower when compared to Cr6+. Pb is a bluish-white metal that is both soft and highly flexible and ductile.

It is also resistant to corrosion [28]. The mineral matter types (Table 1) consist of intricate compounds that are primary components of both clay and coal.

Table 1. Mineralogy of a tailing dump in South Africa [29].

Fraction	Name	Chemical formula
Gauge	Quartz	SiO_2
Mineral	K-feldspar	$KAlSi_3O_8$
	Na-feldspar	$NaKAlSi_3O_8$
	Ca-feldspar	$CaKAl_2Si_2O_8$
	Serisite	KAl (AlSi)(F,OH)
	Chlorite	$(Mg,Fe)_3(Si,Al)_4O_{10}(OH)_2(OH)_6)$
	Calcite	$CaCO_3$
	Dolomite	$Ca,Mg(CO_3)_2$
Sulfide-oxide	Pyrite/pyrrhotite	FeS_2/FeS
	Arsenopyrite	FeAsS
	Galena	PbS
	Sphalerite	ZnS
	Chalcopyrite	$CuFeS_2$
	Magnetite	Fe_3O_4

The oxidation of pyrite causes the water that passes through the dumps to become more acidic. This acidic water then seeps into the groundwater system underneath the dumps [30]. It is hypothesized that this water with high acidity is flowing into streams along the Witwatersrand in South Africa [30, 31]. In addition to direct pollution by metals, tailings can contribute to environmental damage. The Gauteng Province in South Africa has a significant concentration of tailings, which release metal contaminants into the environment through leaching during periods of heavy rainfall [30]. Fig. 3 depicts a tailing dump.

Fig. 3. Photograph of gold mine tailing dumps in Johannesburg, South Africa taken by George Steinmetz.

Tailings typically exhibit a pyrite level of up to 6%, high salinity, low nutritional value, and a low organic matter content [32, 33, 34]. The presence of a significant amount of sulfide leads to elevated levels of acidity and increased concentrations of metals in the groundwater near the tailings [32]. Rafiei et al. [35] observed a pH level of 7.35 in gold mine tailings in Iran, while Mitileni et al. [36] documented pH levels ranging from 3.25 to 6.28 in South Africa. Harish et al. [37] documented a pH range of 3.48–8.12 in India. The pH level is crucial for the survival of aquatic organisms, as various species exhibit varying responses. In the case of mine pollution, elevated soil acidity can lead to the destruction of microbes that play a vital role in the decomposition of organic matter into nutrients. Furthermore, acidity has the ability to dissolve aluminum, resulting in the production of harmful organic substances that are toxic to plant roots. Additionally, acidity can also decrease the levels of vital nutrients. Gold mine tailings are characterized by high quantities of harmful heavy metals, specifically arsenic (As), cadmium (Cd), nickel (Ni), phosphorus (P), copper (Cu), zinc (Zn), cobalt (Co), and mercury (Hg) [38].

3. THE IMPACT OF CHEMICAL TREATMENT ON THE ECOSYSTEM

The primary factors contributing to the growing accumulation of heavy metals are industrial and population expansions. This increase in heavy metal concentration is inevitable and difficult to regulate due to their natural and human-caused origins [39]. Gold mining is known to generate numerous large tailings dumps that are dispersed throughout the countries, posing a possible threat to the ecology [40]. Nuclear power generating produces harmful waste due to the utilization of uranium as fuel. The challenge associated with nuclear power is in effectively managing and disposing of radioactive waste materials. This review study proposes that an effective approach to addressing the issue of dangerous minerals is through the implementation of mineral processing techniques such as mineral liberation-classification. However, it acknowledges that mineral sliding is an unavoidable occurrence. While mineral liberation-classification does not yield all minerals in their pure state, certain minerals exist in intricate complexes that necessitate additional separation.

A significant quantity of trash is produced during several processes, particularly in gold mining, where almost 99% of the waste consists primarily of ore [41]. Another option involves the utilization of bacteria, although a major challenge is in controlling their production rate and ensuring optimal growth conditions, which are primarily influenced by pH and temperature. The bacteria that are reliant on pH, such as acidophilic, chemolithotrophic iron- and sulfur-oxidizing bacteria, help in dissolving the sulfides found in the gold ore deposits by releasing gold from the deposit [42]. While biomining is considered to be less environmentally hazardous than some physicochemical extraction methods [42], it is nevertheless seen as unsustainable. Bacteria, such as Acidithiobacillus, get energy by oxidizing Fe2+ ions to Fe3+ ions or elemental sulfur (SO).

Eqs. (1)–(3) depict the process of oxidation-reduction involving iron and sulfur in pyrite, a mineral that is commonly exposed during gold and coal mining. The acidithiobacvillus bacteria derive energy through the oxidation of ferrous iron (Fe^{2+}) to ferric iron (Fe^{3+}) or the reduction of elemental sulphur (SO) to generate sulphuric acid [45, 46]. This process ultimately leads to the creation of acid mine drainage (AMD).

$$2\,S^0 + 3\,O_2 + 2\,H_2O \rightarrow 4\,H^+ + 2\,SO_4^{2-} \tag{1}$$

$$FeS_2\,(Au) + 2\,Fe^{3+} \rightarrow 3\,Fe^{2+} + 2\,S^0 + (Au) \tag{2}$$

$$FeS_2\,(Au) + 14\,Fe^{3+} + 8\,H_2O \rightarrow 15\,Fe^{2+} + 2\,SO_4^{2-} + 16\,H^+ + Au \tag{3}$$

One notable feature of the bacteria involved in gold purification is their capacity to secrete ligands that stabilize gold by creating gold-rich complexes and/or colloids [43,44,45]. This research examines the efficacy and reliability of utilizing bacteria for gold purification. However, the presence of residual heavy metals poses a significant challenge, since they contribute to the formation of waste that poses a threat to the ecosystem. Mercury, a poisonous metal, is employed in the process of gold purification. This is achieved by combining it with mineral ore deposits obtained from the ground or stream beds to create an amalgam. The combustion of the latter causes the conversion of elemental mercury into a poisonous cloud, leading to the extraction of gold from the ore deposit. The practice is widely recognized as the second most significant contributor to atmospheric mercury pollution, following coal burning [46]. In addition to mercury, cyanide is also a hazardous metal used in the gold purification process (Eq. (4)). This procedure has two stages: extraction and recovery.

Eq. (4) shows dissolution of gold:

$$4\,Au(s) + 8\,NaCN(aq) + O_2(g) + 2\,H_2O(l) \rightarrow 4\,NaAu(CN)_2(aq) + 4\,NaOH(aq) \quad (4)$$

Cyanide, a highly poisonous element, is used in the process of extracting gold. In addition, the procedure employs zinc, which is another hazardous element, to solidify the cyanide solution [46]. Environmental deterioration arises from cyanide solution spillages, while activated carbon can be employed in cases of significant gold concentration. The ground and crushed ore deposits are contained within spacious tanks equipped with agitators to facilitate the dissolution of gold, which then adheres to activated carbon particles. The activated carbon and gold are extracted from the solution and then discarded along with the leached ore [46]. The lingering presence of trace heavy metals from the procedure poses possible risks to the ecology. The sludge generated from gold and coal operations has been seen to contain several substances, including mercury amalgam, cyanide, uranium, and mineral debris. Due to incomplete gold recovery from ore deposits, there are residual amounts of gold in the tailings. Over time, these residual amounts may break down and create different complexes or react with other harmful substances, resulting in the formation of new toxic compounds. In previous investigations, lime has been employed to control the pH of the AMD solution [47]. The main issue associated with this technology is the production of an excessive amount of sludge, which necessitates additional treatment of sludge contaminated with heavy metals. Alternatively, the sludge is transferred by pumping mechanisms to expansive repositories known as slimes dumps.

4. THE IMPACT OF METAL SALTS ON THE ELIMINATION OF SUSPENDED PARTICLES

Mine wastewater and acid mine drainage (AMD) are the primary vehicles for transporting the heavy metals generated by mining activities. Several studies have been conducted to explore cost-effective methods for treating mine wastewater. The usage of inorganic coagulants is considered an outdated strategy in this regard. The study in this review focuses on the traditional treatment technique that utilizes inorganic coagulants. Their value lies in their ability to undergo hydrolysis and produce flocs that act as adsorbents. For iron (Fe), there are at least four different forms that exist together in a watery solution with a pH between 1 and 5. These forms are Fe^{3+}, $Fe(OH)^{2+}$, $Fe(OH)^{2+}$, and a very little amount of the dimer Fe2(OH)24+ [48]. Conversely, Al species have a preference for a pH that is somewhat higher than that preferred by Fe, within the range of 3-8. In addition to its favorable effects, naturally occurring iron serves as a source for the production of AMD (acid mine drainage). Through a geological process, it undergoes a reaction with sulfur to produce FeS_2, which has detrimental effects on the environment. Pyrite is non-toxic in its natural state as a mineral. However, when it comes into contact with oxygen in a watery environment, it undergoes oxidation and transforms into sulfuric acid, which is a harmful substance that causes environmental degradation (as shown in Equation 5).

$$2\,FeS_2 + 7\,O_2 + 2\,H_2O \rightarrow 2\,Fe^{2+} + 4\,SO_4^{\;2-} + 4\,H^+ \quad (5)$$

Fe2+ ions may be oxidized to form Fe3+ ions (Eq. (5)), which oxidize excess pyrite:

$$FeS_2 + 14\,Fe^{3+} + 8\,H_2O \rightarrow 15\,Fe^{2+} + 2\,SO_4^{2-} + 16\,H^+ \tag{6}$$

Eq. (6) yields unstable Fe2+ ions species which are further oxidized by excess Fe3+ ions that are subjected to hydrolysis when added to wastewater.

The catastrophe associated with the AMD (Fig. 4) is not only about the surface or ground water pollution and degradation of the quality of the soil, but endanger it poses to aquatic sediments and fauna, allowing heavy metals to seep into the environment.

Fig. 4. Diagram of AMD pumped from the underground workings [48].

Prolonged exposure to drinking water contaminated with acid mine drainage (AMD) can result in elevated incidences of diseases such as cancer, reduced cognitive abilities, and the development of skin sores. The presence of high levels of heavy metals in drinking water can hinder the growth of fetal neurons, leading to cognitive impairment. Furthermore, AMD is responsible for the depletion of stream beds, the desiccation of rivers, the restricted water resources available for expanding industrialization and population, and the curtailment of water supplies for agriculture purposes. The discharge of acidified water (AMD) from the mine tailing dumps, which is highly concentrated with heavy metals, leads to significant environmental deterioration, particularly in the surrounding stream flow [49, 50]. The condition causes a decrease in pH and an increase in metal concentrations in the water, leading to the formation of excessive amounts of iron hydroxide in the stream. This poses a threat to aquatic organisms. The potential for groundwater evaporation from the capillary zone above the water table could result in the contamination of the surface soil layer with heavy metals. The dissolution of metals in the Earth's crust and surface soil can contribute to the accumulation of metals in the stream. The introduction of iron into the stream by seepage can act as a redox buffer, regulating the pH of the stream water. Raising the pH of acid mine drainage (AMD) above 3, either by exposing it to fresh water or by neutralizing minerals, causes Fe^{3+} ions to form iron hydroxide species through precipitation. Additional forms of iron precipitates can occur, such as iron oxides and oxyhydroxides [48]. These species all lead to the water being discolored and suffocate aquatic plant and animal life, hence altering stream ecosystems. According to the literature, it is commonly believed that the best way to remove turbid materials is in an alkaline environment. However, the findings of this review present a contrasting perspective. Although this review has discovered certain information, the studies conducted on AMD samples using inorganic coagulants, metal hydroxide, and bentonite clay have shown that the elimination of turbid materials and heavy metals relies on the physicochemical characteristics of both the colloidal suspension and the reagents, as well as the structural morphology of the flocs (as observed in SEM micrographs).The hypothesis of pH adjustment relies not only on the presence of metal hydroxide, but also on the release of OH- ions resulting from the breakage of bipolar water molecules. The dehydroxylation of hydrated metal ions is facilitated by the high ionic strength of the AMD and the presence of charges in the solution. This leads to the release of OH- ions, which then combine with metal ions to create precipitates of metal hydroxides.

5. VARIABLES THAT IMPACT THE ELIMINATION OF HEAVY METALS

Extensive research has been carried out to explore effective methods for removing heavy metals. However, many of these methods are expensive and complex to implement. The most influential factors during treatment are dosage, destabilization-hydrolysis, time, concentration of the heavy metals, type of colloidal suspension, the reagents, temperature, retention time, sorption capacity, and structural morphology of a substrate, among others. While pH is commonly believed to be a factor in the elimination of cloudy substances, as mentioned later, research conducted by Ntwampe et al. [53, 54] on paint wastewater and AMD using conventional inorganic coagulants, metal hydroxide, or bentonite clay showed that the pH of the colloidal suspension has little impact on the removal of cloudy material or heavy metals. The studies have shown that the most effective elimination of cloudy substances is a physicochemical process that is influenced by the rate of destabilization-hydrolysis. Although there are two reactions involved, this review treats them as a single stage due to their simultaneous occurrence. Specifically, destabilization takes place in the colloidal suspension, while hydrolysis happens in the metal ions [55].

5.1. Destabilization-hydrolysis reaction

The equilibrium state of a colloid is disrupted by the reagent in this reaction, as a result of the opposing forces of van der Waals attraction and electrostatic repulsion [56, 57]. The reaction, a physico-chemical process, governs the efficiency of subsequent reactions. The colloid undergoes separation into two distinct ionic zones, known as the diffuse and Stern layers, in order to create an electrical double layer (EDL) [58]. The effectiveness of a reagent in treatment is assessed by its capacity to compress the electrical double layer (EDL), leading to a decrease in particle radius. This reduction promotes particle-particle collisions, resulting in the formation of agglomerates or flocs.

5.2. Type of a colloid

Colloids are categorized based on their affinity for water, either being hydrophilic (having an affinity for water) or hydrophobic (having a repulsion towards water). Treating hydrophilic colloids is challenging because they readily dissolve in the solvent, making the process of flocculation (clumping together) problematic. Conversely, hydrophobic colloid is composed of solid particles that are suspended and some of them settle naturally because of gravity [58].

5.3. Type of a reagent

A coagulant is considered effective when it has a strong ability to destabilize substances due to its high valence and electronegativity. This phenomenon becomes evident when metal ions from a salt are added to a colloidal suspension, causing them to spread inside the colloidal particle and disrupt the stability of the system, leading to the production of flocs. Nucleation takes place, leading to the formation of tiny flocs that subsequently aggregate into bigger flocs, resulting in the removal of turbid materials [58].

5.4. Structural morphology

The crystal shape observed in the scanning electron microscope (SEM) micrographs demonstrates the reagent's capacity to adsorb the absorbate. The majority of the scanning electron microscope (SEM) images that represent ideal adsorption show tightly packed aggregates connected together with minimal or no empty spaces [51]. The SEM micrograph indicates that there is no penetration of turbid materials (heavy metals). Conversely, the SEM micrograph showing flocs and voids has a lower ability to adsorb. [51]

5.5. pH of the solution

Several studies [59, 60, 61] indicate that the adsorption rate of turbid materials and heavy metals is more advantageous in an acidic colloidal suspension, specifically at a low pH. This conclusion aligns with previous research that have shown that treating AMD with inorganic coagulants without adjusting the pH results in the most effective adsorption. An acidic solution has a high concentration of hydronium ions (H^{+}), which have a tendency to increase the electrostatic forces of attraction between negatively charged heavy metal oxyanions and protonated sorbent. This reaction follows equation (7).

$$Fe^{0} + 2\,H^{+} + Fe^{2+} + H_2 \qquad (7)$$

Fe^{2+} can be further oxidized to form Fe3+.

$$2\,Fe^{2+}{}_{(aq)} + 2\,H^{+}{}_{(aq)} + 0.5\,O_{2(aq)} \rightarrow 2\,Fe^{3+}{}_{(aq)} + O \qquad (8)$$

6. REMOVAL OF HEAVY METALS IN THE TREATMENT OF AMD

The removal of heavy metals from industrial effluent discharge has been a serious challenge for decades due to the complex compounds formed during chemical reactions. The wastewater consists of metals as pure and trace elements, where the latter always disperse throughout the solution and sorbed onto other compounds or form other compounds during nucleation. The problem associated with trace elements or compounds consisting of trace elements is a plausibility to slip through the adsorbents. The residual heavy metals are caused by the existence of the trace elements in the system [62], a problem caused by low adsorption potential of an adsorbent in a medium with low concentration of absorbate. Technologies such as membrane separation, electrochemical, precipitation, ion exchange and adsorption were investigated to treat wastewater/AMD [63]. The results showed that they are expensive and incur high waste disposal costs, apart from high costs, some studies [64] revealed that some of the new technologies have a tendency of regenerating pollutants back to the environment by producing toxic chemicals such as sodium borohydride ($NaBH^{4}$). Chemical reaction between iron salt and sodium borohydride to release flammable hydrogen gas [65] is shown in Eq. (9).

$$4Fe^{3+} + 3BH_4^{+} + 9H_2O \rightarrow 4F^{0}\downarrow + 3H_2BO_3^{-} + 12H^{+} + H_2\uparrow \qquad (9)$$

Some studies were conducted [51, 54] to reduce turbid materials from AMD using a combination of bentonite clay and inorganic coagulants. Among their studies, a combination of bentonite clay and Na_2CO_3 [53] was employed to remove the heavy metals from the AMD sample. Inorganic coagulants, mainly the metal ions (Mn^{+}), have exhibited a significant influence in wastewater and AMD treatment [51, 52, 53], including the pivotal role played by the bentonite clay. There are more studies which have been conducted in the treatment of AMD sample using a variety of metal salts without or with a combination of bentonite clay [51, 52, 53]. The results show that the turbid materials removal efficiency of inorganic coagulants is optimal when dosed alone or in combination. The observations show that they are all subjected to hydrolysis to form species which are adsorption substrates. Additionally, bentonite clay demonstrates ionic exchange, sorption, and intercalation as the primary processes responsible for eliminating turbid materials and heavy metals [52, 53]. Fig. 5 displays the mineralogy of the AMD sample, including its pH and conductivity, after treatment with a mixture of bentonite clay and Na_2CO_3, with the addition of mixing.

Table 2. ICP-OES analyses of untreated AMD sample with clay (5× dilution).

Element	Conc. (ppm)
Al	1.171
Ca	182.1
Co	39.7
Cu	0.14
Fe	28.3
K	4.5
Mg	67.3
Mn	35.3
Na	44.5
Ni	27.6
Pb	35.2
As	45.2
Se	0.71
Zn	31.7

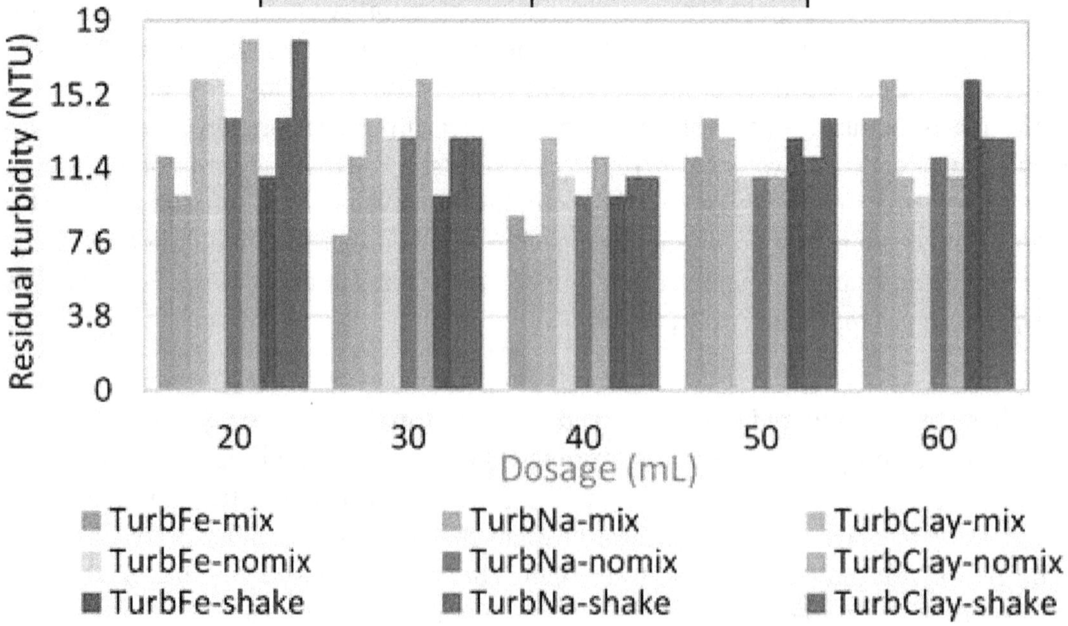

Fig. 5. Residual turbidity of AMD with FeCl3, Na2CO3 and a combination of bentonite clay and Na2CO3 with and without mixing and shaking.

Table 1 shows that the concentration of metals, specifically Ca, Mg, and Na, in the AMD is high, with values of 182, 67, and 46 ppm, respectively. The content of non-toxic heavy metals, Fe and Mn, is 28 and 35 ppm, respectively. The concentration of dangerous metals, including Co, Ni, Pb, As, and Zn, is significantly elevated. These metals are particularly harmful to the ecology. The turbid substances found in the AMD (Fig. 5) are classified as turbidity, encompassing both suspended and soluble particles, inorganic matter, heavy metals, and any trace chemicals. The treatment involving agitation and the absence of blending is not included in this review study and will not be further explained. Due to the high turbidity of the AMD sample, residual turbidity levels below 10 NTU are considered highly acceptable, levels between 10 and 15 NTU are considered acceptable, levels between 15 and 20 NTU are considered moderately acceptable, and levels over 20 NTU are considered unsatisfactory. The experimental results for the samples containing a mixture of $FeCl_3$ and Na_2CO_3 fall within the range of 8.4–14.0 and 8.2–16.5 NTU, respectively. Despite their similar efficiency, the cost of using the pure form of the dosage is prohibitively expensive compared to the dosage indicated in the study by Ntwampe et al. [51], which examines the effectiveness of bentonite clay in treating AMD. Table 3 displays the weight percentage of the mineral content in the bentonite clay.

Table 3. Weight % of the mineral content in the bentonite clay.

Element	Wt.%	Wt.%	Atomic%	Compd%	Formula	No. of ions
C K	9.07	0.46	16.55	33.22	CO_2	2.15
Mg K	0.91	0.08	0.82	1.50	MgO	0.11
Al K	1.56	0.08	1.26	2.94	Al_2O_3	0.16
Si K	5.06	0.13	3.95	10.82	SiO_2	0.51
S K	1.48	0.10	1.01	3.69	SO_3	0.13
Ca K	1.14	0.08	0.62	1.60	CaO	0.08
Ti K	0.53	0.09	0.24	0.89	TiO_2	0.03
Fe K	35.25	0.49	13.84	45.34	FeO	1.79
O	45.02	0.58	61.70			8.00
Total	100.00					

Where Wt is weight and Compd is compound.

The weight % mineral content (Table 3), where CO_2 and FeO representing siderite ($FeCO_3$) form a larger weight % (33.2 and 45.3 wt. %) of the total complex compounds contained in both coal/gold mineral composition, and the rest are trace elemental compounds. The particle size of the bentonite clay is 180 µm mesh size with intensive mineral liberation (Table 3). The mineralogy of the tailing dump (Table 1) shows complex compounds of which some are original and others reacted to form new complex compounds; however, the chemical compounds are also representative of the clay minerals. Pulverizing the mineral compound (Table 1) results in minerals liberation to form simple compounds (Table 3). The elements/ metals with high weight percentage include Fe, K and O, an indication of oxidation-reduction of ferric and ferrous ions during chemical reactions. A simple chemical composition of simple compounds (Table 3) indicates a high rate of chemical reactions by the colloid or the reagent(s).

The residual turbidity in the samples with bentonite clay mixing is in a range of 9.8–12.0 NTU, without mixing in a range of 10.5–19.2 NTU and shaking in a range of 10.4–13.3 NTU. As indicated, the method of chemical dispersion, i.e. mechanical agitation, is insignificant in this review, and the focus is on the efficiency of the combination of bentonite clay and Na2CO3 during AMD treatment. The mineral matter composition and structural configuration of the bentonite clay play a pivotal role during treatment process. The former identifies the metals and content present in the colloidal suspension, the information which are indicative of inevitable electrochemical reactions. On the other hand, the latter is distinguished by the planar layers, which include octahedral and tetrahedral formation to form T-O-T. According to mineralogy of the bentonite clay (Table 1), dehydroxylation and processes are plausible. On the other hand, the structural configuration (T-O-T) is effective in sorption by intercalation of the turbid materials (heavy metals) onto the sheets and porous surface of the bentonite clay. In addition, the mineralogy (Table 1) is indicative of a series of physicochemical reactions in the system, especially on the serisite and chlorites, and oxidation of Fe on the iron-bearing compounds (chlorite, pyrite, pyrrhotite, arsenopyrite, chalcopyrite and magnetite). The formation of CO_2 from the decomposition of dolomite is inevitable, including oxidation of metals and other physico-chemical reactions. These form the main chemical reactions occurring in the bentonite clay in the system to form new simple compounds suitable for further reactions or sorption. Na_2CO_3 component in the flocculent also plays a pivotal role which includes increasing of the pH when it solubilizes to $NaCO_3^-$ and CO_3^{2-}. A slight pH rise decreased the solubility of the metals in the colloidal suspension, thus ameliorating the rate of agglomeration. Table 4 illustrates the ICP-OES results of treated AMD samples.

Table 4. ICP-OES analyses of treated AMD sample with clay (5× dilution).

Element	Conc. (ppm)
Al	1.023
Ca	142.1
Co	3.2
Cu	0.14
Fe	19.2
K	4.14
Mg	55.7
Mn	23.1
Na	27.4
Ni	6.6
Pb	5.5
As	4.1
Se	0.71
Zn	4.8

The findings of the AMD treatment using a mixture of bentonite clay (as shown in Table 4) demonstrate a significant reduction in the concentrations of heavy metals, namely Arsenic (As), Cobalt (Co), Nickel (Ni), Lead (Pb), and Zinc (Zn). The concentrations of these metals decreased from 45.5, 39.7, 27.6, 35.2, and 31.7 parts per million (ppm) to 4.1, 3.2, 6.6, 5.4, and 4.8 ppm, respectively. Fig. 6 displays scanning electron microscope (SEM) images of the sludge from the samples of acid mine drainage (AMD) mixed with FeCl3 (Fig. 6A) and bentonite clay (Fig. 6B), respectively.

Fig. 6. SEM of samples the sludge of AMD sample with FeCl3 and bentonite clay respectively (25,000×) [53].

The scanning electron microscope (SEM) images display the dehydrated sediment of the acid mine drainage (AMD) together with the addition of FeCl3 (Fig. 6A) and bentonite clay (Fig. 6B). The former displays flocs that are aggregated to create clusters of distinct, compact, and small irregular porous structures with voids interspersed on the surface. The structural morphology suggests that shear stress was applied during the treatment. In contrast, the micrograph in Fig. 6B displays denser clusters of flocs with a consistent crystal structure, exhibiting fewer empty spaces compared to the ones shown in Fig. 6A.

7. CONCISE SUMMARY

Heavy metals have demonstrated both positive and negative consequences, with economic benefits and environmental degradation, respectively. Extracting heavy metals from the solvent is not economically viable due to their presence as trace elements that are difficult to eliminate. Conversely, the trace elements present in tailings gradually build up over time and combine to create novel, harmful chemicals that impact the ecosystem. The remediation of tailings is a highly costly endeavor that has not been thoroughly considered, resulting in environmental degradation, particularly during rainy events when the hazardous substances dissolve and spread throughout the surroundings. The bulk of modern technologies that have been investigated are not economically viable and environmentally sustainable, as determined by the use of advanced and imported equipment. Furthermore, developing and impoverished nations are unable to manage such financial outlays. The sole resolution to the current predicament about environmental deterioration caused by the mine's wastewater revolves around selecting a cost-efficient treatment method that is uncomplicated to operate and utilizes the most plentiful and affordable reagent(s) available. Industrial wastewater treatment has utilized natural inorganic coagulants. However, a significant issue arises when these coagulants are administered in their pure form, as it leads to detrimental effects on the plant and equipment. This includes corrosion, scaling, and the introduction of more sludge. This review examines the effectiveness of using a mixture of bentonite clay and inorganic coagulants (flocculants) to treat the AMD sample with a lower mass fraction. The selection of bentonite clay was based on its chemical composition and crystal shape, both of which suggest its ability to effectively adsorb physicochemical substances. However, inorganic coagulants also utilize a physicochemical process to eliminate cloudy substances, as shown by SEM micrographs. Nevertheless, bentonite clay is more effective due to its ability to engage in ion exchange, adsorption, and intercalation. The utilization of a combination of bentonite clay and inorganic coagulants at a reduced molarity of 50% (m/m %) offers the benefit of minimizing their negative impact while also addressing issues of excessive or insufficient dose, which can lead to restabilization and deflocculation, respectively. The present review utilized a flocculent composed of bentonite clay and Na2CO3, which shown an excellent ability to remove heavy metals. While certain inorganic coagulants effectively remove turbid materials, the treatment process is influenced by changing air conditions, necessitating ongoing modifications. While it is possible for this phenomena to occur with the flocculent utilized in this work, bentonite clay consistently acts as the primary substrate due to its various adsorption qualities. Another benefit of this procedure is that it eliminates the need for pH modification due to the reagents' strong destabilization-hydrolysis potential, which causes the elimination of turbid materials. The elimination of the murky substances from the Acid Mine Drainage (AMD) suggests that bentonite clay and Na_2CO_3 (flocculent) work well together.

REFERENCES

1. Barakat MA. New trends in removing heavy metals from industrial wastewater. Arabian Journal of Chemistry. 2011;4(4):361-377
2. Coetzee H et al. An Assessment of Sources, Pathways, Mechanisms and Risks of Current and Potential Future Pollution of Water and Sediments in Gold-Mining Areas of the Wonderfonteinspruit Catchment. Report, WRC, Council for Geosience. 2004. Report No 1214/1/06. 2006
3. Winde F. Uranium pollution on water – A global perspective on the situation in South Africa. Epidemiological data related to persons living around the gold mine tailings are unavailable. Inaugural Lecture. 22 February 2013. ISBN 978-1-86822-629-0
4. Bambas-Nolen L, Birn AE, Cairncross E, Kisting S, Liefferink M, Mukhopadhyay B, Shroff FM, van Wyk D. Case study on extractive industries prepared for the lancer commission on global governance: Report from South Africa. The Lancet. 2013;383(9917):630-667
5. Fergusson JE. The Heavy Elements: Chemistry, Environmental Impact and Health Effects. Oxford: Pergamon Press; 1990
6. Duffus JH. Heavy metals-a meaningless term? Pure and Applied Chemistry. 2002;74(5):793-807
7. He ZL, Yang XE, Stoffella PJ. Trace elements in agroecosystems and impacts on the environment. Journal Trace Elemental Medical Biology. 2005;19(2–3):125-140 [PubMed]
8. Goyer RA. Toxic effects of metals. In: Klaassen CD, editor. Cassarett and Doull's Toxicology: The Basic Science of Poisons. New York: McGraw-Hill Publisher; 2001
9. Herawati N, Suzuki S, Hayashi K, Rivai IF, Koyoma H. Cadmium, copper and zinc levels in rice and soil of Japan, Indonesia and China by soil type. Bulletin of Environmental Contamination and Toxicology. 2000;64:33-39 [PubMed]
10. Shallari S, Schwartz C, Hasko A, Morel JL. Heavy metals in soils and plants of serpentine and industrial sites of Albania. Science Total Environment. 1998;19209:133-142
11. Nriagu JO. A global assessment of natural sources of atmospheric trace metals. Nature. 1989;338:47-49

12. Arruti A, Fernández-Olmo I, Irabien A. Evaluation of the contribution of local sources to trace metals levels in urban PM2.5 and PM10 in the Cantabria region (Northern Spain). Journal Environmental Monitoring. 2010;12(7):1451-1414 [PubMed]

13. Sträter E, Westbeld A, Klemm O. Pollution in coastal fog at Alto Patache, Northern Chile. Environmental Science Pollution Research International. 2010. [Epub ahead of print] [PubMed]

14. Pacyna JM. Monitoring and assessment of metal contaminants in the air. In: Chang LW, Magos L, Suzuli T, editors. Toxicology of Metals. Boca Raton, FL: CRC Press; 1996

15. Poguberović SS, Krčmar DM, Dalmacija BD, Maletić SP, Tomašević-Pilipović DD, Kerkez DV, Rončević SD. Removal of Ni (II) and Cu (II) from aqueous solutions using 'green'zero-valent iron nanoparticles produced by oak and mulberry leaf extracts. Water Science and Technology. 2016;74(9):2115-2123

16. Fu F, Wang Q. Removal of heavy metal ions from wastewaters: A review. Journal of Environmental Management. 2011;92(3):407-418

17. Duan J, Gregory J. Coagulation by Hydrolysing Metal Salts. UK: Elsevier B. V; 2002

18. Swaine DJ. The Significance of Trace Elmer: in Solving Petrogenetic, Problems and Controversies. Athens, Greece: Theophrastus PublicJtii 3s S.A.; 1983. pp. 521-532

19. Ren D. Mineral matter in coal. In: Dexin H, editor. Coal petrology of China. Xuzhou: 1996. pp. 67-78

20. Singh R, Gautam N, Mishra A, Gupta R. Heavy metals and living systems: An overview. Indian Journal of Pharmacology. 2011;43:246. DOI: 10.4103/0253-7613.81505. [PMC free article] [PubMed]

21. Piotrowska-Seget Z, Cycoń M, Kozdroj J. Metal-tolerant bacteria occurring in heavily polluted soil and mine spoil. Applied Soil Ecolology. 2005;28:237-246. DOI: 10.1016/j.apsoil.2004.08.001

22. Gadd GM. Metals, minerals and microbes: Geomicrobiology and bioremediation. Microbiology, 2010;156:609-643. DOI: 10.1099/mic.0.037143-0. [PubMed]

23. Golyshina OV, Yakimov MM, Lünsdorf H, Ferrer M, Nimtz M, Timmis KN, Wray V, Tindall BJ, Golyshin PN. Acidiplasma aeolicum gen. Nov., sp. Nov., a Euryarchaeon of the family ferroplasmaceae isolated from a hydrothermal pool, and transfer of Ferroplasma cupricumulans to Acidiplasma cupricumulans comb. International Journal of Systematic and Evolutionary Microbiology. 2009;59:2815-2823. DOI: 10.1099/ijs.0.009639-0. [PubMed]

24. Chevrel S, Croukamp L, Bourguignon A, Cottard T. A remote-sensing and GIS-based integrated approach for risk-based prioritization of gold tailings facilities–Witwatersrand, South Africa. In: Fourie AB, Tibbett M, Weiersbye IM, editors. Proceedings of the Third International Seminar on Mine Closure, 14–17 October. Johannesburg, South Africa: Australian Centre for Geomechanics; 2008. pp. 639-650

25. Ezzatahmadi N, Ayoko GA, Millar GJ, Speight R, Yan C, Li J, Li S, Zhu J, Xi Y. Clay supported nanoscale zero-valent iron composite materials for the remediation of contaminated aqueous solutions: A review. Chemical Engineering Journal. 2016

26. Fazlzadeh M, Rahmani K, Zarei A, Abdoallahzadeh H, Nasiri F, Khosravi R. A novel green synthesis of zero valent iron nanoparticles (NZVI) using three plant extracts and their efficient application for removal of Cr(VI) from aqueous solutions. Advanced Powder Technology. 2017;28(1):122-130

27. Jiang W, Cai Q, Xu W, Yang M, Cai Y, Dionysiou DD, O'Shea KE. Cr (VI) adsorption and reduction by humic acid coated on magnetite. Environmental Science and Technology. 2014;48(14):8078-8085

28. Aliyah NA. Adsorption of lead using rice husk. [Doctoral dissertation] UMP. 2012

29. Lottermoser B.Mine Wastes: Characterization, Treatment and Environmental Impacts. New York, NY, USA: Springer; 2007. pp. 1-290

30. Marsden DD. The current limited impact of Witwatersrand gold mine residues on water pollution in the Vaal River system. Journal of the South African Institute of Mining and Metallurgy. 1986;86:481-504

31. Jones GA, Brierly SE, Geldenhuis SJJ, Howard JR. Research on the Contribution of Mine Dumps to the Mineral Pollution Load at the Vaal Barrage. Water Research Commission Report No. 136/1/89. Pretoria: Water Research Commission; 1988

32. Vega F, Covelo E, Andrade M, Marcet P. Relationships between heavy metals content and soil properties in minesoils. Analytica Chimica Acta. 2004;524:141-150. DOI: 10.1016/j.aca.2004.06.073

33. Vega FA, Covelo EF, Andrade M. Competitive sorption and desorption of heavy metals in mine soils: Influence of mine soil characteristics. Journal Colloid Interface Science. 2006;298:582-592. DOI: 10.1016/j.jcis.2006.01.012

34. Vega FA, Covelo EF, Andrade M. Competitive sorption and desorption of heavy metals in mine soils: Influence of mine soil characteristics. Journal of Colloid and Interface Science. 2006;298:582-592. DOI: 10.1016/j.jcis.2006.01.012. [PubMed]

35. Rafiei B, Bakhtiari NM, Hashemi M, Khodaei A. Distribution of heavy metals around the Dashkasan Au mine. Internationa Journal of Environmental Research. 2010;4:647-654

36. Mitileni C, Gumbo J, Muzerengi C, Dacosta F. The distribution of toxic metals in sediments: Case study of new union gold mine tailings, Limpopo, South Africa. In: Mine Water – Managing the Challenges. Aachen, Germany: IMWA; 2011

37. Harish E, David M. Assessment of potentially toxic cyanide from the gold and copper mine ore tailings of Karnataka, India. International Journal of Science and Technology. 2015;3:171-178

38. Da Silva EF, Zhang C, Pinto LSS, Patinha C, Reis P. Hazard assessment on arsenic and lead in soils of Castromil gold mining area, Portugal. Applied Geochemistry. 2004;19:887-898. DOI: 10.1016/j.apgeochem.2003.10.010

39. Akabzaa TM. Boom and Dislocation: A Study of the Social and Environmental Impacts of Mining in the Wassa West District of Ghana. Accra, Ghana: Third World Network, Africa Secretariat; 2000

40. Dold B. Basic Concepts in Environmental Geochemistry of Sulfidic Mine-Waste Management. Rijeka, Croatia: InTech Open Access Publisher; 2010

41. Adler R, Rascher JA. Strategy for the Management of Acid Mine Drainage from Gold Mines in Gauteng. Pretoria, South Africa: CSIR; 2007

42. Acevedo F. The use of reactors in biomining processes. Electronic Journal of Biotechnology. 2000. DOI: 10.2225/vol3-issue3-fulltext-4

43. Van Hille RP, van Wyk N,Froneman T, Harrison ST. Dynamic Evolution of the Microbial Community in Bio Leaching Tanks. Pfaffikon, Switzerland: Trans Tech Publications; 2013. pp. 331-334 (Advanced Materials Research)

44. Brehm U, Gorbushina A, Mottershead D. The role of microorganisms and biofilms in the breakdown and dissolution of quartz and glass. Palaeogeography, Palaeoclimatology, Palaeoecology. 2005;219:117-129. DOI: 10.1016/j.palaeo.2004.10.017

45. Baker BJ, Banfield JF. Microbial communities in acid mine drainage. FEMS Microbiological Ecology. 2003;44:139-152. DOI: 10.1016/S0168-6496(03)00028-X. [PubMed]

46. Telmer K, Stapper DA. Practical Guide: Reducing Mercury Use in Artisanal and Small-Scale Gold Mining. Nairobi, Kenya: Geneva, Switzerland: United Nations Environment Programme; 2012

47. Maree JP. Treatment of industrial effluent for neutralization and sulphate removal. A thesis submitted for PhD at the North West University, RSA. 2004

48. Flynn CM. Hydrolysis of inorganic iron(III). Journal of American Chemical Society. 1984;84:31-41

49. Coetzee H et al. WRC, Council for Geosience. Report No 1214/1/06. 2004

50. Tchounwou PB, Yedjou CG, Patlolla AK, Sutton DJ. Heavy Metals Toxicity and the Environment. The publisher's final edited version of this article is available at EXS. 2005

51. Ntwampe IO, Waanders FB, Bunt JR. Reaction dynamics of iron and aluminium salts dosage in AMD using shaking as an alternative technique in the destabilization-hydrolysis process. International Journal of Science. 2016;4:5-23

52. Ntwampe IO, Waanders FB, Bunt JR. Destabilization dynamics of clay and acid-free polymers of ferric and magnesium salts in AMD without pH adjustment. Water Science and Technology. 2016;74(4):861-875

53. Ntwampe IO, Waanders FB, Bunt JR. Turbidity removal efficiency of clay and a synthetic af-PACl polymer of magnesium hydroxide in AMD treatment. International Journal of Science. 2015;4:88-104

54. Ntwampe IO, Moothi K. Determination of the efficiencies of a flocculant of clay and PFCl/PACl with Na2CO3 in the removal of TSS from the AMD. Accepted by Journal of Environmental Science and Technology. 2017

55. Ntwampe IO, Jewell LL, Hildebrandt D, Glasser D. The effect of mixing on the treatment of paint wastewater with Fe3+ and Al3+ salts. Journal of Environmental Chemistry and Ecotoxicology. 2013;5(1):7-16

56. Coulson JM, Richardson JF. Chemical Engineering. 6th ed. Vol. 1. London: Elsevier, Butterworth-Heinemann; 1999

57. Sincero AP, Sincero GA. Physical-Chemical Treatment of Water and Wastewater. Londao, USA: IWA Publishing; 2003

58. Thomas PR, Allen D, McGregor DL. Evaluation of combined chemical and biological nutrients removal. Water Science Technology. 1996;34:285-292

59. Pratt C, Shilton A, Pratt S, Haverkamp RG, Elmetri I. Effects of redox potential and pH changes on phosphorus retention by melter slag filters treating wastewater. Environmental Science and Technology. 2007;4(18):6583-6590

60. Nozaic DJ, Freeze SD, Thompson P. Water Science and Technology: Water Supply. 2001;1:43-50

61. Ghaly AE, Snow A, Faber BE. Treatment of grease filter washwater by chemical coagulation. Canadian Biosystem Engineering. 2006;48:6.13-6.22

62. Kabata-Pendia 3rd A, editor. Trace Elements in Soils and Plants. Boca Raton, FL: CRC Press; 2001

63. Niyogi DK, William Jr ML, McKnight DM. Effects of stress from mine drainage on diversity, biomass, and function of primary producers in mountain streams. Ecosystems. 2002;5:554-567

64. Saif S, Tahir A, Chen Y. Green synthesis of iron nanoparticles and their environmental applications and implications. Nanomaterials. 2016;6(11):209

65. Demirci UB, Akdim O, Andrieux J, Hannauer J, Chamoun R, Miele P. Sodium borohydride hydrolysis as hydrogen generator: Issues, state of the art and applicability upstream from a fuel cell. Fuel Cells. 2010;10(3):335-350

Chapter 4 DEVELOPING EFFECTIVE HEAVY METAL REMOVAL AGENTS USING ZETA POTENTIAL MEASUREMENTS

1. INTRODUCTION: AN OVERVIEW

The issue of environmental pollution is escalating, necessitating more environmental regulations to manage and mitigate the detrimental effects of industrial operations on the environment. One of the most significant scientific and technological concerns is the preservation of water quality, which arises from the challenges of limited supply and contamination by various chemical elements, including heavy metals [1]. Major sources of heavy metals include mineral processing, pulp and paper industry, graphic industry printing, metallurgical processes, printed board manufacturing, nuclear industry, mining operations, battery manufacturing, leather tanning, smelting, petrochemical industry, metal finishing and plating, semiconductor manufacturing, textile dyeing, ceramic industry, and other water-intensive industries. These industries generate wastewater with a high concentration of heavy metals (>200 mg/L), which cannot be discharged into water bodies without prior treatment [2, 3]. Common heavy metals frequently found in industrial effluent include nickel, zinc, silver, lead, iron, chromium, copper, arsenic, cadmium, and uranium [1]. One of the primary justifications for avoiding the release of heavy metals into water bodies is their nonbiodegradable nature, toxicity, and ability to accumulate in living organisms, particularly humans. This accumulation can lead to severe health issues such as cancer, damage to the nervous system, and kidney failure. At high concentrations, heavy metals can be fatal [4]. Various technologies are available for treating wastewater that contains heavy metals. These include adsorption using different types of adsorbent materials such as polymers, carbon, nanomaterials, clays, and zeolites. Other methods include electrodeposition, membrane filtration using techniques like ultrafiltration, nanofiltration, and reverse osmosis, coagulation-flocculation, chemical precipitation using hydroxides, sulfides, and chelating precipitations, electrocoagulation, electrodialysis, ion exchange, biological treatment, photocatalysis, and electroflotation. Every technology possesses specific benefits and drawbacks. In light of stringent environmental regulations, companies require more effective wastewater treatment methods to remove metals that are either suspended or dissolved [5, 6, 7, 8, 9, 10, 11, 12, 13, 14, 15, 16, 17, 18, 19, 20]. This study proposes the utilization of waste materials from the food industry and renewable sources, such as shrimp shell waste, nopal mucilage, nejayote, mesquite seeds, coffee, and tule residues, as biodegradable agents for coagulation, flocculation, and chelation [21, 22]. In this study, zeta potential measurements were employed to investigate the chemical interactions between the functional groups of biopolyelectrolytes (BPE) and the heavy metals found in wastewater at the molecular level. The zeta potential is a crucial factor in determining the surface charge density of the BPE, as well as its isoelectric point (IEP). It also helps in formulating the approach for introducing the BPE into wastewater containing heavy metals.

2. EXPERIMENTAL SECTION

2.1. Materials and methods

2.1.1. Materials

The heavy metal wastewater model was prepared with the nitrate salts of each metal ion: $Ca(NO_3)_2$, $Zn(NO_3)_2$, $Cd(NO_3)_2$, $Cu(NO_3)_2$, $Ni(NO_3)_2$, $Pb(NO_3)_2$, and $Cr(NO_3)_3$. Sodium alginate (AG) with viscosity (5–40 cP) were purchased from Sigma Aldrich.

2.2. Methods

2.2.1. Extraction and physico-chemical analysis of biopolyelectrolytes

Chitosan is generated from discarded shrimp shells using the modified technique suggested by Goycoolea et al. [23]. The process of obtaining maize gum involved fractional separation using hexane, ethanol, hydrochloric acid, isopropanol, acetone, and methanol. This was achieved through the phases of desalmidonado, deproteinization, delipidation, and delignification, which were adapted from the methods provided in references [24, 25, 26]. The extraction of mesquite gum from mesquite seeds was conducted following the method described by Sciarini et al. [27]. The technology for extracting nopal mucilage (NM) was disclosed by Cárdenas et al. [28]. The delignification process was used to extract lignin from the dark liquors of coffee and tule [29]. The procedure involves the precipitation of lignins through acidification of black liquors derived from tule and coffee dregs. The FTIR spectra were obtained using an FT-IR Spectrometer manufactured by Perkin Elmer. The biopolyelectrolyte was analyzed using the Universal Attenuated Total Reflectance (UATR Two) attachment.

2.2.2. $\zeta = f$ (pH) profiles of the alginate, maize gum, mesquite gum, nopal mucilage, coffee and tule lignin and chitosan

The zeta potential of the anionic, cationic, and neutral biopolyelectrolytes was quantified using a Stabino Particle Charge Mapping instrument (Microtrac). The measurements were conducted at the prevailing temperature using cuvettes made of teflon. The impact of pH on the zeta potential behavior of each biopolyelectrolyte was investigated throughout a pH range of 2-11 using 0.1 M NaOH and 0.1 M HCl. This study aimed to identify the isoelectric point (IEP) and the maximum degree of ionization of the biopolyelectrolytes [30].

2.2.3. Evaluation of the chelating performance of the BPE

The performance of the BPE in the separation of heavy metals present in the wastewater model was evaluated using stability analysis to determine the kinetics of coagulation-flocculation, morphology and elemental composition of the formed flocs, content of metals remaining in the treated water, in each test of dosage of BPE to synthetic heavy metals wastewater. Samples of flocs obtained were examined in a SEM ZEISS EVO-MA15, equipped with an EDS (energy dispersive spectroscopy) BRUKER detector for heavy metals-relative estimations. Agilent Technologies 4200 MP-AES were used to determine metals concentration. Stability tests were carried out, performing a transmittance scan every 25 s for 30 min using Turbiscan Lab® Expert.

3. RESULTS AND DISCUSSION

3.1. Physicochemical characterization of biopolyelectrolytes

Considering that the surface functional groups of the BPE are responsible for the molecular interactions between the biopolyelectrolytes (AG, Ch, Mag, nopal mucilage, Meg, CL, and TL) (see Fig. 1) and the heavy metals contained in the wastewater, they were studied by FTIR [31]. The most common BPE functional groups are carbonyl, phenolic, acetamido, alcoholic, amido, amino, and sulfhydryl, which have an affinity for heavy metal ions to form complexes or metal chelates [32].

In Fig. 2, the FTIR spectra corresponding to the anionic biopolyelectrolytes that have higher density of negative surface charge are shown. In Fig. 2a and c, the characteristic signals are shown to the three main compounds present in the coffee and tule lignin: p-coumaryl alcohol, coniferyl alcohol, and sinapyl alcohol. Assignment bands shown at the wavelength 3500–3100 cm^{-1} are attributed to the stretching vibrations of the OH, the stretching vibrations at 2910–2835 cm^{-1} are attributed to the CH group, and the stretching vibrations at 1700–1550 cm^{-1} are attributed to the C—C group. There are also other signals at 1230 cm^{-1} and 1012 cm^{-1}, corresponding to the vibration bands of the C—O.

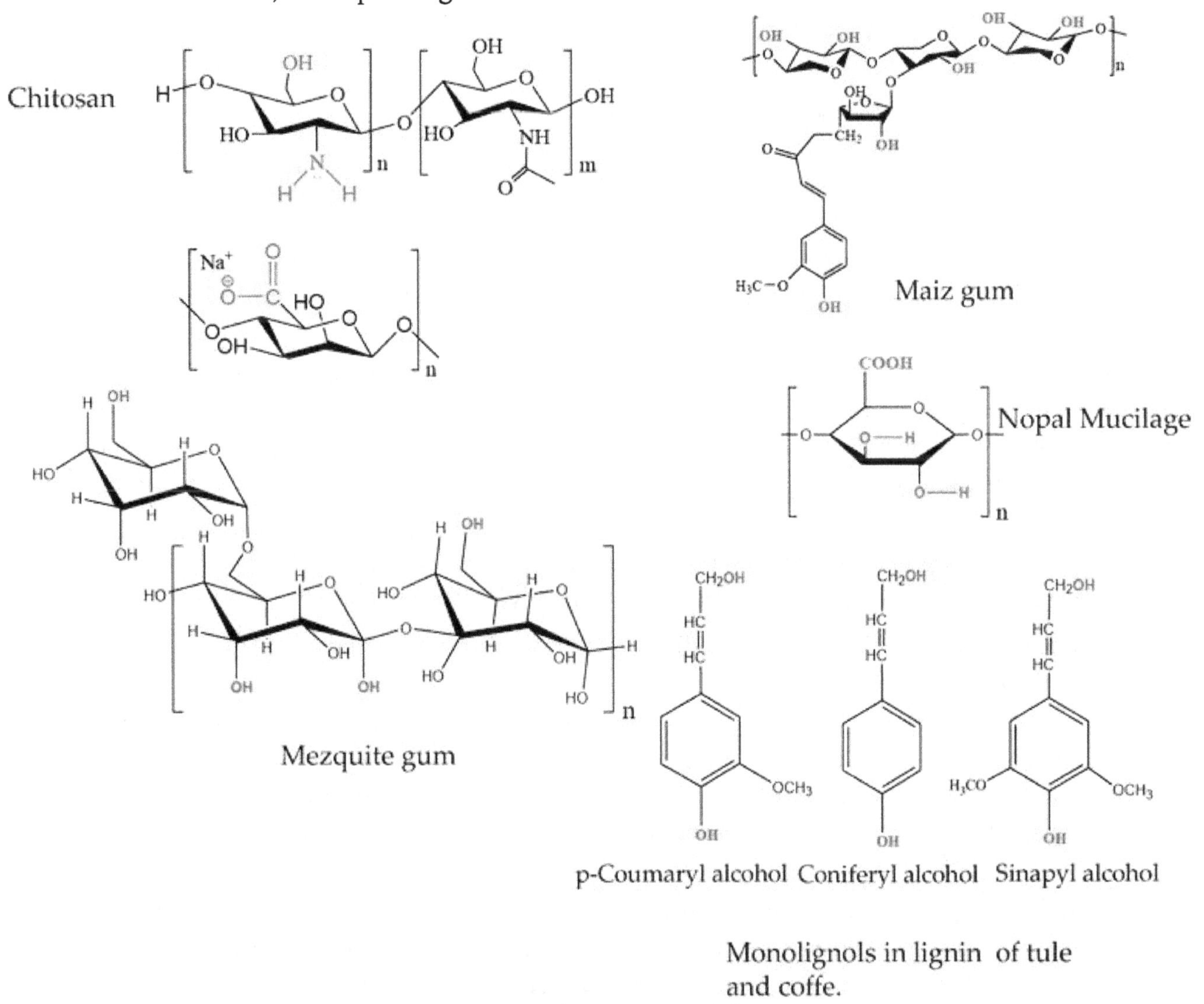

Fig. 1.Chemical structure of the BPE.

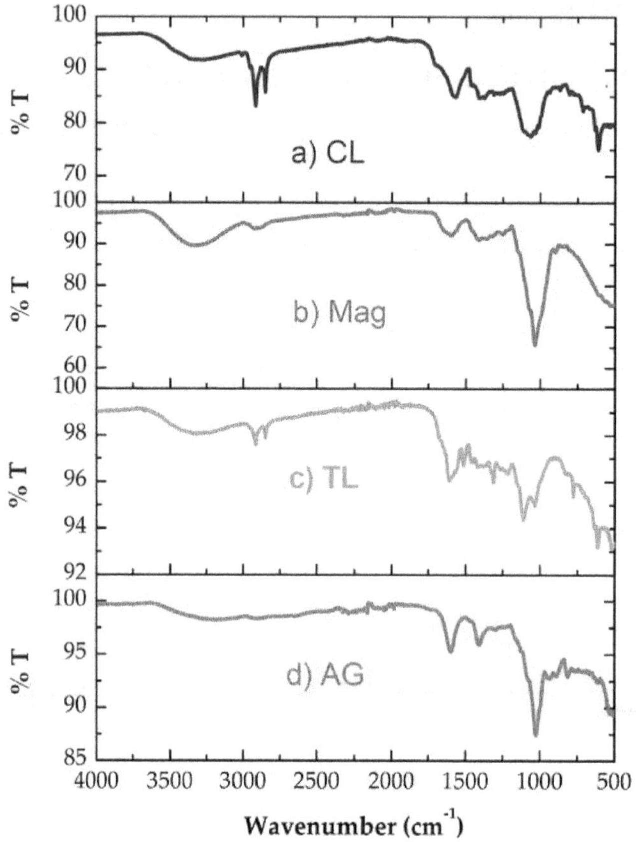

Fig. 2. FTIR spectra of biopolyelectrolytes with higher density of negative surface charge: a) Coffe lignin, b) Maize gum, c) Tule lignin, d) Alginate.

Fig. 2b shows the FTIR spectrum of the anionic biopolyelectrolyte (maize gum) obtained from nejayote. The signal observed at 2923 cm^{-1} corresponds to the vibrations of the CH bonds of the methyl groups and the wide band that appears at 3421 cm^{-1} is due to the stretches of the HO bonds of the hydroxyl groups present in the macromolecule of the maize gum. The OH groups are those that give anionic character to this biopolyelectrolyte and that favors its interaction with metal ions present in the wastewater.

Fig. 2d shows the characteristic signals of the alginate macromolecule, in which the band at the 3500–3200 cm^{-1} region is attributed to the hydroxyl groups and the signal at 1500 cm^{-1} corresponds to the carboxyl groups found distributed along the anionic biopolyelectrolyte chain.

Fig. 3 shows the FTIR spectra of the BPE of cationic, non-ionic and slightly anionic character, considering the zeta potential values.

In the spectrum of Ch (Fig. 3a), the characteristic bands at 3450 cm^{-1} (—OH group), at 3292 cm^{-1} (NH group), at 2919 and 2862 cm^{-1} (CH group), at 1655 cm^{-1} (amide I), at 1580 cm^{-1} (double group —NH2), at 1313 cm^{-1} (amide III), at 1154 cm^{-1} (antisymmetric stress of the COC bridge), at 1082 and 1032 cm^{-1} (skeletal vibrations characteristic of the pyranotic structure) and 896 cm^{-1} (CH tension of the anomeric groups) are shown.

The absorption bands that are common for mesquite gum (Fig. 3b) correspond to the OH-group centered around 3400 cm^{-1} and the C—H found in the band at 2900 cm^{-1}. Other characteristic absorption bands are observed in the bands at 1150–950 cm^{-1}, which are associated with the signals of C—O and carbohydrates (C—O—H). Also, in the bands at 870 cm^{-1} and 813 cm^{-1} indicate the presence of units of β-D mannopyranose and units of α-D galactopyranose.

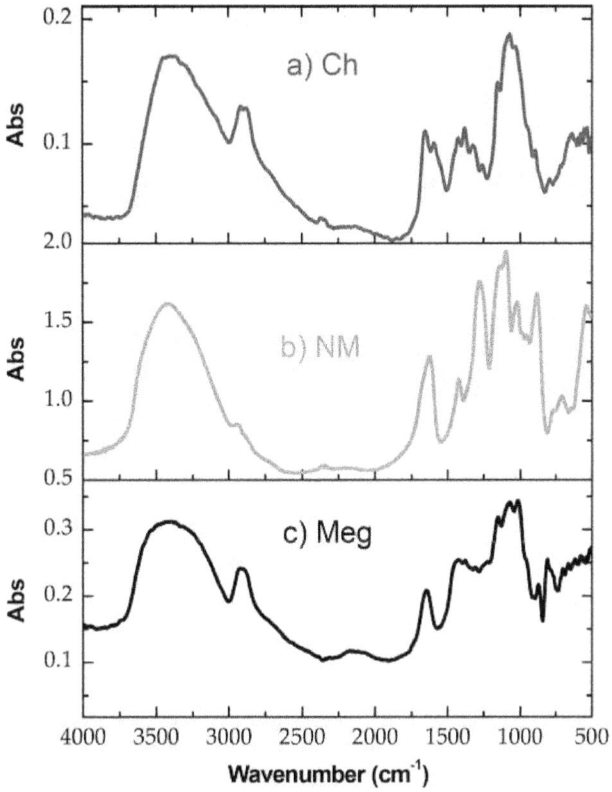

Fig. 3.FTIR spectra of the BPE of cationic, non-ionic, and slightly anionic character: a) Chitosan, b) Nopal mucilage, c) Mezquite gum.

The nopal mucilage spectrum is shown in Fig. 3c, and the spectral region at 1350 cm-1 to 1750 cm-1 assigned to the vibrations of the bands of the nonionized COOH carboxylic groups is observed. Likewise, ionization leads to its disappearance and to the appearance of new bands associated with the symmetric and asymmetric vibrations of the COO— groups, centered approximately between at 1600–1650 cm-1 and at 1400–1450 cm-1.

3.2. ζ = f (pH) profiles of the BPE

In Fig. 4, the ζ = f(pH) plot shows the charge density variation for AG, NM, Meg, CL, TL, Mag and Ch with respect to pH. The change in pH had a distinct effect with each biopolyelectrolyte because of the difference between the functional groups present in the chains of the macromolecules. Sodium alginate is a strong anionic biopolyelectrolyte and maintains negative zeta potential values (ζ = - 480.0 mV) between pH 3.5 and 10, attributed to the carboxylate group (—COO-Na^+) and was not observed their IEP.

Chitosan has a cationic character at pH < 6.0, reaching a maximum value in zeta potential ζ = 55 mV at pH 3.0. This behavior is attributed to the amine groups that are protonated at pH < pka = 6.3. At pH = 6.5, the isoelectric point of the chitosan was reached and at pH > IEP, the values of zeta potential remained close to neutrality (ζ = 0), and this corresponds to the hydrophobic character acquired by the chitosan macromolecule due to the deprotonation of the amine groups.

The zeta potential values of the NM indicate the acid-base behavior of the carboxyl groups present in the macromolecule. The NM has its isoelectric point at pH = 4.0, observing the formation of NM aggregates at pH < IEP. At pH > IEP, it had an anionic character, and this derived that the carboxyl groups are ionized at pH > pka = 3.5. The behavior zeta potential vs. pH of anionic BPE obtained of nejayote is shown in Fig. 4, which has a high negative charge density (ζ = -35 mV) in the pH range 6–12, having the isoelectric point close to pH = 2.0.

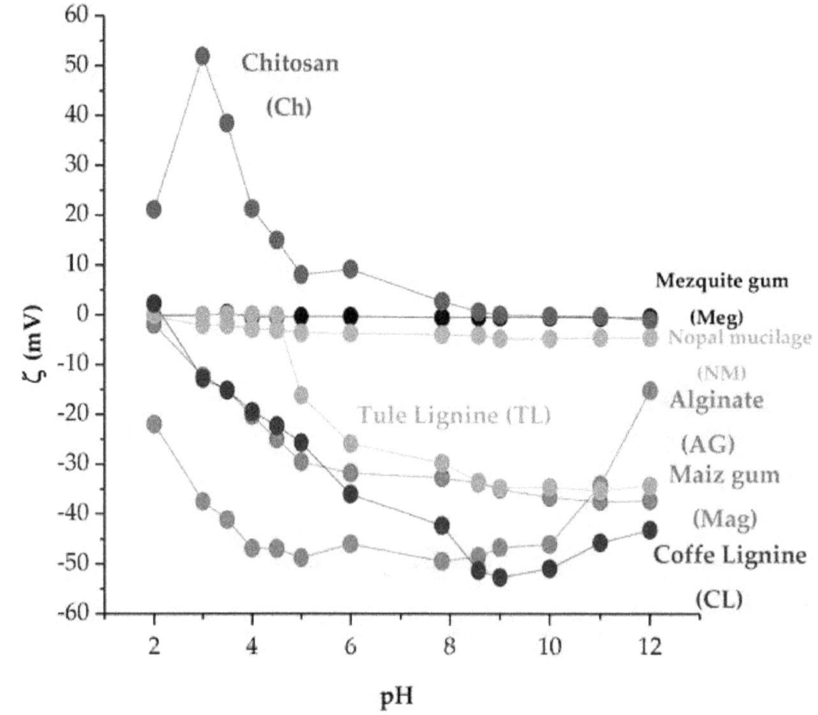

Fig. 4. **ζ = f (pH) profiles of the BPE.**

The coffee lignin showed an average negative zeta potential value (ζ = -40 mV) at a pH of 6–12, decreasing linearly according to the low pH (ζ = -40 mV to ζ = -10 mV). The lignin chains of coffee present their IEP at pH = 2.0.

Similarly, tule lignin showed this tendency, but with a lower negative charge density (ζ = -30 mV) at pH 2–12, and its IEP at pH = 4.0. This behavior is attributed to the difference in the composition of the three main monomers that contain the lignin and which is a function of its source of origin.

The profiles of ζ = f (pH) of the BPE allow to determine and predict the physicochemical behavior of the macromolecules in the coagulation processes of flocculation of residual water with heavy metals. Through these profiles, it is possible to determine the pH at which the BPE have their highest surface charge density (negative or positive), define a strategy to make additions of BPE mixtures sequentially, and orient the dose needed to carry out the separation of heavy metals from waste water.

3.3. Evaluation of the chelating performance of the BPE

The BPE have a great potential to get to replace the synthetic coagulant-flocculant-chelating agents, considering their chemical structure and their surface charge density in aqueous solution. The BPE contain in their structure different functional groups that give rise to the surface charge of the macromolecules and that can interact with the heavy metals frequently found in industrial wastewater. As shown in Fig. 5, the evaluation of the performance of each BPE can be performed for the separation of metals in a natural way as well as to carry out strategies for modifying the macromolecules to make them more efficient in terms of coagulation-flocculation kinetics and removal of heavy metals. In this chapter, the evaluation of chitosan for the separation of heavy metals in wastewater model is presented.

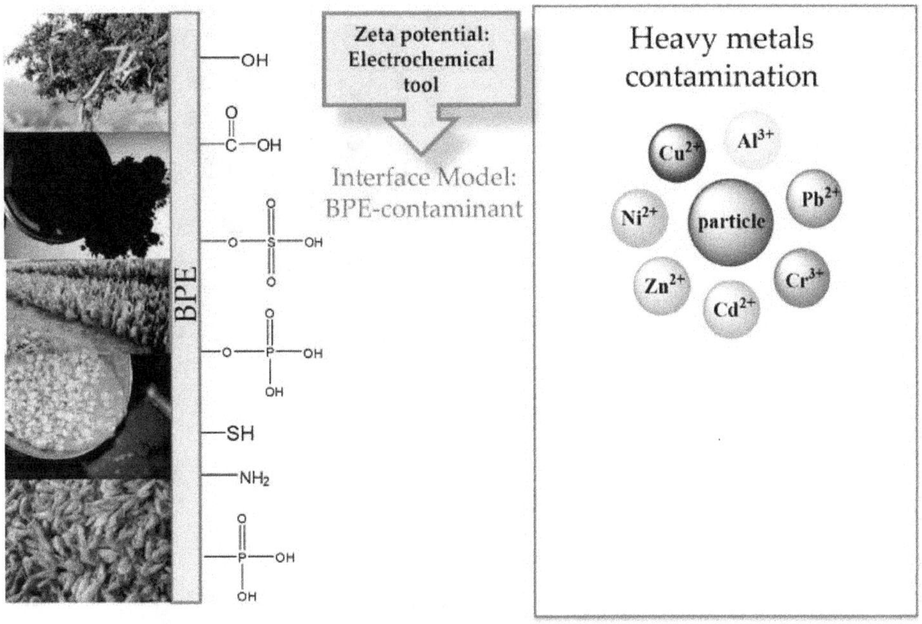

Fig. 5. Outline of BPE-contaminant interface model.

In Fig. 6, the profile of $\zeta = f$ (Ch dose) is shown with the residual water model with heavy metals. The model wastewater has a zeta potential value ($\zeta = -2.8$ mV) at pH = 5.0, as the Ch dose increases, the zeta potential of the wastewater rises linearly ($\zeta = -2.8$ mV to $\zeta = -0.5$ mV), in a dose range of Ch 0–1.5 mg. The isoelectric point of the wastewater model was reached with a dose of 2.5 mg of Ch.

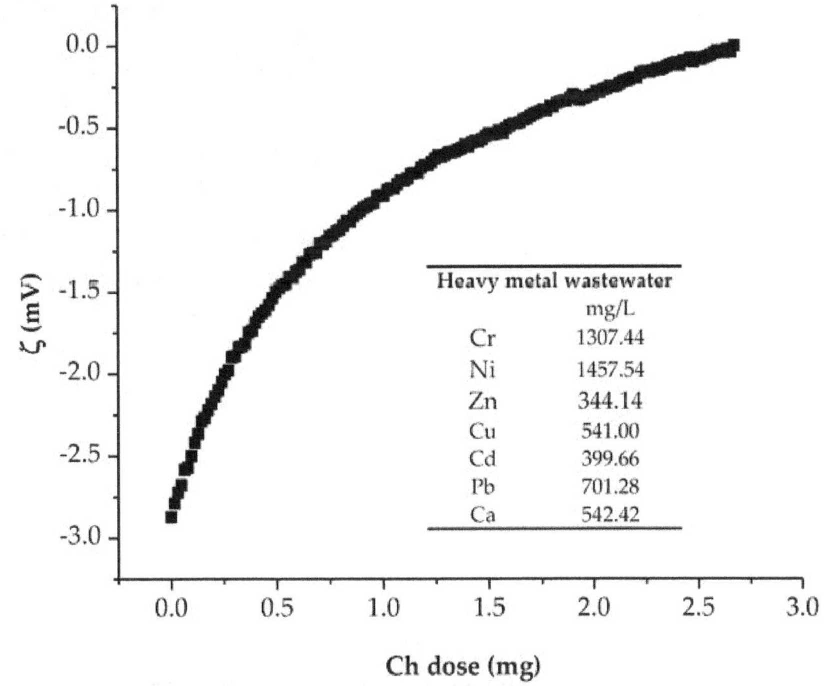

Fig. 6. $\zeta = f$ (Ch dose) profile with heavy metals wastewater model.

The superficial chemical analysis of the flocs formed in the coagulation-flocculation process using the best dose of Ch (2.5 mg), at pH = 5.0, Fig. 7a shows the morphology of the chitosan without interacting with the heavy metals. The EDS spectrum of chitosan shows that it does not contain heavy metals in its natural form. In Fig. 7b, the morphology of flocs formed with heavy metals and chitosan is shown.

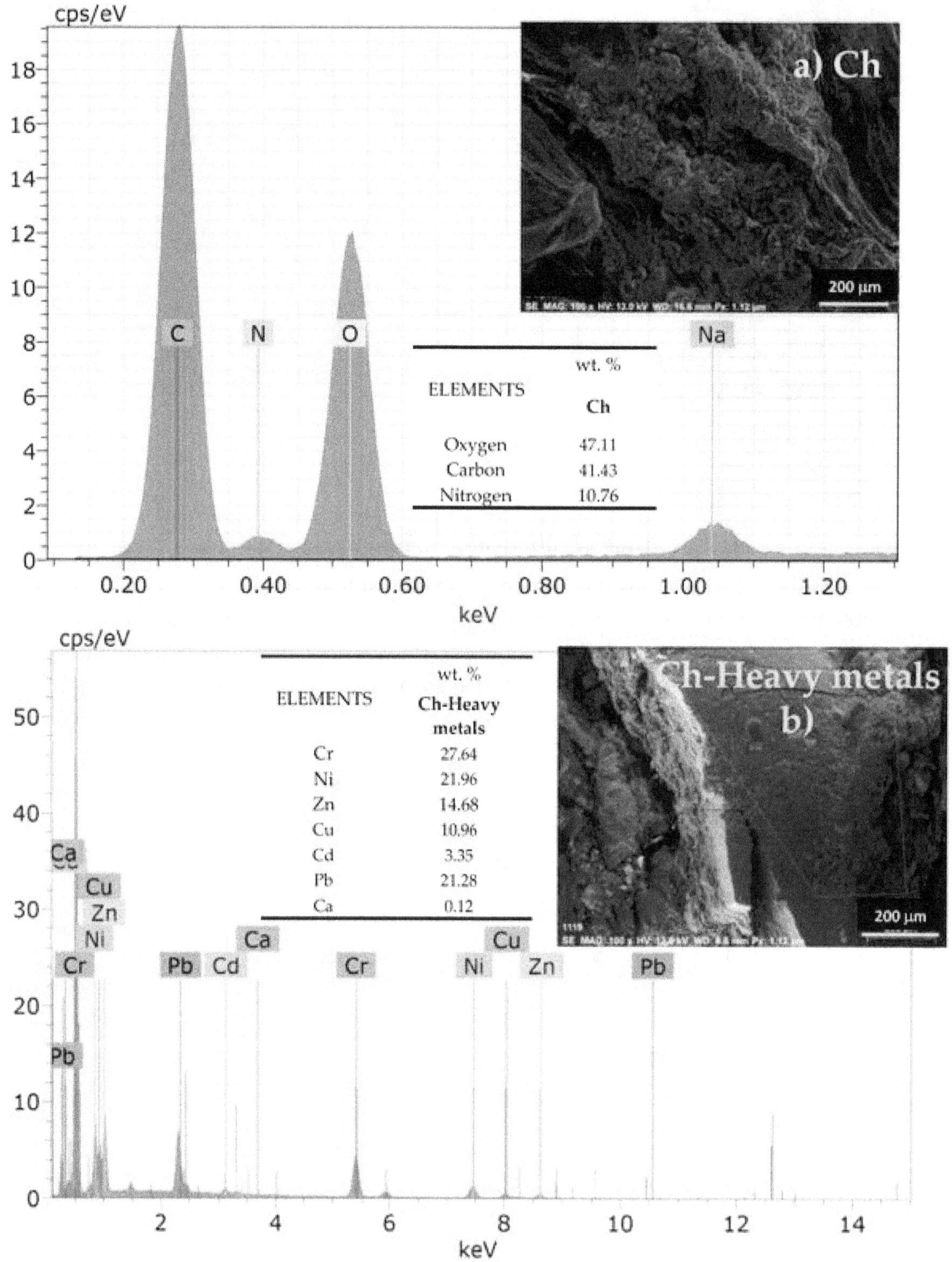

Fig. 7. SEM micrograph and EDS spectrum of (a) Ch and (b) flocs of Ch-heavy metals.

The efficacy of chitosan in separating heavy metals is evidenced by an 85% reduction in heavy metal content in the treated wastewater. Fig. 8 displays the relationship between the backscattering profile (% BS) and the height of the vial. The optimal dose of chitosan for clarity kinetics is 187.49%T/h, while for sediment kinetics it is 93.96 mm/h.

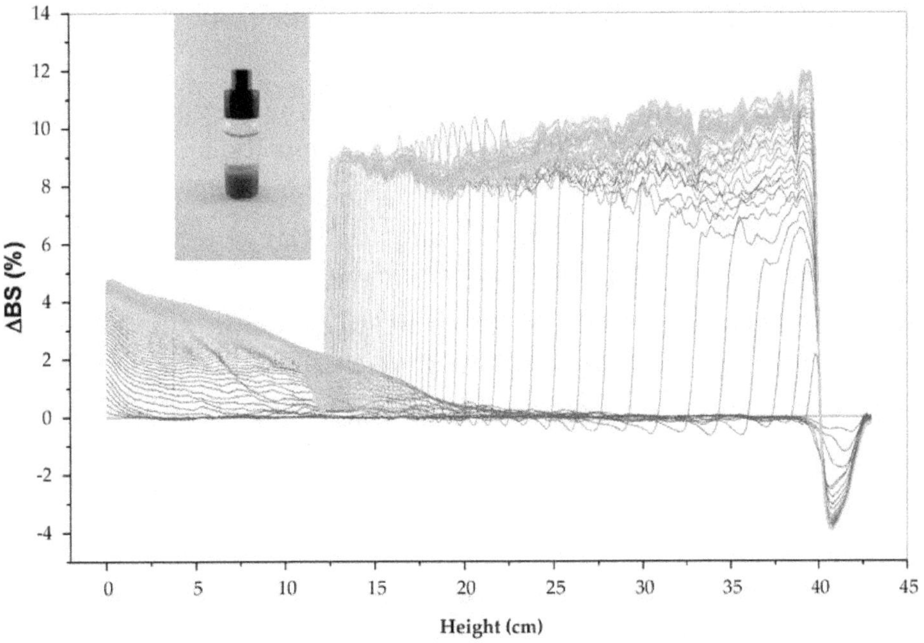

Fig. 8. Backscattering profile of NP at optimum dose; data are reported as a function of time (0–120 min).

4. CONCISE SUMMERY

An alternative for treating wastewater with a high concentration of heavy metals involves utilizing BPE derived from food sector waste, thereby enhancing its worth. The zeta potential enables the targeted application of coagulant-flocculant-chelating agents to effectively separate heavy metals. It helps determine the optimal physicochemical conditions to enhance the efficiency of coagulation-flocculation processes. Achieving an 85% separation efficiency of heavy metals and a clarity kinetics of 187.49% T/h, as well as a sediment kinetics of 93.96 mm/h, was seen in wastewater treatment using a 2.5 mg dosage of chitosan.

REFERENCES

1. Khan A, Badshah S, Airoldi C. Dithiocarbamated chitosan as a potent biopolymer for toxic cation remediation. Colloids and Surfaces B: Biointerfaces. 2011;88:88-95
2. Kurniawan TA, Chan GYS, Lo WH, Babel S. Physico-chemical treatment techniques for wastewater laden with heavy metals. Chemical Engineering Journal. 2006;118:83-98
3. Babel S, Kurniawan TA. Cr (VI) removal from synthetic wastewater using coconut shell charcoal and commercial activate carbon modified with oxidizing agents and/or chitosan. Chemosphere. 2004;54:951-967
4. Blais JF, Djedidi Z, Cheikh RB, Tyagi RD, Mercier G. Metals precipitation from effluents: Review. Practice Periodical of Hazardous, Toxic, and Radioactive, Waste Management. 2008;12:135-149
5. Chen B, Qu R, Shi J, Li D, Wei Z, Yang X, Wang Z. Heavy metal and phosphorus removal from water by optimizing use of calcium hydroxide and risk assessment. Environmental Pollution. 2012;1:38-54
6. Charerntanyarak L. Heavy metals removal by chemical coagulation and precipitation. Water Science and Technology. 1999;39:135-138
7. Tünay O, Kabdasli NI. Hydroxide precipitation of complexed metals. Water Research. 1994;28:2117-2124

8. Andrus ME. A review of metal precipitation chemicals for metal-finishing applications. Metal Finishing. 2000:20-23

9. Li YJ, Zeng XP, Liu YF, Yan SS, Hu ZH, Ya M. Study on the treatment of copper-electroplating wastewater by chemical trapping and flocculation. Separation and Purification Technology. 2003;31:91-95

10. Lazaridis NK, Matis KA, Webb M. Flotation of metal-loaded clay anion exchangers. Part I: The case of chromate. Chemosphere. 2001;42:373-378

11. Doyle FM, Liu ZD. The effect of triethylenetraamine (trien) on the ion flotation of Cu2+ and Ni2+. Journal of Colloid and Interface Science. 2003;258:396-403

12. Kongsricharoern N, Polprasert C. Electrochemical precipitation of chromium (Cr^{6+}) from an electroplating wastewater. Water Science Technology. 1995;31:109-117

13. Holt PK, Barton GW, Mitchell CA. The future for electrocoagulation as a localized water treatment technology. Chemosphere. 2005;59:355-367

14. Pansini M, Colella C, De Gennaro B. Chromium removal from water by ion exchange using zeolite. Desalination. 1991;83:145-157

15. Álvarez E, García A. Purification of metal electroplating wastewater using zeolites. Water Research. 2003;37:4855-4862

16. Abdel-Halim SH, Shehata AMA, El-Shahat MF. Removal of lead ions from industrial waste water by different types of natural materials. Water Research. 2003;37:1678-1683

17. Algarra M, Jiménez MV, Rodríguez E, Jiménez A, Jiménez J. Heavy metals removal from electroplating wastewater by aminopropyl-Si MCM-401. Chemosphere. 2005;59:779-786

18. Karabulut S, Karabakan A, Denizli A, Yûrûm Y. Batch removal of copper (II) and zinc (II) from aqueous solutions with low-rank Turkish coals. Separation and Purification Technology. 2000;18:177-184

19. Juang R, Shiau RC. Metal removal from aqueous solutions using chitosan-enhanced membrane filtration. Journal Membrane Science. 2000;165:159-167

20. Ahn KH, Song KG, Cha HY, Yeom IT. Removal of ions in nickel electroplating rinse water using low-pressure nanofiltration. Desalination. 1999;122:77-84

21. Chang Q, Wang G. Study on the macromolecular coagulant PEX which traps heavy metals. Chemical Engineering Science. 2007;62:4636-4643

22. Chitanu GC, Carpov A. Ecologically benign polymers: The case of maleic polyelectrolytes. Environmental Science and Technology. 2002;36:1856-1860

23. Goycoolea F, Agullo E, Mata R. Sources and processes of obtaining. In: Pastor de Abram A, editor. Chitin and Chitosan: Obtaining, Characterization and Applications. 1st ed. Fondo editorial de la Pontificia Universidad Catolica del Perú; 2004. pp. 103-154

24. Niño-Medina G, Carvajal-Millán E, Lizardi J, Rascon-Chu A, Marquez-Escalante J, Gardea A, Martínez-López A, Guerrero V. Maize processing wastewater arabinoxylans: Gelling capability and cross-linking content. Food Chemistry. 2009;115:1286-1290

25. Zhang Z. Extraction and modification technology of arabinoxylans from cereal byproducts: A critical review. Food Research International. 2014;65:423-436

26. Simpson J. A survey of microorganisms for the production of enzymes that attack the pentosans of wheat flour. Canadian Journal of Microbiology. 1954;1:131-139. DOI: 10.1139/m55-017

27. Sciarini LS, Maldonado F, Ribotta PD, Pérez GT, León AE. Chemical composition and functional properties of Gleditsia triacanthos gum. Food Hydrocolloids. 2009;23:306-313

28. Cárdenas A, Goycoolea FM, Rinaudo M. On the gelling behaviour of 'nopal' (Opuntia ficus indica) low methoxyl pectin. Carbohydrate Polymers. 2008;73:212-222

29. Sarwar M, Nasima D, Islam M, Iqbal S. Characterization of lignin isolated from some nonwood available in Bangladesh. Bioresource Technology. 2007;98:465-469

30. López-Maldonado EA, Oropeza-Guzman MT, Jurado-Baizaval JL, Ochoa-Terán A. Coagulation-flocculation mechanisms in wastewater treatment plants through zeta potential measurements. Journal of Hazardous Materials. 2014;279:1-10

31. Abiola Oladoja N. Headway on natural polymeric coagulants in water and wastewater treatment operations. Journal of Water Process Engineering. 2015;6:174-192

32. Bhatnagar A, Sillanpää M, Witek-Krowiak A. Agricultural waste peels as versatile biomass for water purification—A review. Chemical Engineering Journal. 2015;270:244-271

5 Chapter

IMMEDIATE PLANT RESPONSE TO HIGH LEVELS OF HEAVY METAL TOXICITY

1. INTRODUCTION: AN OVERVIEW

Cadmium (Cd) is one of the components of the earth's crust and present in several places or in different The ecosystems on Earth include terrestrial, aquatic, and other types. According to Benavides et al. [1], cadmium is a highly dangerous heavy metal found in the atmosphere, soil, and aquatic systems. It enters the food chain and causes significant environmental issues, leading to health hazards in living organisms such as mutagenesis, lung cancer, convulsions, and brain damage. The mitigation or suppression of cadmium in plants has, consequently, garnered significant interest from society as a whole [2, 3]. It is essential to have knowledge about cadmium, including its physical and chemical properties, as well as isotopic research. Cadmium (Cd) can be introduced into the atmosphere through the combustion of coal, metal mining, and refining processes. This can result in an increase in Cd concentrations in the soil due to atmospheric deposition. The atmospheric deposition of Cd from the air varies in the following sequence: Remote places < Rural areas < Urban areas. As a result of the prolonged impact of cadmium, the countries established a maximum acceptable level. The European Economic Committee introduced the notion of Potential Toxic Elements (PTE). The permissible threshold level for Cadmium in soil is 1.0–3.0 mg/kg of dry soil [4]. When considering the aging of metal in the soil, it is important to differentiate between the long-term presence of all metals in the soil and the long-term presence of metals that are available for biological processes. The process of metal aging is contingent upon the level of acidity in the soil. Studies on metal extractability and liability provide evidence of the aging process. Various names are employed to classify and characterize metals, such as trace metals, transition metals, micronutrients, hazardous metals, and heavy metals. Bjerrum's definition of "heavy metals" is based on the natural density of the metal. According to him, "heavy metals" are metals with a basic density above 7 g cm^{-3}. In 1964, the editor of Van Nostrand's International Encyclopedia of Chemical Science and in 1987, the editors of Allow and Hackh's Chemical Dictionary included metals with a density exceeding 4 g cm^{-3}. The presence of several metals such as Chromium (Cr), Nickel (Ni), Copper (Cu), Manganese (Mn), Mercury (Hg), Cadmium (Cd), Lead (Pb), and metalloids like Arsenic (As), Antimony (Sb), and Selenium (Se) in the surrounding surroundings is causing significant concern. This is particularly true around former mining sites, dumps, tailings piles, and impoundments, as well as in urban and industrial areas. The presence of cadmium in soil is attributed to agricultural waste (20%), sediment (38%), fertilizers (2%), and atmospheric residues (40%) [7]. Cadmium is a highly hazardous metal that has been found to have carcinogenic effects in humans [8]. Cadmium and its derivatives have a higher solubility in water and mobility in soils compared to other heavy metals. They are generally more easily absorbed by living organisms and have a tendency to accumulate in their bodies. It causes cellular damage and death by disrupting calcium (Ca) homeostasis in biological systems. According to the IPCS monographs, cadmium does not play a necessary role in the life of plants or animals. Cadmium can be accumulated in the roots of terrestrial plants, where it is typically linked to the cell walls [9]. Cadmium absorption is mostly regulated by the pH level, which is a crucial determinant. The movement of cadmium is notable in the aboveground portions in comparison to copper and lead. The concentration of cadmium in roots is typically only 2-5 times higher than in the above ground sections of the plant. However, the transfer of cadmium to reproductive or storage organs of the plant is challenging. Putrescine (Put), a polyamine (PA), is a crucial regulator in plants for reducing many types of stress [11, 12, 13, 14, 15, 16, 17, 18]. Additionally, it plays a vital role in apoptosis and programmed death in both animals and plants [19]. Chemically, the PAs are small, positively charged aliphatic amines at the pH levels found in cells. As a result, they are able to connect to molecules that have an opposite charge, such as nucleic acids, acid phospholipids, and proteins [14, 20]. The concentration of free amino acids in plants at the cellular level is determined by the processes of synthesis and degradation of PAs, their binding with phenolic acids, and intracellular transport [14, 15, 21, 22, 23]. The presence of Cd results in oxidative stress, leading to a decrease in agricultural output. Polyamines play a role in enhancing plants' ability to withstand abiotic stress. The elevated levels of polyamines in plants under stress have adaptive relevance due to their role in regulating the cellular ionic environment, preserving membrane integrity, preventing chlorophyll loss, and promoting the production of proteins, nucleic acids, and protective alkaloids. The interaction between polyamines and membrane phospholipids suggests that polyamines have a role in maintaining membrane stability during stressful conditions. Polyamines provide additional protection to the membrane by acting as scavengers of free radicals, so preventing oxidative damage. The production of conjugated PAs in plants is associated with their response to abiotic damage and insufficient mineral supplementation. Polyamine compounds undergo modifications in response to exposure to heavy metals. These effectively stabilize and protect the film systems from the detrimental effects of metal particles. The mycorrhiza Glomus acts as a potent inhibitor of the movement of cadmium ions into the soil solution, thereby protecting plants against cadmium toxicity. The presence of AM fungal colonization was reported in soil that had a high level of contamination [24, 25]. Glomus mosseae was documented in locations with high levels of heavy metal contamination [26]. The external mycelium of Glomus mosseae produces a specific protein known as Glycoprotein (Glomalin), which contains binding sites for heavy metals [27]. Cadmium metals concentrate inside these specific binding sites. The presence of arbuscular mycorrhizal (AM) fungi in plants leads to an elevation in antioxidant levels [28, 29, 30]. In the past, certain fungal strains were identified that exhibited resistance to heavy metal contamination. Mycorrhizal fungi are able to mitigate the effects of heavy metal contamination [31, 32].

We are intrigued by the investigation of the sorghum plant due to various factors that we consider particularly pertinent:

- The research of the poisoning effect of the heavy metal Cd on plant mainly focuses on food crops such as rice, wheat and maize, but less on sorghum.
- Sorghum frequently used as animal feed so quality assurance related to heavy metals is not understood very well, hence studies on contamination of heavy metal in fodder crop is required.
- The sensitivity level of sorghum for cadmium varies.

By deciding the sensitivity one can find genes which are responsible for the same. Literature indicates that tolerance level of a heavy metal for sorghum is 1000 ppm onwards. It indicates that it has a wide range of tolerance. The genomic size of the sorghum ranges from 700 to 772 Mbp, and it has been well sequenced. So, search for genes responsible for tolerance is not that difficult now. After finding the gene, it can be transferred into the sensitive crops such as cow pea and other leafy vegetables. Finally, we can get transgenic of the heavy metal tolerance. Da-Lin et al. [33] reported some important changes in growth and physiological characteristics of the sorghum plant under cadmium stress such as height, chlorophyll content, root activities and MDA content. They concluded that Cd effect was the manifestation of oxidative stress. Low concentration of cadmium can promote growth of hard wheat, while under a relatively high concentration, the growth of wheat and tillering were both inhibited and the degree varied among the various species. Liu's [34] research also showed that corn seedling's height under cadmium treatment was reduced significantly as the concentration of cadmium increased with prolonged growth period. These studies showed that lower concentrations of Cd stimulated increase of sorghum height, which may be related to the certain resistance of the sorghum genus plant. The higher concentrations hindered their development; explanations behind the restraining impact of substantial metals to plant development were presumably due to: [A] A progression of physical and chemical responses between the overabundance substantial metal and soil parts which changes soil properties, along these lines influencing soil fruitfulness [35, 36]. For instance, substantial metal contamination can improve the obsession of soil phosphorus, which influenced the plant engrossing phosphorus, consequently influencing the development of plants [37, 38].

- Substantial metal caused a lessening in plant photosynthesis, in this way decreasing the plant water and supplement adsorption, which influenced the ordinary development and improvement of sorghum [39].

There is extensive evidence of significant metal contamination affecting the root activities of grass-like plants. For example, Yang et al. [40] discovered that the direct flooding of sewage had a significant impact on the roots and seedlings of wheat using hydroponics. The results indicated that the concern of sewage flooding hastened the decline of wheat seedlings and root, reducing both the number of roots and their activities significantly. Jiang et al. [41] also showed that contaminated soil caused the roots of rice seedlings to turn yellow and red, increased the growth of rhizomes, and resulted in brown and yellow root coloration. Huang et al. [42] demonstrated that the presence of cadmium in the soil or environment significantly reduced root activity. The aim of our study was to determine the oxidative damage caused by cadmium toxicity in sorghum and evaluate the impact of putrescine and mycorrhiza in reducing the stress responses induced by cadmium in sorghum.

2. MATERIAL AND METHODS

2.1. Selection of crop sorghum (*Sorghum vulgare* L.)

Sorghum is a primary food source for the most economically disadvantaged and food-insecure populations in the semi-arid tropics. Sorghum is a member of the Graminae family, which consists of grasses. This plant is a C4 plant with a short-day photoperiod. The ideal duration of light exposure that triggers the process of flower development is between 10 and 11 hours. Sorghum is a primary food source for the most impoverished and food-insecure populations in the semi-arid tropics. Worldwide, sorghum is grown on 41 million hectares, yielding a total production of 64.20 million tons. The average productivity per hectare is approximately 1.60 tonnes, according to the Directorate of Sorghum Research in Hyderabad [96]. With the exception of few locations, it is primarily produced and consumed by impoverished farmers. India accounts for around 16% of global sorghum production, according to a study conducted by DSR Hyderabad [96]. It ranks fifth in terms of importance among cereal crops in the country. In India, this crop was a significant staple cereal during the 1950s and covered an area of about 18 million hectares. However, it has now decreased to 7.69 million hectares ([96], Directorate of Sorghum Research, Hyderabad). Sorghum necessitates warm climatic conditions, yet it is adaptable to a diverse array of environments. It may be cultivated at elevations ranging from sea level to 1500 m. It exhibits superior heat tolerance throughout its whole life cycle compared to any other crop. It can be cultivated in regions with a mean annual precipitation of 60–100 cm. The seed requires a minimum temperature of 7–10°C for germination. The ideal temperature for growth is approximately 26–30°C. Sorghum has a wide range of genetic material that varies greatly in terms of observable physical and structural characteristics. Sorghum can be categorized into four primary classes based on their production characteristics: grain sorghum, forage sorghum (FS), high-tonnage sorghum (energy), and sweet sorghum. Sorghum cultivars are currently being evaluated as potential candidates in the quest for efficient energy crops, driven by a growing interest in converting biomass into energy.

2.2. Experimental site

The current study was conducted in the polyhouse and laboratory facilities of the Department of Plant Physiology, Institute of Agricultural Sciences, Banaras Hindu University, located in Varanasi, Uttar Pradesh, India. The experimental site is situated in a geographical region between 25°18′ N latitude and 83°03′E longitude. It is roughly 75.7 m above sea level.

2.3. Climatic condition

Varanasi is located in the northern region of India, which has a semi-arid to sub-humid climate. In this region, the monsoon typically begins in the third week of June and lasts until the end of September, perhaps extending into the third week of October. The average annual precipitation is approximately 1100 mm, with 88% occurring during the monsoon season from June to September, 5-7% in the post-monsoon period from October to December, and about 3.3% in the winter season from January to February. The temperature varied between a maximum of 45°C and a minimum of 19°C, with fluctuations between 28°C and 7°C. The average relative humidity in the region is approximately 66%. It reaches a peak of 92% between July and September, and drops to 39% from late April to early June.

2.4. Plant and mycorrhizal origin

The seeds of sorghum cv. CSV15, which were disease-free and in good health, were acquired from the Directorate of Sorghum Research in Hyderabad. The endomycorrhiza Glomus mosseae was acquired from the Tata Energical Research Institute, located in New Delhi.

2.5. Treatments detail

The pot experiment was carried out in the poly house of the Institute of Agricultural Sciences, Banaras Hindu University, Varanasi, using a single genotype of sorghum CSV 15. The experimental pot had a diameter of 30 cm and a height of 25 cm. It had a capacity of 10 kg of soil and a little hole at the bottom. Seeds of sorghum are planted in pots filled with a soil mix consisting of a combination of soil and FYM at a ratio of 3:1. Specific pots were deliberately inoculated with the Endomycorrhiza Glomus sp. The soil was inoculated with 150 inoculants per kilogram, followed by the induction of heavy metal stress in the plant with the external application of cadmium nitrate in the soil. The two most significant quantities of heavy metals, determined during initial screening, were chosen: 70 and 150 ppm per kilogram of soil. Putrescine was administered via foliar spray at concentrations of 2.5 and 5.0 mM, with a 7-day interval, beginning from seven days after sowing (DAS) to one week prior to 90 DAS. The observations were recorded at three different time intervals: 30, 60, and 90 days following seeding in the respective pots. The treatments included in the study were as follows: Control (T0), Control + Mycorrhiza (T1), Control + 2.5 mM Putrescine (T2), Control + 5 mM Putrescine (T3), Control + 2.5 mM Putrescine + Mycorrhiza (T4), Control + 5 mM Putrescine + Mycorrhiza (T5), 70 ppm $Cd(NO_3)_2$ (T6), 70 ppm $Cd(NO_3)_2$ + Mycorrhiza (T7), 70 ppm $Cd(NO_3)_2$ + 2.5 mM Putrescine (T8), 70 ppm $Cd(NO_3)_2$ + 5 mM Putrescine (T9), 70 ppm $Cd(NO_3)_2$ + 2.5 mM Putrescine

+ Mycorrhiza (T10), 70 ppm $Cd(NO_3)_2$ + 5 mM Putrescine + Mycorrhiza (T11), 150 ppm $Cd(NO_3)_2$ (T12), 150 ppm $Cd(NO_3)_2$ + Mycorrhiza (T13), 150 ppm $Cd(NO_3)_2$ + 2.5 mM Putrescine (T14), 150 ppm $Cd(NO_3)_2$ + 5 mM Putrescine (T15), 150 ppm $Cd(NO_3)_2$ + 2.5 mM Putrescine + Mycorrhiza (T16), 150 ppm $Cd(NO_3)_2$ + 5 mM Putrescine + Mycorrhiza (T17).

2.6. Experimental design and layout

The experiment was conducted using a completely randomized design (CRD). There were a total of 18 treatments, including a control group. There were five replicates for each treatment.

2.7. Oxidative damage assessment

The method described by Heath and Packer [43] was used to determine the concentration of Thiobarbituric Acid Reducing Substances (TBARS), which is a measure of Lipid Peroxidation (malondialdehyde (MDA) content). Approximately 0.5 grams of leaf tissues from both the control and treated groups were fragmented into minute fragments and blended by adding 5 milliliters of a 5% trichloroacetic acid (TCA) solution. The homogenates were thereafter transferred into new tubes and subjected to centrifugation at a speed of 12,000 revolutions per minute for a duration of 15 minutes at ambient temperature. A fresh tube was prepared and filled with equal amounts of supernatant and a solution containing 0.5% thiobarbituric acid (TBA) in a 20% TCA solution. The mixture was then heated to 96°C and boiled for 25 minutes. The tubes were placed in an ice bath and thereafter subjected to centrifugation at a speed of 10,000 revolutions per minute for a duration of 5 minutes. The supernatant's absorbance was quantified at a wavelength of 532 nm and adjusted for any non-specific cloudiness by subtracting the absorbance at 600 nm. A blank solution consisting of 0.5% TBA in 20% TCA was utilized for this purpose. The quantity of MDA-TBA complex (red pigment) was determined using the extinction coefficient of 155 M^{-1} cm^{-1}. The MDA contents were derived from measurements. The results were expressed as micromoles of malondialdehyde per gram of fresh weight.

The H_2O_2 content was measured by the method given by Jana and Choudhuri [44]. One hundred mg root samples were extracted using 3 ml of 50 mM sodium phosphate buffer (pH 7.4). The homogenate was centrifuged at 6000 g for 15 min. To determine hydrogen peroxide levels, 3 ml of the extracted solution was mixed with 1 ml of 0.1% titanium sulfate in 20% (w/v) sulfuric acid and then centrifuged at 6000 g for 15 min. The intensity of the yellow color in the supernatant was measured at 410 nm. The hydrogen peroxide level was calculated using extinction coefficient 0.28 μM^{-1} cm^{-1} and was expressed as nmol H_2O_2 g^{-1} tissue FW.

2.8. Statistical analysis

The numerical data collected was examined using the Statistical module of Origin 6.1, which is an advanced software for scientific graphing and data analysis developed by OriginLab Corporation, located at One RoundHouse Plaza, Northhampton, MA 01060. A two-way ANOVA was conducted to analyze the interaction between mycorrhiza and cadmium treatments. A One Way ANOVA was conducted to assess the statistical significance of differences among the means of individual groups.

3. RESULTS

3.1. Malondialdehyde (MDA) [nmole/ml FW] content

The study investigated the impact of polyamine (putrescine), mycorrhiza, and their combination on the levels of MDA (measured in nmole/ml FW) in sorghum variety CSV15 under conditions of cadmium stress. Measurements were taken at three time points: 30, 60, and 90 days after sowing (DAS) (Fig. 1). The average MDA showed a clear and significant rise of 35.64%, 41.39%, and 64.02% when exposed to heavy metal stress (T6) compared to the control (T0) at 30, 60, and 90 DAS of the interval. Similarly, when plants were subjected to higher concentrations of heavy metal (T12), the level of MDA increased dramatically by 62.28%, 59.44%, and 73.67% compared to the control (T0) on the specified dates. The application of endomycorrhiza in the soil (T7) from an external source demonstrated a mitigating impact by reducing the MDA levels by 6.05%, 2.87%, and 7.87% compared to T6 at 30, 60, and 90 DAS. During the treatment, T13 was compared to T12, and the MDA (mean difference of abundance) decreased significantly by 12.92%, 9.85%, and 9.63% at the planned DAS (Days After Sowing). Compared to T6, the external application of putrescine (T8) shown a moderating effect by reducing MDA levels by 1.84, 6.64, and 14.90% on the planned DAS. The mean MDA was significantly decreased by 11.15%, 7.24%, and 29.68% in the high-dose putrescine treatment group (T9) compared to T6, with respect to T8. Comparatively, at the planned DAS, treatment T14 shown a significant reduction in MDA by 21.47%, 17.32%, and 18.45% as compared to treatment T12. The mean MDA level was considerably decreased by 33.17%, 23.72%, and 20.39% when treated with a high dose of putrescine (T15) compared to T12, with respect to T14. The combination of putrescine and mycorrhiza exhibited the most effective reduction in MDA levels in treatment T10, resulting in a decrease of 17.65%, 11.03%, and 36.89% compared to treatment T6 at the suggested DAS. When treatment T11 was compared to treatment T6, the levels of MDA were lowered by 25.51%, 20.21%, and 42.59%, respectively. Treatment T16 shown a comparable impact to treatment T12. In the T16 treatment, the MDA (malondialdehyde) levels were seen to decrease significantly by 43.79%, 30.52%, and 42.96% at the recommended DAS (Days After Sowing). The treatment T17 exhibited superior outcomes, with a notable reduction in MDA levels by 57.32%, 56.76%, and 55.50% compared to T12. The combination of putrescine and mycorrhiza demonstrated the most effective reduction of cadmium toxicity for malondialdehyde.

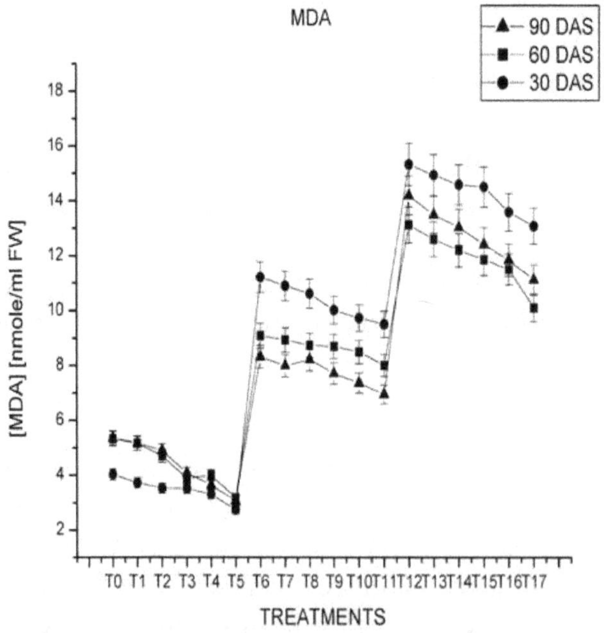

Fig. 1. Effects of putrescine and Glomus on MDA content in sorghum grown in cadmium contaminated soil.

3.2. H_2O_2 [μ mole/g FW] content

The concentration of H_2O_2 [μ mole/g FW] was investigated in the sorghum variety CSV15 during cadmium-induced stress. Measurements were taken at three time points: 30, 60, and 90 days after sowing (DAS) (Fig. 2). The average H_2O_2 levels showed a substantial rise of 34.09%, 32.02%, and 19.37% when exposed to heavy metal stress (T6) compared to the control (T0) at 30, 60, and 90 days after sowing (DAS). Similarly, when plants were subjected to high concentrations of heavy metal (T12), their H_2O_2 levels increased significantly by 30.9%, 37.55%, and 34.66% compared to the control (T0) on the specified dates. The exogenous application of endomycorrhiza in the soil (T7) shown a mitigating effect by reducing the H_2O_2 levels by 32.47%, 14.21%, and 0.54% compared to T6 at 30, 60, and 90 DAS. During the treatment, T13 was compared to T12, and it was shown that the H_2O_2 levels decreased significantly by 1.13%, 2.00%, and 1.25% at the prescribed DAS. Compared to T6, the external application of putrescine (T8) shown a moderating effect by reducing H2O2 levels by 32.74%, 17.24%, and 2.52% on the suggested DAS. The mean concentration of H_2O_2 was considerably decreased by 35.10%, 20.87%, and 4.30% when treated with a high dose of putrescine (T9) compared to T6, with respect to T8. Comparatively, at the planned DAS, treatment T14 showed a substantial reduction of 1.82%, 4.33%, and 2.50% in H2O2 when compared to treatment T12. The mean concentration of H_2O_2 was considerably decreased by 2.5%, 4.98%, and 3.28% when treated with a high dose of putrescine (T15) compared to T12, with respect to T14. The combination of putrescine and mycorrhiza exhibited the most effective reduction in H2O2 levels in treatment T10, resulting in a decrease of 17.65%, 11.03%, and 36.89% compared to treatment T6 at the suggested DAS. Comparing treatment T11 with treatment T6, there was a considerable reduction in H2O2 levels of 42.70%, 34.80%, and 22.80%, respectively. At the intended DAS, treatment T16 showed a comparable effect to treatment T12, where H_2O_2 levels decreased significantly by 8.41%, 9.66%, and 10.21%, respectively. The treatment T17 exhibited superior outcomes, with a notable reduction in H_2O_2 levels by 12.70%, 19.24%, and 23.89% compared to T12. The combination of putrescine and mycorrhiza demonstrated the most effective synergy in reducing cadmium toxicity in the presence of hydrogen peroxide.

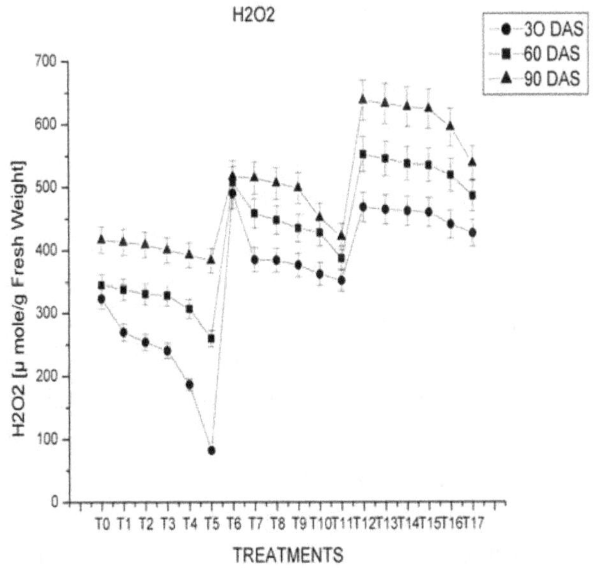

Fig. 2. Effects of putrescine and Glomus on H_2O_2 content in sorghum grown in cadmium contaminated soil.

4. DISCUSSION

The ROS might potentially cause harm to cellular layers, resulting in the oxidation of lipid membranes (lipid peroxidation). Malondialdehyde (MDA) is a byproduct of lipid peroxidation and can be used as an indicator of lipid peroxidation in living organisms. Prior research has indicated that the absorption of cadmium leads to oxidative stress, which in turn causes an elevated production of hydrogen peroxide and lipid peroxidation [45, 46]. Plants have an antioxidant system that becomes active when they are exposed to cadmium stress, in order to deal with increased levels of oxidative stress [46, 47]. Ascorbate peroxidase, peroxidase, and catalase are essential in the degradation process of the plant sample cultivated under cadmium-induced stress. The statistical analysis of treatment T17 revealed a noteworthy decrease in MDA content. This can be attributed to the prominent feature of PA conjugates with phenolic acids, which is their antioxidant characteristic. This trait is essential for plants to adapt to stressful conditions. Bors et al. [48] demonstrated that conjugates containing coffeic, cinnamic, and ferulic acids had greater binding affinity to reactive oxygen species (ROS). In other words, unbound polyamines are less effective at scavenging radicals compared to their conjugates. The conjugation of polyamines by transglutaminases, particularly to Rubisco, appears to play a crucial role in safeguarding this protein from protease activity, hence preserving its photosynthetic efficacy [49]. Thus, it is probable that PAs have a significant impact on photosynthesis as they possess the ability to reverse harm caused by stress to the photosynthetic machinery. The process of cross-linking polyamines by transglutaminases may have a substantial impact on the bioactivity of polyamines in flower development and reproductive compatibility [50]. Conjugates play a role in defending against cadmium toxicity and also regulate specific growth and developmental events [51, 52]. Similarly, the mycorrhiza Glomus functioned as an effective inhibitor of the movement of cadmium ions into the soil solution, so safeguarding the plants against cadmium toxicity. The presence of arbuscular mycorrhizal (AM) fungi was detected in soil with significant levels of contamination [24, 25]. Researchers [26] have also documented the occurrence of Glomus mosseae in locations contaminated with heavy metals. The mycelium of certain arbuscular mycorrhizal (AM) fungus produce an external protein known as Glycoprotein (Glomalin), which has binding sites for heavy metals [53]. These binding sites serve as sites of accumulation for heavy metals. The presence of arbuscular mycorrhizal (AM) fungi in plants leads to an elevation in the antioxidant content [28, 29, 30]. In the past, certain fungal strains were identified that exhibited resistance to heavy metal contamination. Several mycorrhizal fungi are able to mitigate the effects of heavy metal exposure [31, 32]. In this study, it was shown that when plants were exposed to cadmium stress, the level of lipid peroxidation was lower in plants treated with mycorrhizal fungi compared to plants that did not get mycorrhizal treatment, under the same level of cadmium stress. A comparable pattern was observed when the plant was subjected to putrescence. Optimal outcomes were achieved when mycorrhiza and putrescine were administered simultaneously to mitigate the harmful effects of Cd toxicity in sorghum. Mycorrhizal plants, in conjunction with putrescine, exhibited reduced levels of lipid peroxidation and diminished formation of hydrogen peroxide [40, 54, 55, 56]. It is commonly acknowledged that decreasing levels are one of the strategies via which arbuscular mycorrhizal (AM) fungi safeguard plants against various challenges [57, 58, 59]. Research has proposed that peroxidase has the potential to effectively remove H2O2 in plant vacuoles when phenolics and reduced ascorbate are present [60]. The methanol extracts of rhizome polyphenols from Nymphaea were found to directly bind to Cr, Pb, and Hg. This chelation process was seen in multiple studies, with references ranging from [61] to [96]. Phenolic antioxidants prevent lipid peroxidation by capturing the lipid alkoxyl radical. The activity of this process is contingent upon the molecular structure, as well as the quantity and arrangement of hydroxyl groups inside the molecules. Santiago et al. [62] propose a cycle in which H2O2 is eliminated by phenolics using a peroxidase enzyme, and phenolics are then converted to phenoxyl radicals. These radicals can be subsequently reduced by amino acids [97, 98, 99, 100, 101, 102, 103, 104, 105, 106, 107, 108, 109, 110, 111, 112, 113, 114, 115, 116, 117, 118]. The current evidence indicating a positive correlation between metal exposure and the accumulation of proline suggests a protective role of this amino acid against heavy metal toxicity. Proline has been recommended as a means to maintain water balance in this particular scenario, as it can counteract the disruption caused by heavy metals [63, 64]. It can also scavenge hydroxyl radicals [65], chelate heavy metals in the cytoplasm [66, 67, 68], and reduce metal uptake [69]. Many experts believe that the accumulation of proline is merely a byproduct of various stresses and is not involved in protecting against metal and other burdens [70, 71, 72]. Wu et al. [73] showed that the efflux of K^+ in Anacystis nidulans, prompted by Cu, was restricted in the presence of proline. This suggests that proline potentially protected the plasma membrane from Cu toxicity, or it may have complexed Cu^+ thereby reducing the concentration of free Cu+ in the external environment. In a subsequent study, Wu et al. [69] demonstrated that cells with high concentrations of proline displayed reduced uptake of Cu. This suggests that proline acts as a stabilizer for the plasma membrane and also decreases the uptake of Cu by algal cells. A study was conducted to investigate the effects of lead, copper, cadmium, and mercury on the proline content of 17-day-old bean seedlings (*Phaseolus vulgaris* L.) grown in a Hoagland solution spiked with varying concentrations of Pb, Cu, Cd, and Hg. Control and metal-treated plants were cultivated for 10 days in a Hoagland solution. A significant increase in proline levels was seen in primary leaves after a ten-day exposure to heavy metals. The highest influence on proline content was observed in plants exposed to mercury, followed by the order of $Cd^{++} > Cu^{++} > Pb^{++}$ [74].

Early investigations indicated that peroxidase serves not only as a defense protein, but also as a crucial antioxidant enzyme that plays a role in responding to environmental challenges [75]. It encompasses a diverse range of heme-containing enzymes that play a role in numerous physiological processes in plants. According to reports, it is implicated in the response to both living and non-living stressors, the breakdown of auxin, and the formation and breakdown of cell wall lignin [76]. Peroxidase participates in the removal of Reactive Oxygen Species (ROS). The presence of a large amount of cadmium in sorghum led to the suppression of certain enzymes and the enhancement of others. The accumulation of cadmium in the cellular compartment of the enzyme is necessary for the inhibition of the enzyme in living organisms [77, 78]. The upregulation of peroxidase is thought to be associated with oxidative reactions occurring at the biomembrane. The phytotoxicity of metal polluted soils was assessed by evaluating the isoperoxidase pattern, which is a measure of enzymatic alterations generated by metals in plants. Extensive evidence suggests that peroxidase plays a significant role in developmental processes, perhaps because this enzyme is involved in auxin metabolism (specifically auxin oxidation) and in the creation of cross-links between cell wall components. The cadmium treatment resulted in a more prominent peroxidase activity in sorghum [79, 80, 81]. The plants possess a significant level of peroxidase activity, which allows them to defend against oxidative stressors, as indicated by Scalet et al. [82]. Peroxidase in plants serves to safeguard cells against detrimental levels of hydroperoxides by stimulating the production of peroxidases, particularly anionic peroxidase. These enzymes have been observed to play a role in the response to both abiotic and biotic stressors. Mycorrhiza demonstrated the most effective reduction of cadmium toxicity in relation to peroxidase activity. The current investigation found that the presence of Glomus dramatically decreased the activity of peroxidase. This was because the fungus trapped the cadmium ion in the soil solution, which prevented its movement and transportation in plants [83]. The movement of cadmium is contingent upon various circumstances. The availability of cadmium to sorghum roots from the soil solution is determined by its concentration in the soil and is influenced by the presence of organic matter, pH, and temperature. The presence of chelating organisms, such as mycorrhiza, also has an impact on it. Although metal ions in plants may have varying levels of mobility, the concentration of cadmium was consistently higher in the roots compared to the above ground tissue [84, 85, 86]. Cadmium (Cd) typically reaches the roots in a majority of environmental situations, making them the primary site for potential Cd harm [87]. Putrescine, like-wise, is a promising substance that can help reduce the harmful effects of heavy metals in sorghum due to its distinct characteristics. Multiple studies have demonstrated that the formation of polyamines (PA) takes place in response to abiotic stressors such as drought, salt, low temperature, heavy metals, and PQ [21, 84, 88, 89, 90].

5. CONCISE SUMMARY

In conclusion, it has been determined that polyamines such as putrescine and mycorrhiza Glomus play a crucial role in reducing the harmful effects of cadmium toxicity in sorghum. This is achieved by the activation of both enzymatic and non-enzymatic antioxidants, which serve as a defense mechanism in plants. The levels of MDA and hydrogen peroxide were much higher in plants treated with cadmium compared to plants treated with putrescine and mycorrhiza.

REFERENCES

1. Benavides MP, Gallego SM, Tomaro ML. Cadmium toxicity in plants. Brazilian Journal of Plant Physiology. 2005;17:21-34. DOI: 10.1590/S1677-04202005000100003

2. Uraaguchi S, Mori S, Kuramata M, Kawasaki A, Arora T, Ishikawa S. Root-to shoot Cd translocation via the xylem is the major process determining shoot and grain cadmium accumulation in rice. Journal of Experimental Botany. 2009;60:2677-2268. DOI: 10.1093/jxb/erp119

3. Wang L, Zhou QX, Ding LL, Sun Y. Effect of cadmium toxicity on nitrogen metabolism in leaves of Solanum nigrum L. as a newly found cadmium hyperaccumulator. Hazardous Materials. 2008;154(1-3):818-825. DOI: 10.1016/j.jhazmat.2007.10.097

4. Smith CJ, Hopmans P, Cook FJ. Accumulation of Cr, Pb, Cu, Ni, Zn and Cd in soil following irrigation with untreated urban effluents in Australia. Environmental Pollution. 1996;94(3):317-323. PMID:15093492

5. Bjerrum J. Metal amine formation in aqueous solution. In: Elving J, editor. Treatise on Analytical Chemistry. Vol. I. New York: The Interscience Encyclopedia; 1959

6. Adriano DC. Chromium. In: Trace Elements in the Terrestrial Environment. New York: Springer; 1986. pp. 58-76

7. Juste C, Mench M. Long term application of sewage sludge and its effect on metal uptake by crops. In: Adriano DC, editor. Biogeochemistry of Trace Metals. Ann. Arbor, London, Tokyo: Leuwis publishers; 1992. pp. 159-193

8. Goering PL, Waalkes MP, Klaassen CD. Toxicity of cadmium. In: Goyer RAC, Herian MG, editors. Handbook of Experimental Pharmacology: Toxicity of Metals, Biochemical Effects. New York: Springer Verlag; 1994. pp. 189-213

9. AMAP. AMAP Assessment. Heavy Metals in the Arctic-Pre-Print Files. Oslo, Norway: Arctic Monitoring and Assessment Programme; 2002. p. 870

10. Mench M, Amans V, Sappin-Didier V, Fargues S, Gomez A, Loffler M, Masson P, Arrouays D. A study of additivies to reduce availability of Pb in soils to plants. In: Iskandar A, Adriano DC, editors. Remediation of Soils Contaminated with Metals. Northwood, UK: Science Reviews; 1997. pp. 202-185

11. Baron K, Stasolla C. The role of polyamines during in vivo and in vitro development. In Vitro Cellular & Developmental Biology. Plant. 2008;44:384-395. DOI: 10.1007/s11627-008-9176-4

12. Dossantos RW, Schmidt EC, Depmartins R, Latini A, Maraschin M, Horta PA, Bouzon ZL. Effects of cadmium on growth, photosynthetic pigments, photosynthetic performance, biochemical parameters and structure of chloroplasts in the agarophyte Gracilaria domingensis (Rhodophyta, Gracilariales). American Journal of Plant Sciences. 2012;3:1077-1084. DOI: 10.4236/ajps.2012.38129

13. Kakkar RK, Sawhney VK. Polyamine research in plants–a changing perspective. Physiologia Plantarum. 2002;116:281-292. DOI: 10.1034/j.1399-3054.2002.1160302.x

14. Kuznetsov V, Radyukina NL, Shevyakova NI. Polyamines and stress: Biological role, metabolism and regulation. Russian Journal of Plant Physiology. 2006;53:583-604. DOI: 10.1134/S1021443706050025

15. Kusano T, Berberich T, Tateda C, Takahashi Y. Polyamines: Essential factors for growth and survival. Planta. 2008;228:367-381. DOI: 10.1007/s00425-008-0772-7

16. Steiner N, Santa-Catarina C, Silveira V, Floh EIS, Guerra MP. Polyamine effects on growth and endogenous hormones levels in Araucaria angustifolia embryogenic cultures. Plant Cell, Tissue and Organ Culture. 2007;89:55-62. DOI: 10.1007/s11240-007-9216-5

17. Santa-Catarina C, Silveira V, Scherer GFE, Floh EIS. Polyamine and nitric oxide levels correlate with morphogenetic evolution in somatic embryogenesis of Ocotea catharinensis. Plant Cell, Tissue and Organ Culture. 2007;90:93-101. DOI: 10.1007/s11240-007-9259-7

18. Tun NN, Santa-Catarina C, Beghum T, Silveira V, Handro W, Floh EIS, Scherer GFE. Polyamines induce the rapid biosynthesis of nitric oxide (NO) in Arabidopsis thaliana seedlings. Plant & Cell Physiology. 2006;47:346-354. DOI: 10.1093/pcp/pci252

19. Kuhen GD, Phillips GC. Role of polyamines in apoptosis and other recent advances in plant polyamines. Critical Reviews in Plant Sciences. 2005;24:123-130. DOI: 10.1080/07352680590953161

20. Yoda H, Hamaguchi R, Sano H. Induction of hypersensitive cell death by hydrogen peroxide produced through polyamine degradation in tobacco plants. Plant Physiology. 2003;132:1973-1981. DOI: 10.1104/pp.103.024737

21. Bouchereau A, Aziz A, Larher F, Martin-Tanguy J. Polyamines and environmental challenges: Recent development. Plant Science. 1999;140:103-125. PMCID: PMC2835953

22. Bhatnagar P, Minocha R, Minocha SC. Genetic manipulation of the metabolism of polyamines in poplar cells: The regulation of putrescine catabolism. Plant Physiology. 2002;128:1455-1469. DOI: 10.1104/pp.010792

23. Minocha SC, Minocha R. Role of polyamines in somatic embryogenesis. In: Bajaj YPS, editor. Biotechnolgogy in Agriculture and Forestry. Vol. 30. Berlin: Spring-Verlag; 1995. pp. 53-70. DOI: 10.12691/plant-1-2-1

24. Leung HM, Ye ZH, Wong MH. Survival strategies of plants associated with arbuscular mycorrhizal fungi ontoxic mine tailings. Chemosphere. 2007;66:905-915. DOI: 10.1016/j.chemosphere.2006.06.037

25. Pawlowska TE, Blaszkowski J, Ruhling A. The mycorrhizal status of plants colonizing a calamine spoil mound in southern Poland. Mycorrhiza. 1996;6:499-505. DOI: 10.1007/s005720050154

26. Debiane D, Garcon G, Verdin A, Fontaine J, Durand R, Grandmougin-Ferjani A, Shirali P, Lounces-Hadj Sahraui A. In vitro evaluation of the oxidative stress and genotoxic potenitials of anthracene on mycorrhizal chicory roots. Environmental and Experimental Botany. 2008;64:120-127. DOI: 10.1016/j.envexpbot.2008.04.003

27. Vivas A, Azcon R, Biro B, Barea JM, Ruiz Lozano JM. Influence of bacterial strains isolated from lead polluted soil and their interactions with arbuscular mycorrhizae on the growth of Trifolium pertense L.under lead toxicity. Canadian Journal of Microbiology. 2003a;49:577-588. DOI: 10.1139/w03-073

28. Gopi R, Jaleel CA, Sairam R, Lakshamanam GMA, Gomathinayagam M, Paneerselvam R. Differential effects of hexaconazole and paclobutrazol on biomass, electrolyte leakage, lipid peroxidation and antioxidant potential of Daucus carota L. Colloids and Surfaces Biointerfaces. 2007;60:180-186. DOI: 10.1016/j.colsurfb.2007.06.003

29. Zhang FQ, Wang YS, Lou ZP, Dong JD. Effect of heavy metal stress on antioxidative enzymes and lipid peroxidation in leaves and roots of two mangrove plant seedlings (Kandelia candel and Bruguiera gymnorrhiza). Chemosphere. 2007a;67:44-50. DOI: 10.1016/j.chemosphere.2006.10.007

30. Zhang LZ, Wei N, Wu QX, Ping ML. Antioxidant response of Cucumis sativus L. to fungicide carbendazin. Pesticide Biochemistry and Physiology. 2007b;89:54-59. DOI: 10.1016/j.pestbp.2007.02.007

31. Hildebrandt U, Regvar M, Bothe H. Arbscular mycorrhiza and heavy metal tolerance. Phytochemistry. 2007;68:139-146. DOI: 10.1016/j.phytochem.2006.09.023

32. Joschim HJ, Makoi R, Ndakidemi PA. The agronomic potential of vesicular-arbuscular mycorrhiza (AM) in cereals-legume mixtures in Africa. African Journal of Microbiology Research. 2009;11:664-675

33. Da-Lin L, Shu-pan Z, Zheng C, Wei-wei Q. Soil cadmium regulates antioxidases in sorghum. Agricultural Sciences in China. 2011;9(10):1475-1480

34. Liu JX. Effects of cadmium and zinc interaction on corn seedlings physiological and biochemical characteristics. Journal of Yichun College. 2004;26(6):55-57. DOI: 10.5897/AJB11.848

35. Cieslinski G, Neilser GH, Hogue EJ. Effect of soil cadmium application and pH on growth and cadmium accumulation in roots, leaves and fruit of strawberry plants. Plant and Soil. 1996;18:267-271. DOI: 10.1080/01904169409364791.

36. Chang ZM, Wu XH. Difference comparison of three alfalfa varities resisitant to cadmium pollution. Pratac. Sci. 2005;22(2):20-23

37. Li J, Gao XH, Guo SR, Zhang RH, Wang X. Effects of exogenous spermidine on photosynthesis of salt-stressed Cuellmis sativus seedlings. Chinese Journal of Ecology. 2007;26(10):1595-1599. DOI: 10.1007/s10265-014-0653-z

38. Zhang EH, Zhang XH, Wang HZ. Adaptation effects of phosphorus stress on different genotypes of faba-bean. Acta Ecologica Sinica. 2004;24(8):1589-1593

39. Qin TC, Ruan J, Wang LJ. Effects of cadmium on plant photosynthesis. Environmental Science and Technology. 2000;13:33-35. DOI: 10.5897/AJB11.848

40. Yang Y, Han X, Liang Y, Ghosh A, Chen J, Tang M. The combined effects of arbuscular mycorrhizal fungi (AMF) and lead (Pb) stress on Pb accumulation, plant growth parameters, photosynthesis, and antioxidant enzymes in *Robinia pseudoacacia* L. PLoS One. 2015;10:e0145726. DOI: 10.1371/journal.pone.0145726

41. Jiang Y, Liang WJ, Zhang YG, Xu YF. Research on effect of sewage irrigation on soil heavy metal environmental capacity and rice growth. Chinese Journal of Eco-Agriculture. 2004;12(3):124-127

42. Huang CC, Chen MW, Hsieh JL, Lin WH, Chen PC, Chien LF. Expression of mercuric reductase from Bacillus megaterium MB1 in eukaryotic microalga Chlorella sp. Applied Microbiology and Biotechnology. 2006;72:197-205. DOI: 10.1007/s00253-005-0250-0

43. Heath RL, Packer L. Photoperoxidation in isolated chloroplasts: I. Kinetics and stoichiometry of fatty acid peroxidation. Archives of Biochemistry and Biophysics. 1968;125:189-198. DOI: 10.1016/0003-9861(68)90654-1

44. Jana S, Choudhuri MA. Glycolate metabolism of three submerged aquatic angiosperms during aging. Aquatic Botany. 1981;12:345-354

45. Liu J, Macarisin D, Wisniewski M, Sui Y, Droby S, Norelli J, et al. Production of hydrogen peroxide and expression of ROS-generating genes in peach flower petals in response to host and non-host fungal pathogens. Plant Pathology. 2013;62:820-828. DOI: 10.1111/j.1365-3059.2012.02683

46. Singh HP, Batish DR, Kohli RK, Arora K. Arsenic-induced root growth inhibition in mung bean (*Phaseolus aureus* Roxb.) is due to oxidative stress resulting from enhanced lipid peroxidation. Plant Growth Regulation. 2007;53:65-73. DOI: 10.1007/s10725-007-9205

47. Sobrino-Plata J, Meyssen D, Cuypers A, Escobar C, Hernández LE. Glutathione is a key antioxidant metabolite to cope with mercury and cadmium stress. Plant and Soil. 2014;377:369-381. DOI: 10.1007/s11104-013-2006-4

48. Bors W, Langebartels C, Michel C, Sandermann H. Polyamines as radical scavengers and protectants against ozone damage. Phytochemistry. 1989;28:1589-1595

49. Serafini-Fracassini D, Del Duca S, Beninati S. Plant transglutaminases. Phytochemistry. 1995;40:355-365

50. Serafini-Fracassini D, Del Duca S. Transglutaminases: Widespread cross-linking enzymes in plants. Annals of Botany. 2008;102:145-152. DOI: 10.1093/aob/mcn075

51. Bagni N, Tassoni A. Biosynthesis, oxidation and conjugation of aliphatic polyamines in higher plants. Amino Acids. 2001;20:301-317. PMID:11354606

52. Martin-Tanguy J. Metabolism and function of polyamines in plants: Recent development (new approaches). Plant Growth Regulation. 2001;34:135-148

53. Vivas A, Voros I, Biro B, Campos E, Barea JM, Azcon R. Symbiotic efficiency of autochtonous arbuscular Mycorrhizal fungus (G. mosseae) and Brevibacillus brevis isolated from cadmium polluted soil under increasing cadmium levels. Environmental Pollution. 2003b;126:179-189. DOI: 10.1016/S0269-7491(03)00195-7

54. Evelin H, Kapoor R. Arbuscular mycorrhizal symbiosis modulates antioxidant response in salt-stressed Trigonella foenum-graecum plants. Mycorrhiza. 2014;24:197-208. DOI: 10.1007/s00572-0130529-4

55. Jiang QY, Zhuo F, Long SH, Zhao HD, Yang DJ, Ye ZH, et al. Can arbuscular mycorrhizal fungi reduce Cd uptake and alleviate cd toxicity of Lonicera japonica grown in Cd-added soils? Scientific Reports. 2016;6:21805. DOI: 10.1038/srep21805

56. Tan SY, Jiang QY, Zhuo F, Liu H, Wang YT, Li SS, et al. Effect of inoculation with Glomus versiforme on cadmium accumulation, antioxidant activities and phytochelatins of Solanum 977 photeinocarpum. PLoS One. 2015;10:e0132347. DOI: 10.1371/journal.pone.0132347

57. Garg N, Singla P. The role of Glomus mosseae on key physiological and biochemical parameters of pea plants grown in arsenic contaminated soil. Scientia Horticulturae. 2012;143:92-101. DOI: 10.1016/j.scienta.2012.06.010

58. Hajiboland R, Aliasgharzadeh N, Laiegh SF, Poschenrieder C. Colonization with arbuscular mycorrhizal fungi improves salinity tolerance of tomato (Solanum lycopersicum L.) plants. Plant and Soil. 2010;331:313-327. DOI: 10.1007/s11104-009-0255-z

59. Ruiz-Sánchez M, Aroca R, Muñoz Y, Polón R, Ruiz-Lozano JM. The arbuscular mycorrhizal symbiosis enhances the photosynthetic efficiency and the antioxidative response of rice plants subjected to drought stress. Journal of Plant Physiology. 2010;167:862-869. DOI: 10.1016/j.jplph.2010.01.018

60. Zancani M, Nagy G. Phenol-dependent H2O2 breakdown by soybean root plasma membrane-bound peroxidase is regulated by ascorbate and thiols. Journal of Plant Physiology. 2000;156:295-299. DOI: 10.1016/S0176-1617(00)80064-4

61. Lavid N, Schwartz A, Yarden O, TelOr E. The involvement of phenylpropanoids and peroxidase activities in heavy metal accumulation by epidermal glands of the waterlily (Nymphaeaceae). Planta. 2001;212:323-331

62. Santiago LJM, Louro RP, De Oliveira DE. Compartmentation of phenolic compounds and phenylalanine ammonia-lyase in leaves of Phyllanthus tenellus Roxb. And their induction by copper sulphate. Annals of Botany. 2000;86:1023-1032. DOI: 10.1006/anbo.2000.1271

63. Costa G, Morel JL. Water relations gas exchange and amino acid content in cd treated lettuca. Plant Physiology and Biochemistry. 1994;32:561-570

64. Schat H, Sharma SS, Vooijst R. Heavy metal-induced accumulation of free proline in a metal-tolerant and nontolerant ecotype of Silene vulgaris. Physiologia Plantarum. 1997;101:477-482. DOI: 10.1034/j.1399-3054.1997.1010304. x

65. Smirnoff N, Cumber QJ. Hydroxyl radical scavenging activity of compatible solutes. Phytochemistry. 1989;28:1057-1060. DOI: 10.1016/0031-9422(89)80182-7

66. Farago ME, Mullen WA. Plants which accumulate metals. A possible copper-proline complex from the roots of *Armeria maritima.* Inorganica Chimica Acta. 1979;32:L93-L94. DOI: 10.1016/0031-9422(89)80182-7

67. Hemalatha S, Anburaj A, Francis K. Effect of heavy metals on certain biochemical constituents and nitrate reductase activity in *Orzya sativa* L. seedlings. Journal of Environmental Biology. 1997;18:313-319. DOI: 10.2478/eko-2014-0012

68. Pandey P, Tripathi AK. Effect of heavy metals on morphological and biochemical characteristics of *Albizia procera* (Roxb. Benth.) seedling. International Journal of Environmental Sciences. 2011;1(5):1009-1018. DOI: 10.2478/eko-2014-0012

69. Wu JT, Hsieh MT, Kow LC. Role of proline accumulation in response to toxic copper in Chlorella sp. (Chlorophyceae) cells. The Journal of Physiology. 1998;34:113-117. DOI: 10.1046/j.1529-8817.1998.340113.x

70. Bhaskaran S, Smith RH, Newton RJ. Physiological changes in cultured sorghum cells in response to induced water stress. I. Free proline. Plant Physiology. 1985;79:266-269. PMCID: PMC1075388

71. Lutts S, Kinet JM, Bouharmont J. Effects of various salts and of mannitol on ion and proline accumulation in relation to osmotic adjustment in rice (*Oryza sativa* L.) callus cultures. Journal of Plant Physiology. 1996;149:186-195. DOI: 10.1016/S0176-1617(96)80193-3

72. Perez-Alfocea F, Santa-Cruz A, Guerrier G, Bolarin MC. NaCl stress-induced organic solute changes on leaves and calli of *Lycopersicon esculentum*, L. pinnellii and their interspecific hybrid. Journal of Plant Physiology. 1994;143:106-111

73. Wu JT, Chang SC, Chen KS. Enhancement of intracellular proline level in cells of Anacystis nidulans (cyanobacteria) exposed to deleterious concentrations of copper. Journal of Phycology. 1995;31:376-379. DOI: 10.1111/j.0022-3646.1995.00376.x

74. Rodriguez-Serrano M, RomeroPuertas MC, Pazmino DM, Testillano PS, Risueno MC, Del Rio LA, et al. Cellular response of pea plants to cadmium toxicity: Cross talk between reactive oxygen species, nitric oxide, and calcium. Plant Physiology. 2009;150:229-243. DOI: 10.1104/pp.108.131524

75. Flohé L, Ursini F. Peroxidase: A term of many meanings. Antioxidants & Redox Signaling. 2008;10:1485-1490. DOI: 10.1089/ars.2008.2059

76. Wang LK, Hung YT, Shammas NK. Physicochemical Treatment Processes. Vol. 3. New Jersey: Humana Press; 2004, 2004. pp. 141-198

77. Rabie MH, Eleiwa ME, Aboseoud MA, Khalil KM. Effect of nickel on the content of carbohydrate and some mineral in corn and broad bean plant. Journal of King Saud University – Science. 1992;4:37

78. Sharma SS, Dietz KJ. The significance of amino acids and amino acid-derived molecules in plant responses and adaptation to heavy metal stress. Journal of Experimental Botany. 2006;57(4):711-726. DOI: 10.1093/jxb/erj073

79. Ahmad P, Sharma S, Srivastava PS. In vitro selection of NaHCO3 tolerant cultivars of Morus alba (Local and Sujanpuri) in response to morphological and biochemical parameters. Horticultural Science (Prague). 2007;34:114-122

80. Chugh LK, Sawhney SK. Photosynthetic activities of *Pisum sativum* seedlings grown in the presence of cadmium. Plant Physiology and Biochemistry. 1999;37:297-303. DOI: 10.1016/S0981-9428(99)80028-X

81. Deniz B, Merve H, Sermin E. Evaluation of Lead Removal onto Black Cumin by Using Multi Linear Regression. Ohrid, Republic of Macedonia: BALWOIS 2012; 2012. pp. 1-6

82. Scalet M, Federico R, Guido MC, Manes F. Perioxidase activity and polyamines changes in response to ozone and simulated acidrain in Aleppo pine needles. Environmental and Experimental Botany. 1995;35:417-425. DOI: 10.1016/0098-8472(95)00001-3

83. Ernst WHO, Verkleij JAC, Schat H. Metal tolerance in plants. Acta Botanica Neerlandica. 1992;41:229-248

84. Jentschke G, Godbold DL. Metal toxicity and ectomycorrhizas. Physiologia Plantarum. 2000;109:107-116

85. Leyval C, Turnau K, Haselwandter K. Effect of heavy metal pollution on mycorrhizal colonization and function: Physiological, ecological and applied aspects. Mycorrhiza. 1997;7:139-153. DOI: 10.1007/s005720050174

86. Schützendübel A, Polle A. Plant responses to abiotic stresses: Heavy metal-induced oxidative stress and protection by mycorrhization. Journal of Experimental Botany. 2002;53:1351-1365. DOI: 10.1093/jexbot/53.372.1351

87. Santia di Toppi L, Gabbrielli R. Response to cadmium in higher plants. Environmental and Experimental Botany. 1999;41:105-130 PII: S0098-8472(98)00058-6

88. Groppa MD, Tomaro ML, Benavides MP. Polyamines as protectors against cadmium or copper induced oxidative damage in sunflower leaf discs. Plant Science. 2001;161:481-488. DOI: 10.1007/s00726-006-0343-9

89. Marschner H. Mineral Nutrition of Higher Plants. 2nd ed. London: Academic Press; 1995. 889 pp

90. Pang XM, Zhang ZY, Wen XP, Ban Y, Moriguchi T. Polyamine, all-purpose players in response to environment stresses in plants. Plant Stress. 2007;1:173-188

91. Aebi HE. Catalase. In: Bergmeyer HU, editor. Methods of Enzymatic Analysis. 3rd ed. Weinheim, Florida: Verlag Chemie; 1983. pp. 273-286

92. Asada K. The water –water cycle in chloroplast: Scavenging of active oxygen and dissipation of excess photons. Annual Review of Plant Physiology and Plant Molecular Biology. 1999;50:601-639. DOI: 10.1146/annurev.arplant.50.1.601

93. Abdul Jaleel C, Jayakumar K, Chang-Xing Z, Iqbal M. Low concentration of cobalt increases growth, biochemical constituents, mineral status and yield in Zea mays. Journal of Scientific Research. 2009;1:128-137. DOI: 10.3329/jsr.vlil.1226

94. Bates L, Waldren RP, Teare ID. Rapid determination of free proline for water-stress studies. Plant and Soil. 1973;39:205-207. DOI: 10.1007/BF00018060

95. Cho UH, Seo NH. Oxidative stress in Arabidopsis thaliana exposed to cadmium is due to hydrogen peroxide accumulation. Plant Science. 2005;168:113-120. DOI: 10.1016/j.plantsci.2004.07.021

96. DSR Annual Report (2011). Directorate of Sorghum Research, Hyderabad; 2010-11

97. El-Sayed E, Omran, Afaf A, Abd ER. Mapping and screening risk assessment of heavy metals concentrations in soils of the Bahr El-Baker region, Egypt. Journal of Soil Science and Environmental Management. 2010;6(7):182-195. DOI: 10.5897/JSSEM12.010.

98. Gonzalez-Chavez MC, Carrillo-Gonzalez R, Wright SF, Nichols KA. The role of glomalin, a protein produced by arbuscular mycorrhizal fungi, in sequestering potentially toxic elements. Environmental Pollution. 2004;130:317-323. DOI: 10.1016/j.envpol.2004.01.004

99. Gohre V, Paszkowski U. Contribution of arbuscular mycorrhizal symbiosis to heavy metal phytoremediation. Planta. 2006;223:1115-1122. DOI: 10.1007/s00425-006-0225-0

100. Garg N, Manchanda G. ROS generation in plants: Boon or bane? Plant Biosystems. 2009;143(1):81-96. DOI: 10.1080/11263500802633626

101. Ivanova J, Toncheva-Panova T, Chernev G, Samuneva B. Effect of Ag+, Cu2+ and Zn2+ containing hybrid nano matrixes on the green algae *Chlorella keissleri*. General and Applied Plant Physiology. 2008;34:339-348

102. Joner EJ, Briones R, Leyval C. Metal-binding capacity of arbuscular mycorrhizal mycelium. Plant and Soil. 2000;226:227-234. DOI: 10.1023/A:1026565701391.

103. Kar M, Mishra D. Catalase, peroxidase, and polyphenoloxidase activities during rice leaf senescense. Plant Physiology. 1976;57:315-319. PMID: 16659474

104. Kaldorf M, Kuhn AJ, Schroder WH, Hildebrandt U, Bothe H. Selective element deposits in maize colonized by a heavy metal tolerance conferring arbuscular mycorrhizal fungus. Journal of Plant Physiology. 1999;154:718-728. DOI: 10.1016/S0176-1617(99)80250-8

105. Kukreja S, Nandval AS, Kumar N, Sharma SK, Unvi V, Sharma PK. Plant water status, H2O2 scavenging enzymes, ethylene evolution and membrane integrity of Cicer arietinum roots as affected by salinity. Biologia Plantarum. 2005;49:305-308

106. Mehlhorn H, Lelandais M, Korth HG, Foyer CH. Ascorbate is the natural substarte for plant perioxidase. FEBS Letters. 1996;378:203-206. SSDI: 0014-5793(95)01448-9

107. Mozgawa W, Król M, Bajda T. Application of IR spectra in the studies of heavy metal cations immobilization on natural sorbents. Journal of Molecular Structure. 2009;924-926:427-433. DOI: 10.1016/j.molstruc.2008.12.028

108. Nakano Y, Asada K. Hydrogen peroxide is scavenged by ascorbate specific peroxidase in spinach chloroplasts. Plant & Cell Physiology. 1981;22:867-880. DOI: 10.1093/oxfordjournals.pcp.a076232

109. Pooria G, Tahereh TM, Bijan R. Differential scanning calorimetry techniques: Applications in biology and nanoscience. Journal of Biomolecular Techniques. 2010;21(4):167-193. PMCID: PMC2977967

110. Stoeppler M. Cadmium. In: Merian E, editor. Metals and their Compounds in the Environment: Occurrence, Analyses and Biological Relevance. New York: VCH; 1991. pp. 803-851

111. Sadasivam S, Manickam A. Biochemical methods for agricultural sciences; 1992. pp. 12-13. Sumner JB, Gjessing EC. Arch. Biochem. 1943;2:291

112. Shah K, Kumar RG, Verma S, Dubey RS. Effect of cadmium on lipid perioxidation, superoxide anion generation and activities of antioxidants enzymes in growing rice seedlings. Plant Science. 2001;161:1135-1144. DOI: 10.1016/S0168-9452(01)00517-9

113. Smeets K, Cuypers A, Lambrechts A, et al. Induction of oxidative stress and antioxidative mechanisms in *Phaseolus vulgaris* after cd application. Plant Physiology and Biochemistry. 2005;43(5):437-444. DOI: 10.1016/j.plaphy.2005.03.007

114. Sbartai H, Rouabhi R, Sbartai I, Berrebbah H, Djebar RM. Induction of anti-oxidative enzymes by cadmium stress in tomato (*Lycopersicon esculentum*). African Journal of Plant Science. 2008;2(8):72-76. Article Number - E4FC5413005

115. Thomet M, Vogel E, Krahenbuhl U. The uptake of cadmium and zinc by mycelia and their accumulation in mycelia and fruiting bodies of edible mushrooms. European Food Research and Technology. 1999;209(5):317-324. DOI: 10.1155/2013/1499120

116. WHO. Lead Environmental Health Criteria. Geneva: WHO; 1995

117. Yang X, Baligar VC, Martens DC, Clark RB. Cadmium effects on influx and transport of mineral nutrients in plant species (Sedum alfredii Hance). Plant and Soil. 1996;259:181-189. DOI: 10.1080/01904169609365148

118. Zhang H. Chromium contamination in the soil from an alloy steel factory in Nanjing. China Environmental Science. 1997;17(2):80-82

Chapter 6 THE EFFECTS OF HEAVY METALS ON FOREST ECOSYSTEMS IN THE NORTHERN EUROPEAN REGION OF RUSSIA

1. INTRODUCTION: AN OVERVIEW

The metallurgy industries, both ferrous and non-ferrous, are the most environmentally harmful sectors. The primary pollutants released into the atmosphere are sulfur dioxide and polymetallic dust, which contain high levels of heavy metals such as copper, lead, cadmium, zinc, iron, nickel, cobalt, and others. The companies belonging to Norilsk Nickel Mining and Metallurgical Company (Severonickel, Pechenganickel, Norilsk Nickel) are extensively researched as major contributors to environmental contamination in Russia. Within the area affected by their presence, terrestrial ecosystems experience disruption ranging from partial deterioration to complete destruction, resulting in the creation of man-made wastelands [1, 2, 3, 4, 5, 6, 7, 8, 9, 10]. In recent decades, there has been a decrease in the amount of air pollution released by industrial companies in economically advanced and some developing nations. This reduction is due to advancements in production and treatment technologies, as well as the relocation or downsizing of production activities. The non-ferrous metallurgy enterprises had the highest levels of atmospheric emissions between 1970 and the late 1980s, with volumes ranging from 140,000 to 275,000 tons per year. However, by the early 2000s, these emissions decreased significantly, by a factor of 3 to 5. Furthermore, there was a further reduction in atmospheric emissions by the end of the 2000s. Currently, the total volume of emissions is no more than 5000 to 35,000 tons per year [10, 11]. The reduction in air emissions has led to a decrease in the impact of human activities on ecosystems. This has allowed for the observation of changes in the condition of both individual components of ecosystems and the overall communities. Over the past ten years, a limited number of these studies have emerged in the scientific literature, namely references [12, 13, 14, 15, 16]. Prior research has demonstrated that the investigation of the buildup of toxic metals found in the dusty emissions from a non-ferrous metallurgy plant in soils, plants of different varieties, mosses, and lichens accurately indicates the detrimental impact of the combined presence of sulfur dioxide and polymetallic dust on the forest ecosystems in the northern regions of Russia [1, 4, 8]. Field tests were conducted to identify the harmful impact of polymetallic dust on ecosystems in the north-taiga forests. The influence of heavy metals on forest ecosystems in the European North of Russia.

2. MATERIAL AND METHODS

Laboratory of Ecology of Plant Communities of the Botanical Institute named after V.L. Komarov of the Russian Academy of Sciences for more than 35 years has been monitoring the state of forest ecosystems in the impact and buffer zones of the non-ferrous metallurgy complex Severonickel and conducted experimental studies of the effect of heavy metals on various components of forest ecosystems in the Kola Peninsula (European North of Russia).

Studies of the ecosystems of young (formed after fires and felling with a prescription of 40–50 years) and middle-aged (with the prescription of fire or felling of 80–90 years) pine forests of lichen-green-moss site type were carried out in 1980–2017 on a series of permanent sampling sites, 0.1–0.15 ha in size, located at different distances (8–15, 30–35, 60–80 km) from Severonickel smelter (the Kola Peninsula, Russia) within the impact, buffer and background zones.

On the sampling site, the categories of the vital state of all individuals of the tree layer were determined using two parameters: the relative density of the crown (in % of the density of the crown of healthy individuals in the stand of the background zone) and the degree of damage to the assimilation organs by chlorosis and necrosis. Five categories of vitality for trees and undergrowth were identified: I—healthy, II—weakened, III—severely weakened, IV—dying and V—dead individuals. To calculate the vitality index [17, 18], the percentage of individuals of each category was multiplied by the corresponding coefficient (I—1, II—0.71, III—0.43, IV—0.14 and V—0), and then summarized the values obtained. To study the growth dynamics of trees by diameter, increment cores were selected from 20 trees on the sampling site. The registration and determination of the vitality of small seedling (individuals less than 1.3 m in height) was carried out at 40–200 plots with a size of 1 m^2. The characteristics of the ground cover were determined on regularly located marked plots of 1 m^2 (from 20 to 50 on each sampling site).

Analysis of the content of acid-soluble forms of Ni and Cu (extract 1.0 N HCl) in samples of the organogenic horizon of Al-Fe-humus podzols (albic rustic podzol) was carried out by atomic absorption spectrometry, according to accepted methods [19]. An analysis of the content of Ni and Cu in assimilating organs of plants *(Pinus sylvestris, Vaccinium myrtillus, V. vitis-idaea, V. uliginosum, Empetrum hermaphroditum, Arctostaphylos uva-ursi, Avenella flexuosa and Solidago lapponica)* was performed by atomic absorption spectrometry after dry ashing of plant material and transfer of ash to the solution [20].

In 1992, four permanent sampling sites (PSSs) were laid in middle-aged pine forests of the background area of the Kola Peninsula, remote from the Severonickel Smelter beyond 80 km, where no visually observed damages to the plants were registered. One half of the PSS was left as a control. On the surface of the snow cover the other half of PSS was dispersed polymetallic dust identical to emitted into the atmosphere by smelter "Severonickel" in the amount of 352–563 kg/ha. The composition of dust included: Ni—1.3–2.1%, Cu—1.3–1.8% and Co—0.06–0.09% [21]. Disintegration of polymetallic dust led to a spatially very uneven destruction of the ground cover and contamination of the Al-Fe-humus podzol of experimental sites.

The study of the dynamics of the ground cover of the control and experimental plots was performed according to the procedure described above. Chemical analysis of samples of the organogenic horizon of Al-Fe-humus podzol, leaves (needles) of plants *(Pinus sylvestris, Vaccinium myrtillus, V. vitis-idaea, Empetrum hermaphroditum)*, living parts of lichen thallus (*Cladonia stellaris, Cl. rangiferina, Cl. mitis and Cl. uncialis*) and live parts of moss (*Pleurozium schreberi*) collected in experimental plots were carried out according to the procedure described above.

3. RESULTS AND DISCUSSION

3.1. The effects of airborne pollution on forest ecosystems surrounding the "Severonickel" smelter complex.

Between 1981 and 1990, the Severonickel Smelter released an average of over 220,000 tons of SO^2 and 16,000 tons of solids into the atmosphere each year. Since then, there has been a gradual reduction in pollutant emissions. Currently, the annual volume of SO^2 and solid matter emissions at the Severonickel Smelter has decreased by 8 and 5 times, respectively, compared to their highest recorded levels (Fig. 1).

The fine-dispersed polymetallic dust is mostly composed of sulfides and metal oxides, as well as metallic Ni and Cu [21].

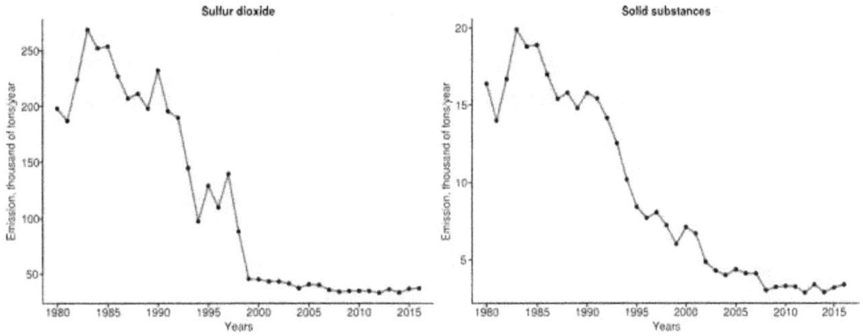

Fig. 1. Volume of atmospheric emissions of the Severonickel Smelter (Kola Peninsula, Russia) (after data [22]).

From 1981 to 2014, in the background area of the Kola Peninsula, the average content of acid-soluble forms of nickel (Ni) and copper (Cu) in the organogenic horizon of the soil was 10.0 ± 0.9 and 10.5 ± 1.1 mg/kg of soil, respectively. These values were considered as the regional background levels of heavy metals in forest litter. The average concentrations of acid-soluble forms of Nickel (Ni) and Copper (Cu) in the organogenic horizon of the soil within the buffer zone were 78.1 ± 5.9 and 137 ± 21 mg/kg, respectively, for the entire study period. These concentrations are 7.8 and 13.7 times higher than the regional background values. Within the area affected by the impact, the average concentrations of Ni and Cu for the entire period of observation were 530 ± 30 and 1035 ± 95 mg/kg soil, respectively. These values are 53 and over 100 times higher than the background values in the surrounding region [22]. There is a strong negative correlation ($r = -(0.63–0.70)$, $p < 0.05$) between the levels of acid-soluble forms of Ni and Cu in the organogenic horizon of the soil and the distance from the source of pollution. This correlation has been confirmed by multiple specialists who conducted studies in the impact zone of the Severonickel and Pechenganickel Smelters [1, 4, 5, 6, 7, 9, 10, 13, 23, 24]. The relationship between the concentration of acid-soluble forms of nickel (Ni) and copper (Cu) in the organogenic horizon of Al-Fe-podzols and the distance from the contamination source can be well described by the exponential equation (Fig. 2). It is important to mention that the amount of Cu in acid-soluble forms in the soil is consistently higher than that of Ni.

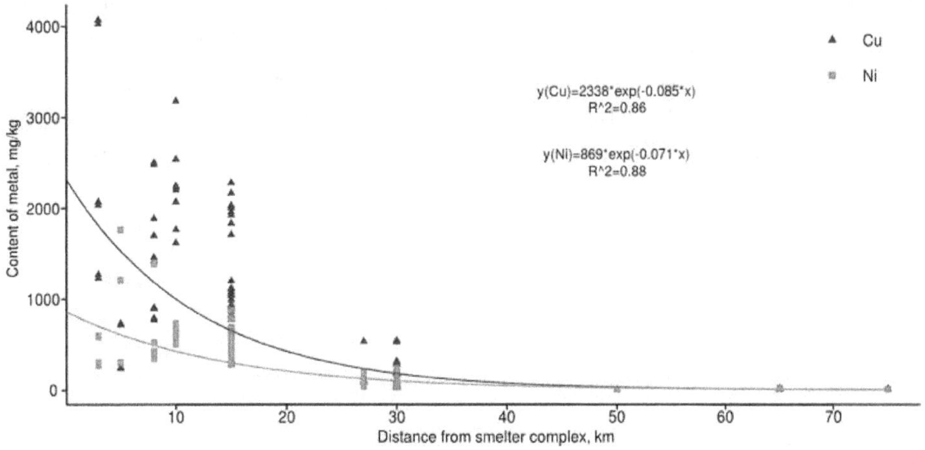

Fig. 2. The content of heavy metals in the organogenic horizon of podzols along the profile from the Severonickel Smelter (Kola Peninsula, Russia).

The technogenic load index is employed to assess the pollution level of habitats. It measures the surplus of acid-soluble forms of Ni and Cu in the polluted organogenic horizon of podzols compared to their regional background values. Between 1981 and 2017, the average technogenic load index was 11.2 rel. units in the buffer zone and 72 rel. units in the impact zone. The analysis of the man-caused load index dynamics (Fig. 3) reveals that, despite the significant reduction in atmospheric emissions from the Severonickel Smelter (Fig. 1), the pollution level of the upper soil horizon continues to rise within the buffer zone. Furthermore, within the impact zone, the pollution level remains extremely high, as previously observed [22].

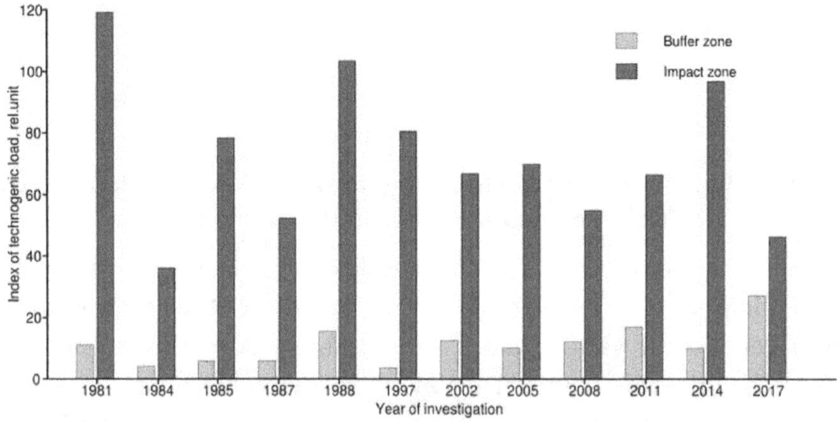

Fig. 3. The dynamic trend of the technogenic load index within the buffer and impact zones of the Severonickel Smelter (Kola Peninsula, Russia).

The concentration of heavy metals in the assimilation organs of all plant species being studied (*Pinus sylvestris, Vaccinium myrtillus, V. vitis-idaea, V. uliginosum, Empetrum hermaphroditum, Arctostaphylos uva-ursi, Avenella flexuosa, and Solidago lapponica*) increases as the source of contamination approaches. Fig. 4 displays the concentrations of Ni and Cu in the needles of *Pinus sylvestris*, a tree species found at varying distances from the Severonickel Smelter. During the study period, the average nickel (Ni) content in the assimilation organs of all species was found to be 4.9 ± 0.3 mg/kg in the background region and 22.3 ± 1.9 mg/kg in the buffer zone. Similarly, the average copper (Cu) content was 4.8 ± 0.5 mg/kg in the background region and 8.8 ± 0.6 mg/kg in the buffer zone. These values in the buffer zone were 2-4.5 times higher than the background values. In the impact zone, the Ni content was found to be 137 ± 10 mg/kg and the Cu content was 42 ± 3 mg/kg, which were 9-27 times higher than their background concentrations [25]. It is important to highlight the distinctiveness in the accumulation of heavy metals by various plant species. The plant species at a single level of soil pollution, such as within the impact zone, exhibited a variation in content by a factor of 3-9. It is important to observe that, unlike the soil, the Ni concentration is consistently higher than the Cu concentration in all plant species. This is attributed to the rapid migration of Ni ions from the contaminated soil to the above-ground plant organs.

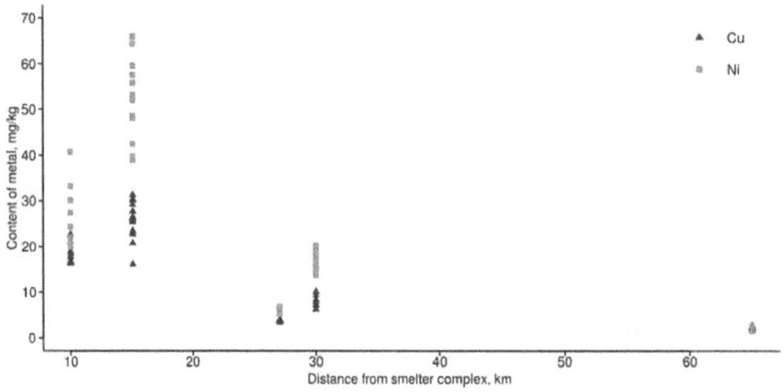

Fig. 4. The content of heavy metals in the needles of *Pinus sylvestris* L. along the profile from the Severonickel Smelter (Kola Peninsula, Russia).

Within the buffer zone, there was a significant build-up of heavy metals in leaves (needles) from 1981 to 1997 (Fig. 5). However, from 2008 to 2014, it was observed that the accumulation of heavy metals decreased by 1.5 to 2.5 times compared to the highest recorded value. The average concentration of Ni and Cu in the assimilation organs within the impact zone consistently decreased from the highest levels (300 and 85 mg/kg) to the lowest levels (47 and 16 mg/kg). This means that over the entire observation period, there was a significant 5-6 fold decrease in the average content of heavy metals in the assimilation organs of all the plant species studied within the impact zone. The reduction in the accumulation of heavy metals by plant leaves (needles) is mostly caused by the severity of air pollution and is not strongly influenced by the absorption of heavy metals from polluted soil. Previously, it was demonstrated that there is a strong positive correlation ($r = 0.89–0.99$, $p < 0.05$) between the level of heavy metal contamination in the upper soil horizon and their accumulation in the leaves (needles) of plants. It was found that up to 80% of these metals enter the plant through contaminated air, potentially in the form of dust deposits on the surface of the leaf blade.

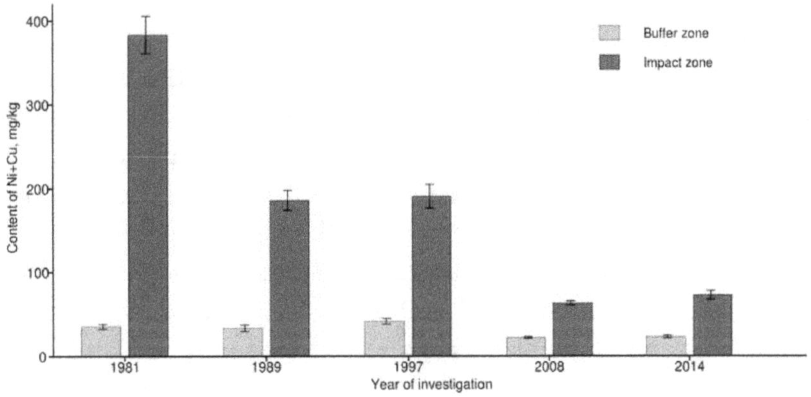

Fig. 5. Dynamics of the average content of heavy metals in all species under investigation within the buffer and impact zones of the Severonickel Smelter (Kola Peninsula, Russia).

Therefore, there is a notable disparity in the trajectory of dynamic patterns regarding the contamination levels of heavy metals in the organogenic horizon of Al-Fe-humus podzols and the concentration of Nickel (Ni) and Copper (Cu) in the assimilation organs of plants. Within the buffer zone, the level of technogenic load index is steadily rising, while in the impact zone, the concentration of heavy metals in the soil is persistently high. The Severonickel Smelter has achieved a significant reduction in atmospheric emissions, resulting in a 2–16-fold decrease in the concentration of nickel (Ni) and copper (Cu) in plant leaves (needles). This decrease is attributed to a reduction in the amount of solid particles deposited on the leaf surface from the contaminated air. Long-term studies conducted on permanent sample plots have enabled the assessment of the overall trends in the dynamics of the condition of trees and stands of *Pinus sylvestris* in the pine forests of the Kola Peninsula, both in terms of spatial distribution and over time. The condition of trees and stands can be evaluated based on the development of their assimilation apparatus. This is important because the degree of tree crowning, the vitality of needles, and the lifespan of trees determine the productivity of forest communities and their ability to withstand environmental stressors. Throughout the study conducted in the pine forests, the lifespan of pine needles ranged from 5.7 to 6.7 years and remained very stable over the course of the study period (Fig. 6). Within these habitats, a minimal portion (not exceeding 5%) of pine needles displayed chlorosis and/or necrosis. This affected less than 5% of the overall surface and was limited to only the older needles (5 years and older). However, in some cases, the extent of chlorosis and necrosis reached up to 20% of the surface area. This occurrence is linked to natural changes in the pine tree's organs responsible for assimilation, which are influenced by age [4, 26].

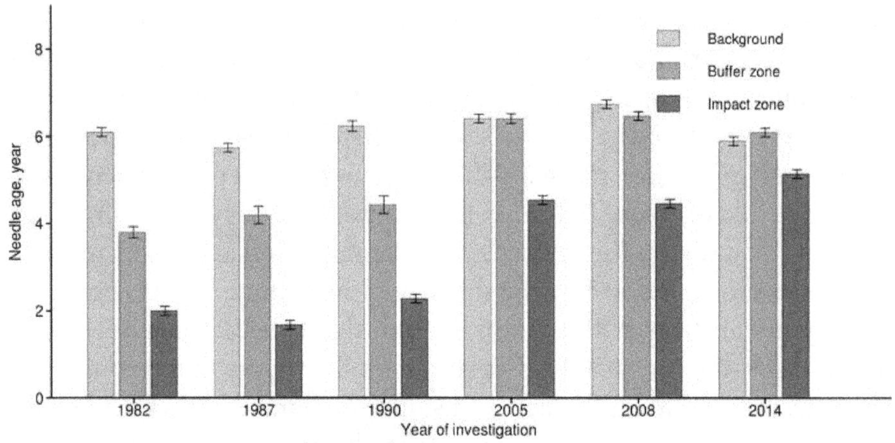

Fig. 6. Dynamics of the life expectancy of *Pinus sylvestris* needles in the background, buffer and impact zones of the Severonickel Smelter (Kola Peninsula, Russia).

In the buffer zone in the period of high aerial emission (1982–1990), the life expectancy of pine needles was 3.8–4.4 years, which is on average 2–3 years less than in the background area (Fig. 6). Later (2005–2014), there was a significant increase in the life expectancy of the needles to 6.1–6.4 years, and the duration of its life ceased to differ significantly from its background value. At the beginning of the research (1988), only 66% of the 1-year-old needles were classified as healthy, more than 30% of the needles of this age had traces of chlorosis and necrosis, with increasing needle age, proportion of healthy needles decreased and the area of injuries increased, 2–3% of needles had necrosis of red-brown color. Studies in 2008 showed that the bulk (93–98%) of P. sylvestris needles of 1-year-old age had no traces of damage. Spots of chlorosis and pinpoint necrosis were found only in a small part (6%) of needles of 1–3 years of age [4, 26].

On the territory of the impact zone during the initial period of research (1982–1987), the average life expectancy of pine needles did not exceed 2 years. Against the backdrop of a sharp decrease in the volume of atmospheric emissions of the Severonickel Smelter experienced a twofold increase in the life expectancy of needles, and now its value is approaching the background value of this parameter (Fig. 6). In 1988, only 25% of 1-year-old needles were classified as healthy in this area, the remaining 75% of 1-year-old needles were covered with chlorosis and necrosis, all needles of 2- and 3-year-old age were damaged. In addition to spotted chlorosis and point necrosis, spotted, girdle and marginal necrosis was noted on the needles, as well as apical necrosis of needles. A re-examination of 2008 showed that the needles of the current year had no visible traces of damage, the proportion of 1-year-old healthy needles increased threefold and was 74%, with the increase in the age of the needles, the degree of its damage increased, but the area of the damaged needle surface was significantly smaller compared with the data of 1988 [4, 26]. An analysis of the vitality patterns of the tree stand reveals not only the natural processes of its formation, development and self-sustainment, but also reflects the impact of stress factors such as fires, felling and airborne pollution. During the period of research, the vital state index very slowly decreased from 0.96 to 0.83 (Fig. 7), which is due to the natural process of differentiation of individuals in the categories of vitality in middle-aged stands [4, 18, 26].

In the buffer zone in 1982, the proportion of healthy individuals did not exceed 50%, the proportion of the weakened ones was 31% and the severely weakened and dying – 19%, then the state of the *Pinus sylvestris* stands worsened. The vitality index of pine stands in 1982 was 0.76, in 1987–1990s it decreased on average to 0.55. Then, with a reduction of the degree of needle damage, an increase in the proportion of healthy individuals up to 60% occurred, which led to an increase in the vitality index of the stand to 0.72–0.78 (Fig. 7).

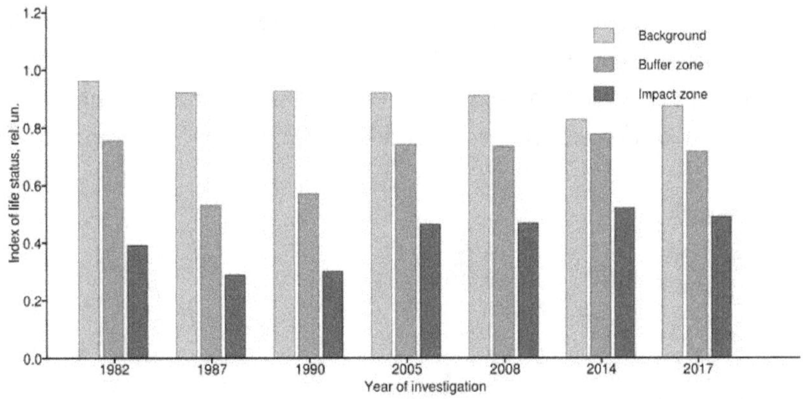

Fig. 7. Dynamic trend of the vitality index of *Pinus sylvestris* stands in the background, buffer and impact zones of the Severonickel Smelter (Kola Peninsula, Russia).

In pine stands of the impact zone, in 1982, severely weakened and dying individuals predominated, vital state index was 0.39, which is 2.5 times less than its background value (Fig. 7). In 1987–1990 the state of the stands in this zone continued to deteriorate, the index of the vital state decreased to its minimum value of 0.30. Since 2005, the vitality spectrum has become fuller due to the appearance of healthy individuals of Scots pine, whose share was almost a quarter (23%) of all individuals in the stand, the proportion of weakened individuals somewhat decreased to 22–23%. The index of the vital state has significantly increased in relation to the beginning of the period of research, and in the period 2005–2017 its value was in the range 0.47–0.52, however, this is more than 1.5 times less than the background values. Fig. 8 shows that the annual growth of *Pinus sylvestris* tree trunks has different patterns in the background and impact zone. Throughout the monitoring period, the area of yearly rings in the pine forests remained rather consistent, ranging from 0.7 to 2.0 cm^2. From the mid-1990s till now, the size of annual rings in the impact zone has fluctuated between 0.2 and 0.7 cm2. However, since 2000, there has been a significant increase in their size.

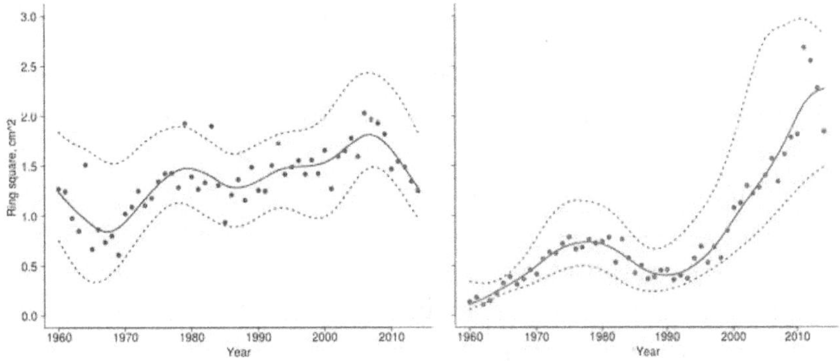

Fig. 8. The annual increment of *Pinus sylvestris* trunks in the background area and impact zone of the Severonickel Smelter (Kola Peninsula, Russia).

Analyzed the annual rings of *Pinus sylvestris* in both the background and impact zone. Found that the direction of changes in this parameter differed significantly during two observation periods: the period of high aerial emission (1985-1999) and the period of 5-8-fold decrease in atmospheric emission (2000-2014). During the second observation period, there was a significant and consistent increase in the size of yearly rings ($z = 2.47–4.67$, $p = 0.000–0.01$) compared to the first observation period. This increase was observed in both the background and impact zones. However, the size of the ring rose by an average factor of 1.3 in the background area, and by an average factor of 3.4 in the hit zone. From 2000 to 2014, there was no significant difference in the size of annual rings between the background and impact zone ($z = 0.28$, p

= 0.78). Consequently, when the Severonickel Smelter reduced its atmospheric emissions by 5-8 times in both the buffer and impact zones, the *Pinus sylvestris* stands responded positively. The needles of the trees had a longer lifespan and there was a decrease in the area of needle damage caused by chlorosis and necrosis. As a result, the vigor of the stands improved and the yearly increment area increased.

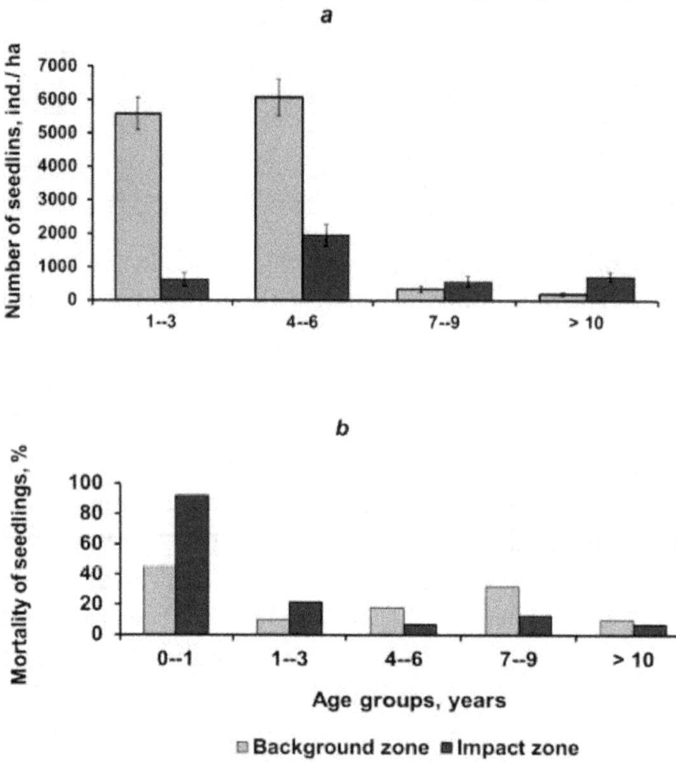

Fig. 9. Number (a) and average annual mortality (b) of different age groups of *Pinus sylvestris* seedlings in the background area and impact zone of the Severonickel Smelter (Kola Peninsula, Russia) at the beginning of the research.

Studying the process of regeneration of the primary tree species is crucial in evaluating the stability of pine forests and determining their potential for restoration in regions that have been disrupted. A comparative research was conducted to examine the abundance and vitality structure of *Pinus sylvestris* seedlings. Initially, the combined number of seedlings in the background region and the buffer zone was between 10,000 and 12,000 individuals per hectare. There was no notable difference in the number of seedlings between these two areas. However, after a period of 5 years, the number of seedlings reduced to a range of 2500-3000 individuals per hectare. Within the area affected by the influence, the quantity of young plants exhibiting a comparable pattern was notably reduced, roughly thrice (Fig. 9a).

The primary factors contributing to the decline in the quantity of pine seedlings within the impact zone, in comparison to the surrounding area, are the reduction in the generation of viable seeds and the harmful effects of heavy metal accumulation in the soil. Within the pine forests of the impact zone, the quantity of female cones on the trees reduces by a factor of 8-10, while the number of fully developed seeds within each cone decreases by a factor of 3. Additionally, the laboratory germination rate of seeds created under favorable climatic conditions is reduced by half. The results of a field experiment conducted in the background area of the Kola Peninsula [27] showed that the germination of *Pinus sylvestris* seeds on soils with a high level of contamination (index of technogenic load 120 rel. units) was 1.5-2 times lower than in the control area. Specifically, the germination rate was 10.7% in the contaminated soil compared to 19.0% in the control soil. An examination of the mortality patterns of sprouts and seedlings during their growth revealed that the highest mortality rate occurred in the first year of life for both the uncontaminated and contaminated soil conditions (Fig. 9b). However, in the contaminated soil, the mortality rate was twice as high as in the uncontaminated soil, with rates of 45% and 92% respectively. Throughout the following three years of the seedlings' lives, an analogous link was sustained, resulting in a notable decrease in the average annual death rate. The high level of heavy metal contamination in soils leads to a significant decrease in the growth of pine sprouts. This is primarily due to the disruption in the development of their root systems. Compared to the control group, the average length of the main root decreases by 7.5-8 times (12 ± 7 mm compared to 93 ± 18 mm). Additionally, the number of first-order lateral roots decreases by 10 times (3 ± 4 pieces compared to 37 ± 13 pieces). Many other species of coniferous woody plants have a comparable response in their root systems to high levels of soil pollution by heavy metals [9, 28, 29, 30].

Within the age range of pine seedlings, specifically from 4–6 to 10–15 years, the rate of individual extinction on uncontaminated soils is 1.5–2 times greater than on contaminated soils (Fig. 9a). However, by the end of the experiment, the density of 15-year-old seedlings on both uncontaminated and contaminated soils equalizes and reaches an average of 1–2 individuals per square meter. Based on our findings, we can infer that soil contamination is the primary factor influencing the survival of pine seedlings under the age of 3 years in the impact zone. However, its significance diminishes as the seedlings continue to mature.

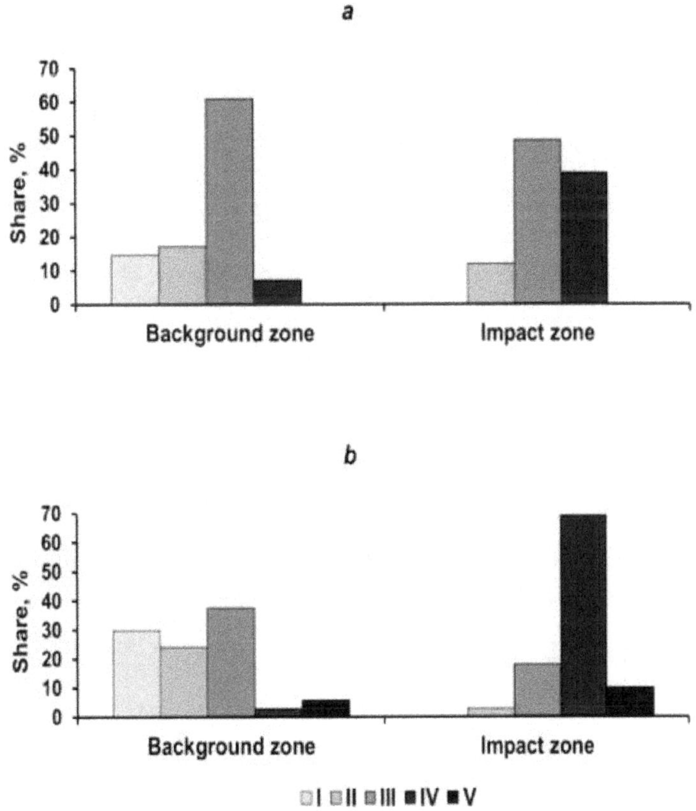

Fig. 10. The vitality structure of the 5-year (a) and 10-year (b) *Pinus sylvestris* seedlings in the background area and the impact zone of the Severonickel Smelter (Kola Peninsula, Russia). Status categories: I – healthy; II – weakened; III – severely weakened; IV – dying; V – dead.

In the vitality spectrum of the 5-year-old seedlings, there is a prevalence of severely impaired people in the background area (Fig. 10a). The combined percentage of healthy and weakened individuals is approximately 30%. In the area affected by the impact, there is a noticeable absence of healthy individuals. The majority of pine trees in this area, namely those that are 5 years old, can be categorized as badly weakened and in the process of dying. The overall condition of the 10-year-old pine seedlings in the background region is much healthier, but in the impact zone it is comparatively poorer than the 5-year-old seedlings (Fig. 10). In 10-year-old pine seedlings, the overall percentage of healthy and weakened people in the background rises to 55%, but the number of seriously impaired individuals falls. Within the impact zone, the rate of decay for the most vulnerable specimens intensifies, resulting in a nearly twofold rise in the proportion of dead individuals. Conversely, the number of weaker and seriously weakened individuals drops by a factor of 3 to 4.

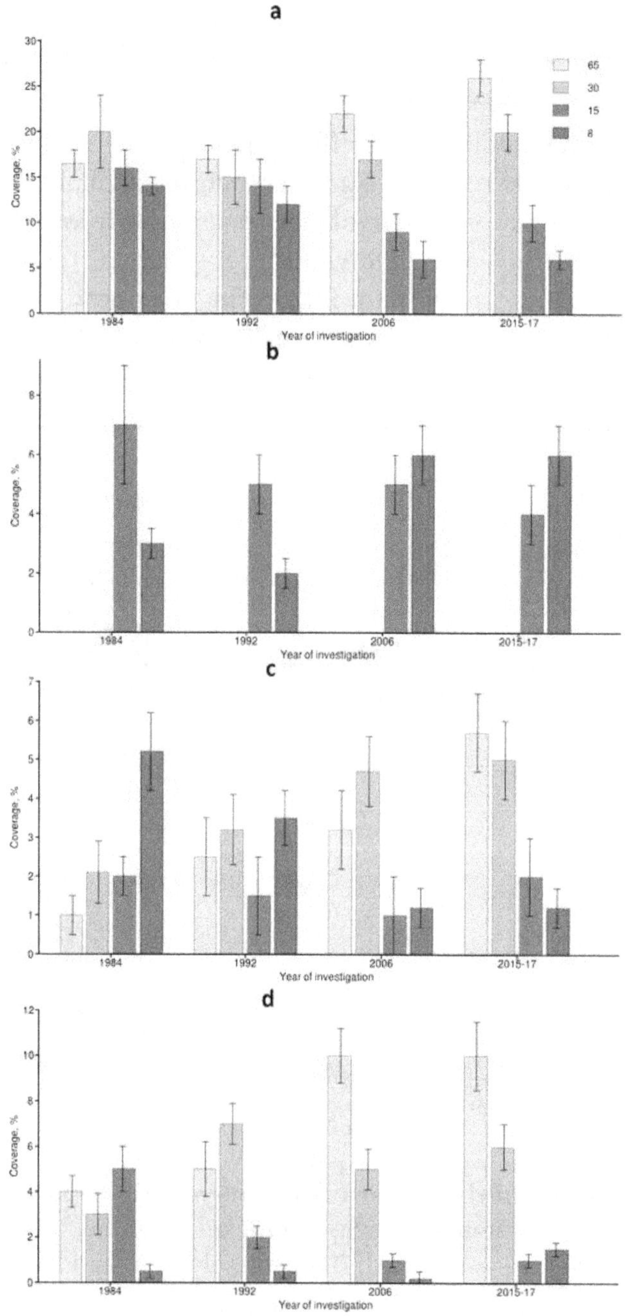

Fig. 11. The 33-year dynamics of the total coverage and coverage of some species of the dwarf-shrub and herb layer at various distances from the Severonickel Smelter (Kola Peninsula, Russia): a – total coverage of the layer; b – coverage of Arctostahyllos uva-ursi; c – coverage of Empetrum hermaphroditum; d – coverage of Vaccinium vitis-idaea.

The reduction in atmospheric emissions and the enhancement of the health of pine stands in the affected area contribute to an augmentation in the production and viability of pine seeds. Consequently, despite the persistent high level of soil contamination by heavy metals, there is an increase in the number of newly emerging pine sprouts. In certain highly favorable germination years, such as 2006 and 2007, the number of sprouts and 1-year-old pine seedlings in the impact zone was similar to the background values observed during the past decade. Nevertheless, the death rate of young plants in the area affected by the impact remains exceedingly high during the 1 to 3–5 year period. Therefore, despite a notable reduction in the level of pollution caused by aviation activities, there was no substantial enhancement in the regeneration of Scots pine trees within the affected area.

Ground cover plays a crucial role in boreal forests. The moss-lichen layer responds to daily changes in humidity and temperature by shedding and scattering its debris. Air pollution causes the degradation of the lower levels and the impairment of their ability to create and sustain their natural environment [31, 32].

During the observation period, there was a recorded rise in the overall projective coverage of the dwarf-shrub and herb layer from 17% (1984–1992) to 24% (2006–2017) in the pine woods where the fire occurred 60 years ago (Fig. 11). The increase in coverage was attributed to the expansion of all the primary species: *Empetrum hermaphroditum, Vaccinium vitis-idaea, and Vaccinium myrtillus.*

Throughout the observation period, the overall extent of the moss-lichen layer remained constant, with an average coverage of 70-75% (Fig. 12). Significant reorganization is shown in the composition of the moss-lichen layer. During the initial time of analysis in 1984, when the fire had been burning for 60 years, the primary dominant species on the stage were fruticose lichens belonging to the genus Cladonia. These lichens covered around 50-60% of the area. The proportion of bryophytes in the overall projective cover was 10%, with the dominating moss species *Pleurozium schreberi* accounting for 4% of the cover in the later stages of succession. During the period of 80-90 years (2006-2017), the presence of fruticose lichens declined by 1.5-2 times, whereas the presence of *Pleurozium schreberi* increased to 35-40% (Fig. 12). The observed alterations in the ground cover are a result of the increased thickness of the forest litter [33] and the modification in the wind patterns beneath the canopy as the height of the stand grows.

During the specified era, 140 years ago, pine forests in the backdrop were set on fire. However, all aspects of the dwarf-shrub and herb layer, as well as the moss-lichen layer, remained consistent [4]. The dwarf-shrub and herb layer had a coverage of 25-26%, with *Empetrum hermaphroditum, Vaccinium vitis-idaea, and V. myrtillus* co-dominating the cover. The moss-lichen layer covered around 85% of the area. The dominant lichens in the cover were fruticose lichens belonging to the genus Cladonia, whereas the moss species present was *Pleurozium schreberi.*

In the buffer zone in 1984, the projected coverage of the dwarf-shrub and herb layer was 18% and remained relatively stable from 1984 to 2017 (Fig. 11). The dominant dwarf-shrubs in this area were Vaccinium and Empetrum hermaphroditum. Overall, there is no substantial difference in the overall coverage, dominating species composition, and long-term dynamics between the dwarf-shrub and herb layer of pine forests in the buffer zone and the background region.

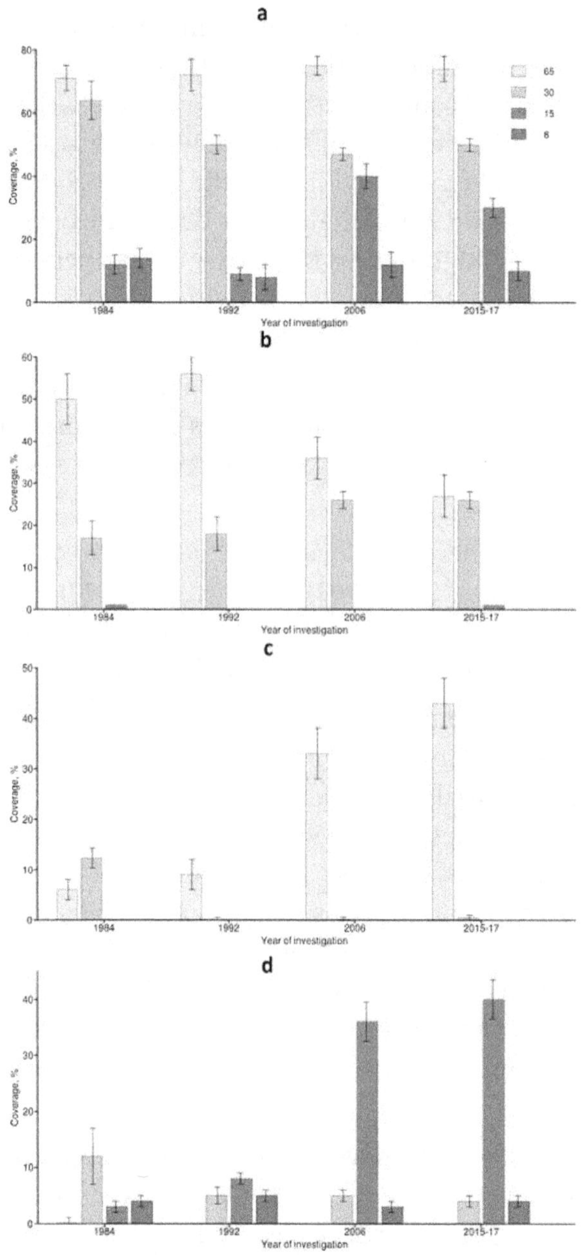

Fig. 12. The 33-year of characteristics of the moss-lichen layer at various distances from the Severonickel Smelter (Kola Peninsula, Russia): a – total coverage of the layer; b – coverage of fruticose lichens; c – coverage of *Pleurozium schreberi*; d – total coverage of the crustaceous lichens and primary thallus of Cladonia sp.

In 1984, the extent of the moss-lichen layer in pine forests of the buffer zone, which had experienced a fire 50 years ago, was found to be similar to the coverage in the background communities, with a value of 64% (Fig. 12). However, the dominant species in this layer were early succession lichens belonging to the Cladonia genus (25%) and *Trapeliopsis granulosa* (10%). In 1992, there was a significant reduction in the extent of the moss-lichen layer, with a decrease in coverage to 50%. The dominant moss species, *Pleurozium schreberi*, virtually disappeared. Additionally, there was a decrease in the coverage of crustose forms of lichens from 14% to 5%, and an increase in the coverage of fruticose lichens of the genus Cladonia from 18% to 26%. The structural modifications of the moss-lichen layer that have been recorded persist up until the present day.

Within pine forests that had experienced a fire 140 years ago, the overall extent of the moss-lichen layer declined from 54% to 37% between 1984 and 2006. This decline can be attributed to the ongoing reduction in the covering of Cladonia, a type of fruticose lichen, as well as the removal of *Pleurozium schreberi*. From 2006 to 2015, there was a notable positive trend in the ecosystem. The overall projective cover of the layer increased to 50%. Additionally, there was an emergence of a dominant species in the final stage of succession, *Pleurozium schreberi*, which accounted for 0.1-0.2% of the cover. Furthermore, the coverage of fruticose lichens increased from 10% to 18%. During the study period, crustose lichens (covering approximately 7% of the area) and Cladonia species with unbranched podecia (also 7% of the area) played a significant role in the ground cover. This is highly unusual considering the prescribed fire, suggesting that the moss-lichen layer has been suppressed for an extended period of time. One notable characteristic of the populations in the buffer zone is their involvement in the growth of liverworts, which typically cover around 10-15% of the area and have been present for over 140 years. This phenomenon is atypical for the context.

The results depicted in Fig.s 11 and 12 describe the pine communities in the impact zone, indicating the age of fire at the start of the experiment in 1984 as 130 years and about 50 years. The projective covering of the dwarf-shrub and herb layer in the impact zone in 1984 and 1992 was similar to the background and had an average value of around 16%. During the last stages of the study, conducted in 2006 and 2017, it was seen that the total coverage of the layer decreased to 10% and 6% correspondingly. The loss in coverage found was mostly caused by a decline in the coverage of Vaccinium species (130 years after the fire), as well as Empetrum hermaphroditum and Calluna vulgaris (50 years after the fire), whose coverage decreased from 4-6% to 1%. The dwarf-shrub and herb layer in the impact zone is primarily composed of the dwarf-shrub Arctostaphylos uva-ursi, which has a coverage of 2-6%. It is important to mention that this differs from the background and buffer zones. This phenomenon is caused by two factors. Firstly, the organogenic horizon of the soil, where the majority of roots and subterranean shoots of dwarf-shrubs are found, contains a significant amount of heavy metals. Secondly, the destruction of the moss-lichen layer and the degradation of forest litter, which are crucial for maintaining the moisture capacity of the upper soil horizons, contribute to this phenomenon.

The moss-lichen layer's overall coverage in the impact zone differed significantly from the background throughout the entire research, ranging from 10% to 40% (Fig. 12). In this instance, the prevailing species commonly found in unaffected regions, namely Cladonia fruticose lichens and *Pleurozium schreberi* moss, were nearly entirely gone. The stratum consisted solely of cruticose lichens and main thalli from various lichen species. Distinct variations were seen in the patterns of lichen mat growth in populations with varying fire ages. From 1984 to 2017, in a neighborhood with a fire age of 50 years, the overall projected coverage remained relatively stable at approximately 10%. The lack of changes is attributed to the community's positioning in the higher portion of the slope, where soil erosion processes are actively occurring and consistent removal of the forest litter (which measured 1-2 cm in 1984) is carried out by rain and meltwater. As a result, the mineral layers of the soil and big stones were exposed, inhibiting the growth of lichen cover. Lichen cover was only able to develop in certain limited locations. Within a community that has existed for 130 years, the proportion of lichen cover grew significantly from approximately 10% to 40% between 2006 and 2017. This change can be attributed to a shift in the condition of the surface on which the lichen cover develops. The long-term decline in the decline of inflow (caused by the destruction of the stand by the early 1990s) and the accumulation of forest litter enabled an increase in the prevalence of lichen species that are resistant to pollution. However, currently, the condition of the moss-lichen layer in the pine forests of the impact zone is assessed as completely devastated. The overall projective coverage is 2-8 times lower compared to the unaffected areas, and all the primary dominant species that are typically found in communities with a fire age of over 70 years are absent.Comparison of dynamics of ground vegetation in pine forests of the impact zone of and rate of its recovery in field experiment with the complete removal of ground vegetation conducted in the background area [27], allows to separate the effects of soil pollution and comprehensive aerial pollution by sulfur dioxide together with polymetallic dust. Under the conditions of the field experiment (unpolluted air – polluted soils taken from the impact zone), a significant slowing of the restoration of the dwarf-shrub and herb layer as well as of moss-lichen layer was recorded. This explains the lack of positive dynamics of the state of the ground cover of pine forests with a significant decrease in the level of aerial pollution.

Consequently, in the atmosphere contaminated with airborne pollutants, the normal progression of the ground cover's ecological changes is disturbed. The ground cover of communities in the buffer zone closely resembles that of woods burned approximately 30-40 years ago, whereas in the impact zone it resembles that of forests burned around 10 years ago. Within the impact zone, when there is forest litter present, a protective layer is created by the crustose lichens and the main body of lichens belonging to the Cladonia genus. There is no moss covering. The reduction of atmospheric emissions had no impact on the condition of the ground cover in both the buffer and impact zones. This is because the upper layer of the soil remains heavily contaminated with heavy metals.

3.2. An empirical investigation of the effects of soil pollution caused by heavy metals on forest ecosystems

The acid-soluble forms of Nickel (Ni) and Copper (Cu) in the organogenic horizon of Al-Fe-humus podzols at the control sample sites had an average total level of no more than 20 mg/kg. The amounts of heavy metals, specifically nickel (Ni) and copper (Cu), ranged within the range of 9.4-260 mg/kg and 22-615 mg/kg, respectively, in the experimental plots. Fig. 13 displays the average concentrations of acid-soluble forms of heavy metals in the forest litter of the experimental sites. The technogenic load index ranged from 2.0 to 45.5 relative units, with an average of 15.0 relative units. There was no significant difference in the index between the experimental locations. There is no relationship between the quantity of polymetallic dust introduced and the technogenic load index. There was no observed link between the quantity of deposited polymetallic dust and the technogenic load index.

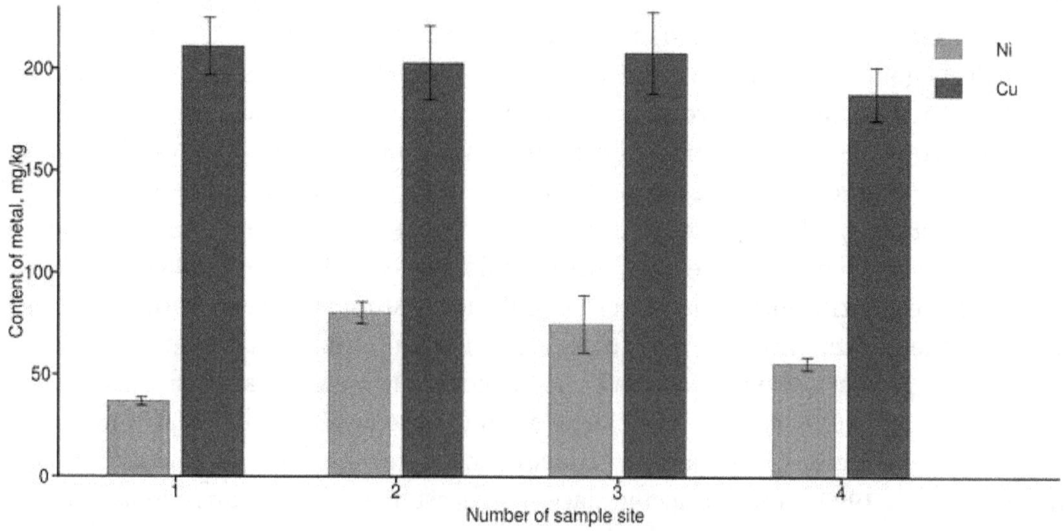

Fig. 13. The average content of acid-soluble forms of Ni and Cu in the organogenic horizon of Al-Fe-humus podzol in the experimental sites.

The concentration of heavy metals in the needles of *Pinus sylvestris* trees and the leaves of dwarf-shrubs (*Vaccinium myrtillus, V. vitis-idaea, Empetrum hermaphroditum*) in the experimental plots was found to be significantly higher than their natural levels, but only by a factor of 1.3-3. The concentration of heavy metals in the living components of lichens (*Cladonia stellaris, Cl. rangiferina, Cl. mitis, and Cl. uncialis*) and moss (*Pleurozium schreberi*) collected from both control and experimental sites did not show any significant differences, except for a minor increase in nickel (Ni) and copper (Cu) levels compared to their natural levels. The slight elevation in the levels of Ni and Cu found in the plant material at the experimental sites suggests a limited movement of heavy metals from the polluted soil to the above-ground parts of plant organisms. Furthermore, the concentrations of these metals in the tissues remain below the threshold at which they become toxic to plants, mosses, and lichens.

Therefore, based on the field experiment, it can be concluded that the higher levels of heavy metals found in the plant assimilation organs and live parts of lichens and mosses do not exceed the minimum tolerance threshold and do not cause death or harm to their essential functions.

Fig. 14 illustrates the changes in the projective covering of the dwarf-shrub, herb, and moss-lichen layers at the control and experimental sites. In 1992, both locations showed identical characteristics in terms of the dwarf-shrub and herb layer, but there was a modest variation in the total projective covering of the moss-lichen layer.

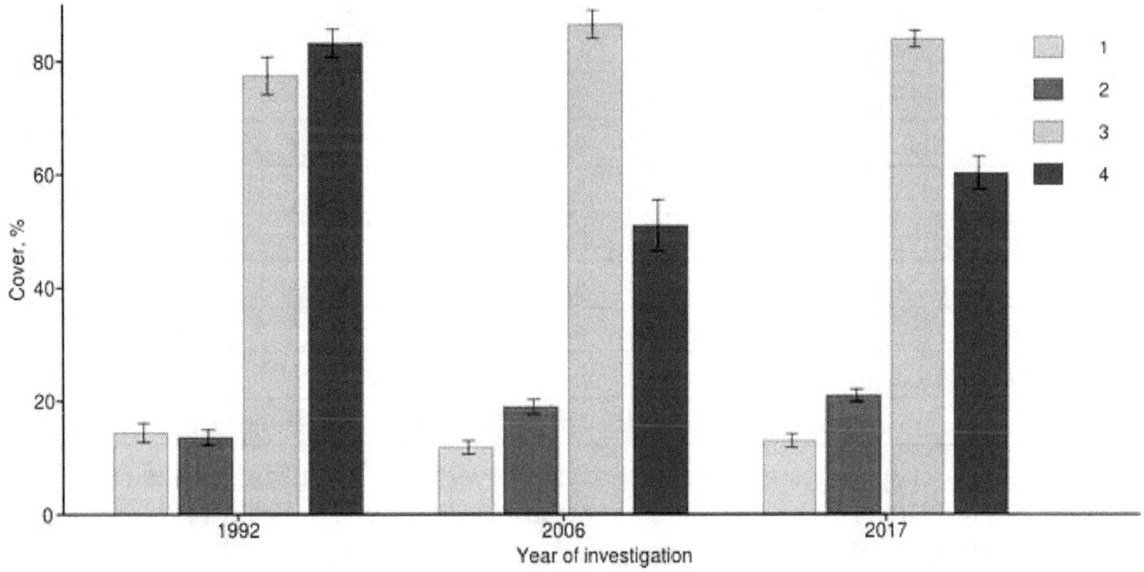

Fig. 14. Dynamics of covearages (%) of dwarf-shrub and herb (1, 2) and moss-lichen (3, 4) layers in control (1, 3) and experiment (2, 4) sites.

Within the control community, the changes in the vegetation on the ground followed the principles of post-fire succession. No significant change in the overall coverage of dwarf-shrub and herb layer was observed between 1992 and 2017 (Fig. 14). This lack of change can be attributed to the varying trends of individual species: an increase in the coverage of Vaccinium vitis-idaea and V. myrtillus, and a decrease in the coverage of Arctostaphyllos uva-ursi and Calluna vulgaris (Table 1). In 2017, the overall extent of the moss-lichen layer grew by 7% compared to 1992. However, the total coverage of lichens remained rather stable, whereas the bryophytes had a threefold rise. In 2017, the primary species of lichen covering was Cladonia stellaris, while the moss cover consisted of *Pleurozium schreberi*. The coverage of *Pleurozium schreberi* rose by more than 12 times compared to the start of the experiment.

Table 1. Dynamics of the most active species in the control and experimental sites.

Species and characteristics	Control site			Experimental site		
	Year of investigation					
	1992	2006	2017	1992	2006	2017
	a	b	c	d	e	f
Dwarf-shrubs, total cover	14.4	11.8^{e}	13.0^{f}	$13.6^{e,f}$	$19.0^{b,d}$	$21.0^{c,d}$
Vaccinium vitis-idaea L.	3.2^{c}	3.4^{c}	$5.8^{a,b}$	$1.8^{e,f}$	$4.2^{d,f}$	$7.5^{d,e}$
Vaccinium myrtillus L	0.2^{c}	1.1	1.8^{a}	0.2^{f}	1.0	2.3^{d}
Empetrum hermaphroditum Hagerup	0.3	0.2	0.9	0.9	0.3	0.6
Arctostaphylos uva-ursi (L.) Spreng.	$1.0^{b,c}$	$<0.1^{a}$	0.1^{a}	0.5^{f}	0.4	0.1^{d}
Calluna vulgaris (L.) Hull	9.7^{c}	7.1^{e}	$4.3^{a,f}$	10.0	13.0^{b}	10.4^{c}
Lichens, total cover	73.4	81.3	70.7	80.2	50.0	58.3
Cladonia mitis Sandst.	$40.0^{b,c,d}$	$21.5^{a,c}$	$3.4^{a,b,f}$	$52.0^{a,e,f}$	$16.4^{d,f}$	$7.3^{c,d,e}$
Cladonia rangiferina (L.) Weber ex F.H. Wigg.	8.8^{b}	$25.5^{a,c,e}$	$16.5^{b,f}$	9.5^{f}	$14.4^{b,f}$	$23.7^{c,d,e}$
Cladonia stellaris (Opiz) Pouzar & Vězda	$9.2^{b,c}$	$26.6^{a,c,e}$	$48.4^{a,b,f}$	7.1^{f}	$10.6^{b,f}$	$19.5^{c,d,e}$
Cladonia uncialis (L.)Weber ex F.H. Wigg.	$8.4^{b,c}$	$4.8^{a,c}$	$1.4^{a,b}$	$6.6^{e,f}$	$3.1^{d,f}$	$1.6^{d,e}$
Cladonia deformis (L.) Hoffm.	0.1	<0.1	$<0.1^{f}$	0.1	0.9	0.2^{c}
Cladonia gracilis (L.) Willd.	0.3^{d}	0.2^{e}	0.1	0.5^{a}	0.5^{b}	0.3
Cladonia crispata (Ach.) Flot.	1.8^{c}	0.9	$0.5^{a,f}$	1.1	1.5	1.2^{c}
Cladonia cornuta (L.) Hoffm.	$1.0^{b,c}$	$0.1^{a,e}$	$0.1^{a,f}$	0.8	0.5^{b}	1.1^{c}
Stereocaulon paschale (L.) Hoffm.	3.2^{c}	1.4	0.3^{a}	1.6	1.3	1.5
Mosses, total cover	4.1^{c}	$5.3^{c,e}$	$13.3^{a,b,f}$	3.0	1.1^{b}	2.0^{c}
Pleurozium schreberi (Willd. ex Brid.) Mitt.	0.9^{c}	$4.7^{c,e}$	$12.2^{a,b,f}$	0.1^{f}	$0.1^{b,f}$	$1.4^{c,d,e}$
Polytrichum spp.	$2.4^{b,c}$	0.4^{a}	0.7^{a}	$2.4^{e,f}$	0.7^{d}	0.3^{d}

Notes: The names of species are given in accordance with the latest (on November 25, 2017) data on the systematics of plants [34] and fungi [35]. The letters indicate significant (at $p \leq 0.05$) differences between the characteristics of the ground cover.

The disturbance has been seen in the succession dynamics of the dwarf-shrub and herb layer at the experimental location. The absence of a drop in the coverage of Calluna vulgaris, coupled with a considerable twofold rise in the coverage of Vaccinium vitis-idaea, led to a 1.5 times increase in the overall projective cover of the layer. This is illustrated in Fig. 14 and Table 1.

The most notable disparities in the dynamics of species coverage were observed in the moss-lichen layer. After fifteen years since the experiment began in 2006, the experimental site experienced a decrease in the projective coverage of Cladonia mitis that was twice as large as the decrease observed in the control site. On the other hand, the increase in *Cladonia rangiferina and Cl. stellaris* was only half as much as the increase observed in the control site. As a result, there was a significant 40% decrease in the total projective cover of the layer as a whole in the experimental site (Fig. 14, Table 1). During this period in the control areas, the overall extent of the layer had a 10% increase. After 25 years of dust dispersion, a partial repair of the moss-lichen layer was seen. The disparity with the control was diminished by 30%.

The direction of the changes in the coverage of two prominent species of lichen mat, *Cladonia mitis and Cl. stellaris*, was consistent between the control and experimental groups. However, in the first scenario, the changes were significantly more pronounced, with the disparities in the coverings of these species reaching a range of 7-11 times. This is in contrast to the changes observed in the experimental area, where the discrepancies did not surpass a range of 3-7 times. The dynamics of the cover of the dominant lichen mat *Cl. rangiferina* differed between the control and experimental areas. In the control area, the maximum projective cover of this species was recorded in 2006, after which it decreased. In contrast, the experimental site showed a gradual increase in *Cl. rangiferina* coverage.

At the experimental site, the presence of heavy metals in the soil hindered the recovery of both the overall growth of mosses and the dominance of *Pleurozium schreberi*. After twenty-five years of using polymetallic dust, the amount of bryophytes covered by the projective covering was seven times lower compared to the control group (Table 1).

Consequently, the presence of heavy metals in the soil caused a disturbance in the progression of plant growth in pine forests during the intermediate stages of recovery following a fire. Typically, the deceleration of the succession processes can be approximated to occur within a timeframe of 15–20 years.

4. CONCISE SUMMARY

The long-term (1980–2017) monitoring of the north taiga pine forests in the area affected by atmospheric emissions from the non-ferrous metallurgy Smelter, along with field experiments conducted in the Kola Peninsula (Russia), enable us to analyze and summarize the results. This analysis provides insights into the spatial and temporal changes in the condition of the forest ecosystems and their components. Despite a significant decrease of 5–8 times in atmospheric emissions from the Severonickel Smelter in the Kola Peninsula, Russia, the pollution level in the upper soil layer in the buffer zone continues to rise, and remains extremely high within the affected area. Throughout the study period, the technogenic load index of the organogenic horizon of Al-Fe-humus podzols varied between 4 and 30 in the buffer zone, and between 35 and 120 relative units in the impact zone. The dynamics of heavy metal content in plant assimilation organs varies greatly from the pollution level in the higher soil horizon. The atmospheric emissions of Severonickel Smelter decreased by a factor of 5–8, resulting in a decrease of Ni and Cu in the plant assimilation organs by a factor of 2–16. This decrease can be attributed to a reduction in the quantity of dust particles originating from the polluted air. The level of heavy metal accumulation in assimilation organs is correlated with the physiological characteristics of plant species. The leaves of Empetrum hermaphroditum exhibit the highest concentration of Ni and Cu, while the leaves of Arctostaphylos uva-ursi have the lowest concentration. The atmospheric emissions from the Severonickel Smelter have significantly decreased in both the buffer and impact zones, leading to a noticeable improvement in the condition of *Pinus sylvestris* stands. The overall condition of the stands improved, resulting in a reduction in the extent of pine needle damage caused by chlorosis and necrosis. As a result, there was a 1.5 to 2.5 times longer lifespan and a 3 to 4 times larger yearly tree trunk growth area. In order to restore the normal renewal process of *Pinus sylvestris* in the impact zone, it is necessary to decrease the pollution level of heavy metals in the upper layer of soil. Due to airborne pollution, the natural progression of ground cover dynamics is disrupted. The parameters of pine forests in the buffer zone, independent of their stage of succession, resemble forests that have experienced a fire approximately 30-40 years ago. In the impact zone, the parameters resemble forests that have had a fire around 10 years ago. The reduction in atmospheric emissions by the Severonickel Smelter did not result in an improvement in the condition of the ground cover in both the buffer zone and impact zones. This is owing to the ongoing high level of soil contamination with heavy metals.

The introduction of pollutants into the soil, with a technogenic load index ranging from 2 to 45 relative units, resulted in a reduction in overall vegetation coverage and a modification of the species composition in both the dwarf-shrub and herb and moss-lichen layers. The technogenic load index had an average value of 15 relative units. As a result, the coverage of the dwarf-shrub and herb layer grew by 1.5 times, while the moss-lichen layer fell by 1.5 times. The alterations are a result of a shift in the relative abundance of dominating species of dwarf-shrubs (*Calluna vulgaris, Vaccinium vitis-idaea*) and lichens (*Cladonia mitis, Cl. rangiferina, Cl. stellaris*). Consequently, there is a 15-20 year delay in the succession processes observed in the ground cover of pine forests that are in the intermediate phases of recovery following a fire. The presence of polymetallic dust in the soil, with a technogenic load index ranging from 2 to 45 relative units, resulted in a 1.5 to 3 times higher concentration of nickel (Ni) and copper (Cu) in the plant material at the experimental sites. The elevated concentration of heavy metals suggests a limited transfer of these substances from the soil to the aerial sections of the plant. The concentrations of Nickel (Ni) and Copper (Cu) in plant assimilation organs and live parts of lichens and mosses, ranging from 1.5 to 15 mg/kg, do not above the toxicity threshold for heavy metals in plants.

REFERENCES

1. Norin B, Yarmishko V, editors. Influence of Industrial Air Pollution on Pine Forests of the Kola Peninsula. Leningrad: Botanical Institute of Academy of Sciences of Soviet Union; 1990. 195 p. (in Russian)
2. Vodyanitskii Y. Contamination of soils with heavy metals and metalloids and its ecological hazard (analytic review). Eurasian Soil Science. 2013;46(7):793-801. DOI: 10.1134/S1064229313050153
3. Vodyanitskii Y, Plekhanova I, Prokopovich E, Savichev A. Soil contamination with emissions of nonferrous metallurgical plants. Eurasian Soil Science. 2011;44(2):217-226. DOI: 10.1134/S1064229311020177
4. Yarmishko V, editor. Dynamics of Forest Communities of the Northwestern Russia. St. Petersburg: VVM; 2009. 276 p. (in Russian)
5. Evdokimova G, Kalabin G, Mozgova N. Contents and toxicity of heavy metals in soils of the zone affected by aerial emissions from the Severonickel Enterprise. Eurasian Soil Science. 2011;44(2):237-244. DOI: 10.1134/S1064229311020037
6. Lukina N, Nikonov V. Biogeochemical Cycles in Northern Forests Affected by Technogenic Air Pollution. Apatity: Kola Scientific Center of Academy of Sciences of Soviet Union. 1996. Vol. 1, 213 p. Vol. 2, 192 p. (in Russian)
7. Lukina N, Nikonov V. Nutrient Balance of the Forests of Northern Taiga: Natural and Technogenic Aspects. Apatity: Kola Scientific Center of Academy of Sciences of Soviet Union; 1998. 316 p. (in Russian)
8. Yarmishko V, editor. Environmental Problems of the Northern Plant Communities. St. Petersburg: VVM; 2005; 450 p. (in Russian)
9. Kozlov M, Zvereva E, Zverev V. Impacts of Point Polluters on Terrestrial Biota. New York: Springer-Verlag, 2009; 466 p
10. Kozlov M, Zvereva E. Industrial barrens: Extreme habitats created by non-ferrous metallurgy. Reviews in Environmental Science and Biotechnology. 2007;6:231-259

11. Environmental Monitoring of the Zone Affected by Emissions of the Kola Smelter [Internet]. Available from: http://www.kolagmk.ru
12. Lyanguzova I. Dynamics of nickel and copper concentrations in plants of pine forests of the Kola Peninsula under conditions of industrial air pollution. Rastit Resursy. 2008;44(4):91-98 (in Russian)
13. Evdokimova G, Mozgova N. Dynamics of the state of forest ecosystems exposed to emissions from a copper-nickel smelter over 30 years. In: Proceedings of the Conference on Ecological Problems in Northern Regions and Ways to Solve Them; 14-16 October 2008; Apatity: Kola Scientific Center of RAS; 2008. Part 1. pp. 68-73 (in Russian)
14. Zverev V. Mortality and recruitment of mountain birch (Betula pubescens ssp. czerepanovii) in the impact zone of a copper-nickel smelter in the period of significant reduction of emissions: The results of 15-year monitoring. Russian Journal of Ecology. 2009;40(4):254-260
15. Sukhareva T, Lukina N. Mineral composition of assimilative organs of conifers after reduction of atmospheric pollution in the Kola peninsula. Russian Journal of Ecology. 2014;45(2):95-102. DOI: 10.1134/S1067413614020088
16. Chernen'kova T, Basova E, Kabirov R. Regeneration successions of northern taiga spruce forests under reduction of aerothechnogenic impact. Contemporary Problems of Ecology. 2011;4(7):742-757. DOI: 10.1134/S199542551107006X
17. Alexeyev V. Impact of air pollution on far North forest vegetation. Science of the Total Environment. 1995;160/161:605-617
18. Yarmishko V, Gorshkov V, Stavrova N. *Pinus sylvestris* L. vital state structure in the tree layer of pine forest with different degree and type of anthropogenic disturbance (Kola peninsula). Rastit Resursy. 2003;39(4):1-19 (in Russian)
19. Fomin G, Fomin A. Control of Quality and Environmental Safety According to International Standards: A Handbook. Moscow: Protektor; 2001. 304 p. (in Russian)
20. Yarmishko V, Lyanguzova I, editors. Methods of the Investigation of forest communities. St. Petersburg: SRI of Chemistry; 2002. 240 p. (in Russian)
21. Barcan V. Nature and origin of multicomponent aerial emissions of the copper-nickel smelter complex. Environmental International. 2002;28:451-456
22. Lyanguzova I, Goldvirt D, Fadeeva I. Spatiotemporal dynamics of the pollution of Al-Fe-humus podzols in the impact zone of a nonferrous metallurgical plant. Eurasian Soil Science. 2016;49(10):1189-1203. DOI: 10.1134/S1064229316100094
23. Kashulina G, Pereverzev V, Litvinova T. Transformation of the soil organic matter under the extreme pollution by emissions of the Severonikel smelter. Eurasian Soil Science. 2010;43(10):1174-1183
24. Barcan V. Leaching of nickel and copper from a soil contaminated by metallurgical dust. Environmental International. 2002;28(1-2):63-68
25. Lyanguzova I. Dynamic trend of heavy metal contents in plants and soil under different industrial air pollution regimes. Russian Journal of Ecology. 2017;48(4):311-320. DOI: 10.1134/S1067413617040117
26. Yarmishko V. Scots Pine and Aerial Pollution in the European North. St. Petersburg: Research Institute of Chemistry of St. Petersburg State University; 1997. 210 p. (in Russian)

27. Stavrova N, Lyanguzova I, Gorshkov V, Bakkal I. Dynamics of forest ecosystems components recovery in field experiment of soil pollution by heavy metals. Rastit Resursy. 2007;43(3):48-65 (in Russian)

28. Wotton D, Jones D, Phillips S. The effect of nickel and copper deposition from a mining and smelting complex on coniferous regeneration in the boreal forest of Northern Manitoba. Water, Air and Soil Pollution. 1986;31:349-358

29. Arduini I, Godbold D, Onnis A. Infuence of copper on root growth and morphology of *Pinus pinea* L. and *Pinus pinaster* Ait. seedlings. Tree Physiology. 1995;15:411-415

30. Veselkin D. Reduction of absorbing root length in Siberian fir and Siberian spruce under heavy metal pollution. Lesovedenie. 2003;3:65-68 (in Russian)

31. Kershaw K. Studies of lichen-dominated systems. XV. The temperature and humidity profiles in a Cladina alpestris mat. Canadian Journal of Botany. 1975;53(22):2614-2620

32. Likens G. Biogeochemistry of a Forested Ecosystem. 3rd ed. NY, Heidelberg, Dordrecht, London: Springer, 2013. 208 p

33. Gorshkov V, Bakkal I, Stavrova N. Postfire recovery of forest litter in scots pine forests in two different regions of boreal zone. Silva Fennica. 1996;30(2-3):209-2019. DOI: 10.14214/sf.a9233

34. The Plant List. Version 1.1. [Internet]. Available from: http://www.theplantlist.org (Accessed: Nov 25, 2017)

35. Index Fungorum. Version 1. [Internet]. Available from: http://www.indexfungorum.org/Names/names.asp (Accessed: Nov 25, 2017)

7 Chapter

ENVIRONMENTAL POLLUTION CAUSED BY HEAVY METALS

1. INTRODUCTION: AN OVERVIEW

The term "environment" can be defined as the immediate surroundings in which humans live. The environment encompasses the land, water, and atmosphere of the earth, as well as microorganisms, plant and animal life. It includes any combination of these elements and the interrelationships among them, as well as the physical, chemical, aesthetic, and cultural properties and conditions that affect human health and well-being. Additionally, it is distinguished by several domains that impact its actions and inherent worth. The biosphere is the most crucial component of the environment as it serves as a habitat for all living organisms. The biosphere is the realm in which living species, including plants and animals, interact with their abiotic environment, consisting of soil, air, and water. Over the past few centuries, the processes of industrialization and globalization have negatively impacted natural habitats and their capacity to support life. This has added elements that undermine the overall functioning of the environment and its inherent worth [1].

1.1. Environmental Pollution

An environment has the potential to become polluted or contaminated. Pollution and contamination are distinct concepts, however contaminants can be considered pollutants and have a harmful effect on the environment. Pollution, as defined in literature, refers to the deliberate or unintentional introduction of substances or energy into the environment by humans. This leads to harmful effects such as damage to living resources, risks to human health, obstacles to environmental activities, deterioration of environmental quality, and a decrease in amenities. Contamination, in contrast, refers to the existence of higher concentrations of substances in the environment that exceed the normal background level specific to the place and organism. Environmental pollution is the term used to describe the negative and undesirable alteration of the physical, chemical, and biological properties of the air, water, and soil. This alteration poses danger to both animal and plant life. Pollution can manifest as chemical substances or various forms of energy, such as noise, heat, or light [2]. Pollutants are chemicals or energy that can cause pollution, and they can be either alien or naturally occurring contaminants.

1.1.1. Different Types of pollutants

Environmental contaminants remain a global concern and a significant challenge for the global community. Pollutants can encompass both naturally occurring substances and foreign matter that, upon interacting with the environment, result in detrimental alterations. Various categories of contaminants exist, specifically inorganic, organic, and biological. Regardless of the categorization of pollutants, they all attract significant attention due to the environmental damage they cause. The correlation between environmental pollution and world population is undeniably positive, as seen by the escalating discharge of potentially hazardous compounds into the environment, which corresponds to the worrisome expansion of the global population. This dilemma has resulted in pollution being a prominent environmental issue.

1.1.1.1. Inorganic pollutants

Industrial, agricultural, and home waste all contribute to environmental pollution, resulting in detrimental effects on both human and animal health. Inorganic pollutants are discharged from these particular sources. Inorganic pollutants often consist of minerals, metals, salts, and other compounds derived from minerals [2]. Research has documented the presence of inorganic pollutants, which are substances that occur naturally but have been modified by human activities, resulting in an elevated concentration of these pollutants in the environment. Inorganic substances are introduced into the environment by various human activities, including mine drainage, smelting, metallurgical and chemical operations, as well as natural processes. These contaminants are hazardous because they build up in the food chains [3].

1.1.1.2. Organic pollutants

Organic pollution is the presence of biodegradable pollutants in an environment. These sources of pollution occur naturally and are generated by the environment, but human activity has also been adding to their significant production in order to suit human requirements. Several organic pollutants that have been identified as particularly concerning include human waste, food waste, polychlorinated biphenyls (PCBs), polybrominated diphenyl ethers (PBDEs), polycyclic aromatic hydrocarbons (PAHs), pesticides, petroleum, and organochlorine pesticides (OCPs) [4]. Organic pollutants have garnered significant attention due to their emergence as a prominent environmental issue. Organic pollutants are classified as persistent due to their notable characteristics, including high lipid solubility, stability, lipophilicity, and hydrophobicity. Organic contaminants possess features that facilitate their bioaccumulation in several environmental realms, resulting in toxicological effects [5, 6].

1.1.1.3. Biological pollutants

Biological pollutants are contaminants that arise from human activities and have a detrimental effect on the quality of both water and land environments. These pollutants encompass bacteria, viruses, molds, mildew, animal fur and cat saliva, home dust, mites, cockroaches, and pollen. Various studies have recorded diverse origins of these pollutants, such as pollens derived from plants, viruses transferred by humans and animals, and bacteria transported by humans, animals, soil, and plant detritus [7].

2. HEAVY METALS

While there is no precise definition of a heavy metal, it is generally described in literature as a naturally occurring element with a high atomic weight and density that is five times more than that of water [8]. Heavy metals have garnered significant attention from environmental chemists due to their hazardous nature, making them one of the most prominent contaminants. Heavy metals are typically found in small quantities in natural bodies of water, but even at extremely low levels, many of them can be harmful [9]. Arsenic, lead, cadmium, nickel, mercury, chromium, cobalt, zinc, and selenium are extremely poisonous, even in little amounts. The escalating presence of heavy metals in our natural resources is currently a significant cause for alarm, particularly due to the indiscriminate release of metal-laden waste by numerous enterprises into freshwater bodies without proper treatment [3].

Heavy metals become poisonous when they are not metabolized by the body and accumulate in the soft tissues. Pathogens can infiltrate the human body via ingestion, inhalation, or dermal contact when individuals interact with humans in agricultural, manufacturing, pharmaceutical, industrial, or residential environments. Industrial exposure is a frequently seen method of exposure for adults. Children are most commonly exposed to substances through ingestion. The environment and its resources are being contaminated by both natural and human activities, resulting in the discharge of a volume that exceeds the ecosystem's capacity to handle it [9, 10] (Fig. 1).

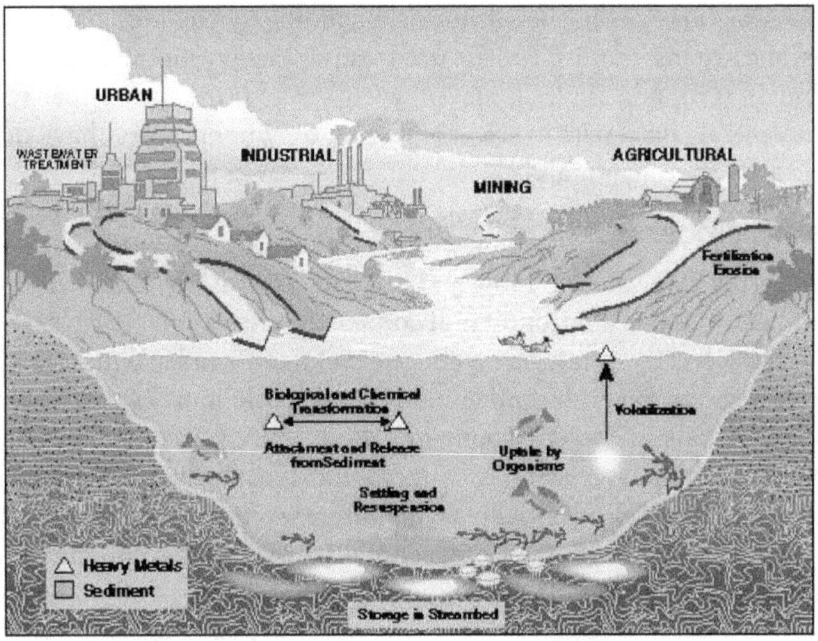

Fig. 1. Sources and sinks of heavy metals [11].

2.1. Origins of metallic elements with high atomic weights

Heavy metals can be released through both natural and human activities and can accumulate in several environmental components such as soil, water, air, and their interfaces (Fig. 2).

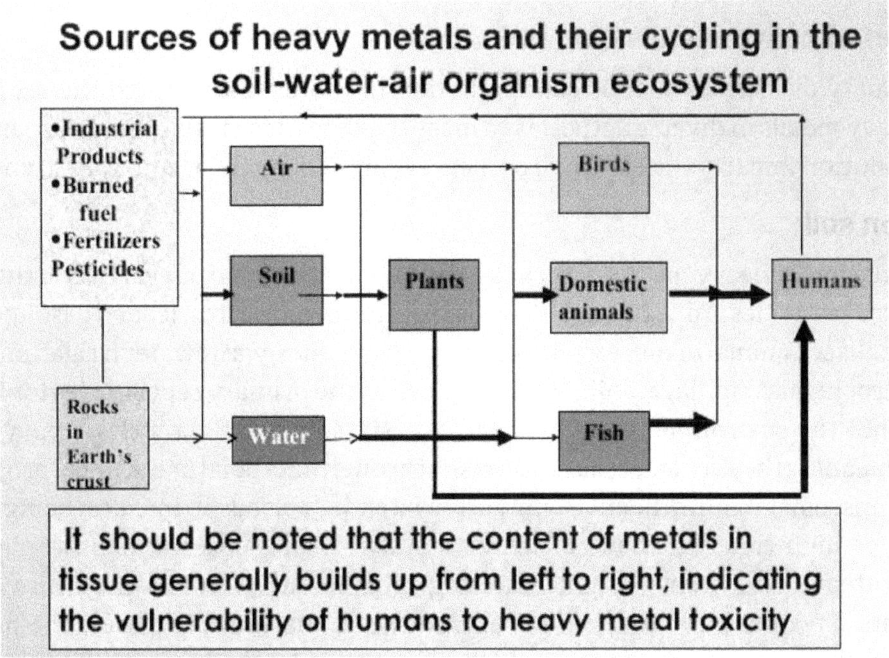

Fig. 2. Sources of heavy metals and their cycle in the environment [12].

2.1.1. Natural processes

Various investigations have documented diverse natural sources of heavy metals. Heavy metals are naturally emitted under specific environmental circumstances. These emissions consist of volcanic eruptions, sea-salt sprays, forest fires, rock weathering, biogenic sources, and wind-borne soil particles. Weathering processes can cause the release of metals from their original locations into various parts of the environment. Heavy metals exist in various forms, including hydroxides, oxides, sulphides, sulphates, phosphates, silicates, and organic molecules. The predominant heavy metals include lead (Pb), nickel (Ni), chromium (Cr), cadmium (Cd), arsenic (As), mercury (Hg), zinc (Zn), and copper (Cu). Even though the aforementioned heavy metals may exist in small amounts, they nonetheless have detrimental effects on the health of humans and other mammals [9].

2.1.2. Anthropogenic processes

Industries, agriculture, wastewater, mining, metallurgical operations, and runoffs all contribute to the emission of pollutants into various environmental compartments. Human activities have been observed to cause heavy metals to exceed their natural levels. The metals emitted in wind-blown dusts are primarily derived from industrial regions. Several significant anthropogenic sources contribute significantly to the contamination of heavy metals in the environment. These sources include automobile exhaust, which releases lead; smelting, which releases arsenic, copper, and zinc; insecticides, which release arsenic; and the burning of fossil fuels, which releases nickel, vanadium, mercury, selenium, and tin. Human activities have been identified as the primary cause of environmental contamination, mostly due to the constant production of goods to suit the needs of the growing population [10].

2.2. Ecological consequences of high concentrations of metallic elements

Heavy metals in the environment have numerous negative effects. These impacts have a wide-ranging effect on all aspects of the ecosystem, including the hydrosphere, lithosphere, biosphere, and atmosphere. Health and mortality issues arise and food networks are disrupted until the repercussions are addressed. Fig. 3 provides a concise overview of the health effects caused by heavy metals.

2.3. The impact of heavy metals contamination

The global worry over heavy metals contamination has intensified due to the escalating usage and processing of heavy metals in diverse activities to meet the demands of the quickly expanding population. Heavy metals pollution impacts the key environmental compartments of soil, water, and air.

2.3.1. Impact on soil

Soil contamination by heavy metals occurs as a result of emissions from various activities and sources, including industrial activities, mine tailings, disposal of high metal wastes, leaded gasoline and paints, land application of fertilisers, animal manures, sewage sludge, pesticides, wastewater irrigation, coal combustion residues, and petrochemical spillage. Soils are recognized as the primary repositories for heavy metals that are discharged into the environment by the previously stated human activities. Heavy metals are often nondegradable, meaning they do not break down easily through microbial or chemical processes. As a result, their concentrations remain high in the environment for extended periods of time after being discharged [5,14]. The occurrence of high concentrations of heavy metals in soils poses a significant concern as they become incorporated into food chains, hence causing detrimental effects on the entire ecosystem. While organic pollutants have the potential to be broken down by natural processes, the presence of heavy metals in the environment reduces the pace at which they can be biodegraded. This leads to a doubling of environmental pollution, as both organic pollutants and heavy metals are present. Heavy metals pose dangers to humans, animals, plants, and ecosystems in multiple ways. These methods include directly consuming, being absorbed by plants, entering the food chain, consuming contaminated water, and changing the pH, porosity, color, and natural chemistry of the soil, all of which affect its quality [15].

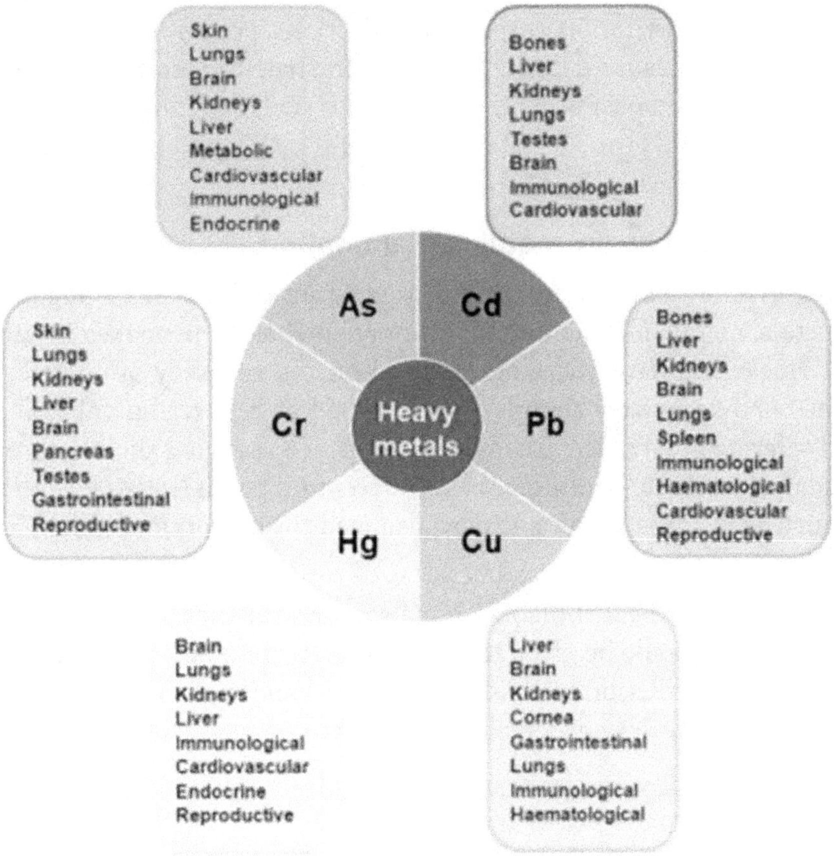

Fig. 3. Impacts of heavy metals on the environment [13].

2.3.2. Impacts on water

Although there are many sources of water contamination, industrialisation and urbanisation are two of the culprits for the increased level of heavy metal water contamination. Heavy metals are transported by runoff from industries, municipalities and urban areas. Most of these metals end up accumulating in the soil and sediments of water bodies [15].

Heavy metals can be found in traces in water sources and still be very toxic and impose serious health problems to humans and other ecosystems. This is because the toxicity level of a metal depends on factors such as the organisms which are exposed to it, its nature, its biological role and the period at which the organisms are exposed to the metal. Food chains and food webs symbolise the relationships amongst organisms. Therefore, the contamination of water by heavy metals actually affects all organisms. Humans, an example of organisms feeding at the highest level, are more prone to serious health problems because the concentrations of heavy metals increase in the food chain [16].

2.3.3. Impacts on air

Industrialisation and urbanisation, due to rapid world population growth, have recently made air pollution as a major environmental problem around the world. The air pollution was reported to have been accelerated by dust and particulate matters (PMs) particularly fine particles such as PM2.5 and PM10 which are released through natural and anthropogenic processes. Natural processes which release particulate matters into air include dust storms, soil erosion, volcanic eruptions and rock weathering, while anthropogenic activities are more industrial and transportation related [17].

Particulate matters are important and require special attention as they can lead to serious health problems such as skin and eyes irritation, respiratory infections, premature mortality and cardiovascular diseases. These pollutants also cause deterioration of infrastructure, corrosion, formation of acid rain, eutrophication and haze [9]. Amongst others, heavy metals such as group 1 metals (Cu, Cd, Pb), group 2 metals (Cr, Mn, Ni, V and Zn) and group 3 metals (Na, K, Ca, Ti, Al, Mg, Fe) originate from industrial areas, traffic and natural sources, respectively [17, 18].

2.4. Methods for mitigating heavy metal contamination

Treatment processes for acid mine water typically generate high-density sludge that is heterogeneous due to variety of metals, metalloids and anionic components, and this makes it difficult to dispose the sludge [19]. Recent researches have therefore focused on the recovery of chemical species from acid mine drainage (AMD) and secondary sludge. This is aimed at recovering valuable resources and also enabling easier and safer disposal of the treated sludge, hence reducing their environmental footprints. Disposal of metal ladened waste to landfills and waste retention ponds/heaps lead to secondary pollution of surface and subsurface water resources. It may also lead to soil contamination, hence affecting their productivity [19].

In order to protect the human health, plants, animals, soil and all the compartments of the environment, proper and careful attention should be given to remediation technologies of heavy metals. Most physical and chemical heavy metal remediation technologies require handling of large amounts of sludge, destroy surrounding ecosystems and are very expensive [19] (Fig. 4).

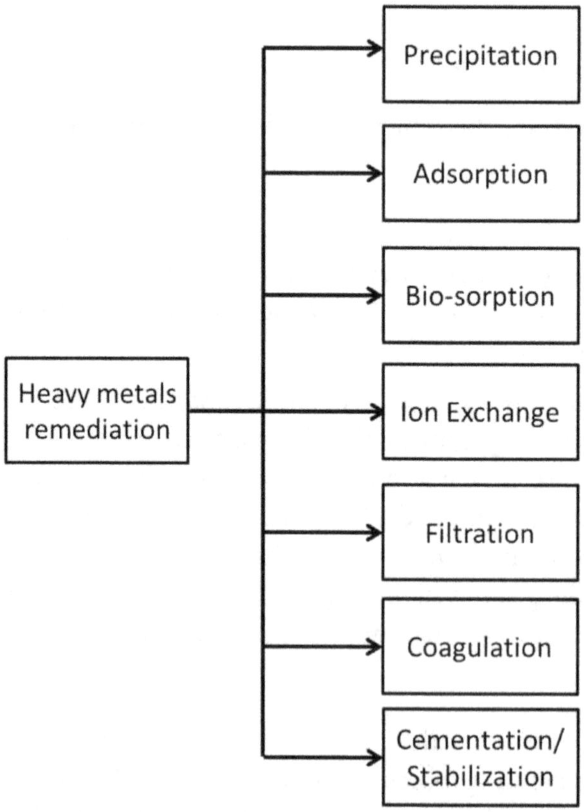

Fig. 4. Mechanisms for the removal of heavy metals [20].

2.4.1. Precipitation

A variety of alkaline chemical reagents have been used over the years for neutralisation of acid mine drainage (AMD) in order to increase the pH and consequently precipitate and recover the metals. The most common alkaline reagents used for sequential recovery of minerals resources from AMD are limestone ($CaCO_3$), caustic soda (NaOH), soda ash (Na_2CO_3), quicklime (CaO), slaked lime ($Ca(OH)_2$) and magnesium hydroxide ($Mg(OH)_2$) [21]. Some processes have recovered metals at varying pH regimes (Table 1) and synthesised commercially valuable materials such as pigments and magnetite [22]. Some minerals are recovered and sold to metallurgical industries, hence off-setting the treatment costs [19].

Table 1. pH values at which metals in AMD precipitate [21].

Metal ion	pH	Metal ion	pH	Metal ion	pH
Al^{3+}	4.1	Hg^{2+}	7.3	Cd^{2+}	6.7
Fe^{3+}	3.5	Na^{+}	6.7	Fe^{2+}	5.5
Mn^{2+}	8.5	Pb^{2+}	6.0	Cu^{2+}	5.3
Cr^{3+}	5.3	Zn^{2+}	7.0		

2.4.2. Adsorption

Adsorption is the process in which an adsorbate molecule sticks to the surface of an adsorbent material. Adsorption is considered the most efficient and cost-effective method for removing metals from water due to its ability to be reversed and its capacity for desorption. While adsorption is generally efficient, it becomes less effective when dealing with highly concentrated solutions because the adsorbent quickly becomes saturated with the adsorbate. It is only practical for solutions with extremely low concentration, involves a lot of effort due to the need for frequent regeneration, and is not specific in terms of reducing the amount of metal [21]. Adsorption is not often used for large-scale metal cleanup.

2.4.3. Ion exchange

Ion exchange refers to the process of exchanging ions between two or more electrolyte solutions. It can also denote the process of ion exchange between a solid substrate and a solution in the soil. Clay and resins with a high cation exchange capacity are frequently employed to remove metals from water-based solutions. Nevertheless, this approach necessitates significant manual labor and is restricted to specific levels of metal concentration in the solution. This system functions within defined temperature and pH parameters. Metals in wastewater have been effectively eliminated and reduced using natural and synthetic clays, zeolites, and synthetic resins [19, 23].

2.4.4. Biosorption

Biosorption is the process of eliminating contaminants from water systems by utilizing biological substances. It involves the absorption, adsorption, ion exchange, surface complexation, and precipitation. Biosorbents possess the advantageous qualities of being easily obtainable, very effective, and having a large capacity. This method is simply accessible and straightforward. Regeneration is a simple process, which makes it highly desirable. Nevertheless, if the concentration of the feed solution is exceedingly high, the process quickly approaches a point where no further pollutant removal can occur, hence restricting its effectiveness [24].

2.4.5. Membrane technologies

Membrane technologies are highly efficient in the treatment of acid mine drainage with high levels of contaminants. It utilizes the phenomenon of concentration gradients or reverse osmosis. Various membranes are employed for mine water treatment, such as ultrafiltration, nano-filtration, reverse osmosis, microfiltration, and particle filtration [19, 25, 26].

3. ANALYSIS OF ACID MINE DRAINAGE IN SOUTH AFRICA

South Africa possesses abundant mineral reserves, which has led to its significant reliance on natural resources for its gross domestic product and overall economy. Nevertheless, the consequences of coal and gold mining have resulted in significant environmental issues. The primary issue at hand is the occurrence of acid mine drainage. Acid mine drainage (AMD) is created through the hydro-geochemical process of sulphide-bearing rocks (such as pyrite, arsenopyrite, and marcasite) reacting with water and oxygen [23, 27]. Iron (Fe) and sulphur-oxidising bacteria [28, 29] also act as catalysts for this process. To put it concisely, the development of AMD can be summarized as follows [19, 23, 30, 31]:

$$2FeS_{2(s)} + 7O_{2(g)} + 2H_2O_{(l)} \rightarrow 2Fe^{2+}_{(aq)} + 4SO^{2-}_{4(aq)} + 4H^+_{(aq)} \quad (1)$$

The oxidation of sulphide to sulphate solubilises the ferrous iron (Fe(II)), which is subsequently oxidised to ferric iron (Fe(III)):

$$4Fe^{2+}_{(aq)} + O_{2(g)} + 4H^+_{(aq)} \rightarrow 4Fe^{3+}_{(aq)} + 2H_2O_{(l)} \quad (2)$$

Either these reactions can occur spontaneously or can be catalysed by microorganisms (sulphur- and iron-oxidising bacteria) that derive energy from the oxidation reaction [26]. The ferric cations produced can also oxidise additional pyrite into ferrous ions:

$$FeS_{(s)} + 14Fe^{3+}_{(aq)} + 8H_2O_{(l)} \rightarrow 15Fe^{2+}_{(aq)} + 2SO^{2-}_{4(aq)} + 16H^+_{(aq)} \quad (3)$$

The net effect of these reactions is to produce H+ and maintain the solubility of the ferric iron [32]. Because of the high acidity and elevated concentration of toxic and hazardous metals, AMD has been a prime issue of environmental concern that has globally raised public concern [33].

The discharge of metalliferous drainage from mining activities has rendered the environment unfit to foster life [22]. Pragmatic approaches need to be developed to counter for this mining legacy that is perpetually degrading the environment and its precious resources [21]. Researches and piloted studies have indicated that active and passive approaches can be successfully adopted to treat acid mine drainage and remove potentially toxic chemical species [23, 31]. The presence of Al, Fe, Mn and sulphates is a prime concern in addition to the trace of Cu, Ni, Pb and Zn [29]. Metalloids of As and earth alkali metal (Ca and Mg) are also present in significant levels [33]. Several studies have shown the feasibility of treating acid mine drainage to acceptable levels as prescribed by different water quality guidelines, but the resultant sludge has been an issue of public concern due to its heterogeneous and complex nature loaded with metal species [23, 34].

Research investigations have strongly focused on the recovery of important minerals from AMD, based on the available evidence [19, 23]. Various methods are employed to recover chemical components from AMD, including as precipitation [35], adsorption [36], biosorption [24], ion exchange [19, 25, 26], desalination [37], and membrane filtration [38, 39]. Among these strategies, precipitation has emerged as a promising technology because it can effectively manage vast amounts of water using minimal dosage [35]. The processes of adsorption and ion exchange have the difficulty of low effectiveness when dealing with high concentrations and rapid saturation rates. Membrane methods face the issue of producing brine, which in turn poses an additional environmental concern. Desalination is hindered by the production of impure salts, rendering them unsuitable for use. Cryogenic desalination has shown potential as a technology, but it has not yet been used on a big scale [19, 23, 34].

3.1. Effects of high concentrations of heavy metals in South Africa

ISouth Africa possesses abundant coal and mineral deposits that encompass valuable commodities like gold, platinum, and copper. Mining plays a crucial role in the economic development and expansion of the country due to its substantial reserves of minerals and coal. This is apparent from the extensive proliferation of mines discovered throughout the nation. Nevertheless, mining has been observed to have detrimental effects on human health, creatures, and the ecosystem as a whole. Water resources are particularly susceptible to contamination caused by mining [40].

The mining of coal and gold for multilateral uses exposes pyrite to oxidising agents. Iron hydroxide and sulphuric acid are toxic chemical species to living organisms when introduced into water resources (both surface and underground). This deteriorates the natural form of the water bodies and its ability to foster life. Acid mine drainage has very low pH of about <1.4 to >3 [41, 42]; high TDS, EC and other metals in toxic concentration. Previous studies documented the following concentrations in AMD: < 75 ppm to >47,800 acidity; <3560 to >41,700 SO42- ppm; <460 to >12,270 ppm total Fe; <17,400 to 37,700 μg/L Zn; <270 to >13,000 μg/L Cu; <520 to >1500 μg/L Co; <75 to >360 μg/L Ni; <8 to >30 μg/L Pb and 6 to 30 μg/L Cd [41, 42, 43, 44].

However, the concentrations listed above are dependent on the pH of the AMD. As the pH raises, the concentrations fall. Organisms are very susceptible to mortality and diseases when they are exposed to such environments, along with other health issues [45]. Furthermore, AMD devastates ecosystems and has a detrimental effect on the country's economy. The presence of heavy metals in current and abandoned mines in South Africa has had a detrimental effect on both surface and subsurface water sources.

3.2. Legal requirements of water quality

The National Environmental Management Act (NEMA) 108 of 1998, stipulates that everyone has the right to live in an environment which is safe and unlikely to pose any deleterious effects to their health. The legislative requirements for industrial effluents are primarily governed by the Department of Water Affairs DWS Water Quality Guidelines [46]. This purpose requires that any person who uses water for industrial purposes shall purify or otherwise treat such water in accordance with requirements of DWA [41, 46, 47, 48]. The relevant criteria for discharge of acidic and sulphate-rich water are given in Table 2.

Table 2. The relevant criteria for discharge of acidic and sulphate-rich water as compared to DWS water quality guidelines.

Parameter	Gold AMD*	Coal AMD**	Neutral drainage†	DWS industrial	DWS irrigation
pH	2.3	2.5	6.5	5.0–10.0	6.5–8.4
EC	22,713	13,980	500	0–250	>540
Na	248.4	70.5	20.1	—	430–460
K	21.6	34.2	29.1	—	—
Mg	2.3	398.9	861.8	—	—
Ca	710.8	598.7	537.5	—	—
Al	134.4	473.9	0.01	—	5.0–20
Fe	1243	8158.2	0.07	0.0–10	5.0–20

Parameter	Gold AMD*	Coal AMD**	Neutral drainage†	DWS industrial	DWS irrigation
Mn	91.5	88.2	25.0	0.0–10.0	0.02–10.0
Cu	7.8	—	—	—	0.2–5.0
Zn	7.9	8.36	0.16	—	1.0–5.0
Pb	6.3	—	—	—	0.2–2.0
Co	41.3	1.89	0.29	—	0.05–5.0
Ni	16.6	2.97	0.21	—	0.2–2.0
SO_4^{2-}	4635	42,862	4603	0–500	—

*Gold mining AMD [44].

**Coalmining AMD.

†Neutral drainage water [40, 42, 45, 47, 48, 49, 50, 51].

As shown in Table 2, mine effluents in South Africa are dominated by dissolved Fe, Al, Mn, Ca, Na, Mg and traces of Cu, Co, Zn, Pb and Ni. These concentrations are far above the legal requirements.

4. DELETERIOUS EFFECTS OF ACID MINE DRAINAGE ON TERRESTRIAL AND AQUATIC ECOSYSTEMS

The introduction of effluents from mining activities into receiving streams can severely impact aquatic ecosystems through habitat destruction and impairment of water quality. This will eventually lead to reduction in biodiversity of a given aquatic ecosystem and its ability to sustain life. The severity and extent of damage depends on a variety of factors including the frequency of influx, volume and chemistry of the drainage and the buffering capacity of the receiving stream [22, 52, 53, 54, 55, 56, 57, 58].

4.1. Acidity

When metals in AMD are hydrolysed, they lower the pH of the water making it unsuitable for aquatic organisms to thrive [52]. AMD is highly acidic (pH 2–4), and this promotes the dissolution of toxic metals [44]. Those toxic species exert hazardous effects on terrestrial and aquatic organisms [23]. Also, if the water is highly acidic, only acidophile microorganisms will thrive on such water with the rest of aquatic organisms migrating to other regions which are conducive to their survival. Many streams contaminated with AMD are largely devoid of life for a long way downstream. To some aquatic organisms, if the pH range falls below the tolerance range, probability of death is very high due to respiratory and osmoregulation failure. Acidic conditions are dominated by H^+ which is adsorbed and pumps out Na from the body which is important in regulating body fluids [23, 52, 53, 56, 57, 58, 59, 60, 61, 62, 63, 64, 65].

4.2. Toxic chemical species

Exposure of aquatic and terrestrial organisms to potentially toxic metals and metalloids can have devastating impacts to living organisms [44, 66, 67]. Toxic chemical species present in AMD have been reported to be toxic to aquatic and terrestrial organisms. They are associated with numerous diseases including cancers. Some of these chemical species may accumulate and be biomagnified in living organisms, hence threatening the life of higher trophic organisms such as birds [68]. Lead causes blood disorders, kidney damage, miscarriages and reproductive disorders and is linked to various cancers. The exposure of living organisms to toxic chemical species in AMD can also lead to nausea, diarrhoea, liver and kidney damage, dermatitis, internal haemorrhage and respiratory problems. Epidemiological studies have shown a significant increase in the risk of lung, bladder, skin, liver and other cancers on exposure to these chemical species. Effects of Al, Fe, Mn, Cu, Mg and Zn on the health of living organisms are summarised in Table 3 [44, 56, 67].

Table 3. Effects of selected AMD metals on the health of living organisms.

Element	DWA limit	Ecological impacts of AMD
Al	<0.5 mg/L	Prolonged exposure to aluminium has been implicated in chronic neurological disorders such as *dialysis dementia* and Alzheimer's disease. Severe aesthetic effects (discolouration) occur in the presence of iron or manganese
Fe	<1 mg/L	Severe aesthetic effects (taste) and effects on plumbing (slimy coatings). Slight iron overload possible in some individuals. Chronic health effects in young children and sensitive individuals in the range of 10–20 mg/L, and occasional acute effects towards the upper end of this range
Mn	<0.2 mg/L	Very severe, aesthetically unacceptable staining. Domestic use unlikely due to adverse aesthetic effects. Some chance of manganese toxicity under unusual conditions
Cu	<1 mg/L	Gastrointestinal irritation, nausea and vomiting. Severe taste and staining problems. Severe poisoning with possible fatalities. Severe taste and staining problems
Mg	<200 mg/L	Water aesthetically unacceptable because of bitter taste users if sulphate present. Increased scaling problems. Diarrhoea in most new consumers
Zn	<5 mg/L	Bitter taste; milky appearance. Acute toxicity with gastrointestinal irritation, nausea and vomiting. Severe, acute toxicity with electrolyte disturbances and possible renal damage

REFERENCES

1. Sands P. Principles of International Environmental Law. 2nd ed. London: Cambridge; 2003
2. Wong MH. Environmental Contamination: Health Risks and Ecological Restoration. United States of America: Taylor & Francis Group; 2012
3. Salomons W, Forstner U, Mader P. Heavy Metals: Problems and Solutions. Berlin, Germany: Springer-Verlag; 1995
4. El-Shahawi MS, Hamza A, Bashammakhb AS, Al-Saggaf WT. An overview on the accumulation, distribution, transformations, toxicity and analytical methods for the monitoring of persistent organic pollutants. Talanta. 2010;80:1587-1597
5. Lepp NW. Effect of heavy metal pollution on plants. Metals in the Environment, Pollution Monitoring Series, Applied Science Publishers. Department of Biology. Liverpool, United Kingdom: Liverpool Polytechnic; 2012;2
6. Van Ael E, Covaci A, Blust R, Bervoets L. Persistent organic pollutants in the Scheldt estuary: Environmental distribution and bioaccumulation. Environmental International. 2012;48:17-27
7. Elliott M. Biological pollutants and biological pollution—an increasing cause for concern. Marine Pollution Bulletin. 2003;46:275-280
8. Banfalvi G. Cellular Effects of Heavy Metals. Netherlands, London, New York: Springer; 2011
9. Herawati N, Suzuki S, Hayashi K, Rivai IF, Koyoma H. Cadmium, copper and zinc levels in rice and soil of Japan, Indonesia and China by soil type. Bulletin of Environmental Contamination and Toxicology. 2000;64:33-39
10. He ZL, Yang XE, Stoffella PJ. Trace elements in agroecosystems and impacts on the environment. Journal of Trace Elements in Medicine and Biology. 2005;19(2–3):125-140

11. Garbarino JR, Hayes HC, Roth DA, Antweiler RC, Brinton TI, Taylor HE. Contaminants in the Mississippi River: Heavy metals in the Mississippi River. Reston, Virginia: U.S. GEOLOGICAL SURVEY CIRCULAR; 1995. pp. 1133

12. Brady D, Stoll AD, Starke L, Duncan JR. Bioaccumulation of metal cations by Saccharomyces cerevisiae. Applied Microbiology and Biotechnology. 1994;41:149-154

13. García-Niño WR, Pedraza-Chaverrí J. Protective effect of curcumin against heavy metals-induced liver damage. Food and Chemical Toxicology. 2014;69:182-201

14. Athar M, Vohora SB. Heavy Metals and Environment. New Delhi: New Age International (P) Limited; 2001

15. Musilova J, Arvay J, Vollmannova A, Toth T, Tomas J. Environmental contamination by heavy metals in region with previous mining activity. Bulletin of Environmental Contamination and Toxicology. 2016;97:569-575

16. Lee G, Bigham JM, Faure G. Removal of trace metals by coprecipitation with Fe, Al and Mn from natural waters contaminated with acid mine drainage in the Ducktown Mining District, Tennessee. Applied Geochemistry. 2002;17(5):569-581

17. Soleimani M, Amini N, Sadeghian B, Wang D, Fang L. Heavy metals and their source identification in particulate matter (PM2.5) in Isfahan City, Iran. Journal of Environmental Sciences. 2018. In press

18. Ventura LMB, Mateus VL, de Almeida ACSL, Wanderley KB, Taira FT, Saint'Pierre TD, Gioda A. Chemical composition of fine particles (PM2.5): Water-soluble organic fraction and trace metals. Air Quality, Atmosphere and Health. 2017;10:845-852

19. Nleya Y, Simate GS, Ndlovu S. Sustainability assessment of the recovery and utilisation of acid from acid mine drainage. Journal of Cleaner Production. 2016;113:17-27

20. Yusuf M, Elfghi FM, Zaidi SA, Abdullah EC, Khan MA. Applications of graphene and its derivatives as an adsorbent for heavy metal and dye removal: A systematic and comprehensive overview. RSC Advances. 2015;5:50392-50420

21. Masindi V, Gitari MW, Tutu H. Passive Remediation of Acid Mine Drainage. LAP Lambert Academic Publishing; 2016

22. Masindi V. A novel technology for neutralizing acidity and attenuating toxic chemical species from acid mine drainage using cryptocrystalline magnesite tailings. Journal of Water Process Engineering. 2016;10:67-77

23. Simate GS, Ndlovu S. Acid mine drainage: Challenges and opportunities. Journal of Environmental Chemical Engineering. 2014;2:1785-1803

24. Silvas FPC, Buzzi DC, Espinosa DCR, Tenório JAS. Biosorption of AMD metals using Rhodococcus opacus. Revista Escola de Minas. 2011;64:487-492

25. Buzzi DC, Viegas LS, Rodrigues MAS, Bernardes AM, Tenório JAS. Water recovery from acid mine drainage by electrodialysis. Minerals Engineering. 2013;40:82-89

26. Park S-M, Shin S-Y, Yang J-S, Ji S-W, Baek K. Selective recovery of dissolved metals from mine drainage using electrochemical reactions. Electrochimica Acta. 2015;181:248-254

27. Nordstrom DK, Blowes DW, Ptacek CJ. Hydrogeochemistry and microbiology of mine drainage: An update. Applied Geochemistry. 2015;57:3-16

28. Baker BJ, Banfield JF. Microbial communities in acid mine drainage. FEMS Microbiology Ecology. 2003;44:139-152

29. Hallberg KB. New perspectives in acid mine drainage microbiology. Hydrometallurgy. 2010;104:448-453

30. Amos RT, Blowes DW, Bailey BL, Sego DC, Smith L, Ritchie AIM. Waste-rock hydrogeology and geochemistry. Applied Geochemistry. 2015;57:140-156

31. Johnson DB, Hallberg KB. Acid mine drainage remediation options: A review. Science of the Total Environment. 2005;338:3-14

32. Candeias C, Ávila PF, Ferreira da Silva E, Ferreira A, Salgueiro AR, Teixeira JP. Acid mine drainage from the Panasqueira mine and its influence on Zêzere river (Central Portugal), Journal of African Earth Sciences, 99, Part 2. 2014. pp. 705-712

33. Akinwekomi V, Maree JP, Zvinowanda C, Masindi V. Synthesis of magnetite from iron-rich mine water using sodium carbonate, Journal of Environmental Chemical Engineering. 2017

34. Kefeni KK, Msagati TAM, Mamba BB. Acid mine drainage: Prevention, treatment options, and resource recovery: A review. Journal of Cleaner Production. 2017;151:475-493

35. Seo EY, Cheong YW, Yim GJ, Min KW, Geroni JN. Recovery of Fe, Al and Mn in acid coal mine drainage by sequential selective precipitation with control of pH, Catena, 148, Part 1. 2017. pp. 11-16

36. Masindi V, Gitari MW, Tutu H, DeBeer M. Efficiency of ball milled south African bentonite clay for remediation of acid mine drainage. Journal of Water Process Engineering. 2015;8:227-240

37. Mulopo J. Continuous pilot scale assessment of the alkaline barium calcium desalination process for acid mine drainage treatment. Journal of Environmental Chemical Engineering. 2015;3:1295-1302

38. Meschke K, Herdegen V, Aubel T, Janneck E, Repke J-U. Treatment of opencast lignite mining induced acid mine drainage (AMD) using a rotating microfiltration system. Journal of Environmental Chemical Engineering. 2015;3:2848-2856

39. Mhamdi M, Elaloui E, Trabelsi-Ayadi M. Adsorption of zinc by a Tunisian Smectite through a filtration membrane. Industrial Crops and Products. 2013;47:204-211

40. Gitari M, Petrik L, Etchebers O, Key D, Iwuoha E, Okujeni C. Treatment of acid mine drainage with fly ash: Removal of major contaminants and trace elements. Journal of Environmental Science and Health - Part A Toxic/Hazardous Substances and Environmental Engineering. 2006;41:1729-1747

41. Van der Linde M, Feris L. Compendium of South African Environmental Legislation. Pretoria, South Africa: Pretoria University Law Press; 2010

42. Madzivire G, Gitari WM, Vadapalli VRK, Ojumu TV, Petrik LF. Fate of sulphate removed during the treatment of circumneutral mine water and acid mine drainage with coal fly ash: Modelling and experimental approach. Minerals Engineering. 2011;24:1467-1477

43. Africa S, Van der Linde M, Feris L. Compendium of South African Environmental Legislation. Pretoria University Law Press; 2010

44. Tutu H, McCarthy TS, Cukrowska E. The chemical characteristics of acid mine drainage with particular reference to sources, distribution and remediation: The Witwatersrand Basin, South Africa as a case study. Applied Geochemistry. 2008;23:3666-3684

45. Madzivire G, Petrik LF, Gitari WM, Ojumu TV, Balfour G. Application of coal fly ash to circumneutral mine waters for the removal of sulphates as gypsum and ettringite. Minerals Engineering. 2010;23:252-257

46. Biswas AK, Tortajada C, Izquierdo R. Water Quality Management: Present Situations, Challenges and Future Perspectives. New York, USA: Taylor & Francis; 2014

47. Madzivire G, Maleka P, Lindsay R, Petrik LF. Radioactivity of mine water from a gold mine in South Africa. WIT Transactions on Ecology and the Environment. 2013;178:147-158

48. Madzivire G, Maleka PP, Vadapalli VRK, Gitari WM, Lindsay R, Petrik LF. Fate of the naturally occurring radioactive materials during treatment of acid mine drainage with coal fly ash and aluminium hydroxide. Journal of Environmental Management. 2014;133:12-17

49. Masindi V, Gitari MW, Tutu H, De Beer M. Application of magnesite–bentonite clay composite as an alternative technology for removal of arsenic from industrial effluents. Toxicological & Environmental Chemistry. 2014:1-17

50. Gitari WM, Petrik LF, Etchebers O, Key DL, Iwuoha E, Okujeni C. Passive neutralisation of acid mine drainage by fly ash and its derivatives: A column leaching study. Fuel. 2008;87:1637-1650

51. Masindi V, Gitari MW, Tutu H, De Beer M. Application of magnesite–bentonite clay composite as an alternative technology for removal of arsenic from industrial effluents. Toxicological & Environmental Chemistry. 2014;96:1435-1451

52. Torres E, Ayora C, Jiménez-Arias JL, García-Robledo E, Papaspyrou S, Corzo A. Benthic metal fluxes and sediment diagenesis in a water reservoir affected by acid mine drainage: A laboratory experiment and reactive transport modeling. Geochimica et Cosmochimica Acta. 2014;139:344-361

53. Šucha V, Dubiková M, Cambier P, Elsass F, Pernes M. Effect of acid mine drainage on the mineralogy of a dystric cambisol. Geoderma. 2002;110:151-167

54. Peretyazhko T, Zachara JM, Boily JF, Xia Y, Gassman PL, Arey BW, Burgos WD. Mineralogical transformations controlling acid mine drainage chemistry. Chemical Geology. 2009;262:169-178

55. Netto E, Madeira RA, Silveira FZ, Fiori MA, Angioleto E, Pich CT, Geremias R. Evaluation of the toxic and genotoxic potential of acid mine drainage using physicochemical parameters and bioassays. Environmental Toxicology and Pharmacology. 2013;35:511-516

56. Mohapatra BR, Douglas Gould W, Dinardo O, Koren DW. Tracking the prokaryotic diversity in acid mine drainage-contaminated environments: A review of molecular methods. Minerals Engineering. 2011;24:709-718

57. Martins M, Santos ES, Faleiro ML, Chaves S, Tenreiro R, Barros RJ, Barreiros A, Costa MC. Performance and bacterial community shifts during bioremediation of acid mine drainage from two Portuguese mines. International Biodeterioration & Biodegradation. 2011;65:972-981

58. Levings CD, Varela DE, Mehlenbacher NM, Barry KL, Piercey GE, Guo M, Harrison PJ. Effect of an acid mine drainage effluent on phytoplankton biomass and primary production at Britannia Beach, Howe Sound, British Columbia. Marine Pollution Bulletin. 2005;50:1585-1594

59. Sun M, Ru X-R, Zhai L-F. In-situ fabrication of supported iron oxides from synthetic acid mine drainage: High catalytic activities and good stabilities towards electro-Fenton reaction. Applied Catalysis B: Environmental. 2015;165:103-110

60. Strosnider WH, Nairn RW. Effective passive treatment of high-strength acid mine drainage and raw municipal wastewater in Potosí, Bolivia using simple mutual incubations and limestone. Journal of Geochemical Exploration. 2010;105:34-42

61. Mitsch WJ, Wise KM. Water quality, fate of metals, and predictive model validation of a constructed wetland treating acid mine drainage. Water Research. 1998;32:1888-1900

62. Mapanda F, Nyamadzawo G, Nyamangara J, Wuta M. Effects of discharging acid-mine drainage into evaporation ponds lined with clay on chemical quality of the surrounding soil and water. Physics and Chemistry of the Earth, Parts A/B/C. 2007;32:1366-1375

63. Mahmoud KK, Leduc LG, Ferroni GD. Detection of Acidithiobacillus ferrooxidans in acid mine drainage environments using fluorescent in situ hybridization (FISH). Journal of Microbiological Methods. 2005;61:33-45

64. Gray NF, Delaney E. Measuring community response of bentic macroinvertebrates in an erosional river impacted by acid mine drainage by use of a simple model. Ecological Indicators. 2010;10:668-675

65. Gray JB, Vis ML. Reference diatom assemblage response to restoration of an acid mine drainage stream. Ecological Indicators. 2013;29:234-245

66. Skoczyńska-Gajda S, Labus K. Acid mine drainage within the abandoned lignite mining area-Muskau Arch. Biuletyn Państwowego Instytutu Geologicznego. 2011:643-650

67. Rubin H, Rubin K, Siodlak A, Skuza P. Assessment of contamination of the bottom sediments of the Stola River with selected metals and metalloids within the urban-industrial area of Tarnowskie Góry. Biuletyn Państwowego Instytutu Geologicznego. 2011:615-624

68. Jooste S, Thirion C. An ecological risk assessment for a South African acid mine drainage. Water Science and Technology. 1999;39:297-303

8 Chapter

ESTIMATING THE CONCENTRATION OF HEAVY METALS IN SOIL USING THE INTEGRATION OF GEOCHEMISTRY AND FIELD SPECTROSCOPY IN THE MIYI MINING AREA

1. INTRODUCTION: AN OVERVIEW

As industries have advanced in recent decades, there has been an increasing demand for mineral resources. Nevertheless, the extraction and subsequent treatment of mineral resources will result in a rise in the concentration of heavy metals that infiltrate and build up in the soil. The presence of these abundant heavy metals, due to their persistence and toxicity, might hinder soil processes and contribute to soil pollution [1, 2, 3]. Excessive accumulation of heavy metals in soil can lead to environmental hazards and endanger human health by contaminating the food chain, water, and air [4, 5, 6]. Therefore, the contamination of soil with heavy metals is seen as a significant environmental issue, and the surveillance of heavy metal levels is crucial for the management of the environment in mining regions. The conventional approach to determining the heavy metal concentrations include collecting field data and conducting laboratory investigation on soil samples. While direct sampling can yield precise measurements of soil pollutants' intensity and diversity at specific locations, these processes are typically time-consuming and expensive, and they can only evaluate individual samples [7, 8, 9]. Regular and frequent monitoring of heavy metals in locations susceptible to contamination is crucial. The advancement of remote-sensing technology, particularly hyperspectral, offers the potential for swift and extensive detection of heavy metals, enabling the collection of a continuous spectrum of the target. Several studies have demonstrated a discernible disparity between the spectral curves of soil contaminated with heavy metals and those of uncontaminated soil [10, 11]. Despite being spectrally passive, soil heavy metals can still be predicted using visible and near-infrared diffuse reflectance spectroscopy (VNIRS) due to their associations with spectrally active soil characteristics like clay and Fe oxides [12]. Therefore, the correlation between the amount of heavy metals and the range of soil properties was utilized to develop a model for predicting and mapping the levels of heavy metals in specific regions [11, 13, 14, 15]. This chapter focuses on the spectral sampling of soil samples using the ASD Fieldspec III spectroradiometer, and the measurement of Mn and Co concentrations through chemical analysis. Subsequently, the feature spectra can be derived from the outcomes of spectroscopic analysis in order to create models for retrieving heavy metals. Subsequently, the characteristics of these models can be utilized to investigate the viability of employing hyperspectral data for the purpose of retrieving soil heavy metals in order to facilitate soil remediation and ecological restoration.

2. MATERIALS AND METHODS

2.1. Study area

Miyi Country is situated in the northern region of Panzhihua City, in the southwestern part of Sichuan Province. Its geographical coordinates are around 26°42′–27°10′ N and 101°44′–102°15′ E. The topography is elevated in the southern region and rather flat in the northern region. The study regions for this chapter are the Bai-ma, De-sheng, and Yuan Bao-shan mining areas in Miyi County, as depicted in Fig. 1. The study area is situated in the eastern region of the Yalong River and the northwestern region of the Anning River. The mining area and its surrounds have been polluted by heavy metals, such as manganese (Mn) and cobalt (Co), which enter the soil through discharge and infiltration. These metals are present in the waste residue and waste liquor produced during the mining process.

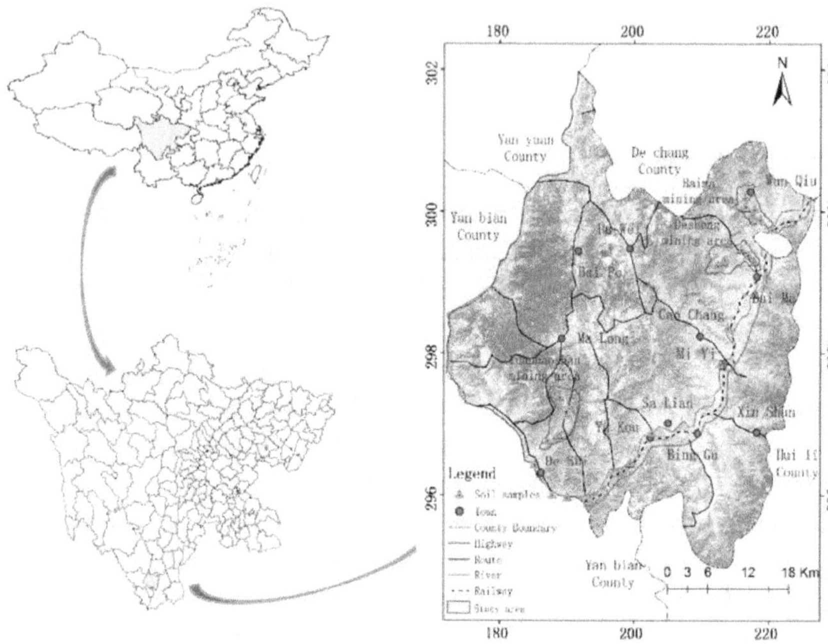

Fig. 1. The study area locations of sampling points.

2.2. Data collection

In June 2015, the spectral data of 55 soil samples were gathered. The geographical coordinates of the sample places can be found in Fig. 1. At the same time, traditional digestion methods were used to chemically test the Mn and Co contents of 32 out of the 55 soil samples using an inductively coupled plasma mass spectrometer (ICP-MS). The ICP-MS is widely regarded as the predominant ion source in the field of analytical chemistry for elemental mass spectrometry. In Inductively Coupled Plasma Mass Spectrometry (ICP-MS), a mass spectrometer is connected to an ICP torch by an interface that consists of sampler and skimmer cones. This interface allows for the transmission of representative samples of the plasma through its orifices to the mass analyzer [16]. The spectrum of the soil sample was acquired using a high spectral resolution ASD Fieldspec III spectroradiometer. This instrument covers the visible and near-infrared band (350–2500 nm) and provides a spectral resolution ranging from 3 to 10 nm, which was then interpolated to 1 nm. During the collection of soil spectra in the field, an ASD high-reflectance probe was utilized to provide illumination. In contrast, a halogen bulb was employed as the light source for collecting water spectra in the laboratory. Three measurements were taken for each sample, and the average value was subsequently determined [11].

2.3. Spectral pre-processing

The soil spectra may include noise or error due to faulty operation of in situ measurement devices or the use of a poorly calibrated *in situ* measurement instrument [17]. Furthermore, heavy metals lack distinct spectral features in the visible and near-infrared regions of the electromagnetic spectrum. Therefore, a significant challenge arises when attempting to analyze overlapping bands of the analytes and interferences, which complicates the extraction of qualitative and quantitative data [18]. Pre-processing of the spectrum is frequently necessary to diminish the impact of noise and amplify the spectral signature. Meanwhile, the Savitzky-Golay differentiation approach is frequently employed as a spectrum pre-treatment technique. In practical applications, it effectively removes the interference caused by the baseline or background, enhances sensitivity, and detects and amplifies small or subtle spectral characteristics [18, 19]. The obtained spectra underwent continuum removal and normalization to enhance the spectral absorption properties. The convex hull of straight-line segments is applied to a reflectance spectrum and then eliminated by dividing or ratioing it with the total reflectance spectrum [8].

2.4. Analysis of spectral data and creation of mathematical models

To strengthen the correlation between soil spectroscopy and heavy metals, it is possible to utilize the logarithmic treatment of feature bands [8]. This chapter focuses on extracting the reflectivity of all bands in order to generate a matrix that contains the reflectivity values for a single band. The outer loop was chosen to be the band reflectivity from the second band to the last band, while the inner loop was chosen to be the band reflectivity from the first band to the penultimate band. The band ratio matrix is obtained by dividing the outer loop by the inner loop. The band normalization matrix is obtained by dividing the difference of reflectivity by the summation of the outer loop band reflectance and the inner loop. The multivariate analysis matrix of two bands and three bands is obtained by randomly combining two or three output feature bands. Subsequently, the aforementioned matrices were employed to conduct Pearson correlation analysis with the matrices representing the soil heavy metal content. The Pearson correlation coefficient quantifies the degree of linear dependence (correlation) between two variables X and Y. The magnitude of the correlation coefficient indicates the strength of the correlation between the two variables [20, 21]. Therefore, it is feasible to forecast the levels of heavy metals in the soil due to the strong association seen between heavy metal concentration and soil spectrum [22]. This chapter demonstrates the utilization of View Spec Pro and ENVI to carry out techniques such as smoothing, first derivative, second derivative, and continuum removal on spectral data. These approaches are employed to minimize background noise and improve the visibility of spectral features. The techniques of ratio, normalization, and multivariate analysis are employed to optimize the association between the levels of heavy metal content and the characteristics of the spectra. Ultimately, the retrieval model was established using the IBM SPSS program.

3. RESULTS

3.1. Geochemical analysis

Manganese (Mn) and Cobalt (Co) are the primary heavy metals found in mining waste in the research area. Therefore, they were chosen as indicators to assess the environmental consequences of mining activities. Table 1 displays the chemical analysis findings of the soil samples. Table 1 clearly demonstrates that the manganese (Mn) content in the soil had a significant impact on the cobalt (Co) content. The soil's standard deviation (SD) for Mn was unusually high at 773.6, suggesting that its concentrations exhibit minimal variation. In contrast, the Co content in the soil shows little variability.

Table 1. Content of heavy metals in soil statistics (unit: mg/kg).

	Max	Min	Mean	SD
Mn	4114.0	531.3	1536.1	773.6
Co	111.9	6.00	43.5	25.87

Table 2. Feature spectra of four kinds of curves (unit: nm).

	Original spectra	1st spectra	2nd spectra	Continuum removed spectra
Feature spectra	473, 791, 1395, 1413, 1854, 1926, 2136, 2170, 2208, 2243, 2259, 2320,	439, 465, 549, 584, 1382, 1396, 1403, 1421, 1452, 1482, 1766, 1890, 1926, 1993, 2121, 2151, 2172, 2200, 2215, 2234, 2280, 2361, 2375, 2398	550, 590, 625, 995, 1006, 1374, 1393, 1403, 1411, 1425, 1466, 1883, 1906, 2138, 2163, 2196, 2209, 2220, 2240, 2291, 2368, 2387	452, 486, 627, 765, 810, 962, 1029, 1285, 1414, 1698, 1786, 1835, 1918, 1922, 2142, 2205, 2236, 2267, 2296, 2342, 2371, 2386, 2411

3.2. Spectral analysis

Upon visually examining the measured soil spectra using various pre-treatment procedures, a notable disparity was seen. Fig. 2 displays the spectral curve of the soil samplings in the research area, as well as the curves that have undergone pre-processing using the first derivative, second derivative, and continuum removal methods. The feature spectra can be derived from these spectrum curves, as illustrated in Table 2. Fig. 2 shows that the original soil spectral curve has only 12 selected feature spectra at specific wavelengths such as 473, 791, 1395, 1413, and 1926 nm. However, after pre-processing, the soil spectral curves have more selected feature spectra at wavelengths like 584, 1382, 1396, 1403, 1421, 1452, 1890, and 1906 nm.

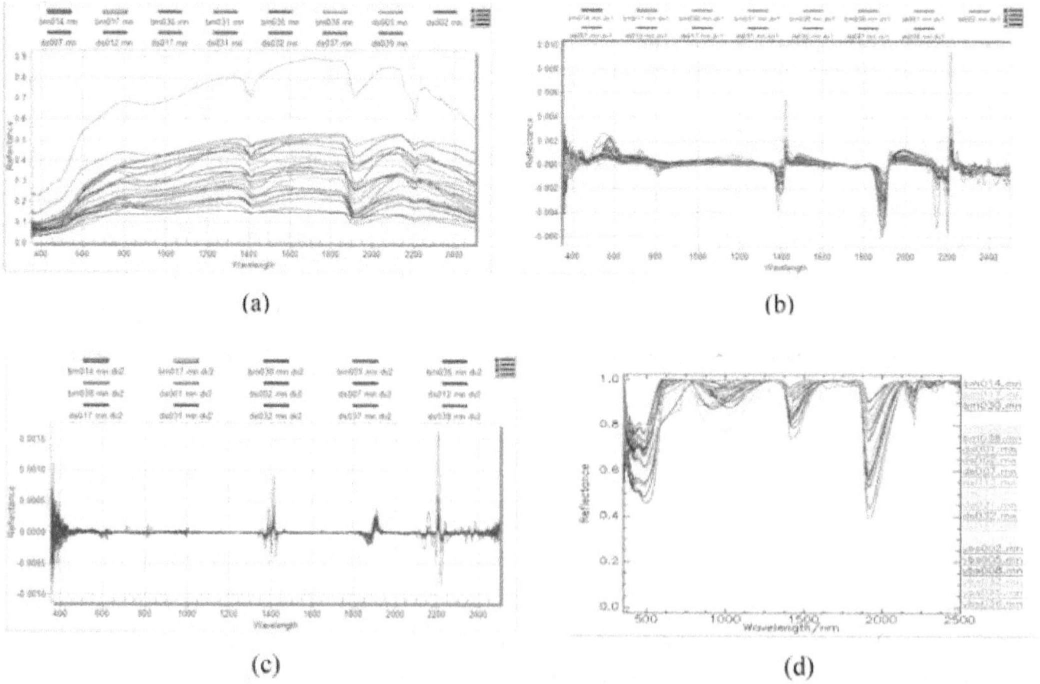

(a) (b) (c) (d)

Fig. 2. Reflectance spectra of soil samples. (a) Raw reflectance spectral, (b) first-derivative reflectance spectral, (c) second-derivative reflectance spectral, (d) continuum removed reflectance spectral.

3.3. Model development

The reflectance spectra are analyzed to identify the spectral features, and the band combination with the highest Pearson correlation coefficient (R) is chosen as the feature band for constructing the inversion models of heavy metals. The regression equations for the amounts of heavy metals in the soil are displayed in Table 3. The regression models for the heavy metal concentrations in the soil have determination coefficients (R2) of 0.645 for Mn and 0.8 for Co. The coefficient of determination (R2) for the regression equations demonstrates a significant correlation between the detected heavy metals and spectral characteristics. The correlation between the bands at 2124 and 2296 nm and the Mn content is significant, while the bands at 1918, 1922, and 2205 nm exhibit high correlations with the Co content. The significance indexes of these regression equations are below 0.05.

Table 3. Spectra parameters and regression equation of Mn and Co.

	Feature bands (nm)	R^2	Regression equations	Significance F
Mn	2142, 2296	0.645	Y = -33749.8 + 34703.04X	2.04507E - 05
Co	1918, 1922, 2205	0.800	Y = -235.03–7507.17X1 + 7452.81X2 + 333.65X3	7.73E - 06
Note: X corresponding to R2142/R2296; X1, X2, X3 corresponding to R2121, R2234, R2398, respectively.				

The relationship of the measured and predicted concentrations of Mn and Co in soil is shown in Fig. 3. From the scatter diagrams, we can see that there is a good linear relationship between the measured and predicted concentrations of Mn and Co.

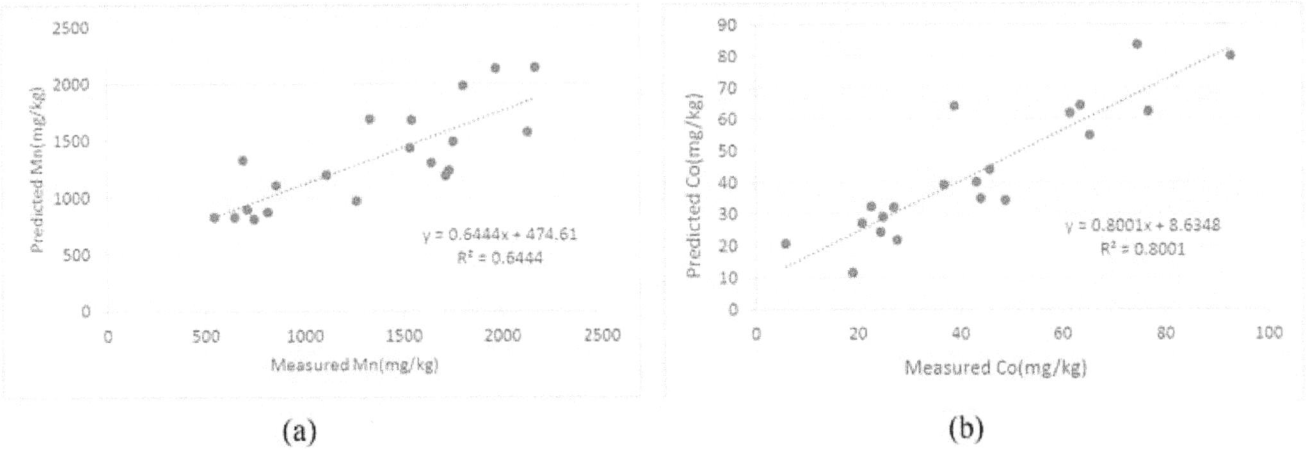

Fig. 3. Scatter plots of the measured values and predicted values: (a) Mn and (b) Co.

The contents of 12 test soil samples of Mn and Co can be calculated by these regression equations. Then the F-test was carried out to validate the feasibility of these regression models for predicting heavy metal contents, as shown in Table 4. From the statistics, we can see that the difference of the mean values of Mn and Co is smaller and the P-value of the F-test between measured and predicted values of Co and Mn in the soil is less than 0.05. This indicates tha Paraphrasing Tool - QuillBot AIt the models can be used to predict the heavy metal contents in the study area.

Table 4. Summary statistics of measured and predicted heavy metal concentration and F-test (concentration unit: mg/kg).

	Measurement			Prediction			P(F ≤ f)
	Max	Min	Mean	Max	Min	Mean	
Mn	4114	531.3	1871.6	2290.1	924.9	1474.7	0.003
Co	111.9	13.1	44.0	82.2	22.1	48.4	0.008

The regression equations can be used to compute the concentrations of Mn and Co in 12 test soil samples. Subsequently, the F-test was conducted to ascertain the viability of these regression models in predicting heavy metal concentrations, as depicted in Table 4. Based on the data, it is evident that the disparity between the average values of Mn and Co is minimal. Additionally, the P-value of the F-test conducted on the measured and anticipated values of Co and Mn in the soil is below 0.05. This suggests that the models have the capability to forecast the levels of heavy metals in the designated region.

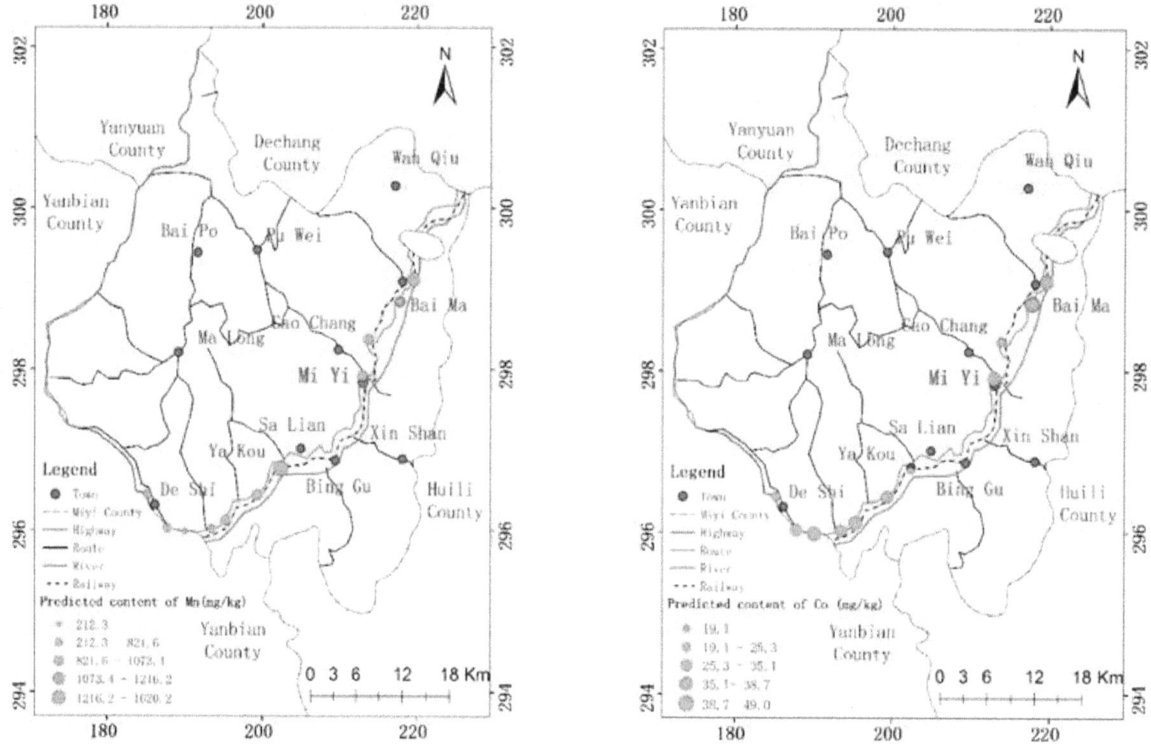

(a) (b)

Fig. 4. Predicted content of Mn and Co. (a) The predicted content of Mn and (b) the predicted content of Co.

4. DISCUSSION

The presence of heavy metal particles adsorbed by the soil organic matter can alter the roughness of the soil surface, as well as molecular vibration and electron transition. This alteration enables the use of soil spectrum analysis to determine the concentration of heavy metals in the soil. The correlation between the spectrum and the concentration of heavy metals was determined through the utilization of spectral analysis and chemical analysis of the soil samples. Subsequently, the concentration of heavy metals cobalt (Co) and manganese (Mn) can be determined to generate predictive data for regional soil quality assessment and remediation. Fig. 4 reveals that the distribution of heavy metal concentration in the research area is uneven due to the effects of the river, vegetation, and mineral transportation. The river's current can facilitate the conversion of slopes into indicators of the presence and magnitude of contamination by heavy metals. Vegetation absorption can decrease heavy metal pollution in the soil, but mineral transportation can cause significant fluctuations in heavy metal pollution levels. The field of hyperspectral approach offers several advantages over existing methods, including rapid, efficient, extensive coverage and nondestructive estimation of heavy metals' concentrations. It can generate predictive data for monitoring the environment in mines to enhance the effectiveness of monitoring and management in mine areas and safeguard the normal quality of life for nearby communities. However, the process of collecting spectrum data and constructing models still requires a significant amount of time. Additionally, the monitoring area is restricted, and mistakes may be present in the collection of field samples and spectral measurements. Hence, the objective of this research is to (1) analyze the impact of temperature, altitude, weather, and other variables on the spectrum in order to enhance the precision of the model, (2) collect spectral data simultaneously with field data collection, and (3) acquire Hyperion or AVIRIS imagery of the study area while taking into account the effects of vegetation, rock, and atmosphere on the soil spectrum.

5. CONCISE SUMMARY

This chapter presents a rapid and efficient approach for determining the levels of heavy metals in the soils of the study area. This method may be used to forecast the eco-remediation of heavy metals in mining areas and ensure the appropriate allocation of human and other resources. Therefore, by shortening the time required for remediating soils polluted with heavy metals, we can effectively avoid the further dispersion of heavy metals in the soil.

REFERENCES

1. Ding Q et al. Effects of natural factors on the spatial distribution of heavy metals in soils surrounding mining regions. Science of the Total Environment. 2017;578:577-585. DOI: 10.1016/j.scitotenv.2016.11.001
2. Li N et al. Effects of double harvesting on heavy metal uptake by six forage species and the potential for phytoextraction in field. Pedosphere. 2016;26(5):717-724. DOI: 10.1016/S1002-0160(15)60082-0
3. Yang Y et al. Risk assessment of heavy metal pollution in sediments of the Fenghe River by the fuzzy synthetic evaluation model and multivariate statistical methods. Pedosphere. 2016;26(3):326-334. DOI: 10.1016/S1002-0160(15)60046-7
4. Cui X et al. Concentrations of heavy metals in suburban horticultural soils and their uptake by Artemisia selengensis. Pedosphere. 2015;25(6):878-887. DOI: 10.1016/S1002-0160(15)30068-0
5. Liu M et al. Monitoring stress levels on rice with heavy metal pollution from hyperspectral reflectance data using wavelet-fractal analysis. International Journal of Applied Earth Observation and Geoinformation. 2011;13(2):246-255. DOI: 10.1016/j.jag.2010.12.006
6. Zheng R et al. Land use effects on the distribution and speciation of heavy metals and arsenic in coastal soils on Chongming Island in the Yangtze River Estuary, China. Pedosphere. 2016;26(1):74-84. DOI: 10.1016/S1002-0160(15)60024-8
7. Schuerger AC et al. Comparison of two hyperspectral imaging and two laser-induced fluorescence instruments for the detection of zinc stress and chlorophyll concentration in bahia grass (Paspalum notatum Flugge.). Remote Sensing of Environment. 2003;84(4):572-588. DOI: 10.1016/S0034-4257(02)00181-5

8. Choe E et al. Mapping of heavy metal pollution in stream sediments using combined geochemistry, field spectroscopy, and hyperspectral remote sensing: A case study of the Rodalquilar mining area, SE Spain. Remote Sensing of Environment. 2008;112(7):3222-3233. DOI: 10.1016/j.rse.2008.03.017
9. Zhuang P et al. Heavy metal contamination in soil and soybean near the Dabaoshan Mine, South China. Pedosphere. 2013;23(3):298-304. DOI: 10.1016/S1002-0160(13)60019-3
10. Trevors J. Water, air, and soil pollution. An International Journal of Environmental Pollution. 2010;(Suppl 1):1
11. Song L et al. Estimate of heavy metals in soil and streams using combined geochemistry and field spectroscopy in Wan-sheng mining area, Chongqing, China. International Journal of Applied Earth Observation and Geoinformation. 2015;34:1-9. DOI: 10.1016/j.jag.2014.06.013
12. St. Luce M et al. Visible near infrared reflectance spectroscopy prediction of soil heavy metal concentrations in paper mill biosolid- and liming by-product-amended agricultural soils. Geoderma. 2017;288:23-36. DOI: 10.1016/j.geoderma.2016.10.037
13. Liu M et al. Estimating regional heavy metal concentrations in rice by scaling up a field-scale heavy metal assessment model. International Journal of Applied Earth Observation and Geoinformation. 2012;19:12-23. DOI: 10.1016/j.jag.2012.04.014
14. Liu Y et al. Heavy metal contamination of agricultural soils in Taiyuan, China. Pedosphere. 2015;25(6):901-909. DOI: 10.1016/S1002-0160(15)30070-9
15. Liu M et al. Regional heavy metal pollution in crops by integrating physiological function variability with spatio-temporal stability using multi-temporal thermal remote sensing. International Journal of Applied Earth Observation and Geoinformation. 2016;51:91-102. DOI: 10.1016/j.jag.2016.05.003
16. Aghaei M, Lindner H, Bogaerts A. Optimization of operating parameters for inductively coupled plasma mass spectrometry: A computational study. Spectrochimica Acta Part B: Atomic Spectroscopy. 2012;76:56-64. DOI: 10.1016/j.sab.2012.06.006
17. John RJ. Introductory Digital Image Processing: A Remote Sensing Perspective. 3rd ed. United States: Pearson Education; 2007
18. Parmar A, Sharma S. Derivative UV-vis absorption spectra as an invigorated spectrophotometric method for spectral resolution and quantitative analysis: Theoretical aspects and analytical applications: A review. TrAC Trends in Analytical Chemistry. 2016;77:44-53. DOI: 10.1016/j.trac.2015.12.004
19. Zheng KY et al. Pretreating near infrared spectra with fractional order Savitzky-Golay differentiation (FOSGD). Chinese Chemical Letters. 2015;26(3):293-296. DOI: 10.1016/j.cclet.2014.10.023
20. Erdeljić V et al. Distributed lags time series analysis versus linear correlation analysis (Pearson's R) in identifying the relationship between antipseudomonal antibiotic consumption and the susceptibility of Pseudomonas aeruginosa isolates in a single Intensive Care Unit of a tertiary hospital. International Journal of Antimicrobial Agents. 2011;37(5):467-471. DOI: 10.1016/j.ijantimicag.2010.11.030
21. Susan Prion ERNC, Katie Anne Haerling PR. Making sense of methods and measurement: Pearson product-moment correlation coefficient. Clinical Simulation in Nursing. 2014;(11):587-588
22. Chen T et al. Identification of soil heavy metal sources and improvement in spatial mapping based on soil spectral information: A case study in northwest China. Science of the Total Environment. 2016;565:155-164. DOI: 10.1016/j.scitotenv.2016.04.163

Chapter 9

STUDY ON THE SPATIOTEMPORAL DISTRIBUTION, GAS-SOLID PARTITION, AND LONG-RANGE TRANSPORTATION OF ATMOSPHERIC MERCURY IN AN INDUSTRIAL CITY AND OFFSHORE ISLANDS

1. INTRODUCTION: AN OVERVIEW

In Title III of the Clean Air Act Amendments (CAAA) of 1990, the United States had previously identified and listed 189 hazardous air pollutants (HAPs) for regulation [1, 2]. Out of the many elements, 11 of them are classified as poisonous heavy metals. These metals include Arsenic (As), Beryllium (Be), Cadmium (Cd), Cobalt (Co), Chromium (Cr), Mercury (Hg), Nickel (Ni), Manganese (Mn), Lead (Pb), Antimony (Sb), and Selenium (Se). In 2005, the U.S.A. Environmental Protection Agency (USEPA) was the first agency to modify the Maximum Achievable Control Technology (MACT) standards by implementing growth control quotas for mercury emissions. This was done with the aim of increasing the initial target of reducing mercury emissions by 30% to a more ambitious objective of 70% by 2018 [2]. Additionally, this was undertaken to incentivize nations globally to reassess their respective targets for reducing mercury emissions. Currently, the chemical technique is still unable to fully eliminate mercury contaminants, posing an ongoing threat to the health of humans and other organisms. Mercury (Hg) is a heavy metal that is persistent, toxic, and bio-accumulative. It is currently regulated by the USEPA and the United Nations Environment Programme (UNEP) [3, 4, 5, 6, 7, 8, 9, 10]. UNEP has identified atmospheric mercury as a global environmental issue, alongside greenhouse gases (GHGs) [1, 2]. Due to its distinctive physicochemical characteristics and ability to be transported over great distances in the atmosphere, it has the potential to be deposited globally [4, 5]. Mercury has the ability to remain and accumulate in the food chain, which can lead to detrimental effects on the brain and nerve systems of humans. Consequently, numerous nations are currently experiencing growing apprehension regarding the issue of atmospheric mercury pollution. This chapter aimed to examine the variation and distribution of atmospheric mercury in an industrial metropolitan area, as well as investigate the long-range transportation path in the marine boundary layer (MBL). Additionally, it sought to compare the concentration of atmospheric mercury with other major cities worldwide.

2. TEMPOSPATIAL VARIATION AND PARTITION OF ATMOSPHERIC MERCURY IN AMBIENT AIR OF AN INDUSTRIAL CITY

2.1. Background

In Taiwan, the primary sources of mercury emissions were combustion sources and metallurgy smelting processes. These sources of combustion include coal-fired boilers, municipal and industrial waste incinerators, petrochemical refineries, and cremators. The processes involved in metallurgy smelting include integrated steel factories, electric arc furnace plants, and secondary metal smelters. The economic growth of metro Kaohsiung was mostly dependent on the presence of heavy industry situated in close proximity to the metropolitan area. The combination of heavy traffic and congestion contributes to the deterioration of the ambient air quality in the Kao-Ping Air Quality Zone in Taiwan. Furthermore, the metro area of Kaohsiung is the largest and most densely populated industrial city in Taiwan. It is responsible for around 70% of the stationary sources of emissions. This area is home to two huge coal-fired power plants, an integrated steel plant, multiple metallurgical smelters, and three petrochemical refineries. The operation of the metro system in Kaohsiung, Taiwan would result in significant mercury emissions. This chapter focuses on the field measurement of atmospheric mercury in Kaohsiung City, located in the coastal region of southern Taiwan. The study aims to investigate the tempospatial variation and partition of gaseous and particulate mercury during both wet and dry seasons. Additionally, the study aims to establish correlations between atmospheric mercury levels and meteorological parameters, as well as criteria air pollutants.

2.2. Selection and description of sampling sites

Field measurements were conducted to determine the speciation and concentration of atmospheric mercury at various locations in Kaohsiung City. These locations include a coastal background site and six sensitivity sites: Nan-zhi (S1) at 22°44′00″ North latitude and 120°19′41″ East longitude, Ren-wu (S2) at 22°41′20″ North latitude and 120°19′57″ East longitude, Zuoying (S3) at 22°40′29″ North latitude and 120°17′34″ East longitude, Cia-jin (S4) at 23°37′56″ North latitude and 120°17′16″ East longitude, Cianjhen (S5) at 22°36′18″ North latitude and 120°18′30″ East longitude, and Hsiao-kang (S6) at 22°33′57″ North latitude and 120°20′15″ East longitude. The coastal background location, situated on the campus of National Sun Yat-Sen University, is placed far away from any emission sources. Its geographical coordinates are 22°37′38″ North latitude and 120°16′01″ East longitude. Fig. 1 displays the precise locations of the coastal background site and six sensitivity sites in Kaohsiung City where sampling for total gaseous mercury (TGM) and particulate mercury (HgP) took place.

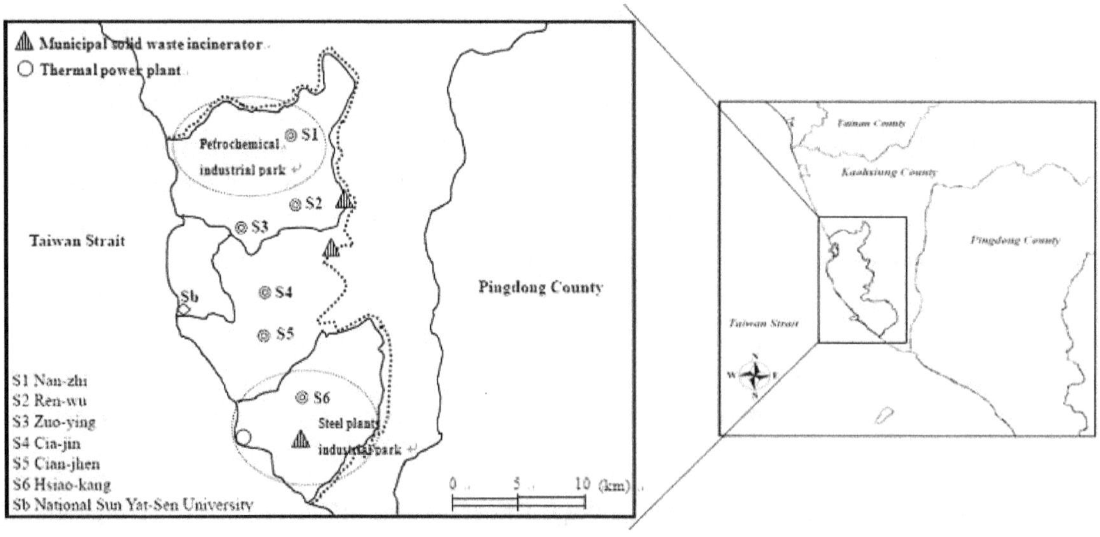

Fig. 1. The location of six sensitivity sites and the coastal background site for TGM and HgP sampling in Kaohsiung City.

The sensitivity sites were primarily situated in the ambient air quality monitoring stations in Kaohsiung City. Among the locations, S5-S6 were situated in close proximity to a steel industrial complex in the southern region of Kaohsiung, while S1 and S2 were located near a petrochemical industrial complex in the northern region of Kaohsiung. Continuous monitoring of total gaseous mercury (TGM) and mercury vapor (HgP) was carried out for 24 hours at each location in Kaohsiung City during both the rainy and dry seasons from June to December 2010. The wet season occurs from June to September, whereas the dry season occurs from October to December. This study aimed to examine the seasonal fluctuations, spatial arrangement, and division of total gaseous mercury (TGM) and mercury in particulate matter (HgP) at a coastal reference location and six locations with high vulnerability to industrial pollution.

2.3. Seasonal variation and partition of TGM and HgP

Fig. 2 depicts the fluctuation of TGM and HgP concentrations in the wet and dry seasons at the northern and southern regions of Kaohsiung, as well as the coastal background location. The meteorological data collected in the field are presented in Table 1. Table 2 provides a summary of the field measurement data, including the mean, standard deviation (SD), and partition of TGM and HgP.

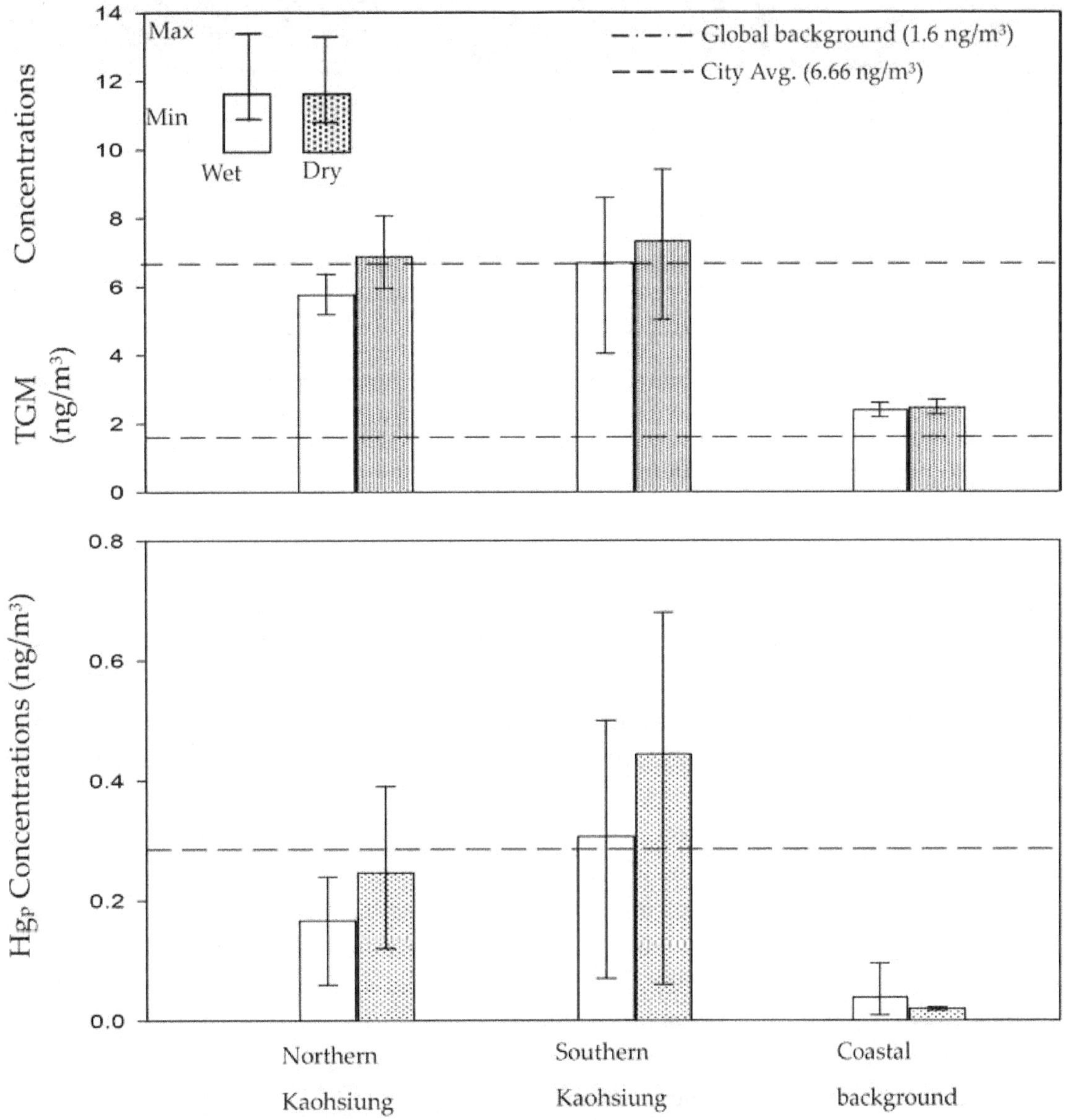

Fig. 2. Tempospatial variation of TGM and HgP concentrations during the wet and dry seasons in the northern and southern Kaohsiung sites and the coastal site.

Table 1. Meteorological parameters monitored in Kaohsiung City during the atmospheric mercury sampling periods for wet and dry seasons.

Seasons	Temp. (°C)	RH (%)	Rainfall (mm)	Rainy days	UV_B (UVI)	WS (m/s)	WD
Wet	28.9 ± 0.7	78.3 ± 1.7	427.9 ± 305.1	50	3.4 ± 0.5	2.2 ± 0.2	SW, NE
Dry	23.7 ± 3.4	73.3 ± 3.1	63.0 ± 97.6	12	2.1 ± 0.3	1.9 ± 0.1	NW, NE

Temp.: ambient temperature; **RH:** relative humidity; **Rainy days:** days with rainfall ≧ 0.01 mm; **WS:** wind speed; **WD:** wind direction; UV_B**:** ultra-violet radiation.

Table 2. Seasonal variation of TGM and HgP concentrations at six sensitivity sites and the coastal background site in Kaohsiung City.

Seasons	Types of Hg	Northern Kaohsiung sites			Southern Kaohsiung sites			Coastal background site
		S1	S2	S3	S4	S5	S6	Sb
Wet season (n = 12)	TGM (ng/m³)	5.72	6.37	5.20	4.04	7.42	8.60	2.38
Hg_P (ng/m³)		0.20	0.24	0.06	0.05	0.35	0.50	0.02
	Mean ± SD (ng/m³)	TGM		6.23 ± 1.62				
		Hg_P		0.23 ± 0.17				
Dry season (n = 12)	TGM (ng/m³)	5.95	8.06	6.59	5.03	7.50	9.41	2.44
	Hg_P (ng/m³)	0.23	0.39	0.12	0.08	0.59	0.68	0.03
	Mean ± SD (ng/m³)	TGM		7.09 ± 1.57				
		Hg_P		0.35 ± 0.25				
Mean ± SD of atmospheric mercury (ng/m³)		TGM		6.66 ± 1.42				2.41 ± 0.04
	Hg_P		0.29 ± 0.21				0.03 ± 0.01	

S1: Nan-zhi site; **S2:** Ren-wu site; **S3:** Zuo-ying site; **S4:** Cia-jin site; **S5:** Cian-jhen site; **S6:** Hsiao-kang site; **Sb:** coastal site.

In the wet season, the ambient temperature, relative humidity, wind speed, and UVB were recorded as 28.9 ± 0.7°C, 78.3 ± 1.7%, 2.2 ± 0.2 m/s, and 3.4 ± 0.5 UVI, respectively. These values were generally higher compared to those observed during the dry season (Table 2). Reported were the prevailing wind directions for the dry and wet seasons: southwest to northeast for the former, and northwest to northeast for the latter. The regular occurrence of sea-land breezes in Kaohsiung City was the key contributing factor. During the wet season, which coincides with the heat convection season (also known as hurricane season), the average rainfall was approximately 427.9 ± 305.1 mm. There were nearly 50 days with measurable rainfall (≧ 0.01 mm). The dry season often brought the northeast monsoon, characterized by significantly lower rainfall and fewer rainy days compared to the wet season. Thus, the weather condition described above is regarded as the contrast between the dry and wet seasons in Kaohsiung City. The field measurements indicated that the levels of TGM and HgP were 6.23 ± 1.62 ng/m³ and 0.23 ± 0.17 ng/m³, respectively, during the rainy season. In contrast, during the dry season in Kaohsiung City, the concentrations of TGM and HgP were 7.09 ± 1.57 ng/m³ and 0.35 ± 0.25 ng/m³, respectively. The levels of atmospheric mercury were significantly greater during the dry season compared to the wet season. The study demonstrated that climatic conditions and air dispersion significantly influenced the seasonal fluctuations in atmospheric mercury levels. Nevertheless, the levels of total gaseous mercury (TGM) and mercury in particulate matter (HgP) did not exhibit substantial fluctuations at the coastal background site. Consequently, the seasonal variations have negligible effects on areas with elevated atmospheric mercury concentrations. Atmospheric mercury concentrations in southern Kaohsiung were consistently greater than in northern Kaohsiung, both during the wet and dry seasons. On average, the concentrations in southern Kaohsiung were 1.12 and 1.79 times higher than those in northern Kaohsiung, respectively. The overall average concentrations of TGM and HgP in Kaohsiung City were approximately 2.94 and 11.7 times higher, respectively, than those at the coastal background site during the rainy season. Similarly, during the dry season, the concentrations were approximately 2.6 and 11.5 times higher, respectively. In Kaohsiung City, the average values of TGM and HgP were 6.66 ± 1.42 and 0.29 ± 0.21 ng/m^3, respectively. The concentration of Total Gaseous Mercury (TGM) in Kaohsiung City was approximately 4.2 times higher than the background TGM concentration in the Northern Hemisphere (1.6 ng/m^3), and roughly 2.8 times higher than the TGM concentration at the coastal background location in Kaohsiung City (2.4 ng/m^3). The data revealed that Kaohsiung City, known for its strong industrial activities, exhibited significant levels of atmospheric mercury pollution. Fig. 3 depicts the division of HgP (presumably referring to a specific substance or element) during both the rainy and dry seasons in Kaohsiung City. The findings indicated that TGM (total gaseous mercury) constituted the predominant mercury species, representing 94.56-99.59% of atmospheric mercury in the rainy season and 92.71-99.37% in the dry season. Moreover, the concentration of HgP exhibited a propensity to rise in proportion to the distance from the emission sources. The highest proportion of HgP partitioned in the ambient air was between 20-40% of the total atmospheric mercury (TAM) [11, 12, 13]. The distribution of HgP varied across different regions, with rural areas having a partition normally below 1% of the total available market (TAM), metropolitan areas ranging from 1-3%, and industrial areas exceeding 5% [14, 15, 16]. According to Fig. 1, site Sb is located in a rural location, while sites S3 and S4 are situated in metropolitan districts. The remaining four sites are found in industrial sectors inside Kaohsiung City. Furthermore, the division of HgP during the arid season was greater than that during the rainy season. It suggests that there may be a correlation between the distribution of HgP and the quantity of rainfall, the frequency of wet days, and the relative humidity.

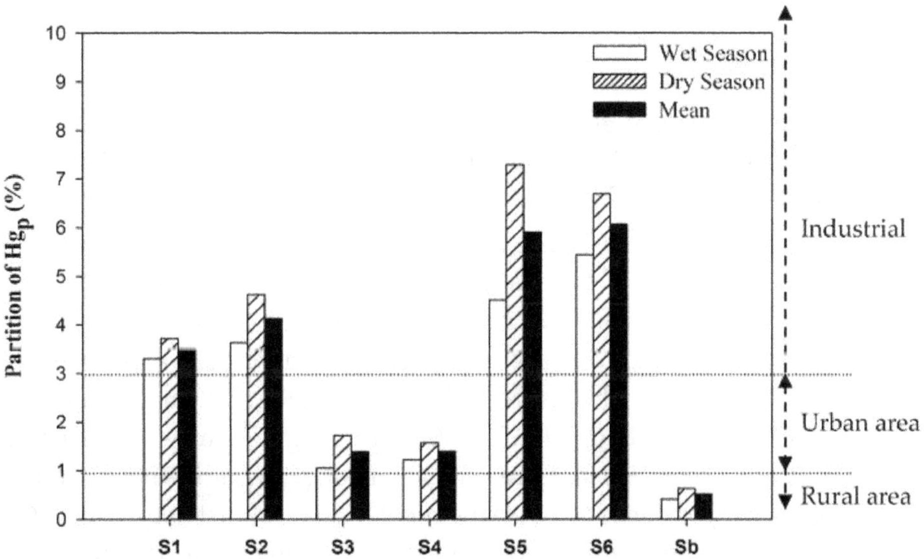

Fig. 3. Partition of HgP during the wet and dry seasons at six sensitivity sites (S1–S6) and the coastal site (Sb).

2.4. Spatial distribution of TGM and HgP

The measured concentrations of total mercury (TGM) and mercury in gaseous form (HgP) at each site are presented in Table 2. In terms of the spatial distribution of atmospheric mercury during the wet and dry seasons in Kaohsiung City, the highest concentrations of atmospheric mercury (TGM of 6.37 ng/m^3; HgP of 0.24 ng/m^3) were found at site S2, followed by sites S1 and S3 in northern Kaohsiung. In southern Kaohsiung, the highest concentrations (TGM of 8.60 ng/m^3; HgP of 0.50 ng/m^3) were found at site S6, followed by sites S5 and S4. During the dry season, the levels of atmospheric mercury at site S2 in northern Kaohsiung were the highest, with a total gaseous mercury (TGM) concentration of 8.06 ng/m^3 and a particulate-bound mercury (HgP) concentration of 0.39 ng/m^3. This was followed by sites S3 and S1 in the same region. In the southern part of Kaohsiung, the highest concentrations were found at site S6, with a TGM concentration of 9.41 ng/m^3 and a HgP concentration of 0.50 ng/m^3.

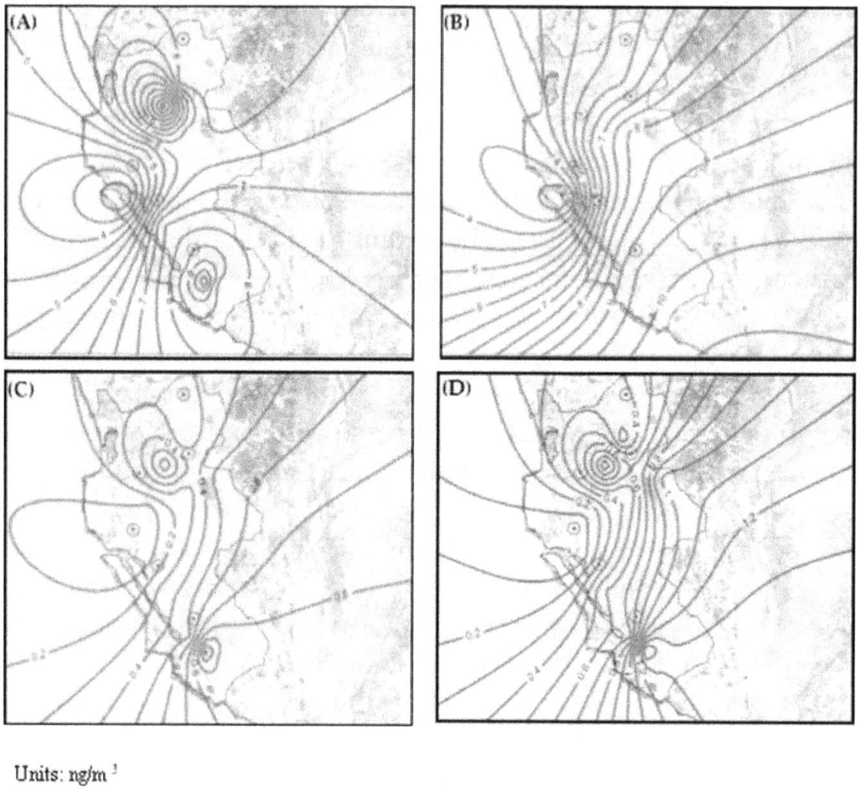

Fig. 4. Spatial distribution of atmospheric mercury in Kaohsiung City. (A) TGM concentration during the dry season, (B) TGM concentration during the wet season, (C) HgP concentration during the dry season, and (D) HgP concentration during the wet season.

This was followed by sites S5 and S4 in the same area. The mapping software SURFER was utilized to generate a concentration contour plot of atmospheric mercury in Kaohsiung City. This software utilizes grid-based interpolation to calculate atmospheric mercury content data gathered from several sampling sites and overlay it onto the city of Kaohsiung. The objective of this study was to investigate the spatial pattern of atmospheric mercury levels in Kaohsiung City. Fig. 4 clearly demonstrates that the air mercury levels in both northern and southern Kaohsiung were significantly influenced by the sources of mercury emissions. There are two significant areas in Kaohsiung, Taiwan, where there are high levels of mercury concentration. These areas are the petrochemical industrial district in northern Kaohsiung and the steel manufacturing complex in southern Kaohsiung. The primary sources of emissions in northern Kaohsiung include a petrochemical manufacturing complex, municipal and industrial waste incinerators, and cremators. In the southern region of Kaohsiung, the primary sources of mercury emissions include steel smelters, electric arc furnaces, cement facilities, petroleum refineries, coal-fired power plants, and municipal garbage incinerators. As a result, the level of mercury in the atmosphere was higher in southern Kaohsiung compared to northern Kaohsiung. This difference can be related to the fact that there was a greater amount of mercury being released into the atmosphere in southern Kaohsiung than in northern Kaohsiung. The overall ambient air quality of Kaohsiung City was significantly impacted by the densely concentrated industries, resulting in a much greater concentration of atmospheric mercury in the metropolitan region compared to the background level during both the wet and dry seasons. The emission sources and fossil fuel consumption in southern Kaohsiung were greater than those in northern Kaohsiung.

2.5. Comparison of TGM and HgP levels with other stationary sources

Fig. 5 presents a comparison of the levels of TGM and HgP at various sources of mercury emissions in southern Taiwan, as documented in references 12 and 13. The steel plant had the greatest levels of TGM concentrations, which were approximately 2.6 times more than those reported in the Tainan Scientific Complex. The HgP concentration at the steel plant was even up to 15 times higher. Similarly, the concentrations of TGM and HgP at other stationary sources were likewise higher than those observed at Tainan Scientific Complex. Fig. 6 depicts the distribution of total gaseous mercury (TGM) and mercury in fine particles (HgP) across different sources of mercury emissions in southern Kaohsiung. The partition of HgP at Tainan Scientific Complex was merely 1.7%, which is comparable to the levels reported in most urban regions (1-3%), and significantly lower than other stationary sources like as coal fired power plants, waste incinerators, and steel mills. Their distribution of HgP varied from 7.3 to 9.5%, resulting in a much higher contribution of HgP to the surrounding air [6, 15, 17, 18]. In southern Taiwan, TGM was the prevailing species of atmospheric mercury.

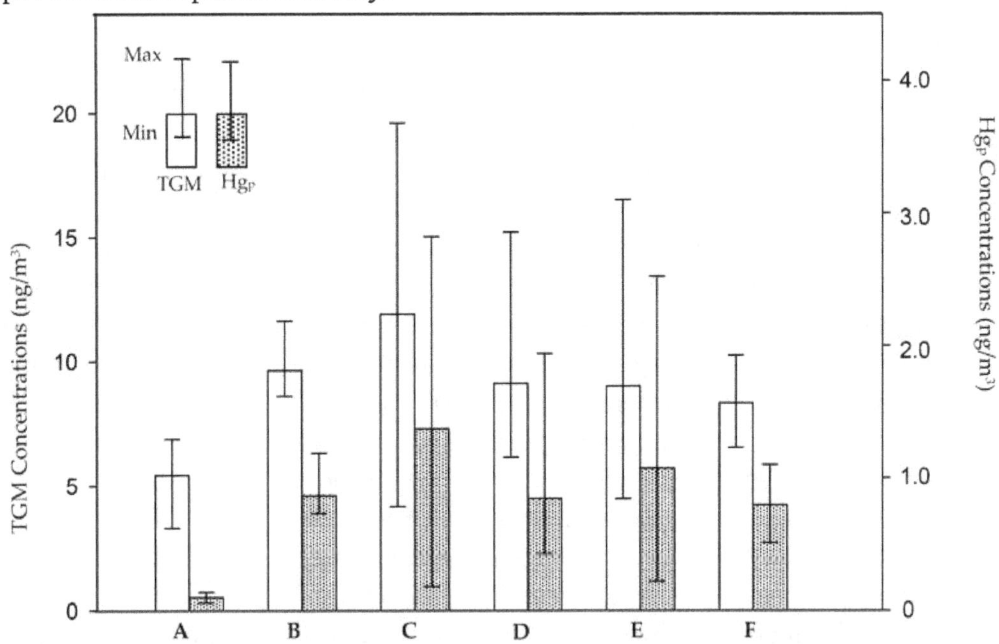

Fig. 5. Comparison of TGM and HgP concentrations with various mercury emission sources in southern Kaohsiung (A: the semi-conductor complex, B: the petroleum refinery, C: the steel plant, D: the coal-fired power plant, E: the electric arc furnace, F: the municipal waste incinerator).

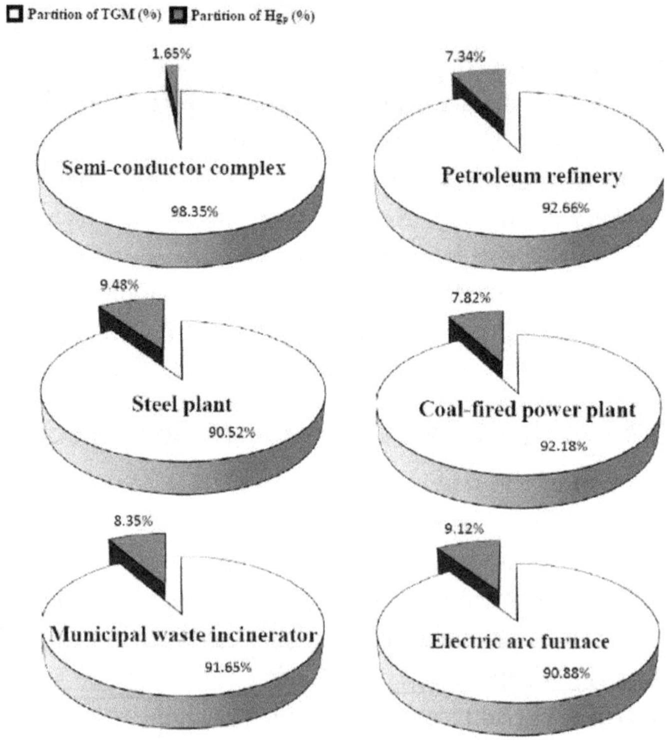

Fig. 6. Location of the TGM monitoring site at the Penghu Islands.

3. THE STUDY FOCUSES ON MEASURING THE CONCENTRATION OF GASEOUS MERCURY IN THE MARITIME BOUNDARY LAYER AND EXAMINING ITS TRANSIT ACROSS GREAT DISTANCES.

3.1. Background

The Penghu Islands are situated in the middle of the Taiwan Strait, between southeastern China and Taiwan. The islands experience a prevailing subtropical climate that is mostly shaped by the East Asian monsoons. Prior research has indicated that the biomass and fossil fuel burning that commonly takes place in Southeast Asia and Southwest China throughout the spring and winter seasons leads to elevated levels of atmospheric mercury in Taiwan due to long-distance transport [11, 18]. Given the absence of notable mercury sources in the Penghu Islands, it can be regarded as the reference site in the area. This chapter presents the findings of a one-year field monitoring study conducted at the Penghu Islands. The study aimed to investigate the seasonal and daily variations of total atmospheric mercury (TAM) concentration. Additionally, the correlation between TAM concentration and meteorological parameters, as well as several criteria air pollutants, was examined and discussed. Furthermore, a retrograde trajectory simulation model was utilized to investigate the transportation of TGM to the Penghu Islands, specifically focusing on the transportation routes seen at an altitude of 500 m above sea level throughout the monitoring seasons. The TGM has been a persistent global concern. The findings of this study would aid in the development of more efficient management strategies to mitigate the negative impacts of TGM on the environment and human health. This is particularly relevant for the Penghu Islands and other areas that may be affected by the long-range transportation of TGM.

3.2. Monitoring sites

Based on meteorological data collected between 1985 and 2011, a dry season with an average rainfall of approximately 800 mm occurs from April to September. Table 3 presents a summary of the meteorological data collected at the Penghu Islands during the TGM monitoring seasons. The data indicates that the dominant winds came from the east, northeast, northwest, and south. The recorded ambient temperatures ranged from 13.4 to 31.2°C, the relative humidity ranged from 53.2 to 91.3%, and the wind speeds ranged from 0.5 to 9.7 m/s. The TGM monitoring station was situated atop a four-story building, positioned at an elevation of roughly 12 meters above ground level. It was set 50 meters away from the beach and 500 meters away from the major roadways. In general, the TGM monitoring site was situated in the marine boundary layer (MBL), which helped to minimize potential interferences caused by mercury (Hg) emission sources. The atmospheric total gaseous mercury (TGM) was consistently measured at the Hsiaomen site (23°38′471″ North latitude, 119°30′316″ East longitude) situated on the northwestern coastline of the Penghu Islands (refer to Fig. 7) over a duration of 15 consecutive days in each season, spanning from March 2011 to February 2012.

Table 3. Meteorological data recorded at the Penghu Islands during the TGM sampling periods.

Seasons	n	Temp. (°C)	RH (%)	WS (m/s)	WD
Spring	360	19.9 ± 2.1	72.5 ± 7.5	4.5 ± 2.2	ENE
Summer	360	29.8 ± 0.6	79.3 ± 2.7	2.2 ± 0.6	S
Fall	360	24.8 ± 0.6	75.1 ± 4.9	6.1 ± 1.5	ENE
Winter	360	16.1 ± 1.5	82.2 ± 4.7	6.2 ± 1.3	ENE
Mean ± SD		22.7 ± 5.9	77.3 ± 4.3	4.8 ± 1.9	—
Max		33.2	91.3	9.7	—
Min		13.4	53.2	0.5	—

Temp.: ambient temperature; **RH:** relative humidity; **WS:** wind speed; **WD:** wind direction; **SD:** standard deviation.

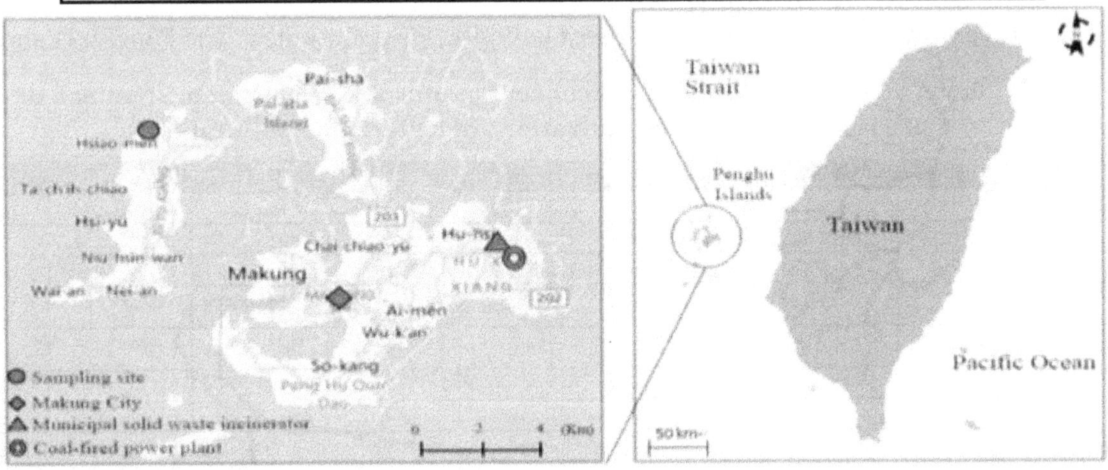

Fig. 7. Variation of TGM concentration during four monitoring seasons at the Penghu Islands.

This chapter compiled the meteorological data and the concentrations of criteria air pollutants from the Makung Air Quality Station during the TGM monitoring periods. It also examined the correlation coefficients between the TGM concentrations and the meteorological parameters as well as the criteria air pollutants. The correlation coefficient (R) was employed to quantify the relationship between TGM concentration, meteorological data, and criterion air pollutant concentrations. The meteorological parameters consisted of the ambient temperature (Temp.), relative humidity (RH), wind speed (WS), and wind direction (WD). The criterion air pollutants of interest were SO_2, NO_X, CO, O_3, PM_{10}, and $PM_{2.5}$. The four seasons are categorized as follows: spring, which spans from March to May; summer, which encompasses June to August; fall, which occurs from September to November; and winter, which lasts from December to February. In this work, the NOAA HYSPLIT model was used to simulate 72-hour backward trajectories. The model input consisted of the National Centers for Environmental Prediction's Global Data Assimilation System (NCEP-GDAS) meteorological dataset. All reverse trajectories originated at an altitude of 500 meters above sea level. The NOAA-HYSPLIT model was utilized to identify the dates with the highest TGM concentration in the Penghu Islands during various seasons. Subsequently, the model was used to simulate the air mass transit pathways towards the Penghu Islands during the monitoring periods. This information was subsequently utilized to analyze the potential transit routes for carrying TGM from the upwind sources to the Penghu Islands. In addition, we utilized the local fire map provided by the FIRMS web platform (http://firms.modaps.eosdis.nasa.gov/firemap/) to determine the potential origins of TGM. This was done by analyzing the moderate resolution imaging spectroradiometer (MODIS) data obtained from the National Aeronautics and Space Administration (NASA). The purpose of this analysis was to gain insight into the geographical distribution and current state of biomass burning in the monitored periods.

3.3. Temporal fluctuations in the concentration of total gaseous mercury (TGM) on both a seasonal and daily basis

Table 4 presents a concise overview of the mean and variability of TGM and criterion air pollutant levels recorded at the Penghu Islands. In spring, the PM10 and PM2.5 levels reached their peak concentrations at 54.60 ± 15.56 and 31.07 ± 10.59 μg/m³, respectively. Conversely, the lowest concentrations of PM10 and PM2.5 were detected in summer, measuring 36.63 ± 8.69 and 17.70 ± 6.48 μg/m³, respectively. Spring and winter are the primary seasons during which Asian dust from northern China is often carried to the Penghu Islands [19], resulting in a large rise in PM^{10} levels. In general, the levels of SO^2 and NOx detected in the Penghu Islands were considerably lower compared to other cities in Taiwan. This suggests that there are no substantial local sources of these pollutants in the Penghu Islands. Furthermore, the concentration of O_3 was comparatively greater than that of other qualifying air pollutants at the Penghu Islands.

Table 4. Continuous monitoring data of TGM concentrations and their mean, standard deviation (SD), range, and the concentrations of criteria air pollutants.

Seasons	n	TGM (ng/m³)	PM_{10} (μg/m³)	$PM_{2.5}$ (μg/m³)	SO_2 (ppb)	NOx (ppb)	O_3 (ppb)	CO (ppm)
Spring	360	4.31 ± 1.87	54.60 ± 15.56	31.07 ± 10.59	1.70 ± 0.72	5.61 ± 0.08	61.58 ± 5.00	0.27 ± 0.06
Summer	360	1.81 ± 0.15	36.63 ± 8.69	17.70 ± 6.48	2.21 ± 0.63	6.15 ± 1.23	33.17 ± 7.70	0.15 ± 0.03
Fall	360	3.03 ± 0.40	45.24 ± 18.49	21.40 ± 6.08	1.68 ± 0.52	3.48 ± 0.46	51.86 ± 5.99	0.18 ± 0.04
Winter	360	3.51 ± 0.67	48.65 ± 9.79	23.81 ± 12.18	2.00 ± 0.08	5.48 ± 1.63	38.59 ± 5.87	0.28 ± 0.08
P-value		3.32×10^{-55}	6.94×10^{-3}	1.93×10^{-3}	1.91×10^{-1}	1.36×10^{-7}	4.00×10^{-18}	3.04×10^{-10}
Mean ± SD		3.17 ± 1.06	46.28 ± 7.51	23.50 ± 5.64	1.90 ± 0.25	5.18 ± 1.17	46.30 ± 12.86	0.22 ± 0.06
Range		1.17–8.63	17.18–77.68	1.32–63.07	1.02–8.63	2.43–14.09	11.35–88.08	0.16–0.57
The concentrations observed among the seasons were significantly different by the analysis of ANOVA (Analysis of Variance) at the confidence level of 95%.								

Field monitoring of TGM at the Penghu Islands showed that the average concentration of TGM was 3.17 ± 1.06 ng/m³ with the range of 1.17–8.63 ng/m³. The concentration of TGM in four monitoring seasons were ordered as: spring (4.34 ± 1.87 ng/m³) > winter (3.51 ± 0.67 ng/m³) > fall (3.03 ± 0.40 ng/m³) > summer (1.81 ± 0.15 ng/m³). Moreover, the average TGM concentration (3.17 ng/m³) at the Penghu Islands was approximately two times of the background TGM concentration of North Hemisphere (1.6 ng/m³). The Penghu Islands are likely to be influenced by long-range transportation of TGM in spring and winter, causing higher TGM concentrations in spring and winter than those in summer and fall. The lowest seasonal average TGM concentration of 1.81 ± 0.15 ng/m³ was observed in summer, which was slightly close to the background TGM concentration of North Hemisphere, suggesting that air masses blown from the South China Sea were relatively clean.

Fig. 8 illustrates the variation of TGM concentrations during the four monitoring seasons at the Penghu Islands. The daily concentration of TGM in spring increased significantly from April 2nd to the peak TGM concentration (8.63 ng/m^3) observed on April 10th. The lowest TGM concentration ranging from 1.50 to 2.70 ng/m^3 was compatibly observed in summer. Moreover, the concentrations of criteria air pollutants were also the lowest in summer compared to other seasons (Table 4), suggesting that the ambient air quality in summer was relatively better than other seasons at the Penghu Islands, and the TGM concentrations was 1.7–2.8 times lower than other seasons.

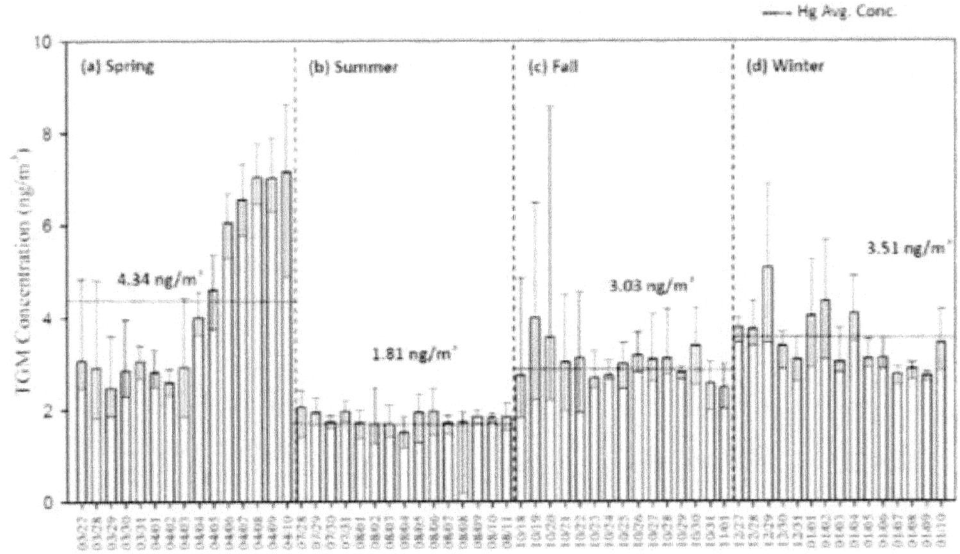

Fig. 8. Hourly variation of TGM concentrations during four monitoring seasons at the Penghu Islands.

In contrast to other seasons, the concentrations of TGM exhibited regular fluctuations during the fall and winter. During the winter season, there was a significant increase in the concentration of a certain substance from 3.74 to 5.08 ng/m^3 on December 28th, followed by a decrease to 3.38 ng/m^3 on December 30th. The peak concentration reached its maximum value of 4.35 ng/m^3 on December 29th and remained constant until January 5th. Monitoring of total gaseous mercury (TGM) for an extended period at the background air quality monitoring station in Mt. Lulin revealed that Taiwan is significantly impacted by atmospheric mercury and gaseous pollutants originating from China and Southeast Asia throughout the spring and winter seasons [11]. The findings of the backward trajectory simulation showed a considerable rise in atmospheric mercury measured in Taiwan. This increase was mostly caused by frequent biomass burning originating from Southeast Asia and industrial emissions from North China between the spring and winter seasons. It is indicated that the levels of TGM (total gaseous mercury) at the Penghu Islands may also be affected by the emission and movement of atmospheric mercury from these areas. The air masses originating from the Pacific Ocean had a comparatively low impact on the concentration of TGM.

3.4. Hourly variation of TGM concentration

Fig. 9 depicts the fluctuation of TGM concentrations on an hourly basis over four seasons at the Penghu Islands. The data indicated that the fluctuation of hourly TGM concentrations remained essentially constant during the summer season. The concentration of TGM exhibited a progressive increase from 6:00 am, reaching its peak at 11:00 am, and subsequently declining after lunch. The increase in TGM concentration likely occurred due to two processes: (a) UV radiation temporarily converting Hg+, Hg^{2}+, and HgP into volatile Hg0, which then entered the atmosphere [6]; (b) the downward mixing of TGM from higher altitudes, which could raise TGM levels by disrupting the nocturnal inversion layer [20]. With the exception of the morning, the concentration of TGM remained quite stable during the whole day. The mean TGM concentrations throughout the daytime (6:00 am–6:00 pm) were generally greater than those during the overnight (6:00 pm–next 6:00 am). The observed results are ascribed to the impacts stemming from the fluctuation in the altitude of the atmospheric boundary layer. Furthermore, it could be impacted by nearby human activity, such as the burning of waste, emissions from fishing boats and automobiles, and other sources.

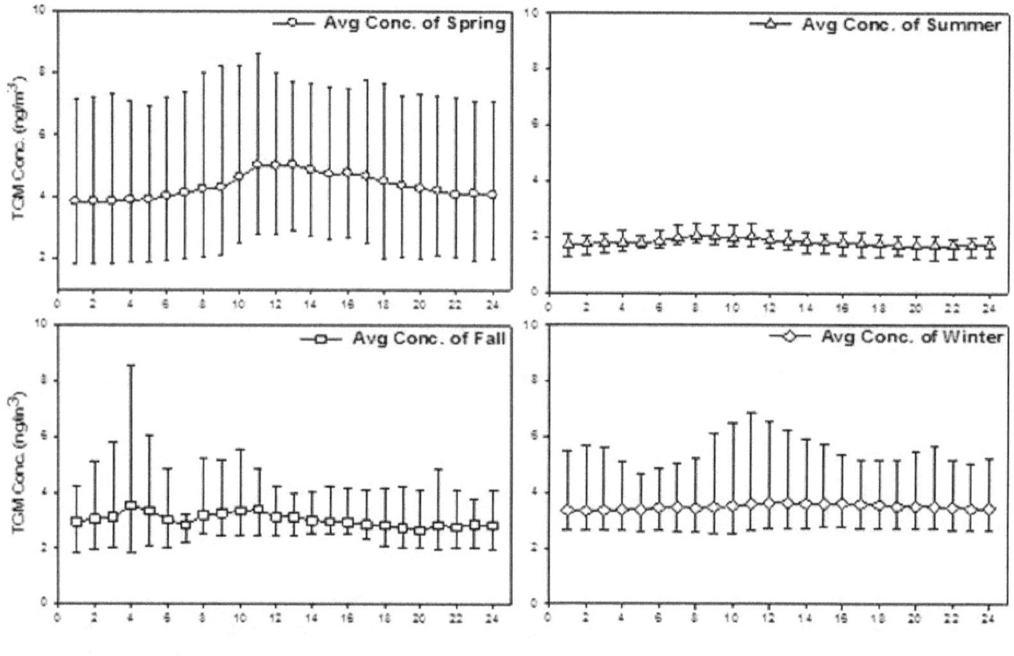

RPD: relative percent difference (%)

Fig. 9. Frequency distribution of TGM concentration covering the four monitoring seasons.

The hourly variation in TGM concentrations was less pronounced throughout the summer and winter, with relative percent differences (RPD) of 61.9% and 88.1% respectively. However, the variance was greater during the spring and fall, with RPDs of 130.7% and 107.8% respectively. Previous studies have demonstrated that the measured concentration of total gaseous mercury (TGM) fluctuates at low elevations in rural areas, as evidenced by research conducted by various sources [11, 21, 22]. The findings of this study indicate that the levels of TGM seen at the Penghu Islands were primarily driven by both local biomass burning and long-range transportation. Fig. 10 depicts the frequency distribution of TGM identified throughout the course of four monitoring seasons. The TGM concentrations exhibited a lognormal distribution pattern within the range of 2.0–4.5 and 6.0–7.5 ng/m^3, constituting approximately 80.0% of the total frequency during the spring season. The readings within the range of 1.5 to 2.5 ng/m3 constituted roughly 89.4% of the total frequency over the summer. In the fall, values ranging from 2.0 to 4.5 ng/m^3 accounted for 93.3% of the entire frequency. In the winter, values ranging from 3.5 to 5.5 ng/m3 accounted for 61.7% of the whole frequency. Nevertheless, occurrences characterized by elevated TGM concentrations (>9.0 ng/m^3) were commonly detected during the seasons of spring, fall, and winter. It is important to mention that the frequency distribution of TGM exhibited two distinct patterns, as illustrated in Fig. 10. The TGM levels in summer and fall had a unimodal distribution, whereas in spring and winter, they displayed a multi-modal distribution.

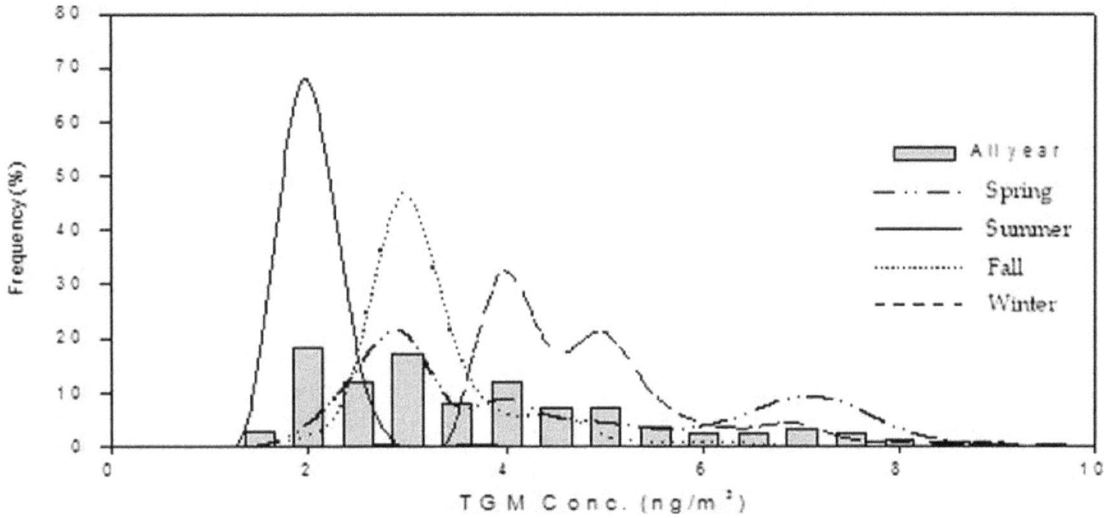

Fig. 10. Pollution rose of TGM in four seasons at the Penghu Islands.

3.5. TGM's transportation routes to the Penghu Islands

Fig. 11 depicts the pollution rise of total gaseous mercury (TGM) during four monitoring seasons at the Penghu Islands. Elevated TGM levels were predominantly detected in wind directions ranging from 0 to 90 degrees, followed by 270 to 360 degrees during the spring season. In the fall and winter, TGM concentrations were highest in wind directions ranging from 60 to 120 degrees. There were no notable sources of mercury emissions at the Penghu Islands in the northwest or southwest, indicating that the elevated levels of total gaseous mercury (TGM) were likely carried from China over a long distance. Multiple studies have demonstrated that biomass burning, industrial combustion, and ocean evaporation are the primary drivers of TGM emissions on a regional scale [10, 23, 24, 25]. Fig. 12 depicts the reverse paths of air masses and the proportion of TGM concentration towards the Penghu Islands across the four seasons.

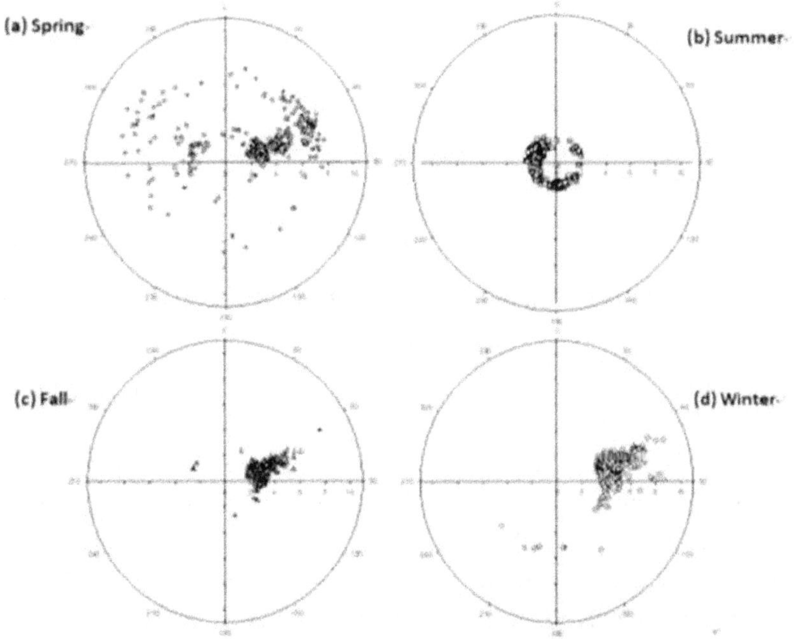

Fig. 11. Backward trajectories of air masses and TGM concentration percentage for different pathways during the monitoring seasons at the Penghu Islands.

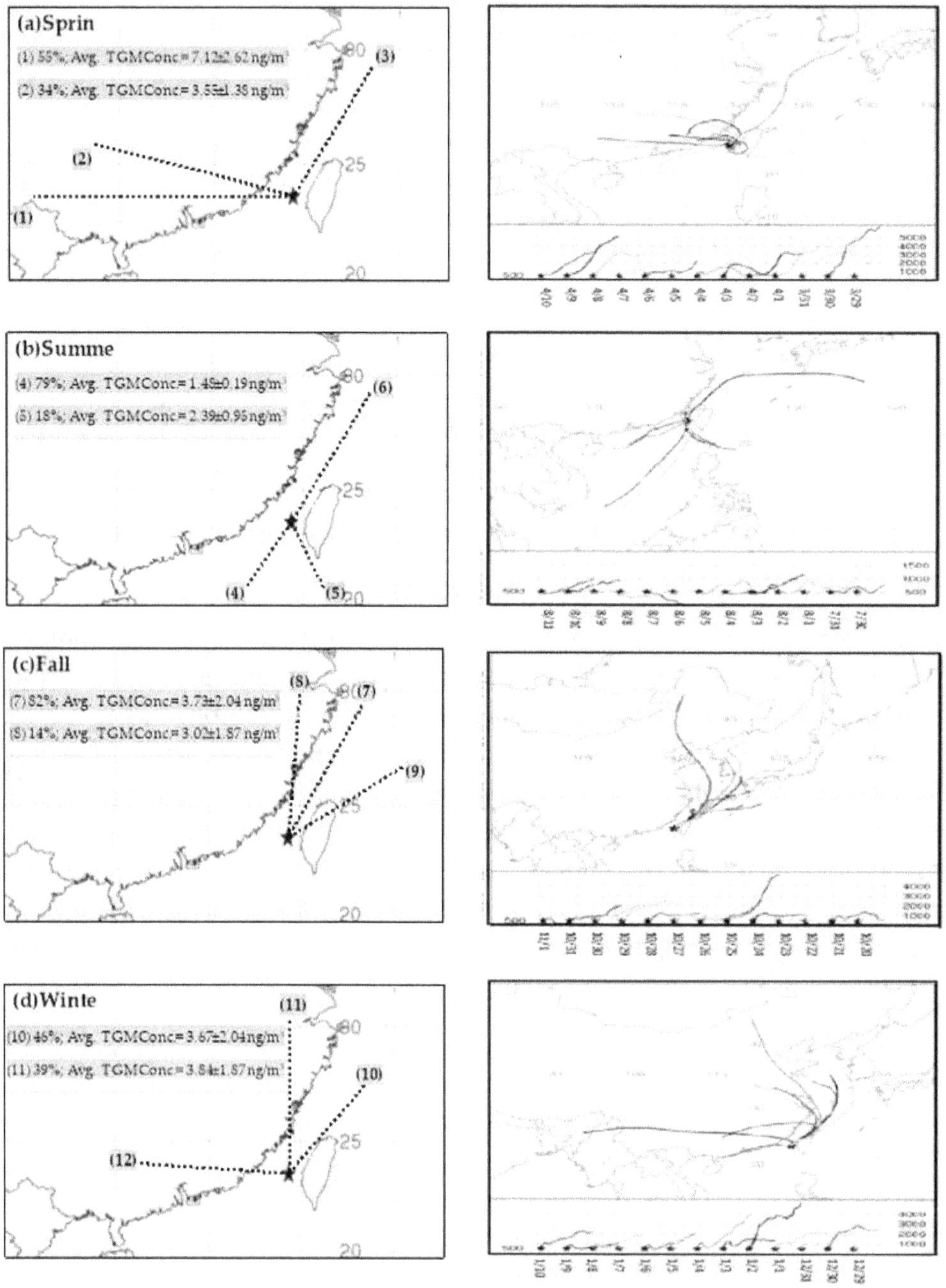

Fig. 12. The percentage of the TGM concentration attributed from long-range transportation at the Penghu Islands.

During the spring season, the concentrations of total gaseous mercury (TGM) in the air masses transported from routes (1) and (2) varied between 3.55 and 7.12 ng/m³. These air masses accounted for approximately 89% of the TGM data. It is likely that the dominant sources of TGM in these air masses were local stationary sources and open burning, with the air masses moving towards the Penghu Islands. Therefore, the southern region of China could have been another potential source that contributed to the amounts of total gaseous mercury (TGM) at the Penghu Islands. During the summer monitoring period, the TGM concentrations for air masses coming from routes (4) and (5) varied from 1.48 to 2.39 ng/m³, making up around 97% of the TGM data. These air masses originated from the South and the East China Sea and were responsible for significantly elevating the TGM concentrations at the Penghu Islands. As a result, the levels of TGM throughout the summer were comparatively lower than in other seasons, and the average TGM concentration was similar to the natural TGM concentration in the Northern Hemisphere, which is roughly 1.6 ng/m^3. During the fall season, the concentrations of total gaseous mercury (TGM) in the air masses coming from routes (7) and (8) varied between 3.02 and 3.73 ng/m^3, representing around 96% of the TGM data. These air masses appeared to originate mainly from northern China, Korea, and Japan, and were transported towards the Penghu Islands. As a result, the TGM concentration in fall was approximately 1.67 times higher than in summer. During the winter season, the concentrations of total gaseous mercury (TGM) in the air masses coming from routes (11) and (12) varied between 3.67 and 3.84 ng/m^3, representing roughly 85% of the TGM data. on winter, the TGM concentrations on the Penghu Islands might increase up to 1.94 times compared to summer due to probable sources from southern Asia, northern China, and Mongolia. The minimal TGM value of 1.5 ng/m^3 observed during the summer monitoring period can be used as the baseline concentration of TGM at the Penghu Islands. Hence, the concentration of TGM resulting from long-range transportation can be determined using Equation (1) and depicted in Fig. 13:

$$\text{Contribution percentage} = (\text{VMH} - \text{BC})/\text{VMH} \times 100\% \quad (1)$$

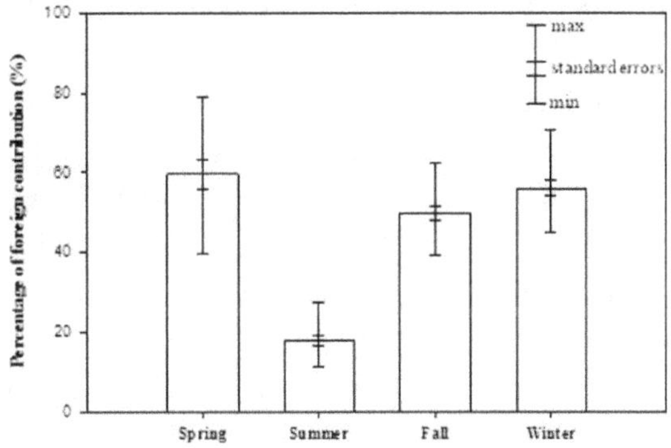

Fig. 13. Fire maps air mass during four monitoring seasons in the East Asia in (a) spring, (b) summer, (c) fall, and (d) winter.

where contribution percentage is the TGM concentration attributed from cross-boundary transportation (%); VMH is the monitored TGM concentrations (ng/m^3); BC is the background concentration of Hg (i.e., 1.5 ng/m^3). The contribution percentage of TGM concentration attributed from cross-boundary transportation at the Penghu Islands were further compared. The percentages of cross-boundary transportation were ordered as: spring (59.6 ± 15.0%) > winter (55.9 ± 7.6%) > fall (49.7 ± 6.2%) > summer (16.8 ± 6.9%). In addition, the maximum percentage of 79.1% was observed in spring. It showed that the TGM concentrations in spring at the Penghu Islands were highly influential, which were mainly affected by the long-range transportation.

The presence of high humidity and frequent rainfall at the Penghu Islands may account for the comparatively lower levels of TGM concentrations seen throughout the winter season. A previous study found that biomass burning, specifically forest fires, in Southeast Asia and the Indochina Peninsula during the months of February to April release a significant quantity of mercury-containing pollutants into the atmosphere [26]. Fig. 14 depicts the fire maps acquired from the FIRMS web fire maps and the routes of air mass transportation in four different seasons. The fire maps provided an explanation for the occurrence of elevated TGM concentration in spring at the Penghu Islands. During the monitoring periods, the concentration of fires was higher in spring compared to other seasons, resulting in a significant release of TGM into the atmosphere. Subsequently, the TGM was carried towards the Penghu Islands.

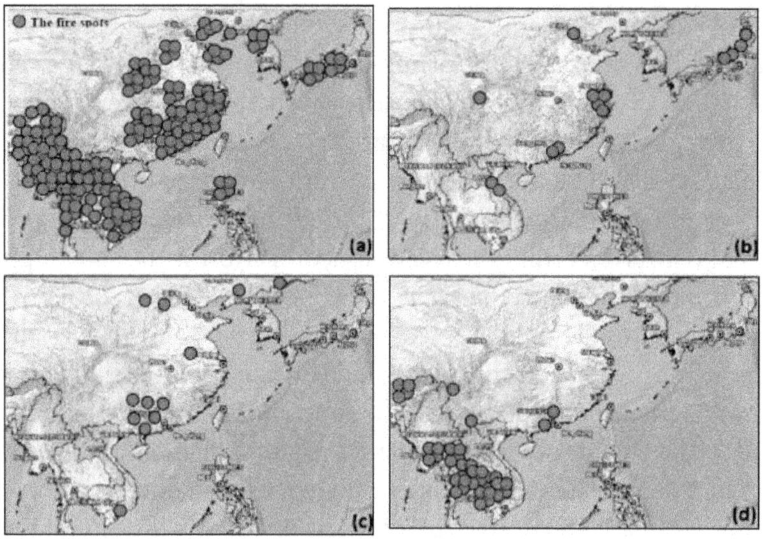

Fig. 14. The concentration map of TGM at islands and seas in East Asia. (a) An-Myun Island (Nguyen et al., 2007); (b) Yellow Sea (Ci et al., 2011); (c) Jeju Island (Nguyen et al., 2010); (d) Okinawa Island (Chand et al., 2008); (e) Penghu Islands (This study); and (f) South China Sea (Fu et al., 2010).

As illustrated in Fig. 15, the TGM concentrations monitored at the Penghu Islands were compared to those observed at islands and seas in East Asia. The TGM concentrations in the ambient air were ordered as: An-Myun (4.61 ± 2.21 ng/m^3) > Jeju (3.85 ± 1.68 ng/m^3) > Penghu (3.17 ± 1.06 ng/m^3) > Yellow Sea (2.61 ± 0.50 ng/m^3) > South China Sea (2.32 ± 2.62 ng/m^3) > Okinawa Islands (2.04 ± 0.38 ng/m^3) [10, 22, 25, 26, 27]. The results showed that the TGM concentrations decreased continuously from the northern islands to the southern islands. The TGM concentrations observed at the offshore islands were generally lower than those close to the continent or main island (except Yellow Sea). It further explained why the high levels of TGM concentration at the offshore islands were much easily influenced by Hg emission sources from the anthropogenic sources in the continent or main island.

Fig. 15. The concentration map of TGM at islands and seas in East Asia. (a) An-Myun Island; (b) Yellow Sea; (c) Jeju Island; (d) Okinawa Island; (e) Penghu Islands; and (f) South China Sea.

To summarize, the concentration of TGM at the Penghu Islands is primarily affected by the long-distance movement of air masses, in addition to local sources and open burning. The prevailing wind direction and routes of air mass transportation are likely to be crucial factors in the fluctuation of TGM concentration in the atmosphere.

4. CONCISE SUMMARY

This chapter examined atmospheric mercury levels by employing modified sampling and analytical techniques in two different scales: small-scale regions and large-scale regions. The study aimed to investigate the tempospatial variation of atmospheric mercury, the partition between gas and particulate forms, the transportation routes of mercury, and the comparison of mercury concentrations in urban areas and stationary sources. The findings of this field study yield various significant conclusions, which can be summarized as follows. The geographical and temporal fluctuation of total gaseous mercury (TGM) and mercury vapor (HgP) in Kaohsiung City were measured to be 6.66 ± 1.42 ng/m^3 and 0.29 ± 0.21 ng/m^3, respectively. The concentration of TGM was roughly 4.1 times more than the background concentration of 1.6 ng/m^3 in the North Hemisphere. There are two areas in Kaohsiung City with significant levels of mercury concentration, which coincide with the petrochemical complex in Northern Kaohsiung and the steel industrial complex in Southern Kaohsiung. During both the wet and dry seasons, the concentrations of TGM and HgP in Kaohsiung City were consistently higher compared to other cities in Taiwan. The levels of total gaseous mercury (TGM) detected in the cities of Taiwan were generally greater than those in Tokyo and Seoul, but lower than those in other cities in China. China's predominant contribution to global mercury emissions stems from the use of coal for winter space heating. Furthermore, Japan, Korea, and Taiwan are located downwind of China, making them one of the primary pathways for atmospheric mercury transmission. During the dominance of the northeastern monsoon, there is a rise in the concentrations of Total Gaseous Mercury (TGM) and Mercury (HgP) in the cities of East Asia. The concentration of TGM (Total Gaseous Mercury) observed at the Penghu Islands was found to be 3.17 ± 1.06 ng/m^3, with a range of 1.17–8.63 ng/m^3. The seasons were ranked in the following order based on TGM concentration: spring > winter > fall > summer. Summer is the sole season that closely matches the background TGM concentration of the Northern Hemisphere at the Penghu Islands. The TGM concentration normally exhibits a rise in the morning (8:00 am–1:00 pm), reaching its maximum concentration, and thereafter decreasing in the late afternoon (after 2:00 pm). Air masses originating from southern and northern China, southern Asia, Korea, Japan, and Mongolia have the potential to influence the levels of Hg (mercury) at the Penghu Islands during the monitoring seasons. The quantities of total gaseous mercury (TGM) may be affected by the long-range transit of air masses contaminated with mercury, originating from either local stationary combustion sources or mobile sources. During the summer, the air masses moved towards the Penghu Islands were primarily from the South China Sea. However, the concentration levels of TGM (Total Gaseous Mercury) at the Penghu Islands were seen to be lower compared to other seasons. The high concentration of TGM observed at the Penghu Islands during spring can be attributed to three main factors: (a) local emissions from field open burning, local stationary combustion, and mobile sources; (b) long-range transportation from biomass burning in Southeast Asia or neighboring Chinese coastal cities; and (c) long-range transportation through Asian dusts originating from North China.

REFERENCES

1. UNEP. Global Mercury Assessment, December. 2002. http://www.chem.unep.ch/mercury/report/gma-report-toc.htm
2. UNEP. Part a: Global Emissions of Mercury to the Atmosphere, UNEP/AMAP 2012 Technical Report, July. 2012
3. Nater EA, Grigal DF. Regional trends in mercury distribution across the Great Lakes states, north central USA. Nature. 1992;358:139-141.
4. Mason RP, Fitzgerald WF, Morel MM. The biogeochemical cycling of elemental mercury: Anthropogenic influences. Geochimica et Cosmochimica Acta. 1994;58:3191-3198.
5. Mason RP, Sheu GR. Role of the ocean in the global mercury cycle. Global Biogeochemical Cycles. 2002;16(4):1093-1107.
6. Schroeder WH, Munthe J. Atmospheric mercury—An overview. Atmospheric Environment. 1998;32:809-822.
7. Lin CJ, Pehkonen SO. The chemistry of atmospheric mercury: A review. Atmospheric Environment. 1999;33:2067-2079
8. Boening DW. Ecological effects, transport, and fate of mercury: A general review. Chemos. 2000;40:1335-1351
9. Clarkson TW, Magos L. The toxicity of mercury and its compounds. Critical Reviews in Toxicology. 2006;36:609-662
10. Fu XW, Feng XB, Dong ZQ, Yin RS, Wang JX, Tang ZR, Zhang H. Atmospheric gaseous elemental mercury (GEM) concentrations and mercury depositions at a high-altitude mountain peak in south China. Atmospheric Chemistry and Physics Discussions. 2010;9:23465-23504
11. Sheu GR, Lin NH, Wang JL, Lee CT, Ou CF, Yang CFO, Wang SH. Temporal distribution and potential sources of atmospheric mercury measured at a high-elevation background station in Taiwan. Atmospheric Environment. 2010;44:2393-2400
12. Jen YH, Yuan CS, Lin YC, Lee CG, Hung CH, Tsai CM, Tsai HH, Ie IR. Partition and tempospatial variation of gaseous and particulate mercury at a unique mercury-contaminated remediation site. Journal of the Air & Waste Management Association. 2011;61:1115-1123
13. Jen YH, Yuan CS, Hung CH, Ie IR, Tsai CM. Tempospatial variation and partition of atmospheric mercury during wet and dry seasons at sensitivity sites within a heavily polluted industrial city. Aerosol and Air Quality Research. 2013;13:13-23
14. Poissant L, Pilote M, Beauvais C, Constant P, Zhang HH. A year of continuous measurements of three atmospheric mercury species (GEM, RGM, and HgP) in Southern Quebec, Canada. Atmospheric Environment. 2005;39:1275-1287
15. Feng XB, Tang SL, Shang LH, Yan HY, Sommar J, Lindqvist O. Total gaseous mercury in the atmosphere of Guiyang, PR China. Science of the Total Environment. 2003;304:61-72
16. Fu X, Feng X, Zhu W, Zheng W, Wang S. Total particulate and reactive gaseous mercury in ambient air in the eastern slope of Mt. Gongga area. Applied Geochemistry. 2008;23(3):408-418
17. Sheu GR, Mason RP. An examination of methods for the measurements of reactive gaseous mercury in the atmosphere. Environmental Science & Technology. 2001;35(6):1209-1216

18. Sheu GR, Lin NH, Wang JL, Lee CT. Monitoring of atmospheric mercury concentration at surface sites in Taiwan. Atmospheric Chemistry and Physics. 2009;30:1-10

19. Yuan CS, Lin HY, Wu CH, Liu MH. Preparing of sulfurized powdered activated carbon from waste tires using an innovative compositive impregnation process.Journal of the Air & Waste Management Association. 2004;54:862-870

20. Stamenkovic J, Lyman S, Gustin MS. Seasonal and diel variation of atmospheric mercury concentrations in the Reno (Nevada, USA) Airshed. Atmospheric Environment. 2007;41:6662-6672

21. Mao H, Talbot R. Speciated mercury at marine, coastal, and inland sites in New England—Part 1: Temporal variability. Atmospheric Chemistry and Physics Discussions. 2011;11:32301-32336

22. Nguyen HT, Kim MY, Kim KH. The influence of long-range transport on atmospheric mercury on Jeju Island, Korea. Science of the Total Environment. 2010;408(6):1295-1307

23. Blanchard P, Froude FA, Martin JB, Clark HD, Woods JT. Four years of continuous total gaseous mercury (TGM) measurements at sites in Ontario, Canada. Atmospheric Environment. 2002;36:3735-3743

24. Chand D, Jaffe D, Prestbo E, Swartzendruber PC, Hafner W, Penzias PW, Kato S, Takami A, Hatakeyama S, Kajii Y. Reactive and particulate mercury in the Asian marine boundary layer. Atmospheric Environment. 2008;42:7988-7996

25. Ci ZJ, Zang XS, Wang ZW, Niu ZC, Diao XY, Wang SW. Distribution and air-sea exchange of mercury (Hg) in the Yellow Sea. Atmospheric Chemistry and Physics Discussion. 2011;11:2881-2892

26. Sigler JM, Lee X, Munger W. Emission and long-range transport of gaseous mercury from a large-scale Canadian boreal forest fire. Environmental Science & Technology. 2003;37:4343-4347

27. Nguyen H, Kim KH, Kim MY, Hong S, Youn YH, Shon ZH, Lee J. Monitoring of atmospheric mercury at a global atmospheric watch (GAW) site on An-Myun Island, Korea. Water, Air, & Soil Pollution. 2007;185:149-164

10 Chapter

PHYTOREMEDIATION OF WATER CONTAMINATED WITH ARSENIC

1. INTRODUCTION: AN OVERVIEW

Over the past few years, the regions experiencing an increase in arsenic poisoning in both the soil and surface waters have been expanding fast across India and its neighboring countries [1, 2, 3]. Arsenic is primarily found in nature as metalloids or chemical compounds, which can lead to various pathological disorders and both skin and internal cancers [4]. Arsenic exhibits toxicity even at low levels of exposure [5] and is responsible for the development of black foot disease [6]. The topic is proving to be a significant challenge for environmental biologists and toxicologists to address. Cost-effective solutions are required to eradicate it from the contaminated water. Phytoremediation is an innovative, economically efficient, and environmentally benign bioremediation technique used for cleaning up the environment. Macrophyte-based bioremediation has proven to be an effective method for removing metal contaminants from different types of polluted effluents [7, 8]. The current assessment evaluated the abilities of three commonly found aquatic macrophytes (Azolla pinnata, Lemna minor, and Hydrilla verticillata) to remove arsenic from water contaminated with arsenic.

2. MATERIALS AND METHODS

The aquatic plants (*A. pinnata*, L. minor, and *H. verticillata*) used in the experiment were obtained from the Agrofarm pond located at Banaras Hindu University in Varanasi, India. To conduct an experiment, a single plant was grown in a controlled environment at the normal temperature of the laboratory, with natural lighting conditions. Before phytoremediation, the objects were carefully washed with tap water (containing dissolved O_2 at a concentration of 6.3 mg L^{-1}, pH of 7.2, water hardness of 23.2 mg L^{-1}, and room temperature of 28 ± 3°C). The tap water used had an arsenic concentration below the detectable limit. The purpose of this washing was to remove any debris present on the objects. Afterwards, they were gradually adapted to tap water for a duration of 2 weeks before being used for disinfection purposes. The test solution was prepared by dissolving 1.0 mg of arsenic in 1.0 L of tap water. The toxicity investigation of arsenic contamination was conducted using the fish bioassay approach, with a concentration of 5% of the LC50 value of Clarias batrachus [9]. 10 grams (in terms of fresh weight) of each macrophyte were put to a 10 L media. To ensure control, equal amounts of each plant were placed in separate aquariums with plain tap water. The growth of each macrophyte was also assessed. During each sample period, distinct experimental and control setups were created. The removal effectiveness of arsenic by aquatic macrophytes was determined by calculating the percentage using the formula provided in Table 1.

$$\% \text{ Removal efficiency} = \frac{C_1 - C_2}{C_1} \times 100 \quad (1)$$

Table 1. Arsenic removal efficiency of experimental macrophytes from media.

Macrophytes	Initial as concentration in media (ppm)	Residual as concentration in media (ppm)	Efficiency (%)
A. pinnata	1.0	0.119 (11.9%)	88.06
L. minor	1.0	0.174(17.4%)	82.56
H. verticillata	1.0	0.224(22.4%)	77.53
Mixture of three macrophytes (*ALH*)	1.0	0.145(14.5%)	85.50
( )Denotes percentage of arsenic residue in media after 10 days of treatment.			

where C1 is the initial concentration of arsenic in the media, and C2 is the final concentration of arsenic in the media.

The bioaccumulation coefficient is a measure of the concentration of arsenic in a plant compared to the concentration of residual arsenic in the surrounding medium where the plants are growing [10]. The bioaccumulation coefficient was determined using the following formula:

$$\text{Bioaccumulation coefficient} = \frac{\text{Arsenic concentration in plant}}{\text{Arsenic concentration in media}} \tag{2}$$

The average bioaccumulation coefficients for the aquatic plants tested here are illustrated in Fig. 1.

The flame photometer was used to analyze magnesium ions. To estimate chlorophyll levels, the total chlorophyll was extracted using Arnon's technique [11] in 80% cooled acetone.

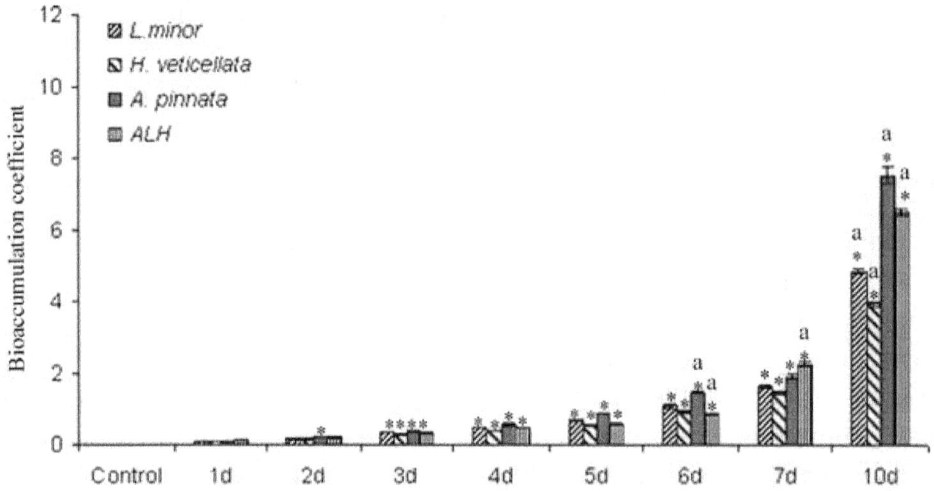

Fig. 1. Bioaccumulation coefficient of arsenic at different exposure day. Data are shown as mean ± SEM. The criterion for significant differences set at p < 0.05. *Indicates significant differences when compared with 1 day exposed controls. aWhen exposed groups were compared with just previous exposed group.

The chlorophyll was then seen using a spectrophotometer. The protein content was determined using the Lowry et al. technique [12]. The water was subjected to bioremediation using macrophytes for a duration of 10 days. A total of 1.0 gram of each plant sample was taken from experimental setups at various time intervals (0, 1st, 2nd, 3rd, 4th, 5th, 6th, 7th, and 10th days). To conduct the arsenic analysis, the plants that were collected were subjected to a drying process at a temperature of 80°C. The desiccated botanical specimens were subjected to acid digestion using a mixture of HNO_3:$HClO_4$ (5:1 v/v) on a heated surface until a transparent solution was achieved. 5 mL of double distilled water was added to get the desired volume. The arsenic buildup was analyzed using a Perkin-Elmer 2380 graphite furnace atomic absorption spectrophotometer. The results pertaining to each plant setting were presented as means accompanied by the standard error of the mean. The treatment effects were assessed using analyses of variance (ANOVA) with the general linear model approach of the standard statistical analysis system. Post hoc comparisons to distinguish treatment differences were conducted using Dunnett's t-test at a significance level of 5%.

3. RESULTS AND DISCUSSION

After being exposed to the arsenic solution, the macrophytes did not show any noticeable changes in their physical structure for a period of up to 4 days. Subsequently, a noticeable decline in the visual condition of all three plants was seen in the four experimental arrangements. The plant's health significantly worsened after a period of 10 days. Nevertheless, the condition of A. pinnata declined rapidly and became evident after a period of 7 days. The types of decay in these plants were essentially the same, whether they were treated individually or together. The concentration of magnesium ions (Mg^{+2}) in these plants remained very stable in all three plants, except after 10 days when there was a statistically significant change. The chlorophyll content of all three plants exhibited a significant decrease after the application of phytoremediation, whether applied individually or in combination with the three macrophytes (Table 2). Decreased levels of chlorophyll have been noticed in different plants that have undergone phytoremediation [13]. Nevertheless, it has been verified that the uptake of heavy metals causes phytotoxic consequences on plants, leading to the suppression of chlorophyll production and biomass generation, often resulting in mortality [14, 15]. Another possible cause for the reduction in chlorophyll content could be the hindered absorption of necessary materials, poor functioning of photosynthetic components, or an elevated chlorophyll activity leading to chlorophyll regeneration [16]. The protein concentration of these plants exhibited a gradual decrease, which reached statistical significance after 10 days in all experimental conditions. Additionally, other noxious metals have been found to decrease the levels of protein in plants [17].

Table 2. Variation in different biomolecule (chlorophylls, magnesium ions, and proteins) contents (μg g^{-1} fresh wt.) of all the three macrophytes after different periods of exposure to 1.0 ppm of sodium arsenite solution.

Exposure period	Biomolecules	*L. minor*	*H. verticillata*	*A. pinnata*	*ALH*
Control (wild)	Chlorophyll	2.767 ± 0.203^a	2.533 ± 0.116^a	3.267 ± 0.392^a	2.946 ± 0.008^a
	Mg^{2+}	0.438 ± 0.0014^b	0.556 ± 0.001^b	0.519 ± 0.001^b	0.607 ± 0.004^b
	Proteins	75.033 ± 1.519^c	94.833 ± 1.476^c	82.867 ± 0.284^c	84.833 ± 1.161^c
1st day	Chlorophyll	2.633 ± 0.088^a	2.173 ± 0.039^a	2.75 ± 0.252^a	2.836 ± 0.024^a
Mg^{2+}		0.434 ± 0.001^b	0.514 ± 0.001^b	0.516 ± 0.001^b	0.584 ± 0.004^b
	Proteins	70.306 ± 0.454^c	90.153 ± 0.608^c	80.42 ± 1.242^c	80.886 ± 0.315^c
2nd day	Chlorophyll	2.093 ± 0.064^a	1.926 ± 0.039^d	2.65 ± 0.078^a	2.473 ± 0.041^a
	Mg^{2+}	0.416 ± 0.001^b	0.511 ± 0.001^b	0.507 ± 0.001^b	0.558 ± 0.007^b
	Proteins	66.383 ± 1.732^d	82.013 ± 0.325^e	72.78 ± 0.015^d	72.667 ± 0.576^d
3rd day	Chlorophyll	1.997 ± 0.097^e	1.786 ± 0.032^d	1.84 ± 0.041^e	1.867 ± 0.029^e
	Mg^{2+}	0.407 ± 0.001^b	0.492 ± 0.001^b	0.504 ± 0.001^b	0.542 ± 0.004^b
	Proteins	62.356 ± 0.305^f	76.82 ± 2.061^f	66.956 ± 2.14^f	66.903 ± 0.214^f

Exposure period	Biomolecules	*L. minor*	*H. verticillata*	*A. pinnata*	ALH
4th day	Chlorophyll	1.873 ± 0.0218^{e}	1.416 ± 0.088^{d}	1.736 ± 0.044^{e}	1.773 ± 0.012^{e}
	Mg^{2+}	0.402 ± 0.001^{b}	0.485 ± 0.001^{b}	0.502 ± 0.001^{b}	0.5073 ± 0.004^{b}
	Proteins	56.946 ± 2.153^{g}	70.053 ± 0.913^{g}	62.703 ± 2.123^{f}	64.623 ± 0.446^{f}
5th day	Chlorophyll	1.713 ± 0.027^{e}	1.166 ± 0.044^{d}	1.573 ± 0.062^{e}	1.156 ± 0.024^{e}
	Mg^{2+}	0.396 ± 0.001^{b}	0.480 ± 0.001^{b}	0.496 ± 0.001^{b}	0.477 ± 0.001^{b}
	Proteins	54.573 ± 1.097^{g}	64.403 ± 1.188^{h}	57.906 ± 1.166^{f}	58.453 ± 1.622^{g}
6th day	Chlorophyll	1.503 ± 0.933^{e}	0.926 ± 0.039^{d}	1.383 ± 0.084^{e}	1.306 ± 0.022^{e}
	Mg^{2+}	0.378 ± 0.002^{b}	0.473 ± 0.001^{b}	0.492 ± 0.001^{b}	0.455 ± 0.004^{b}
	Proteins	48.383 ± 0.661^{h}	56.687 ± 1.567^{i}	44.38 ± 1.603^{g}	50.686 ± 2.391^{h}
7th day	Chlorophyll	1.283 ± 0.116^{e}	0.796 ± 0.029^{d}	1.133 ± 0.060^{e}	1.836 ± 0.059^{e}
	Mg^{2+}	0.368 ± 0.002^{b}	0.465 ± 0.002^{b}	0.489 ± 0.001^{b}	0.438 ± 0.004^{b}
	Proteins	42.233 ± 0.913^{i}	48.226 ± 0.597^{j}	34.86 ± 0.311^{h}	42.576 ± 0.864^{i}
10th day	Chlorophyll	0.673 ± 0.092^{j}	0.503 ± 0.029^{k}	0.693 ± 0.088^{i}	0.653 ± 0.057^{j}
	Mg^{2+}	0.309 ± 0.023^{k}	0.409 ± 0.015^{b}	0.471 ± 0.001^{b}	0.382 ± 0.004^{k}
	Proteins	35.593 ± 1.161^{l}	39.156 ± 2.158^{l}	28.343 ± 1.472^{j}	35.806 ± 1.675^{l}

[a–l]Values followed by different consequent alphabetic (a–l) superscript lowercase letters within the same column are significantly different at $p < 0.05$ level according to Dunnett's *t*-test, and the same letter within the column shows insignificant difference between the mean when compared to preceding value.

During phytoremediation, the concentration of arsenic in the media decreased progressively, and the residual arsenic concentrations were 11.93, 17.49, and 22.46 % in A. pinnata, L. minor, and H. verticillata, respectively. The arsenic concentration remained 14.5% in the medium when it was phytoremediated jointly by all the three macrophytes (Table 1).

Biomass of all the three macrophytes (A. pinnata, L. minor, and H. verticillata) increased with exposure period in the order of A. pinnata (11.55 ± 0.551) > H. verticillata (11.25 ± 0.635) > L. minor (11.00 ± 0.550) after 7 days (Table 3).

Table 3. The changes of macrophytes fresh biomass during the experiment.

Macrophytes	Fresh wt. (g) taken at initial day of experiment	Changes of biomass (g) at the end of 7 days of experiment	Changes of biomass (g) at the end of 10 days of experiment
A. pinnata	10.00 ± 0.00^{a}	11.55 ± 0.551^{a} (+15.5%)	8.95 ± 0.550^{b} (−10.5%)
L. minor	10.00 ± 0.00^{a}	11.00 ± 0.550^{a} (+10%)	9.50 ± 0.635^{b} (−5%)
H. verticillata	10.00 ± 0.00^{a}	11.25 ± 0.635^{a} (+12.5%)	9.65 ± 0.650^{b} (−3.5%)

[a, b]Values refer to the mean followed by standard error of mean. Means followed by the same letter in a column were not significantly different at $p < 0.05$.

( )Denotes percentage change of biomass after 7 and 10 days.

This growth may be attributed, at least in part, to the gradual buildup of arsenic by these plants. The rise became negligible after 7 days and continued exposure. All of these plants consistently showed measurable levels of arsenic, as indicated in Table 4. Nevertheless, over a period of 14 days, they underwent significant degradation. Fig. 1 displays the bioaccumulation coefficients of the aquatic plants examined in this investigation. The results present the bioaccumulation coefficient and demonstrate the variation in arsenic accumulation among different macrophyte species. The decrease in arsenic levels in the media was caused by the bioaccumulation of this metalloid by the macrophytes. This is evident from the presence of arsenic in the plant tissues. A. pinnata accumulated 0.88 mg, L. minor 0.82 mg, H. verticillata 0.77.5 mg, and 0.85 mg per kg dry wt. of the biomass in the combination of all three macrophytes (Table 4). The content of arsenic in all three untreated (control) plants was below the detectable level. The total amount of arsenic found in the medium after phytoremediation, as well as the amount absorbed by the plant tissues, was very similar to the initial concentration of 1.0 mg L^{-1} that was generated for the test solution (Fig.s 2 and 3).

Bharti and Banerjee [13] noted a noticeable disparity between the total amounts of metals remaining in the phytoremediated coal mine wastewater and the metals deposited in the plant tissues during phytoremediation. This phenomenon occurred as a result of the sedimentation process, the adsorption of the substances to the clay particles and organic matter, the co-precipitation with secondary minerals, the exchange of cations and anions, and the formation of complexes [7, 18]. In this instance, the disparity went unnoticed due to the fact that, unlike the effluent from coalmines, the composition and concentration of pollutants in the arsenic solution were uncomplicated. Fig. 2 survey unequivocally demonstrates that the absorption of arsenic in all macrophytes is directly influenced by the duration of treatment, up to a maximum of 10 days. Fig. 3 depicts the correlation between the absorption of arsenic by plants and its removal from the polluted medium. All three macrophytes are effective in the remediation of arsenic contamination. According to this study, A. pinnata is the most effective plant and can be used alone for this purpose. Continuing phytoremediation with the same plant for more than 10 days will not yield any further advantages.

Table 4. Arsenic accumulation in different macrophytes at different periods of exposure.

Period of exposure	Macrophytes			
	L. minor	*H. verticillata*	*A. pinnata*	*ALH*
Control (wild)	ɸ	ɸ	ɸ	ɸ
1st day	0.071 ± 0.004^{a}	0.077 ± 0.004^{a}	0.084 ± 0.005^{a}	0.105 ± 0.001^{a}
2nd day	$0.134 \pm 0.003^{a,b}$	0.133 ± 0.008^{b}	0.176 ± 0.002^{a}	0.169 ± 0.009^{a}
3rd day	0.267 ± 0.012^{b}	0.217 ± 0.002^{c}	0.274 ± 0.002^{b}	0.253 ± 0.020^{b}
4th day	$0.346 \pm 0.008^{b,c}$	0.290 ± 0.003^{d}	0.353 ± 0.022^{b}	0.323 ± 0.001^{c}
5th day	$0.443 \pm 0.012^{c,d}$	0.353 ± 0.002^{d}	$0.466 \pm 0.001^{b,c}$	0.367 ± 0.001^{d}
6th day	0.553 ± 0.008^{e}	0.476 ± 0.005^{e}	$0.593 \pm 0.001^{c,d}$	0.462 ± 0.007^{e}
7th day	0.657 ± 0.024^{f}	0.586 ± 0.001^{f}	$0.654 \pm 0.002^{d,e}$	0.687 ± 0.003^{f}
10th day	0.823 ± 0.020^{g}	0.775 ± 0.005^{g}	0.880 ± 0.003^{g}	0.853 ± 0.006^{g}

[a-g]Values followed by different consequent alphabetic (a–g) superscript lower case letters within the same column are significantly different at $p < 0.05$ level to Dunnett's t-test, and the same letter in a column was not significantly different between the mean when compared to just preceding value at $p < 0.05$. ALH: Mixture of three macrophytes (*L. minor, H. verticillata, and A. pinnata*) in equal (1:1:1) ratio.[ɸ]Below detection limit.

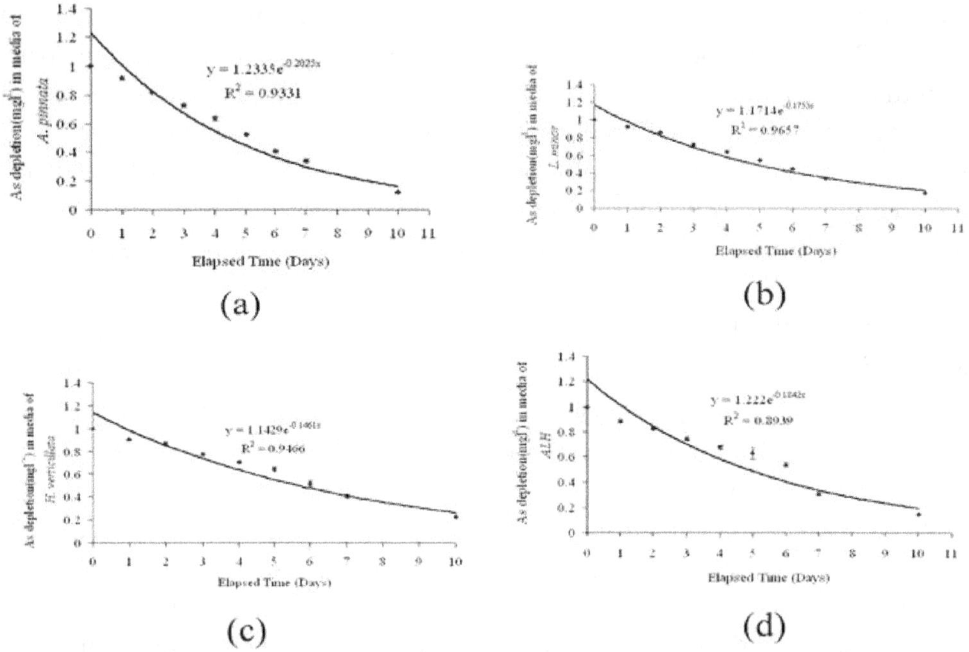

Fig. 2. (a-d) Arsenic depletion curve shown by all three plants (*Azolla pinnata, Lemna minor and Hydrilla verticillata*) separately as well as in their combination (1:1:1) in 10 L of arsenic media (As=1000 µg L^{-1}) for 10 days.

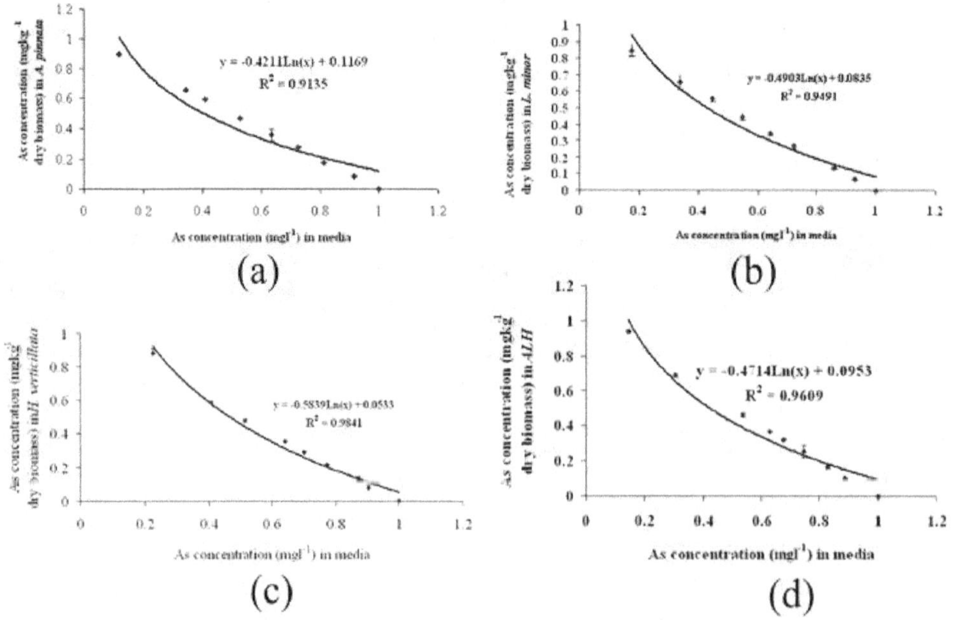

Fig. 3. (a–d) Regression curves between arsenic concentration in plant tissues and media of macrophytes (media = grown in arsenic media) at different exposure day.

REFERENCES

1. Chowdhury TR, Basu GK, Mandal BK, Biswas BK, Samanta G, Chowdhury UK, Chanda CR,Lodh D, Roy SL, Saha KC, Roy S, Kabir S, Quamruzzaman Q, Chakraborti D. Arsenic poisoning in the Ganges delta. Nature. 1999;401:545-546

2. Rahman MM, Chowdhury UK, Mukherjee SC, Mondal BK, Paul K Lodh D, Biswas BK, Chanda CR, Basu GK, Saha KC, Roy S, Das R, Palit SK, Quamruzzaman Q, Chakraborti D.Chronic arsenic toxicity

in Bangladesh and West Bengal, India – A review and commentary. Journal of Toxicology. Clinical Toxicology. 2001;39:683-700

3. Chakraborti D, Hussain A, Allauddin M. Arsenic: Environmental and health aspects with special reference to ground water in South Asia. Journal of Environmental Science and Health, Part A: Toxic Hazard Substance Environmental Engineering. 2003;38:11-15

4. Matsui M, Nishigori C, Toyokuni S, Takada J, Akaboshi M, Ishikawa M. The role of oxidative DNA damage in human arsenic carcinogenesis: Detection of 8-hydroxy-2¢-deoxyguanosine in arsenic-related Bowen's disease. The Journal of Investigative Dermatology. 1999;113:26-31

5. Dikshit AK, Pallamreddy K, Reddy LVP, Saha JC. Arsenic in ground water and its sorption by kimberlite tailings. Journal of Environmental Science and Health, Part A Environmental Science. 2000;35:65-85

6. Lin T-H, Huang Y-L, Wang M-Y. Arsenic species in drinking water, hair, fingernails, and urine of patients with blackfoot disease. Journal of Toxicology and Environmental Health, Part A: Current Issues. 1998;53:85-93

7. Mishra VK, Upadhyay AR, Pathak V, Tripathi BD. Phytoremediation of mercury and arsenic from tropical opencast coalmine effluent through naturally occurring aquatic macrophytes. Water, Air, and Soil Pollution. 2008;192:303-314

8. Rahman MJ, Hasegawa H. Aquatic arsenic: Phytoremediation using floating macrophytes. Chemosphere. 2011;83:633-646

9. Kumar R, Banerjee TK. Arsenic bioaccumulation in the nutritionallyimportant catfish Clarias batrachus exposed to the trivalent arsenic salt,sodium arsenite. Bulletin of Environmental Contamination and Toxicology. 2012;89:445-449

10. Robinson BH, Mills TM, Petit D, Fung LE, Green SR, Clothier BE. Natural and induced cadmium-accumulation in poplar and willow: Implications for phytoremediation. Plant and Soil. 2000;227:301-306

11. Arnon DE. Copper enzyme in isolated chloroplast, polyphenol oxidase in Beta vulgaris. Journal of Plant Physiology. 1949;24:1-15

12. Lowry OH, Rosebrough NJ, Farr AL, Randall RJ. Protein measurement with the Folin phenol reagent. The Journal of Biological Chemistry. 1951;193:265-275

13. Bharti S, Banerjee TK. Phytoremediation of the coalmine effluent. Ecotoxicology and Environmental Safety. 2012;81:36-42

14. Satyakala G, Jamil K. Chromium induced biochemical changes in Eichhornia crassipes (Mart) solms and Pistia stratiotes. Bulletin of Environmental Contamination and Toxicology. 1992;48:921-928

15. Delgado M, Bigeriego M, Guardiola E. Uptake of Zn, Cr and Cd by water hyacinths. Water Research. 1993;27:269-272

16. Sharma P, Dubey RS. Lead toxicity in plants. Brazilian Journal of Plant Physiology. 2005;17:35-52

17. Arvind P, Prasad MNV. Cadmium-zinc interactions in a hydroponic system using Ceratophyllum demersum L.: Adaptive ecophysiology, biochemistry and molecular toxicology. Brazilian Journal of Plant Physiology. 2005;17:3-20

18. Khellaf N, Zerdaoui NM. Growth response of the duckweed Lemna minor to heavy metals pollution. Iran. Journal of Environmental Health Science and Engineering. 2009;6:161-166

11 Chapter

THE STUDY FOCUSES ON THE PROCESS OF PHYTOREMEDIATION AND THE PHYSIOLOGICAL EFFECTS OF MIXED HEAVY METALS ON POPLAR HYBRIDS

1. INTRODUCTION

1.1. Phytoremediation

Phytoremediation has gained considerable interest and support in the last decade. This environment-friendly green technology has gained its popularity over the years in terms of its success with other conventional techniques. The specific definitions of phytoremediation are various however the basic definition involves growing of plants in a contaminated matrix with the intentions of removing, transforming or stabilizing environmental contaminants (Fig. 1) [1, 2, 3, 4]. The generic term phytoremediation originates from the Greek prefix φυτο "phyto" – plant, attached to the Latin root "remedium" – to correct or remove, restoring balance, or remediating [5, 6].

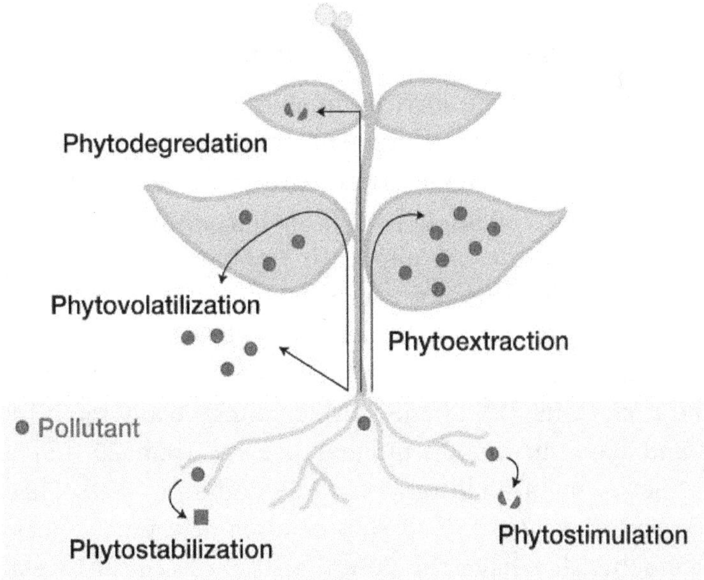

Fig. 1. Possible pollutant fates during phytoremediation: the pollutant (represented by red circles) can be stabilized (phytostabilization) or degraded (phytostimulation) in the rhizosphere, sequestered or degraded (phytoextraction, phytodegradation) inside the plant tissue, or volatilized (phytovolatilization) in the air [4].

There are multiple methods, including physical, chemical, and biological strategies, that can be used to treat soils that have been contaminated with metals. Ex situ methods, also known as classical environmental cleanup methods, are typically costly and damaging. These methods include excavation, thermal treatment, chemical soil washing, soil incineration, volatilization, vitrification, chemical extraction, solidification, and landfills [5, 7]. Their purpose is to chemically or physically detoxify or destroy contaminants. Consequently, the pollutant is subjected to stabilization, solidification, immobilization, incineration, or annihilation [8]. These procedures are both labor-intensive and costly, and they also generate a residue that contains high levels of heavy metals, necessitating additional treatment. In addition, the physiochemical technologies employed for soil remediation cause permanent modifications to soil properties, hence affecting its suitability for plant growth. These technologies eliminate all biological activity and disrupt the natural soil microflora [2, 5]. Therefore, phytoremediation is acknowledged as a cost-efficient and environmentally benign approach to removing heavy metals from polluted areas. Out of the five processes in phytoremediation, phytoextraction has been recognized as the most effective method in which plants remove heavy metals from polluted soil [9]. Phytoextraction, often referred to as phytoaccumulation, phytoabsorption, or phytosequestration, is the process by which plant roots absorb pollutants from the soil and transport and store them in the above-ground parts of the plant [8, 10, 11]. Furthermore, plants that are suitable for phytoremediation demonstrate characteristics such as quick growth with abundant biomass production, an extensive root system, strong survival skills, and the ability to adapt to low-quality soil substrates. Additionally, these plants have a high tolerance for excessive concentrations of heavy metals. Typically, several types of poplar, jatropha, and willow species and hybrids are being utilized for both phytoremediation and energy generation purposes [4, 12, 13, 14]. In general, plants have the ability to uptake metals from the soil, but only a handful can effectively extract, accumulate, and tolerate large concentrations of heavy metals in their tissues. The identification of hyperaccumulator plants, which have the ability to absorb heavy metals at a rate 50-500 times higher than regular plants, has significantly contributed to the groundbreaking progress of phytoextraction technology [15].

1.2. Contamination caused by high levels of heavy metals

The primary cause of pollution in the environment is mostly attributed to the process of global industrialization. Heavy metals are introduced into the environment through both natural and human-caused causes. Notable natural sources of pollution include the breakdown of minerals, erosion, and volcanic activity. On the other hand, human activities such as mining, smelting, electroplating, the use of pesticides and fertilizers in agriculture, the application of biosolids, the dumping of sludge, industrial waste discharge, and atmospheric deposition are anthropogenic sources of pollution [16, 17, 18, 19, 20]. Phytoremediation technology can be used to address a wide variety of pollutants, including metals, radionuclides [21, 22, 23], and chemical molecules such chlorinated solvents, polycyclic hydrocarbons, pesticides, explosives, and surfactants [3]. Chemically speaking, heavy metals are elements that possess metallic characteristics and have an atomic number greater than 20 and a specific gravity higher than 5. The predominant heavy metal pollutants include Cadmium (Cd), Chromium (Cr), Copper (Cu), Mercury (Hg), Lead (Pb), and Zinc (Zn) [2]. In this chapter, the term "heavy metals (HMs)" will be used to describe those elements that have the potential to be harmful to plants and are a significant environmental issue due to their long-lasting presence in the environment and their impact on humans through the food chain. Heavy metals have detrimental impacts on human health, hence necessitating particular focus on heavy metal contamination [5]. Heavy metals (HMs) such as Arsenic (As), Cadmium (Cd), Mercury (Hg), Lead (Pb), and Selenium (Se) are considered non-essential because they do not have any known physiological function in plants [24]. Co, Cu, Fe, Mn, Mo, Ni, and Zn are crucial elements necessary for the regular growth and metabolism of plants [24]. However, when these critical heavy metals are present in excessive amounts, they can cause poisoning [25]. High concentrations of essential metals are hazardous because they induce oxidative stress through the generation of free radicals and interfere with the function of pigments and enzymes by displacing essential metals [8, 26]. Associated with both the concentration of metals in plants and their dry weight.Populus is a genus of 25-35 species of deciduous angiosperms in the Salicaceae family, primarily found in the Northern Hemisphere. Commonly used English names for this tree species include poplar, aspen, and cottonwood. This genus exhibits a significant range of genetic variation and has the capacity to attain heights ranging from 15 to 50 meters (49–164 feet), with trunk diameters reaching up to 2.5 meters (8 feet 2 inches). The genus Populus is classified into six groups based on the characteristics of their leaves and flowers. Four distinct poplar clones were chosen based on prior research and their performance in terms of biomass in the field [27, 28, 29].

- The selected clones exhibited improved responses in conjunction with seasonal variations. The clones comprised the following:
- Eco 28 – Populus euramericana guinier
- DN 034 – Populus deltoides × P. nigra
- TN 074 – Populus trichocarpa × P. nigra
- TD 225 – Populus trichocarpa × P. deltoids

Initial phytoremediation research employed hyperaccumulator species [30, 31], which are plants capable of accumulating abnormally high concentrations of metals in their tissues. Research conducted on willows and poplars from the Salicaceae family demonstrated that their ability to extract metals from soil is significantly lower compared to hyperaccumulators. However, when considering large-scale operations, the overall removal of metals from the soil could still be higher. Zacchini et al. [28] state that Salicaceae plants are capable of flourishing in many soil and climatic circumstances [33]. Additionally, they exhibit a notable tolerance to metals, making them suitable candidates for phytoremediation efforts.

2. POPLAR HYBRIDS USED FOR THE PURPOSE OF PHYTOREMEDIATION

Four poplar hybrids were selected based on their genetic diversity, growth, and wellbeing in the field. Cuttings of approximately 15 cm from the hybrids were planted in 2 L pots filled with sandy loam soil. Two months after sprouting and root stabilization, the plantlets were treated with mixed heavy metals of concentrations ranging from 0 mg L^{-1} as control followed by 5, 50, 100, 200 and 500 mg L^{-1}.

The mixed heavy metals utilized included Chromium III chloride (Cr_3Cl), Copper (II) chloride ($CuCl_2$), Cadmium chloride ($CdCl_2$) and Zinc chloride ($ZnCl^2$). Individual HMs was separately prepared and equal amounts were mixed for each concentration. Each plantlet was treated with 20 ml of mixed HM only once during the 3 month treatment period while the plants were watered regularly. Photosynthesis and transpiration rates were measured using LCi-SD (ADC Bioscientific Ltd., Hoddesdon, UK) portable photosynthesis system before and during the treatment period along with the photosynthetic pigments.

After the 3 month treatment period, the plants were harvested for further heavy metal analysis. Plant leaves, stem, and roots were separated, thoroughly washed with running tap water and finally with distilled water after which morphological characteristics were noted and the plants were prepared for further biochemical and heavy metal analysis. Each concentration constituting of four replicates were analyzed for various parameters and an average was used for statistical analysis. Various plant parts including leaves, stems, and roots were dried at 50°C for 48 hours. These dried plant parts were ground using the 8000 mixer mill (SPEX. SamplePrep). The physical and chemical properties including the levels of heavy metals were determined using the aqua regia method. Soil heavy metal analysis involved, a collection of the soil samples which were air-dried and sieved and digested using the aqua regia method.

Soil samples 1.5 g were digested and the samples were analyzed for various HMs by ICP – AES analysis (iCAP 7000 Series ICP spectrometer, Thermo Fisher Scientific, Waltham, MA, USA).

The biological accumulation coefficient (BAC) was calculated as the ratio of heavy metal in the shoot to that in soil given in Eq. 1 [34]. Biotranslocation coefficient was determined as a ratio of heavy metals in plant shoot to that in plant root given in Eq. 2 [35, 36]. Eq. (3) shows the bioconcentration factor (BCF), calculated as metal concentration ratio in plant roots to soil; Eq. (4) shows the concentration index (CI), calculated as the concentration of heavy metals in the treated plant to control [35, 36].

These factors were used to determine the phytoextraction potential of the studied plants. Shoot in the equations refers to stem + leaves and HM for heavy metal.

$$BAC = HM_{shoot}/HM_{soil} \tag{1}$$

$$BTC = HM_{shoot}/HM_{root} \tag{2}$$

$$BCF = HM_{root}/HM_{soil} \tag{3}$$

$$CI = HM_{treated\ plant}/HM_{normal\ plant} \tag{4}$$

Data are expressed as mean ± standard error of mean (SEM and statistically analyzed by one-way analysis of variance (ANOVA) followed by Duncan's Multiple Range Tests using SPSS Version 21 (IBM Corp., USA). Statistical significance was accepted at $P < 0.05$.

2.1. The impact of heavy metals on the process of photosynthesis and transpiration

Exposure of poplar hybrids to mixed HMs under greenhouse conditions showed an increase in the rate of photosynthesis with increased HM concentrations (Table 1).

Hybrid 1 exhibited the most efficient photosynthetic rate, measuring 14.54 μmol m^{-2} s^{-1} when exposed to a concentration of 500 mg L^{-1} of combined heavy metals. This was recorded as the highest photosynthetic rate among all four hybrids. There were variations in the rates of photosynthesis in hybrid 2 across all doses. The lowest rate was 7.75 μmol m^{-2} s^{-1} at a concentration of 5 mg L^{-1}, while the greatest rate was 12.60 μmol m^{-2} s^{-1} at a concentration of 200 mg L^{-1} of HM. Hybrid 3 had the lowest photosynthetic rate of 2.61 μmol m^{-2} s^{-1} at a concentration of 200 mg L^{-1}, which increased by a factor of 5 to 10.08 at a concentration of 500 mg L^{-1}. However, Hybrid 4 exhibited a substantially greater photosynthetic rate of 8.20 μmol m^{-2} s^{-1} at a concentration of 5 mg L^{-1}, which thereafter reduced to 6.84 and 7.23 at concentrations of 50 and 100 mg L^{-1} respectively.

Table 1. Photosynthesis rates of four poplar hybrids treated with mixed heavy metals (Cd, Cr, Cu, and Zn) under greenhouse conditions.

Heavy metal concentrations (mg L^{-1})	Photosynthesis rate – A (μmol m^{-2} s^{-1})			
	Hybrid 1 (Eco 28)	Hybrid 2 (DN 034)	Hybrid 3 (TN 074)	Hybrid 4 (TD 225)
0	10.21 ± 0.20^{bc}	10.12 ± 0.26^{b}	9.05 ± 0.34^{ab}	9.05 ± 0.17^{a}
5	8.82 ± 1.24^{c}	7.75 ± 0.26^{c}	9.72 ± 0.27^{a}	8.20 ± 0.54^{ab}
50	11.32 ± 0.18^{b}	11.43 ± 0.52^{ab}	7.11 ± 0.42^{c}	6.84 ± 0.63^{b}
100	8.56 ± 0.23^{c}	10.21 ± 0.93^{b}	8.12 ± 0.46^{bc}	7.23 ± 0.50^{b}

Heavy metal concentrations (mg L^{-1})	Photosynthesis rate – A (μmol m^{-2} s^{-1}) Hybrid 1 (Eco 28)	Hybrid 2 (DN 034)	Hybrid 3 (TN 074)	Hybrid 4 (TD 225)
200	8.71 ± 0.21^{c}	12.60 ± 0.52^{a}	2.61 ± 0.50^{d}	8.88 ± 0.53^{a}
500	14.54 ± 0.70^{a}	11.73 ± 0.11^{a}	10.08 ± 0.20^{a}	7.70 ± 0.56^{ab}
Mean values ± SEM, (n = 10).				
Means within columns followed by the same letters are not significantly different at P < 0.05 according to Duncan's multiple range test.				
Values significantly different from control (0).				

An evident correlation can be noticed regarding the photosynthesis rate of all four poplar hybrids. The rate of photosynthesis shows an upward trend as the concentration of mixed heavy metals increases, with the most notable effects occurring at concentrations ranging from 200 to 500 mg L^{-1}. The observed rise in the photosynthetic rate under high HM concentrations may be attributed to the hybrids' capacity to withstand elevated levels of HMs. The basis for this can be found in previous research on higher plants and trees [37]. The effects of chromium stress on photosynthesis are attributed to carbon dioxide fixation, electron transport, photophosphorylation, and enzyme activities [38]. On the other hand, cadmium stress results in a deficiency of Fe(II), which has a significant impact on photosynthesis [39]. Therefore, if the rates of photosynthesis do not drop, it would indicate that the plant is either able to survive in a high concentration of heavy metals or that limited absorption and translocation of heavy metals occur due to competition with other heavy metals and environmental conditions. Transpiration rates varied among all hybrids, with the maximum rate of 11.23 mmol m^{-2} s^{-1} reported at a concentration of 50 mg L^{-1} for hybrid 1, followed by a rate of 8.89 mmol m^{-2} s^{-1} at a concentration of 500 mg L^{-1} (Table 2).

Table 2. Transpiration rates of four poplar hybrids treated with mixed heavy metals (Cd, Cr, Cu, and Zn) under greenhouse conditions.

Heavy metal concentrations (mg L^{-1})	Transpiration rate – E (mmol m^{-2} s^{-1}) Hybrid 1 (Eco 28)	Hybrid 2 (DN 034)	Hybrid 3 (TN 074)	Hybrid 4 (TD 225)
0	7.45 ± 0.24^{c}	10.17 ± 0.41^{a}	8.29 ± 0.39^{b}	5.99 ± 0.16^{c}
5	5.96 ± 0.70^{d}	8.64 ± 0.52^{bc}	9.96 ± 0.1^{b}	6.38 ± 0.15^{bc}
50	11.23 ± 0.05^{a}	9.61 ± 0.26^{ab}	10.87 ± 0.44^{a}	3.55 ± 0.53^{d}
100	7.84 ± 0.28^{c}	8.77 ± 0.78^{abc}	10.14 ± 0.72^{a}	6.24 ± 0.61^{c}
200	5.30 ± 0.11^{d}	9.99 ± 0.38^{ab}	3.42 ± 0.53^{c}	8.14 ± 0.34^{ab}
500	8.89 ± 0.13^{b}	7.64 ± 0.1^{c}	9.53 ± 0.28^{ab}	7.52 ± 0.47^{a}
Mean values ± SEM, (n = 10).				
Means within columns followed by the same letters are not significantly different at P < 0.05 according to Duncan's multiple range tests.				
Values significantly different from control (0).				

Hybrid 2 exhibited the maximum transpiration rate of 9.99 mmol m^{-2} s^{-1} at 200 mg L^{-1}, followed by 9.61 at 50 mg L^{-1}. Distinct variations in transpiration rates were seen for the control group at low mixed heavy metal concentrations of 5 and 100 mg L^{-1}. Hybrid 3 exhibited the greatest transpiration rates among all concentrations, with a rate of 10.87 mmol m^{-2} s^{-1} at 50 mg L^{-1}. This was followed by a rate of 10.14 mmol m^{-2} s^{-1} at 100 mg L^{-1}, and the lowest rate of 3.42 mmol m^{-2} s^{-1} at 200 mg L^{-1}. Hybrid 4 exhibited the lowest transpiration rates at a concentration of 50 mg L^{-1}, measuring 3.55 mmol m^{-2} s^{-1}. Conversely, it displayed the highest transpiration rates at a concentration of 200 mg L^{-1}, measuring 8.14 mmol m^{-2} s^{-1}. There were no unique connections found between transpiration rates and varying concentrations of mixed HM in hybrid 4. However, significant differences in transpiration rates were identified across all four hybrids of poplar. However, the variations in differences were specific to each hybrid; in general, a rise in mixed HM concentrations resulted in a decrease in transpiration rate. Gas exchange measurements are frequently employed by Carlson et al. to identify the most vulnerable site of action [40]. For the hybrids that were examined, although the photosynthetic rate rose at high HM concentrations, the decrease in transpiration rate indicates that the plants experienced stress in specific ways. An in-depth analysis would facilitate the determination of the precise regions impacted by the HMs. Heavy metal stress induces significant changes in various physiological and metabolic processes in plants. The evidence given in this chapter indicates that exposure to a combination of different types of heavy metals (HM) results in a notable reduction in the levels of photosynthetic pigments in poplar hybrids. Chlorophyll content is frequently evaluated to evaluate the effects of environmental stress, as alterations in pigment levels are associated with visible signs of plant disease and photosynthetic efficiency [41]. In this study, the photosynthetic rates of all poplar hybrids reduced at all doses of heavy metals, except for hybrid 1 (Eco 28). Other plants have shown a decrease in photosynthetic rates as a result of heavy metals (HMs) reducing the amount of photosynthetic pigments. Studies have shown that certain plants experience a reduction in chlorophyll levels when exposed to heavy metals such mercury (Hg), copper (Cu), chromium (Cr), cadmium (Cd), and zinc (Zn) [42]. The decrease in photosynthetic pigments is likely caused by the suppression of the reduction processes in the biosynthetic pathways due to the elevated redox potential of many heavy metals (HMs). Furthermore, it is widely recognized that protochlorophyllide reductase, the crucial enzyme responsible for converting protochlorophyll to chlorophyll, is susceptible to inhibition by heavy metals [43]. Several researchers have documented a comparable reduction in chlorophyll levels in cyanobacteria, single-celled chlorophytes (Chlorella), gymnosperms like Picea abies, and angiosperms such as Zea mays, Quercus palustrus, Acer rubrum, sunflower, and almond [44, 45, 46]. There is limited evidence indicating an increase in pigments following exposure to heavy metals [47].

Several studies have also been carried out to investigate the impacts of certain heavy metals on various plant species. The impact of Cd and Pb on Brassica juncea L. resulted in a decrease in growth, chlorophyll content, and carotenoids. However, Cd was discovered to have a greater harmful effect compared to Pb [48]. The investigation on the physiological effects of Cd and Cu on peas (Pisum sativum) found that photosynthetic pigments and photosynthesis rates decreased at all levels of Cd, and only at high doses of Cu [49]. The presence of many heavy metals in the soil-water medium, which are in competition with each other, may contribute to the conflicting portions of the data. The root absorption and leaf accumulation levels differ based on the hybrid, resulting in an anticipated decrease in hybrids 2, 3, and 4. While hybrid 1 exhibited the maximum photosynthetic rate, reductions were found for concentrations of 5, 50, and 100 mg L^{-1}. Nevertheless, a marginal rise, below 0 mg L^{-1}, was detected at a concentration of 500 mg L^{-1} of HM. Hybrid 1 The notable augmentation in photosynthetic and transpiration rates in hybrid 1 is further corroborated by the rise in photosynthetic pigments. The cumulative concentration of hazardous materials (HMs) in hybrid 1 plays a significant role in comprehending hybrid 1's response to elevated levels of mixed HMs. This emphasizes the need for a clearer comprehension of the presence of heavy metal ions in the soil solution, their interaction with plant roots, and ultimately their uptake into the system and transportation to above-ground plant components. Copper (Cu) and zinc (Zn) are crucial heavy metals that are present in several enzymes and proteins, playing a vital role in the proper growth and development of plants. Higher quantities of both necessary and non-essential heavy metals (HMs) can cause toxic symptoms and hinder growth in most plants. In general, a reduction in the efficiency of photosynthesis in plants can contribute to a decline in plant growth and the generation of biomass.

2.2. The phytoextraction potential of poplar hybrids

The tables 3-6 provide the values for the biological accumulation coefficient (BAC), biological translocation coefficient (BTC), bioconcentration factor (BCF), and metal accumulation or concentration index (CI).

Table 3. Biological accumulation coefficient (BAC) of Cu, Cd, Cr and Zn for the four poplar hybrids under greenhouse conditions.

Hybrid	HM conc. (mg L^{-1})	BAC			
		Cu	Cd	Cr	Zn
Eco 28	0	0.10 ± 0.10^{b}	0.23 ± 0.23^{c}	0.83 ± 0.27^{a}	2.38 ± 0.35^{ab}
	5	0.12 ± 0.10^{ab}	nd	0.78 ± 0.19^{a}	2.73 ± 0.16^{ab}
	50	0.17 ± 0.25^{a}	24.01 ± 3.00^{a}	0.18 ± 0.44^{a}	3.16 ± 0.11^{b}
	100	0.13 ± 0.01^{ab}	15.81 ± 5.28^{ab}	0.14 ± 0.44^{a}	2.90 ± 0.15^{ab}
	200	0.12 ± 0.01^{ab}	7.50 ± 1.22^{bc}	0.12 ± 0.36^{a}	2.78 ± 0.26^{ab}
	500	0.11 ± 0.62^{b}	8.13 ± 3.73^{bc}	0.98 ± 0.43^{a}	2.12 ± 0.57^{b}
DN 034	0	0.15 ± 0.10^{a}	nd	0.22 ± 0.12^{a}	1.08 ± 0.19^{a}
	5	0.12 ± 0.01^{ab}	nd	0.09 ± 0.11^{a}	3.51 ± 0.22^{a}
	50	0.88 ± 0.19^{b}	27.72 ± 10.23^{a}	0.85 ± 0.27^{a}	3.10 ± 0.95^{a}
	100	0.96 ± 0.20^{b}	8.59 ± 2.08^{b}	0.10 ± 0.28^{a}	2.40 ± 0.71^{a}
	200	0.12 ± 0.01^{ab}	14.55 ± 6.12^{ab}	0.14 ± 0.38^{a}	3.11 ± 0.11^{a}
	500	0.12 ± 0.01^{ab}	6.52 ± 1.18^{b}	0.09 ± 0.20^{a}	3.18 ± 0.24^{a}
TN 074	0	0.15 ± 0.01^{a}	nd	0.58 ± 0.01^{a}	4.20 ± 0.35^{a}
	5	0.14 ± 0.13^{a}	nd	0.11 ± 0.06^{a}	4.81 ± 0.28^{a}
	50	0.13 ± 0.16^{a}	58.26 ± 26.00^{a}	0.13 ± 0.56^{a}	3.81 ± 0.13^{a}
	100	0.15 ± 0.17^{a}	17.30 ± 2.60^{b}	0.53 ± 0.01^{a}	4.34 ± 0.25^{a}
	200	0.14 ± 0.64^{a}	6.95 ± 1.16^{b}	0.23 ± 0.15^{a}	3.46 ± 0.87^{a}
	500	0.16 ± 0.02^{a}	11.08 ± 4.41^{b}	0.68 ± 0.21^{a}	4.89 ± 0.45^{a}
TD 225	0	0.09 ± 0.02^{ab}	nd	0.13 ± 0.04^{a}	2.25 ± 0.60^{a}
	5	0.12 ± 0.02^{ab}	nd	0.10 ± 0.02^{a}	2.42 ± 0.50^{a}
	50	0.15 ± 0.18^{a}	15.10 ± 2.49 ^{a}	0.19 ± 0.09^{a}	2.83 ± 0.27^{a}
	100	0.98 ± 0.15^{ab}	10.06 ± 1.46^{ab}	0.06 ± 0.01^{a}	1.93 ± 0.42^{a}
	200	0.07 ± 0.02^{b}	8.90 ± 4.28^{ab}	0.12 ± 0.05^{a}	1.48 ± 0.54^{a}
	500	0.93 ± 0.02^{ab}	6.81 ± 2.96^{bc}	0.15 ± 0.04^{a}	1.71 ± 0.49^{a}

Values are Mean ± SE, (n = 4). nd – not detected.

For each metal in each treatment values followed by the same letters are not significantly different at $P \leq 0.05$ according to Duncan's multiple range test.

Table 4. Biological translocation coefficient (BTC) of Cu, Cd, Cr and Zn for the four poplar hybrids under greenhouse conditions.

Hybrid	HM conc. ($mg\ L^{-1}$)	BTC			
		Cu	Cd	Cr	Zn
Eco 28	0	0.35 ± 0.58[ab]	3.40 ± 0.65[a]	0.32 ± 0.72[b]	2.44 ± 0.54[a]
	5	0.28 ± 0.38[b]	1.96 ± 0.44[b]	0.79 ± 0.24[b]	2.54 ± 0.45[a]
	50	0.49 ± 0.10[a]	1.25 ± 0.06[bc]	1.94 ± 0.73[a]	2.81 ± 0.31[a]
	100	0.31 ± 0.56[b]	1.30 ± 0.74[bc]	1.19 ± 0.49[ab]	2.42 ± 0.15[a]
	200	0.26 ± 0.16[b]	1.01 ± 0.26[bc]	0.64 ± 1.44[b]	2.76 ± 0.45[a]
	500	0.90 ± 0.40 [c]	0.43 ± 0.19[c]	0.28 ± 1.67[b]	1.83 ± 1.01[a]
DN 034	0	0.60 ± 0.32[ab]	1.75 ± 0.60[ab]	3.65 ± 1.79[a]	5.25 ± 0.18[a]
	5	0.49 ± 0.40[bc]	2.61 ± 0.42[a]	0.83 ± 0.27[b]	5.20 ± 0.75[a]
	50	0.35 ± 0.10[c]	1.53 ± 0.70[ab]	1.07 ± 0.56[b]	4.28 ± 1.73[a]
	100	0.33 ± 0.69[c]	1.08 ± 0.26[b]	0.78 ± 0.33[b]	2.70 ± 0.70[a]
	200	0.35 ± 0.50[c]	1.08 ± 0.25[b]	1.12 ± 0.33[b]	3.93 ± 0.83[a]
	500	0.30 ± 0.31[c]	0.74 ± 0.70[b]	0.69 ± 0.16[b]	3.82 ± 0.22[a]
TN 074	0	0.70 ± 0.85[a]	4.82 ± 0.60[a]	0.47 ± 0.18[a]	5.97 ± 1.02[c]
	5	0.57 ± 0.92[ab]	2.86 ± 0.56[b]	1.51 ± 2.64[a]	4.47 ± 0.17[c]
	50	0.33 ± 0.23[bc]	1.43 ± 0.65[c]	0.44 ± 0.30[a]	12.80 ± 10.46[c]
	100	0.50 ± 0.96[ab]	1.18 ± 0.30[c]	0.50 ± 0.46[a]	115.97 ± 10.19[a]
	200	0.35 ± 0.15[bc]	0.60 ± 0.32[c]	1.17 ± 0.83[a]	47.53 ± 9.36[b]
	500	0.13 ± 0.16[c]	1.80 ± 0.32[c]	0.06 ± 0.04[a]	11.75 ± 10.62[c]
TD 225	0	0.31 ± 0.0 4[a]	2.53 ± 0.78[a]	1.73 ± 0.85[a]	3.23 ± 1.32[a]
	5	0.29 ± 0.52[a]	2.60 ± 0.93[a]	1.12 ± 0.33[a]	2.26 ± 0.43[a]
	50	0.36 ± 0.58[a]	2.55 ± 0.79[a]	1.32 ± 0.77[a]	2.83 ± 0.38[a]
	100	0.23 ± 0.03[ab]	0.88 ± 0.24[ab]	0.55 ± 0.18[a]	2.87 ± 0.85[a]
	200	0.11 ± 0.24[bc]	0.38 ± 0.07[b]	0.57 ± 0.27[a]	1.23 ± 0.38[a]
	500	0.07 ± 0.03[c]	0.47 ± 0.20[b]	0.59 ± 0.27[a]	0.99 ± 0.37[a]

Values are Mean ± SE, (n = 4). nd – not detected.

For each metal in each treatment values followed by the same letters are not significantly different at $P < 0.05$ according to Duncan's multiple range test.

Table 5. Bio-coefficient factor (BCF) of Cu, Cd, Cr and Zn for the four poplar hybrids under greenhouse conditions.

Hybrid	HM conc. (mg L^{-1})	BCF			
		Cu	Cd	Cr	Zn
Eco 28	0	0.32 ± 0.66[a]	0.58 ± 0.56[b]	0.25 ± 0.58[a]	1.18 ± 0.37[a]
	5	0.45 ± 0.67[a]	nd	0.10 ± 0.01[b]	1.16 ± 0.57[a]
	50	0.30 ± 0.10[a]	13.44 ± 4.80[a]	0.08 ± 0.28[b]	1.12 ± 0.90[a]
	100	0.46 ± 0.78[a]	12.45 ± 1.90[a]	0.13 ± 0.13[b]	1.21 ± 0.89[a]
	200	0.47 ± 0.15[a]	8.27 ± 1.43[a]	0.18 ± 0.15[ab]	1.06 ± 0.98[a]
	500	0.58 ± 0.19[a]	6.99 ± 3.00[ab]	0.17 ± 0.06[ab]	0.62 ± 0.21[a]
DN 034	0	0.24 ± 0.22[c]	nd	0.58 ± 0.01[b]	0.78 ± 0.48[a]
	5	0.25 ± 0.01[c]	nd	0.15 ± 0.49[ab]	0.71 ± 0.84[a]
	50	0.28 ± 0.12[bc]	34.18 ± 15.58[a]	0.12 ± 0.31[ab]	0.86 ± 0.11[a]
	100	0.28 ± 0.12[bc]	8.06 ± 0.81[b]	0.20 ± 0.70[a]	0.83 ± 0.15[a]
	200	0.35 ± 0.32[ab]	12.18 ± 3.74[b]	0.15 ± 0.33[ab]	0.93 ± 0.23[a]
	500	0.42 ± 0.24[a]	8.62 ± 1.03[b]	0.14 ± 0.01[ab]	0.83 ± 0.06[a]
TN 074	0	0.22 ± 0.05[a]	nd	0.33 ± 0.42[a]	0.75 ± 0.11[ab]
	5	0.25 ± 0.06[a]	nd	0.47 ± 0.25[a]	1.08 ± 0.80[a]
	50	0.23 ± 0.16[a]	12.68 ± 8.20[ab]	0.20 ± 0.98[a]	0.57 ± 0.31[c]
	100	0.32 ± 0.53[a]	14.69 ± 2.08 [a]	0.11 ± 0.15[a]	0.04 ± 0.04[b]
	200	0.38 ± 0.74[a]	8.68 ± 2.26[ab]	0.20 ± 0.05[a]	0.06 ± 0.01[a]
	500	0.32 ± 0.38[a]	11.08 ± 8.44[ab]	0.17 ± 0.10[a]	0.38 ± 0.34[bc]
TD 225	0	0.28 ± 0.04[c]	nd	0.10 ± 0.02[b]	0.81 ± 0.14[a]
	5	0.40 ± 0.01[bc]	nd	0.11 ± 0.01[b]	1.05 ± 0.16[a]
	50	0.42 ± 0.02[bc]	8.48 ± 3.05[bc]	0.19 ± 0.03[ab]	1.03 ± 0.12[a]
	100	0.43 ± 0.01[bc]	14.72 ± 5.49[ab]	0.11 ± 0.02[b]	0.75 ± 0.09[a]
	200	0.62 ± 0.15[ab]	19.74 ± 6.48[a]	0.22 ± 0.02[a]	1.16 ± 0.07[a]
	500	0.55 ± 0.18[bc]	4.68 ± 1.71[bc]	0.18 ± 0.06[ab]	0.72 ± 0.25[a]

Values are Mean ± SE, (n = 4). nd – not detected.

For each metal in each treatment values followed by the same letters are not significantly different at $P < 0.05$ according to Duncan's multiple range test.

Table 6. Coefficient index (CI) of Cu, Cd, Cr and Zn for the four poplar hybrids under greenhouse conditions.

Hybrid	HM Conc. (mg L^{-1})	CI			
		Cu	Cd	Cr	Zn
Eco 28	0				
	5	1.04 ± 0.13^{b}	2.88 ± 0.49^{d}	0.55 ± 0.48^{a}	0.89 ± 0.06^{a}
	50	0.78 ± 0.18^{b}	8.92 ± 1.66^{c}	0.82 ± 1.63^{a}	0.93 ± 0.07^{a}
	100	1.02 ± 0.14^{b}	17.29 ± 2.11^{b}	0.81 ± 0.10^{a}	0.97 ± 0.06^{a}
	200	1.20 ± 0.32^{b}	31.13 ± 1.62^{a}	0.96 ± 0.15^{a}	0.91 ± 0.05^{a}
	500	1.79 ± 0.73^{a}	33.65 ± 3.28^{a}	1.05 ± 0.31^{a}	0.78 ± 0.06^{a}
DN 034	0				
	5	1.02 ± 0.03^{b}	1.33 ± 0.03^{d}	1.78 ± 0.35^{a}	0.85 ± 0.04^{a}
	50	1.02 ± 0.10^{b}	5.00 ± 0.66^{cd}	1.30 ± 0.25^{a}	0.77 ± 0.16^{a}
	100	1.11 ± 0.04^{b}	6.89 ± 1.39^{c}	1.95 ± 0.54^{a}	0.82 ± 0.12^{a}
	200	1.25 ± 0.70^{b}	16.49 ± 2.33^{b}	1.97 ± 0.35^{a}	0.86 ± 0.04^{a}
	500	1.50 ± 1.48^{a}	29.62 ± 2.21^{a}	1.72 ± 0.17^{a}	0.83 ± 0.05^{a}
TN 074	0				
	5	1.14 ± 0.10^{b}	1.42 ± 0.24^{c}	3.95 ± 1.23^{a}	1.23 ± 0.18^{a}
	50	1.21 ± 0.85^{b}	4.60 ± 0.54^{c}	2.12 ± 0.66^{a}	0.86 ± 0.72^{b}
	100	1.30 ± 0.71^{b}	12.53 ± 0.89^{bc}	0.94 ± 0.11^{a}	0.87 ± 0.44^{b}
	200	1.55 ± 0.18^{b}	20.49 ± 3.25^{ab}	3.07 ± 1.09^{a}	0.94 ± 0.50^{b}
	500	2.39 ± 0.30^{a}	33.17 ± 8.13^{a}	2.20 ± 0.27^{a}	1.23 ± 0.11^{a}
TD 225	0				
	5	1.39 ± 0.03^{b}	2.94 ± 0.53^{d}	0.89 ± 0.15^{c}	1.52 ± 0.02^{a}
	50	1.49 ± 0.03^{b}	8.30 ± 1.01^{c}	1.85 ± 0.43^{a}	1.43 ± 0.12^{ab}
	100	1.47 ± 0.06^{b}	12.58 ± 1.56^{c}	0.67 ± 0.40^{c}	0.96 ± 0.13^{b}
	200	1.65 ± 0.18^{b}	25.98 ± 1.99^{b}	1.63 ± 0.33^{ab}	0.96 ± 0.21^{b}
	500	2.40 ± 0.19^{a}	35.69 ± 2.87^{a}	1.96 ± 0.20^{a}	1.03 ± 0.09^{b}

Values are Mean ± SE, (n = 4). nd – not detected.

For each metal in each treatment values followed by the same letters are not significantly different at $P < 0.05$ according to Duncan's multiple range test.

2.2.1. The Biological Accumulation Coefficient (BAC)

Individual HMs showed variations in BAC in the studied hybrids. Copper and chromium BAC values were below 1.0 across all treatments in hybrid 1. However, Cd BAC values were at a high of 24.01 at 5 mg L^{-1} followed by 15.81 at 100 mg L^{-1}, decreasing by almost half to 7.50 and 8.13 for 200 and 500 mg L^{-1} respectively. BAC values for Zn were in the range of 2–3 with the highest significant value noted at 50 mg L^{-1}. The BAC values for hybrids 2, 3 and 4, Cu and Cr were all less than 1, whereas Cd and Zn were higher especially at 50 mg L^{-1} decreasing gradually to 500 mg L^{-1}. BAC values for hybrid 3 were higher for Cd and Zn.

2.2.2. Biological translocation coefficient (BTC)

Copper BTC were less than 1.0 for all hybrids across all HM concentrations. Chromium values fluctuated between 0.06 and 1.94. No significant differences were observed at higher heavy metal treatment concentrations. For Eco 28, 50 and 10 mg L^{-1} heavy metal treatment concentrations had BTC values >1.0, whereas for hybrid 2 and 3 the BTC values were greater than 1 at 200 mg L^{-1}. Cadmium BTC values greater than one ranging up to 3.0 for lower heavy metal concentrations only. The highest values were observed at 5 mg L^{-1} for hybrids 3 and 4. Zinc, on the other hand, had the highest BTC for all hybrids.

2.2.3. Bio-coefficient factor (BCF)

BCF values for Cu and Cr for all 4 hybrids were less than 1.0 across all treatments. However, hybrid 1 Zn BCF was slightly above 1.0, with the highest value of 1.21 at 100 mg L^{-1}. For hybrid 2 at 5 mg L^{-1} and hybrid 4 at 5, 50, and 200 mg L^{-1} the BCF value was greater than 1.0.

2.2.4. Concentration index (CI)

The CI values were highest for the heavy metal Cd (Table 6) and increased with increasing heavy metal treatment concentrations across all studied hybrids. However, the highest CI was observed in hybrid 1 at 200 mg L^{-1} and 500 mg L^{-1} followed by hybrids 4, 3, and 2. Copper and chromium CI values were similar, with no significant differences between the two hybrids, whereas CI values for Zn were less than 1.0.

The uptake, storage, and movement of copper (Cu), chromium (Cr), cadmium (Cd), and zinc (Zn) in the four examined poplar hybrids are influenced by the specific characteristics of both the plants and the soil, which ultimately define the presence of heavy metals (HMs). A plant would be classified as a hyperaccumulator if it had BTC and BAC values that are greater than 1, as well as metal concentrations indicated by Baker and Brooks [15]. Regarding the examined hybrids, all four demonstrate the ability to transport and accumulate Cd and Zn, but only hybrid 1 exhibits substantial translocation values for Cr. Conversely, copper exhibits BTC and BAC values below 1, which is corroborated by the high levels of heavy metal concentrations in the tissue (measured in dry weight). The Cd had the highest BCF (the ratio of metal concentrations in the roots to that in soil) for all hybrids, and only Zn in hybrid 1 also had a high BCF. This suggests that the plants have the potential to accumulate the targeted heavy metals from the soil medium. The efficiency of phytoextraction is determined by the concentration of metals in the plant and the amount of dry matter it produces. Therefore, the optimum plant for remedying a contaminated site should have a high yield and be able to withstand and accumulate the specific pollutants being targeted [50]. The selection of poplar hybrids for phytoremediation was determined by their overall growth and performance. Based on the results, the hybrids exhibited varying responses to different types of heavy metals (HMs) and concentrations. The availability of metals in trees and the subsequent accumulation of metals in their tissues might vary significantly depending on the source of metal pollution and the conditions of the location [51]. According to the average HM concentrations in the hybrids, the concentration of Zn is greater in leaves. The greatest concentration of zinc (Zn) was found to be 425.49 mg kg^{-1} dry weight in hybrid 3, followed by 309.52 mg kg^{-1} in hybrid 2 (data not displayed). In general, the hybrids did not exhibit significant accumulation or transport of Cu and Cr, as indicated by the values of BAC, BTC, and BCF. Conversely, Zn was the sole heavy metal that accumulated in significant amounts, particularly in the leaves of all hybrids, regardless of the concentration. The hybrid 3 exhibited the greatest BAC values at a concentration of 500 mg L^{-1}, indicating its superior ability to accumulate substances. Additionally, the hybrid 3 Zn had the best BTC values among all other clones, further highlighting its efficient accumulation and translocation abilities. The bioconcentration factor (BCF) values for Cd and Zn in hybrid one were more than 1 for all concentrations, except for the concentration of 500 mg L^{-1} Zn.

Considering the findings of previous studies [51, 52, 53, 54, 55, 56] and the results reported in this chapter, it can be inferred that trees vary in their capacity to uptake heavy metals from the soil. The translocation capacity from the root to the shoots also varies under different situations.

3. CONCISE SUMMARY

Based on the analysis of physiochemical parameters and the individual heavy metal contents in each hybrid, as well as the phytoextraction potential indices, it can be concluded that hybrid 1 (Eco 28) is a suitable candidate for phytoremediation work, specifically for its ability to extract Cd and Zn. Hybrid 3 (TN 074) exhibits promise as a phytoremediator, however its physiochemical characteristics were significantly impacted when exposed to elevated levels of combined heavy metals. Furthermore, it is imperative to conduct additional research in order to ascertain the viability of employing hybrid 3 as a potential solution for phytoremediation of sites contaminated with a mixture of heavy metals. Phytoextraction is promoted as a highly efficient, environmentally friendly, and cost-effective method for cleaning up soils that are polluted with heavy metals. The efficacy of phytoextraction relies on various elements, such as the level of heavy metal concentration in the soil, the bioavailability of heavy metals for absorption, and the plant's capacity to ingest and store metals in their tissues. The current diversity of plant species should be utilized to identify and choose novel and highly efficient hyperaccumulators. Furthermore, conducting prolonged field-based research is necessary to have a deeper understanding of the absorption and buildup of heavy metals.

REFERENCES

1. Belz KE. Phytoremediation: Overview [Internet]. 1997. Available from: http://www.webapps.cee.vt.edu/ewr/environmental/teach/gwprimer/phyto/phyto.html [Accessed: November 1, 2017]
2. Tangahu BV, Abdullah SRS, Basri H, Idris M, Anuar N, Mukhlisin M. A review on heavy metals (As, Pb, and Hg) uptake by plants through phytoremediation. International Journal of Chemical Engineering. 2011;2011:1-31. DOI: 10.1155/2011/939161
3. Paz-Alberto AM, Sigua GC. Phytoremediation: A green technology to remove environmental pollutants. American Journal of Climate Change. 2013;2:71-86
4. Pilon-Smits E. Phytoremediation. Annual Review of Plant Biology. 2005;56:15-39
5. Ali H, Khan E, Sajad MA. Phytoremediation of heavy metals—Concepts and applications. Chem. 2013;91:869-881
6. Pantola RC, Alam A. Potential of Brassicaceae Burnett (Mustard family; angiosperms) in phytoremediation of heavy metals. International Journal of Scientific Research in Environmental Sciences. 2014;2:120-138
7. Liu X, Li X, Chermaine Ong SM, Chu Z. Progress of phytoremediation: Focus on new plants and molecular mechanisms. Journal of Plant Biology and Soil Health. 2013;1:1-5
8. Ghosh M, Singh SP. A review on phytoremediation of heavy metals and utilization of its by-products. Applied Ecology and Environmental Research. 2005;3:1-18
9. Weerakoon S, Somaratne S. Phytoextractive potential among mustard (Brassica juncea) genotypes in Sri Lanka. Ceylon Journal of Science (Biological Sciences). 2010;38:85-93
10. Kotrba P, Najmanova J, Macek T, Ruml T, Mackova M. Genetically modified plants in phytoremediation of heavy metal and metalloid soil and sediment pollution. Biotechnology Advances. 2009;27(6):799-810

11. Garbisu C, Alkorta I. Basic concepts on heavy metal soil bioremediation. European Journal of Mineral Processing and Environmental Protection. 2003;3(1):58-66

12. Abhilash PC, Powell JR, Singh HB, Singh BK. Plant–microbe interactions: Novel applications for exploitation in multipurpose remediation technologies. Trends in Biotechnology. 2012;30(8):416-420

13. Prasad MN. Phytoremediation of metal-polluted ecosystems: Hype for commercialization. Russian Journal of Plant Physiology. 2003;50(5):686-701

14. Chaudhry TM, Hayes WJ, Khan AG, Khoo CS. Phytoremediation—Focusing on accumulator plants that remediate metal-contaminated soils. Australasian Journal of Ecotoxicology. 1998;4(1):37-51

15. Baker AJ, Brooks RR. Terrestrial higher plants which hyperaccumulate metallic elements. A review of their distribution, ecology and phytochemistry. Biorecovery, 1, 81-126. Bulletin of Environmental Contamination and Toxicology. 1989;64:489-496

16. Modaihsh S, Al-Swailem M, Mahjoub M. Heavy metal contents of commercial inorganic fertilizer used in the Kingdom of Saudi Arabia. Agricultural and Marine Sciences. 2004;9:21-25

17. Chehregani A, Malayeri BE. Removal of heavy metals by native accumulator plants. International Journal of Agricultural Biology. 2007;9:462-465

18. Fulekar M, Singh A, Bhaduri AM. Genetic engineering strategies for enhancing phytoremediation of heavy metals. African Journal of Biotechnology. 2009;8:529-535

19. Sabiha-Javied MT, Tufai M, Irfan N. Heavy metal pollution from phosphate rock used for the production of fertilizer in Pakistan. Microchemical Journal. 2009;91:94-99

20. Wuana RA, Okieimen FE. Heavy metals in contaminated soils: A review of sources, chemistry, risks and best available strategies for remediation. ISRN Ecology. 2011;2011:1-20

21. Saleh HM. Water hyacinth for phytoremediation of radioactive waste simulate contaminated with cesium and cobalt radionuclides. Nuclear Engineering and Design. 2012;242:425-432

22. Saleh HM. Biological remediation of hazardous pollutants using water hyacinth—A review. Journal of Biotechnology Research. 2016;2(11):80-91

23. Saleh HM, Bayoumi TA, Mahmoud HH, Aglan RF. Uptake of cesium and cobalt radionuclides from simulated radioactive wastewater by Ludwigia stolonifera aquatic plant. Nuclear Engineering and Design. 2017;315:194-199

24. Jung MC. Heavy metal concentrations in soils and factors affecting metal uptake by plants in the vicinity of a Korean Cu-W mine. Sensors. 2008;8(4):2413-2423

25. Rascio N, Navari-Izzo F. Heavy metal hyperaccumulating plants: How and why they do it? And what makes them so interesting? Plant Science. 2011;180:169-181

26. Sarma H. Metal hyperaccumulation in plants: A review focusing on phytoremediation technology. Journal of Environmental Science and Technology. 2011;4:118-138

27. Robinson BH, Green SR, Chancerel B, Mills TM, Clothier BE. Poplar for the phytomanagement of boron contaminated sites. Environmental Pollution. 2007;150:225-233

28. Zacchini M, Pietrini F, Scarascia Mugnozza G, Iori V, Pietrosanti L, Massacci A. Metal tolerance, accumulation and translocation in poplar and willow clones treated with cadmium in hydroponics. Water Air and Soil Pollution. 2009;197:23-34

29. Zacchini M, Iori V, Mugnozza GS, Pietrini F, Massacci A. Cadmium accumulation and tolerance in Populus nigra and Salix alba. Biologia Plantarum. 2011;55(2):383-386

30. Baker AJM, Brooks RR. Terrestrial higher plants which accumulate metallic elements: A review of their distribution ecology and phytochemistry. Biorecovery. 1989;1:81-126

31. Baker AJM, Reeves RD, Hajar ASM. Heavy metal accumularion and tolerance in British populations of the metallophyte Thlaspi caerulescens J & C Press (Brassicaeceae). The New Phytologist. 1994;127:61-68

32. Greger M, Landberg T. Use of willow in phytoextraction. International Journal of Phytoremediation. 1999;1:115-123

33. McGee AB, Schmierbach MR, Bazzaz FA. Photosynthesis and growth of populations of Populus deltoids from contrasting habitats. The American Midland Naturalist. 1981;105:305-311

34. Yoon J, Cao X, Zhou Q, Ma LQ. Accumulation of Pb, Cu, and Zn in naïve plants growing on contaminated Florida site. The Science of the Total Environment. 2006;368(2):456-464

35. Cui S, Zhou Q, Chao L. Potential hyperaccumulation of Pb, Zn, Cu and Cd in endurant plants distributed in an old smeltery, Northeast China. Environmental Geology. 2007;51(6):1043-1048

36. Li MS, Luo YP, Su ZY. Heavy metal concentrations in soils and plant accumulation in a restored manganese mine land in Guangxi, South China. Environmental Pollution. 2007;147(1):168-175

37. Van Assche F, Clijsters H. Multiple effects of heavy metal toxicity on photosynthesis. Effects of Stress on Photosynthesis, Springer. 1983:371-382

38. Clijsters H, Van Assche F. Inhibition of photosynthesis by heavy metals. Photosynthesis Research. 1985;7(1):31-40

39. Alcántara E, Romera FJ, Cañete M, De la Guardia MD. Effects of heavy metals on both induction and function of root Fe(III) reductase in Fe-deficient cucumber (*Cucumis sativus* L.) plants. Journal of Experimental Botany. 1994;45(12):1893-1898

40. Carlson RW, Bazzaz FA, Rolfe GL. The effect of heavy metals on plants: II. Net photosynthesis and transpiration of whole corn and sunflower plants treated with Pb, Cd, Ni, and Tl. Environmental Research. 1975;10(1):113-120

41. Parekh D, Puranik RM, Srivastava HS. Inhibition of chlorophyll biosynthesis by cadmium in greening maize leaf segments. Biochemie und Physiologie der Pflanzen. 1990;186:239-242

42. Aggarwal A, Sharma I, Tripati BN, Munjal AK, Baunthiyal M, Sharma V. Metal toxicity and photosynthesis. In: Itoh S, Mohanty S, Guruorasad KN, editors. Photosynthesis: Overviews on Recent Progress & Future Perspectives. 1st ed. I K International Publishing House Pvt. Ltd; 2012. pp. 229-236

43. Filippis LF, Pallaghy CK. Heavy metals: Sources and biological effects. Advances in Limnology[ERGEB. LIMNOL./ADV. LIMNOL.]; 1994

44. Siedlecks A, Krupa Z. Interactions between cadmium and iron and its effects on photosynthetic capacity of primary leaves of *Phaseolus vulgaris*. Plant Physiology. 1996;34:833-841

45. Elloumi N, Ben F, Rhouma ABB, Mezghani I, Boukhris M. Cadmium induced growth inhibition and alternation of biochemical parameters in almod seedlings grown in solution culture. Acta Physiologiae Plantarum. 2007;29:57-62

46. Zengin FK, Munzuroglu O. Toxic effects of cadmium (Cd++) on metabolism of sunflower (*Helianthus annuus* L.) seedlings. Acta Agriculturae Scandinavica Section B Soil and Plant Science. 2006;56:224-229

47. Devi Prasad PV, Devi Prasad PS. Effects of cadmium, lead and nickel on three freshwater green algae. Water, Air, and Soil Pollution. 1982;17:263-268

48. John R, Ahmad P, Gadgil K, Sharma S. Heavy metal toxicity: Effect on plant growth, biochemical parameters and metal accumulation by *Brassica juncea* L. International Journal of Plant Production. 2012;3(3):65-76

49. Hattab S, Dridi B, Chouba L, Kheder MB, Bousetta H. Photosynthesis and growth responses of pea *Pisum sativum* L. under heavy metals stress. Journal of Environmental Sciences. 2009;21(11):1552-1556

50. Turan M, Esringu A. Phytoremediation based on canola (*Brassica napus* L.) and Indian mustard (*Brassica juncea* L.) planted on spiked soil by aliquot amount of Cd, Cu, Pb, and Zn. Plant, Soil and Environment. 2007;53(1):7

51. Pulford ID, Watson C. Phytoremediation of heavy metal-contaminated land by trees—A review. Environment International. 2003;29:529-540

52. Pulford ID, Dickinson NM. 90 Phytoremediation technologies using trees. Trace Elements in the Environment: Biogeochemistry, Biotechnology, and Bioremediation. 2005;14:383

53. Gerhardt KE, Huang XD, Glick BR, Greenberg BM. Phytoremediation and rhizoremediation of organic soil contaminants: Potential and challenges. Plant Science. 2009;176(1):20-30

54. Pulford ID, Watson C, McGregor SD. Uptake of chromium by trees: Prospects for phytoremediation. Environmental Geochemistry and Health. 2001;23(3):307-311

55. Cluis C. Junk-greedy greens: Phytoremediation as a new option for soil decontamination. BioTechology Journal. 2004;2:61-67

56. Raskin I, Ensley BD. Phytoremediation of Toxic Metals. New Jersey, USA: John Wiley and Sons; 2000.

12 Chapter

PHYTOREMEDIATION: THE POTENTIAL OF HALOPHYTES AS EFFECTIVE ACCUMULATORS OF HEAVY METALS

1. INTRODUCTION: AN OVERVIEW

The primary environmental pollutants are heavy metals and other organic compounds. Phytoremediation experiments to remove pollutants from wastewater and contaminated soil have been conducted for centuries, using plants such as Thlaspi caerulescens and Viola calaminaria. These plants have been found to effectively remediate high levels of heavy metals [1]. The growth of these contaminants is mostly caused by anthropogenic activities resulting from industrialization. This can occur either through direct leakage or mishaps during the transportation of solid and liquid wastes from storage and industrial facilities [2, 3]. Methods for remediation of environmental pollutants, including both organic and inorganic substances, can be categorized into physical, chemical, and biological treatments [4, 5]. Nevertheless, physical and chemical procedures are acknowledged to have some drawbacks or constraints, including their high expenses and labor-intensive nature. In addition, chemical procedures contribute to pollution and are particularly expensive because to the large amount of sludge they produce [6]. Given this environment, the need for improved methods to remove metal pollution arose, leading to the investigation of different bio-based technologies. Biological agents are cost-effective, safer, and have minimal or no adverse effects on the environment [7]. Bio-based remediation approaches including bio-augmentation, bioremediation, bioventing, composting, and phytoremediation. Phytoremediation has emerged as the most practical and beneficial alternative, garnering significant attention in recent years [8, 9]. The detrimental and unfavorable consequences linked to these components render them susceptible to phytoremediation [10]. Phytoremediation provides numerous benefits. The benefits of this approach include cost-effectiveness, promotion of biodiversity, erosion reduction, minimal environmental impact, and lower energy usage resulting in fewer carbon dioxide emissions [11]. So far, over 400 plant species have been identified as potential metal hyper-accumulators [12]. However, there is limited research on the toxicity of various metals when combined [13]. While the excessive accumulation of nickel (Ni), cadmium (Cd), manganese (Mn), zinc (Zn), and selenium (Se) in plants has been well-documented, the same cannot be said with certainty for copper (Cu), chromium (Cr), lead (Pb), thallium (Th), and cobalt (Co) metals. Copper (Cu) is a crucial element for the growth and overall physiological functioning of plants. It plays a vital role as a cofactor in several types of enzymes that facilitate the transfer of electrons during metabolic activities, including the tricarboxylic acid (TCA) cycle [14, 15]. Nevertheless, when present in high quantities, copper exhibits toxicity to plants, as seen by inhibited development. While there is some understanding of the physiological effects of copper stress in plants, the specific responses at the functional level remain unclear [16]. Heavy metal deposition in plant tissues leads to many detrimental impacts on growth. While it has been observed that it impacts seed germination, seedling growth, and photosynthetic processes, ultimately inhibiting essential enzyme activity in plants [17, 18], it is crucial to note that plants exhibit varying responses [19]. When faced with heavy metal stress, the root tissue is the initial point of contact with the harmful toxins. The cell wall of the root tissue includes a mechanism that captures and immobilizes the heavy metal ions, preventing them from spreading to other parts of the plant [20, 21]. Multiple studies have documented that numerous plants, especially those found in desert environments, possess the ability to effectively remove pollutants from the soil (known as phytoremediation). However, only a few number of these plants are capable of accumulating high levels of metals, making their selection for successful phytoremediation a difficult task. This is exemplified by the gradual increase in size, the amount of plant material above the ground, the root structure, and the act of gathering crops [22]. Therefore, in order for heavy metal phytoremediation to be successful, it is necessary for candidate plants to have the ability to hyperaccumulate metals. Halophytes are considered suitable phytoremediators due to their hyperaccumulation potential. This is because of their robust stress tolerance mechanism, which allows plants to flourish in saline soil and other desert environments.

2. PHYTOREMEDIATION

Phytoremediation is a method in which plants are used, together with accompanying bacteria, to decrease or eliminate organic and inorganic pollutants from the environment [13]. The remediation of pollutants can vary depending on the method used, which may involve removing, transferring, degrading, or immobilizing them from soil or water [23]. This methodology takes advantage of the plants' roots' ability to initially absorb contaminants and then transport them across the stem to accumulate in the shoot tissue. It is a distinctive method. Phytoremediation is a recently developed approach that has the potential to be a valuable remedy for our declining environment, in contrast to other traditional treatment methods. So far, numerous plant species have been proposed as possible agents for phytoremediation [24].

2.1. Methods of phytoremediation

Phytoremediation is the process in which plants growing in soil or water that is contaminated with little amounts or large amounts of metals can absorb or tolerate these metals in diverse ways. The specific physiological mechanisms and types of metals present play a role in this process [25]. Halder and Ghosh [26] classify phytoremediation techniques into five categories: phytoextraction, phytofiltration, phytovolatilization, phytostabilization, and phytotransformation.

2.1.1. Phytoextraction

Phytoextraction is a method of phytoremediation in which plants absorb metals by translocation and store them in a way that allows for extraction from their tissues [27]. Phytoextraction is a frequently seen kind of phytoremediation, and the terms phytoabsorption, phytoaccumulation, and phytosequestration are commonly used interchangeably to describe it [28]. Phytoremediation is widely regarded as the primary strategy for removing metals from polluted water, sediment, and soil. The effectiveness of this remediation process is contingent upon various parameters, including soil conditions, the bioavailability and speciation of metals, and the specific plant species involved. Typically, a large amount of metals that are taken by plants tends to accumulate in the above-ground parts of the plant, which can be harvested [12]. Several recent investigations have documented several plant species that exhibit the phytoextraction approach from both water and soil media [29, 30, 31, 32]. Plants that can effectively use the phytoextraction technique to remove metals from the environment are known as hyperaccumulators. These plants constantly collect a specific amount of metal in their shoot tissue, which varies depending on the type of metal [22]. In general, hyperaccumulators must exhibit specific traits, including rapid development, extensive branching of shoots, a high capacity for bioaccumulation and translocation, a large amount of above-ground biomass, and ease of cultivation and harvesting [22, 33]. Ali Khan (28) presented two distinct methods for metal phytoextraction in various plant species. One method resulted in lower above ground biomass but a significant accumulation of metals in high concentrations. Conversely, the other method led to higher above ground biomass but lower metal accumulation. The final metal accumulation in both methods aligned with that of hyperaccumulators. Therefore, hyperaccumulation plays a greater role in phytoremediation than the amount of biomass generated. This implies that the use of hyperaccumulators is more favorable due to their benefits, including safe disposal, cost-effectiveness, and ease of handling [28].

2.1.2. Phytofiltration

Phytofiltration or rhizofiltration, as used interchangeably, refers to the absorption or adsorption of contaminants from surface wastewater by plant roots thereby preventing them from leaching to the underground water [34]. It is a type of phytoremediation technique that can be demonstrated in situ by directly growing plants in the polluted water body [24]. Although it is commonly applicable using aquatic plant species [35], there are suggestions that the process may be applied to terrestrial plants, which remediate metals to precipitate with the aid of microbes root bio filter [36]. Indeed, root exudates cause metal precipitation which alters the rhizosphere pH level [37]. Many terrestrial plants including grasses grown in a hydroponic culture were shown to effectively remove metals such via phytofiltration [38]. In the same study, Indian mustard was especially reported to accumulate higher fold of metal concentration far beyond the initial concentration, and the removal is by tissue specific adsorption mediated by root metal concentration.

Quite a number of studies have shown many species of aquatic macrophytes that demonstrate phytofiltration potential. While experimenting for phytoremediation under different water conditions polluted with heavy metals, Liao and Chang [39] found that Eichhonia crassipes absorb and accumulates metal contaminants, it has also exhibit high growth rate and increased biomass production and thus considered a good phytofiltration agent. This plant species absorb high concentrations of Pb, Ni, Zn and Cu which accumulates much higher in the root tissue than the shoot, suggesting the important role of fibrous and tap root system found in the plant, which is one of the key characteristics of potential phytofiltration agent. In a similar study, other aquatic plant species including *Salvinia herzogii, E. crassipes, Pistia stratiotes* and Hydromistia stolonifera were shown to absorb high concentration of Cd with P. stratiotes accumulating higher Cd concentration and exhibiting faster growth rate, a feature attributed to possible complimentary mechanism for the enhanced metal uptake [40]. Absorption of Cd in the root of all the plants relates to the added concentration. In another study by Thayaparan, Iqbal [41] also reported that Azolla pinnata have shown a great potential in the removal of high Pb concentration by phytofiltration from polluted water. As in phytoextraction, potential phytofiltration agents should tolerate high metal concentration, exhibit fast and high growth rate as well as above ground biomass, however, in contrast to phytoextraction, they are expected to show limited translocation capacity of absorbed metals from root to shoot tissues [24]. For efficient phytofiltration, this is an advantage over phytoextraction, since low translocation of contaminants means reduced contamination of other parts of the plant.

2.1.3. Phytostabilization

In this technique, pollutants are converted into a less toxic or bioavailable form by the continuous precipitation of the plant rhizosphere. This is achieved either by surface run off prevention, erosion or leaching [27]. It is applicable in the stabilization of metals in contaminated soil, sediment or water environments, which ensures they are not transferred to the food chain from the soil by translocating to other parts of food crops or to the underground water. This is possible by sorption via the root, precipitation and subsequent metal reduction around the plant rhizosphere, for instance the toxic Cr^{6+} is converted to Cr^{3+}, which is less toxic [42, 43]. Variation exists as to how prone a metal is to phytostabilization and is subject to its chemical character. This is evidenced in a comparative study to evaluate metal accumulation capacity of two aquatic macrophytes Phragmites australis and Typha domingensis, where both are found to stabilize As and Hg but inefficient in the phytostabilization of other metals [44].

Although phytostabilization offer some advantages over other phytoremediation techniques, it is however limited to temporary measure to deal with pollutants contamination owing to the fact that metals are only inactivated and their movement restricted, but still remains in the contaminated environmental compartment [45]. It is useful in emergencies, since it can rapidly immobilize pollutants from soil, water or sediment. Equally important, it ensures that contaminants are not translocated to other plant tissues by trapping most of it in the plant root [46]. Considering the strategies employed in phytostabilization, plants that can appropriately fall under this mechanism is their ability to tolerate and immobilize metals and other contaminants, low translocation capacity from root to plant aerial parts and of course extensive and fibrous tap root system [7]. Among several studies that reported plants species with these characteristics [47, 48, 49] demonstrating the phytostabilization of Zn, Pb, Cu and Cd by different plants in soil and sediment polluted environments.

2.1.4. Phytotransformation/phytodegradation

Phytotransformation or phytodegradation is another technique of phytoremediation where contaminants and other nutrients are chemically modified through plant metabolism and render associated contaminants inactive in both plant root and shoot tissues [6]. Plant metabolic enzymes act on the surrounding contaminants, thereby transforming them to a less toxic form, plants rhizosphere microbes also aid in the transformation process of the compounds [50]. Although this mechanism is mostly against organic contaminants, inorganic compounds such as metals were also suggested, in which case a strategy akin to phytostabilization is employed to convert toxic metals to less toxic form [51]. However, this technique seem less efficient and reliable compared to others in that it requires longer period of time, strict soil characteristic such as depth and underground water availability and often require soil amendments.

2.1.5. Phytovolatilization

In phytovolatilization, contaminants are converted in to a volatile form and released to the air via plants leaves stomata [27, 34]. However, this mechanism merely transfers contaminants from one environmental compartment to another, which may somehow return back to the original source (soil) by precipitation and hence could be less popular to other phytoremediation techniques especially phytoextraction and phytofiltration [34, 52]. It is commonly employed when treating groups of highly volatile metals like Hg and As. Phytovolatilization of As involves the conversion of elemental As to selenoaminoacids, such as selenomethione, which is modified by methylation to a volatile and less toxic form, dimethylselenide [53].

3. METAL HYPERACCUMULATOR PLANTS

Several plants species are known to tolerate high concentration of toxic metals. Tolerant species are best described as excluders, where metal uptake and translocation to different tissue parts are limited. While others that are capable of accumulating higher concentrations with improved translocation from the root to shoot part of the plant, thereby significantly reducing its availability in the soil, and they do so with no visible sign of toxicity effects. To date, heavy metals have no standard definition by recognized bodies in the area. Various researchers use different characteristics and levels in their description such as atomic mass and number, density, chemical character as well as their toxicity; however, there appears no connection between such properties [54]. According to Wang and Chen [55], three categories of heavy metals arising from both natural and artificial sources are of interest, these includes valuable metals e.g. Ag, Au, Pd, Pt, harmful metals e.g., As, Cu, Co, Cd, Cr, Hg, Ni, Pb, and radionuclides such as Am, Th, Ra and so on. The non-biodegradability and stable nature of heavy metals suggests increased exposure to living species including humans [54], periodic reviews of toxic metals effects are documented by many research groups [56, 57, 58].

When determining hyperaccumulators of toxic metal, the most important factor is the concentration of the metal ion threshold. Therefore, plants can be regarded as hyperaccumulators, when capable of accumulating toxic metals concentration to about 50 to 100 times more than non-hyperaccumulator plants [13, 59]. For instance, the threshold for Zn and Mn hyperaccumulation in plant shoot is pegged at 1% of dry biomass, 0.01% for Cd and 0.1% respectively for Ti, Se, Sb, Pb, Ni, Cu, Cr, Co, and As [13, 60]. To date, few plant species are classified as hyperaccumulators, the majority of them (3/4) are tolerant to Ni and belongs to the Brassicaceae family native to Western Asia and Southern Europe, with up to 48 species implicated in Ni accumulation of around 3% dry shoot mass [60, 61, 62]. There is increasing interest in plant hyperaccumulators in recent times, owing to their potential use in metal contaminated soil and water detoxification [25, 63].

3.1. The function of metal chelators in hyperaccumulation

Phytoremediation of heavy metals encompasses several physiological, biochemical, and molecular processes. Phytoextraction is a process that specifically involves the accumulation and transport of heavy metals to plant tissues. Plant metal chelators, such as phytochelatins (PCs) and metallothioneins (MTs), are the primary transporter proteins used in heavy metal phytoremediation. Metallothioneins (MTs) are proteins with a high concentration of cysteine residues. They are well-known for their ability to bind to metals and play a crucial role in the sequestration of metals in their ionic form [64]. PCs, short for phytochelatins, are products of glutathione synthase. They have the ability to bind to heavy metals, which is an important aspect of the phyto-detoxification mechanism [65, 66]. The synthesis of phytochelatins is stimulated by the action of an enzyme called phytochelatin synthase, which is activated by the presence of metal ions [34, 67]. The work aimed to illustrate the function of synthases by studying mutants in the model plant Arabidopsis thaliana. These mutants were shown to have a heightened sensitivity to Hg and Cd due to their inability to make PCs [68]. Conversely, MTs are peptides that are genetically encoded to bind to metals and often have a small molecular weight. Several studies have shown that MTs play a crucial role in safeguarding plants against the harmful effects of heavy metals in soil, sediment, and water [65, 69, 70]. The development of metallothioneins (MTs) and phytochelatins (PCs), in conjunction with organic acid production, plays a crucial role in the absorption of heavy metals by plants and their subsequent transportation to other tissues [42]. To improve the effectiveness of heavy metal accumulation and translocation, the production of these natural chelators could be augmented. Presently, numerous ongoing investigations are focused on defining and identifying biomolecules that are involved in the transportation and detoxification of heavy metals. Understanding the complete process of detoxification in plants [28, 71] might be facilitated by conducting comparative proteome studies, which are of utmost importance.

3.1.1. The shoot proteome

The plant shoot plays a crucial role in the phytoremediation process, particularly in amassing the largest concentration of metals when the plant uses the phytoextraction technique. The effectiveness of this technique depends on the type of metal elements and their bioavailability. Lately, there has been a growing fascination with studying the proteomics of plant hyperaccumulators. The objective is to analyze and discover proteins that have a role in the sequestration and detoxification of metals [72]. Possible applications have been enabled by the progress in modern mass spectrometry techniques, including two-dimensional liquid chromatography matrix-assisted laser desorption/ionization time-of-flight (2D-LC/TOFMS), time of flight/mass spectrometry (TOF/MS), two-dimensional polyacrylamide gel electrophoresis (2D PAGE), and liquid chromatography-tandem mass spectrometry (LC–MS/MS). The proteomes of various plant species, such as Thlaspi caerulescens, Pteris vittata, Helianthus annuus, and Agrostis tenuis, were recently examined to identify proteins involved in detoxifying heavy metals. This search revealed several important functional proteins that help plants defend against oxidative stress and respond to both biotic and abiotic stress conditions, among other functions [12, 73, 74]. Comparisons can be made in the proteomics study of plant metal hyperaccumulators, even when these studies involve distinct plants and metals. In 2005, a study discovered that exposing *Alyssum lesbiacum* to nickel in a controlled experiment resulted in the production of three proteins. One of these proteins, called iron superoxide dismutase (Fe-SOD), was found to have antioxidant properties. The other two proteins were identified as chloroplast phosphoglycerate kinase and a transketolase, both of which play a role in carbohydrate metabolism. When Anemone halleri was exposed to Zn and Cd, the photosynthetic protein (chlorophyll a/b binding) and membrane protein (photosystem II) were consistently produced and increased in quantity. This is associated with the enhanced metabolic energy requirement in this plant, which has the ability to accumulate high levels of metals. At elevated levels of metal concentrations, there is a consistent observation of enhanced protein induction in response to defense against antioxidants and energy metabolism. Examples of these proteins include Renal Epithelial Protein (APX), Superoxide dismutase (SOD), cytochrome P450, and Glutathione S-transferase (GST). These findings indicate that plants need both a working photosynthetic mechanism and the activity of proteins that remove oxygen radical species in order to absorb, transport, and accumulate heavy metals in their shoot tissue [13]. Plant metal tolerance may be influenced by the presence of metabolic energy active proteins. A comparison was made between the proteomes of *T. caerulescens* with different levels of tolerance to Cd and Zn metals. It was shown that the more metal tolerant variety had a higher accumulation of the extrinsic subunit of photosystem II protein, which resulted in its stability compared to the less tolerant variant. Furthermore, with GST and cytochrome P450, other proteins including aspartate aminotransferase and thioredoxin are frequently present and associated with the sequestration of xenobiotics, including metals. Studies have shown that GSTs are increased in many organisms, such as bacteria and fungi, when exposed to metals like Zn, Cu, and Cd [78]. Therefore, it has been proposed that GSTs play a role in protecting these cells from harmful genes.

3.1.2. The root proteome

In plants, the root tissue is the first to be exposed to all potential toxicants whether in the soil or surface water and hence serve as the gateway route through which they can subsequently be translocated to other tissue parts. Plants diversity as to hyperaccumulators and non-hyperaccumulators exist, this is due to the fact that, while some species bear the complete mechanism of enhanced metal uptake and eventual translocation, others have limited sequestration capacity in their root vacuoles [79]. In non-hyperaccumulators plant roots, Zn transporters are only detectable in the absence of Zn, whereas in hyperaccumulators, there is constitutive expression of these proteins such as ZT1 even in Zn deficient condition [80, 81]. In T. caerulescens, the iron transporter protein IRT1 was found to be involved in Zn and Cd uptake [82], similarly, root proteome study of this hyperaccumulator and A. lesbiacum were conducted by Tuomainen, Tervahauta [83] to evaluate peptides involved in Zn and Cd hyperaccumulation. In these studies, various classes of proteins were identified, their availability and or abundance varies relative to metal exposure and accessions. As in the case with similar studies on shoot proteome of hyperaccumulators, ROS scavenging proteins were more abundant in the more metal tolerant accessions compared to the less metal tolerant species. It was concluded that the changes in the enzyme, superoxide dismutase (SOD) availability upon which Zn depends in the different accessions may be linked to ROS increase.

An important organelle, cell wall, in the plant root is also affected by its exposure to heavy metal stress. The putative protein, glycosyl hydrolase family 18, involved in the formation of cell wall structure was shown to be regulated in accordance to treatment conditions and accession. These proteins, which are particularly known to be involved in cell wall expansion, differ in terms of abundance between the root proteome of two accessions, which in turn also affect the capacity of metals uptake; higher Ni and Cd accumulation was observed in the variant with more protein abundance [13]. Despite the recent advancement in proteomics technology, root protein transporters are yet to be differentially identified. Indeed, this is in agreement with transcriptomics studies, with analyzed data suggesting the constitutive expression of metal genes transporters in plant metal hyperaccumulators [13, 81, 84].

3.2. The halophytes found in Qatar show great potential as hyperaccumulators of heavy metals:

Some studies demonstrated the potential of several Qatari plants as good phytoremediation candidates, many among which are heavy metal hyperaccumulators. Examples includes species belonging to the genus Zygophyllum, which are as either metal tolerant or accumulators when tested on both polluted soil and wastewater media [85, 86, 87, 88, 89]. Others include Typha domingensis and Phragmites australis [2]. According to Carvalho and Martin [90], Typha domingensis remediated heavy metals from industrial waste water and solution cultures; similarly, members of the halophytes plant family Brassicaceae were reported to be important phytoremediation agents [3, 59, 91] and the tree plant Prosopis juliflora exhibited phytoremediation of heavy metals potential [92, 93]. Additionally, such as other plants, such as Phragmites australis that were previously shown to clean petroleum-polluted soils may be good candidates for the phytoremediation of typical oil and gas produced wastewater [94], others with similar potentials includes Medicago such as Medicago sativa and Glycine max which also demonstrated strong petroleum polluted soil phytoremediation activity [95, 96]. Some examples of other species tested for phytoremediation studies and their metal uptake capacity are summarized in Table 1 above.

Table 1. Examples of phytoremediation studies using species belong to Qatari flora and/or their relatives.

S/No	Plant species	Metal (s)	Metal accumulation (mg/kg)	References
1	*Atriplex halimus* sub sp. *schweinfurthii*	Cadmium	606.51	[97]
	A. halimus L.	Cadmium	830	[98]
		Zinc	44	
2	*Arthrocnemum macrostachyum*	Lead	620	[99]
3	*Crucianella maritima*	Zinc	390	[99]
4	*Dittrichia viscosa*	Lead	270	[99]
5	*Tamarix smyrnensis* Bunge	Lead	800	[100]
		Cadmium	800	
6	*Typha domingensis*	Selenium	30	[90]
		Lead	59.13	[101]
7	*T. lotifolia* L	Cadmium	210	[102]
8	*Paspalum conjugatum* L. *Prosopis laevigata*	Lead	150	[103]

4. CONCISE SUMMARY

The presence of trace and heavy metals in the environment poses a significant threat to the health of living organisms. These pollutants are linked to a broad spectrum of human illnesses and diverse long-term adverse environmental effects, therefore posing a threat to global sustainable development efforts. Numerous traditional methods are commonly used to address and remove these pollutants. Nevertheless, conventional cleanup methods possess numerous drawbacks, ranging from exacerbating environmental contamination to incurring substantial operational expenses, among other issues. Phytoremediation is a highly promising alternative in this context, as laboratory investigations have shown that numerous plant species have the ability to remediate various heavy metal pollutants. Nevertheless, there are still constraints in the implementation of this nascent technology. This could be associated with exposure to additional stressors in field circumstances, particularly in harsh locations, which could have a substantial impact on physiological function and overall growth. A crucial factor for the efficacy of phytoremediation is the formation of aboveground biomass, which is an essential need for plants. Many new phytoremediation research aim to identify and select plants with a higher potential for metal uptake or hyperaccumulation, even in the presence of additional stress conditions. Furthermore, our insufficient comprehension of the molecular mechanism involved in heavy metal remediation, including the precise function of transporter proteins, is impeding advancements in this field. Several stress response molecules play a crucial role in the tolerance and accumulation of heavy metal pollutants by potential phytoremediators. Halophytes are well-known for their ability to adapt to stressful environmental conditions. As a result, they are considered to be the most promising candidates for identifying heavy metal hyperaccumulators and understanding the mechanisms of metal uptake. Undoubtedly, these processes are crucial for enhancing the efficacy of phytoremediation in large-scale, field, and industrial settings.

REFERENCES

1. Lasat MM. Phytoextraction of toxic metals. Journal of Environmental Quality. 2002;31(1):109-120
2. Yasseen BT, Al-Thani RF. Ecophysiology of wild plants and conservation perspectives in the State of Qatar. In: Stoytcheva M, Zlatev R, editor. Agricultural Chemistry. Croatia: InTech; 2013. pp. 210. ISBN: 978-953-51-1026-2
3. Lone MI et al. Phytoremediation of heavy metal polluted soils and water: Progresses and perspectives. Journal of Zhejiang University. Science. B. 2008;9(3):210-220
4. Baldwin SA et al. The microbial community of a passive biochemical reactor treating arsenic, zinc, and sulfate-rich seepage. Frontiers in Bioengineering and Biotechnology. 2015;3:27
5. Hasegawa H, Rahman IMM, Rahman MA, editors. Environmental Remediation Technologies for Metal-Contaminated Soils. Tokyo: Springer; 2016. 254 p. DOI: 10.1007/978-4-431-55759-3. https://doi.org/10.1007/978-4-431-55759-3
6. Tangahu BV et al. A review on heavy metals (As, Pb, and Hg) uptake by plants through phytoremediation. International Journal of Chemical Engineering. 2011;2011:31. DOI:10.1155/2011/939161
7. Doble M, Kumar A. Biotreatment of Industrial Effluents. Oxford: Elsevier; 2005. 336 p
8. Witters N. Phytoremediation: An alternative remediation technology and a sustainable marginal land management option [Thesis]. Hasselt: Hasselt University; 2011
9. Ullah A et al. Phytoremediation of heavy metals assisted by plant growth promoting (PGP) bacteria: A review. Environmental and Experimental Botany. 2015;117:28-40
10. Monica RC, Cremonini R. Nanoparticles and higher plants. Caryologia. 2009;62(2):161-165
11. Ali-Zade V, Alirzayeva E, Shirvani T. Plant resistance to anthropogenic toxicants: Approaches to phytoremediation. In: Plant Adaptation and Phytoremediation. Netherlands: Springer; 2010. pp. 173-192
12. Walliwalagedara C et al. Differential expression of proteins induced by lead in the dwarf sunflower Helianthus annuus. Phytochemistry. 2010;71(13):1460-1465
13. Visioli G, Marmiroli N. The proteomics of heavy metal hyperaccumulation by plants. Journal of Proteomics. 2013;79:133-145
14. Himelblau E, Amasino RM. Delivering copper within plant cells. Current Opinion in Plant Biology. 2000;3(3):205-210
15. Hall J. Cellular mechanisms for heavy metal detoxification and tolerance. Journal of Experimental Botany. 2002;53(366):1-11
16. Lin C-Y et al. Comparison of early transcriptome responses to copper and cadmium in rice roots. Plant Molecular Biology. 2013;81(4-5):507-522
17. Singh RP et al. Response of higher plants to lead contaminated environment. Chemosphere. 1997;34(11):2467-2493
18. Sharma P, Dubey RS. Lead toxicity in plants. Brazilian Journal of Plant Physiology. 2005;17(1):35-52
19. Liu D et al. Uptake and accumulation of lead by roots, hypocotyls and shoots of Indian mustard [Brassica juncea (L.)]. Bioresource Technology. 2000;71(3):273-277

20. Allan DL, Jarrell WM. Proton and copper adsorption to maize and soybean root cell walls. Plant Physiology. 1989;89(3):823-832

21. Branquinho C, Brown DH, Catarino F. The cellular location of Cu in lichens and its effects on membrane integrity and chlorophyll fluorescence. Environmental and Experimental Botany. 1997;38(2):165-179

22. van der Ent A, et al., Hyperaccumulators of metal and metalloid trace elements: Facts and fiction. Plant and Soil, 2013. 362(1-2): p. 319-334

23. Ahmadpour P et al. Phytoremediation of heavy metals: A green technology. African Journal of Biotechnology. 2012;11(76):14036-14043

24. da Conceição Gomes MA et al. Metal phytoremediation: General strategies, genetically modified plants and applications in metal nanoparticle contamination. Ecotoxicology and Environmental Safety. 2016;134:133-147

25. Pilon-Smits EA, Freeman JL. Environmental cleanup using plants: Biotechnological advances and ecological considerations. Frontiers in Ecology and the Environment. 2006;4(4):203-210

26. Halder S, Ghosh S. Wetland macrophytes in purification of water. International Journal of Environmental Sciences. 2014;5(2):432-437

27. Nwoko CO. Trends in phytoremediation of toxic elemental and organic pollutants. African Journal of Biotechnology. 2010;9(37):6010-6016

28. Ali H, Khan E, Sajad MA. Phytoremediation of heavy metals—Concepts and applications. Chemosphere. 2013;91(7):869-881

29. Li J et al. Cadmium tolerance and accumulation in cultivars of a high-biomass tropical tree () and its potential for phytoextraction. Journal of Environmental Quality. 2010;39(4):1262-1268

30. Simmons R et al. Towards practical cadmium phytoextraction with Noccaea caerulescens. International Journal of Phytoremediation. 2015;17(2):191-199

31. Kacálková L, Tlustoš P, Száková J. Phytoextraction of risk elements by willow and poplar trees. International Journal of Phytoremediation. 2015;17(5):414-421

32. Shaheen SM, Rinklebe J. Phytoextraction of potentially toxic elements by Indian mustard, rapeseed, and sunflower from a contaminated riparian soil. Environmental Geochemistry and Health. 2015;37(6):953-967

33. Sharma P, Pandey S. Status of phytoremediation in world scenario. International Journal of Environmental Bioremediation & Biodegradation. 2014;2(4):178-191

34. Sarma H. Metal hyperaccumulation in plants: A review focusing on phytoremediation technology. Journal of Environmental Science and Technology. 2011;4(2):118-138

35. Olguín EJ, Sánchez-Galván G. Heavy metal removal in phytofiltration and phycoremediation: The need to differentiate between bioadsorption and bioaccumulation. New Biotechnology. 2012;30(1):3-8

36. Salt DE et al. Phytoremediation: A novel strategy for the removal of toxic metals from the environment using plants. Nature Biotechnology. 1995;13(5):468-474

37. Rai PK. Heavy metal pollution in aquatic ecosystems and its phytoremediation using wetland plants: An ecosustainable approach. International Journal of Phytoremediation. 2008;10(2):133-160

38. Khilji S. Rhizofiltration of heavy metals from the tannery sludge by the anchored hydrophyte, Hydrocotyle umbellata L. African Journal of Biotechnology. 2008;7(20)

39. Liao S, Chang W-L. Heavy metal phytoremediation by water hyacinth at constructed wetlands in Taiwan. Photogrammetric Engineering and Remote Sensing. 2004;54:177-185

40. Maine MA, Duarte MV, Suñé NL. Cadmium uptake by floating macrophytes. Water Research. 2001;35(11):2629-2634

41. Thayaparan M et al. Rhizofiltration of Pb by Azolla Pinnata. International Journal of Environmental Sciences. 2013;3(6):1811

42. Wu G et al. A critical review on the bio-removal of hazardous heavy metals from contaminated soils: Issues, progress, eco-environmental concerns and opportunities. Journal of Hazardous Materials. 2010;174(1):1-8

43. Ghosh M, Singh S. A review on phytoremediation of heavy metals and utilization of it's by products. Asian Journal on Energy & Environment. 2005;6(4):18

44. Bonanno G. Comparative performance of trace element bioaccumulation and biomonitoring in the plant species Typha domingensis, Phragmites australis and Arundo donax. Ecotoxicology and Environmental Safety. 2013;97:124-130

45. Vangronsveld J et al. Phytoremediation of contaminated soils and groundwater: Lessons from the field. Environmental Science and Pollution Research. 2009;16(7):765-794

46. Abreu CA et al. Organic matter and barium absorption by plant species grown in an area polluted with scrap metal residue. Applied and Environmental Soil Science. 2012;2012

47. Andreazza R et al. Evaluation of two Brazilian indigenous plants for phytostabilization and phytoremediation of copper-contaminated soils. Brazilian Journal of Biology. 2015;75(4):868-877

48. Najeeb U et al. Enhancing the lead phytostabilization in wetland plant Juncus effusus L. through somaclonal manipulation and EDTA enrichment. Arabian Journal of Chemistry. 2014

49. Zhang X, Zhang X, Huang K. Phytostabilization of acidic soils with heavy metal contamination using three forage grasses in combination with organic and inorganic amendments. Soil and Sediment Contamination: An International Journal, 2016(just-accepted): p. 00-00

50. Dobos L, Carmen P. The most important methods for depollution of hydrocarbons polluted soils. Bulletin of University of Agricultural Sciences and Veterinary medicine Cluj-Napoca Agriculture. 2009;66(1)

51. Kobraee S et al. Influence of micronutrient fertilizer on soybean nutrient composition. Indian Journal of Science and Technology. 2011;4(7):763-769

52. Nikolić M, Stevović S. Family Asteraceae as a sustainable planning tool in phytoremediation and its relevance in urban areas. Urban Forestry & Urban Greening. 2015;14(4):782-789

53. Wang J et al. Remediation of mercury contaminated sites–a review. Journal of Hazardous Materials. 2012;221:1-18

54. Sathya A et al. Cultivation of sweet sorghum on heavy metal contaminated soils by phytoremediation approach for production of bioethanol. In: Bioremediation and Bioeconomy. 2016. pp. 271-292

55. Wang J, Chen C. Biosorption of heavy metals by Saccharomyces cerevisiae: A review. Biotechnology Advances. 2006;24(5):427-451

56. Wongsasuluk P et al. Heavy metal contamination and human health risk assessment in drinking water from shallow groundwater wells in an agricultural area in Ubon Ratchathani province, Thailand. Environmental Geochemistry and Health. 2014;36(1):169-182

57. Duruibe J, Ogwuegbu M, Egwurugwu J. Heavy metal pollution and human biotoxic effects. International Journal of Physical Sciences. 2007;2(5):112-118

58. Nagajyoti P, Lee K, Sreekanth T. Heavy metals, occurrence and toxicity for plants: A review. Environmental Chemistry Letters. 2010;8(3):199-216

59. Baker A, Brooks R. Terrestrial higher plants which hyperaccumulate metallic elements. A review of their distribution, ecology and phytochemistry. Biorecovery. 1989;1(2):81-126

60. Krämer U. Metal hyperaccumulation in plants. Annual Review of Plant Biology. 2010;61:517-534

61. Milner MJ, Kochian LV. Investigating heavy-metal hyperaccumulation using Thlaspi caerulescens as a model system. Annals of Botany. 2008;102(1):3-13

62. Baker AJ et al. Metallophytes: The unique biological resource, its ecology and conservational status in Europe, central Africa and Latin America. Ecology of Industrial Pollution. 2010:7-40

63. Chaney RL et al. Improved understanding of hyperaccumulation yields commercial phytoextraction and phytomining technologies. Journal of Environmental Quality. 2007;36(5):1429-1443

64. Kärenlampi S et al. Genetic engineering in the improvement of plants for phytoremediation of metal polluted soils. Environmental Pollution. 2000;107(2):225-231

65. Fulekar M, Singh A, Bhaduri AM. Genetic engineering strategies for enhancing phytoremediation of heavy metals. African Journal of Biotechnology. 2009;8(4)

66. Yurekli F, Kucukbay Z. Synthesis of phytochelatins in Helianthus annuus is enhanced by cadmium nitrate. Acta Botanica Croatica. 2003;62(1):21-25

67. Cobbett CS. Phytochelatin biosynthesis and function in heavy-metal detoxification. Current Opinion in Plant Biology. 2000;3(3):211-216

68. Memon AR, Schröder P. Implications of metal accumulation mechanisms to phytoremediation. Environmental Science and Pollution Research. 2009;16(2):162-175

69. Jabeen R, Ahmad A, Iqbal M. Phytoremediation of heavy metals: Physiological and molecular mechanisms. The Botanical Review. 2009;75(4):339-364

70. Sheoran V, Sheoran A, Poonia P. Role of hyperaccumulators in phytoextraction of metals from contaminated mining sites: A review. Critical Reviews in Environmental Science and Technology. 2010;41(2):168-214

71. Seth CS. A review on mechanisms of plant tolerance and role of transgenic plants in environmental clean-up. The Botanical Review. 2012;78(1):32-62

72. Visioli G, Marmiroli N. Proteomics of plant hyperaccumulators. In: Metal Toxicity in Plants: Perception, Signaling and Remediation. Berlin, Heidelberg: Springer; 2012. pp. 165-186

73. Bona E et al. Proteomic analysis of Pteris vittata fronds: Two arbuscular mycorrhizal fungi differentially modulate protein expression under arsenic contamination. Proteomics. 2010;10(21):3811-3834

74. Visioli G et al. Comparison of protein variations in Thlaspi Caerulescens populations from metalliferous and non-metalliferous soils. International Journal of Phytoremediation. 2010;12(8):805-819

75. Ingle RA, Smith JAC, Sweetlove LJ. Responses to nickel in the proteome of the hyperaccumulator plant alyssum lesbiacum. BioMetals. 2005;18(6):627-641

76. Freeman JL et al. Increased glutathione biosynthesis plays a role in nickel tolerance in Thlaspi nickel hyperaccumulators. The Plant Cell. 2004;16(8):2176-2191

77. Farinati S et al. Proteomic analysis of Arabidopsis halleri shoots in response to the heavy metals cadmium and zinc and rhizosphere microorganisms. Proteomics. 2009;9(21):4837-4850

78. Waschke A et al. Identification of heavy metal-induced genes encoding glutathione S-transferases in the arbuscular mycorrhizal fungus Glomus intraradices. Mycorrhiza. 2006;17(1):1-10

79. Maestri E et al. Metal tolerance and hyperaccumulation: Costs and trade-offs between traits and environment. Environmental and Experimental Botany. 2010;68(1):1-13

80. Pence NS et al. The molecular physiology of heavy metal transport in the Zn/Cd hyperaccumulator Thlaspi caerulescens. Proceedings of the National Academy of Sciences. 2000;97(9):4956-4960

81. Assunção A et al. Elevated expression of metal transporter genes in three accessions of the metal hyperaccumulator Thlaspi caerulescens. Plant, Cell & Environment. 2001;24(2):217-226

82. Lombi E et al. Influence of iron status on cadmium and zinc uptake by different ecotypes of the hyperaccumulator Thlaspi caerulescens. Plant Physiology. 2002;128(4):1359-1367

83. Tuomainen M et al. Proteomics of Thlaspi caerulescens accessions and an inter-accession cross segregating for zinc accumulation. Journal of Experimental Botany. 2010;61(4):1075-1087

84. Hammond JP et al. A comparison of the Thlaspi caerulescens and Thlaspi arvense shoot transcriptomes. New Phytologist. 2006;170(2):239-260

85. Hu R et al. Physiological responses and tolerance mechanisms to Pb in two xerophils: Salsola passerina Bunge and Chenopodium album L. Journal of Hazardous Materials. 2012;205:131-138

86. Rejeb KB et al. Evaluation of the Cd2+ phytoextraction potential in the xerohalophyte Salsola kali L. and the impact of EDTA on this process. Ecological Engineering. 2013;60:309-315

87. Osman HE, Badawy RK. Effect of pollution on the chemical content and secondary metabolites of Zygophyllum coccineum and Tamarix nilotica. Egyptian Pharmaceutical Journal. 2013;12(1):73

88. De La Rosa G et al. Role of ethylenediaminetetraacetic acid on lead uptake and translocation by tumbleweed (Salsola kali L.). Environmental Toxicology and Chemistry. 2007;26(5):1033-1039

89. Lefèvre I, Corréal E, Lutts S. Cadmium tolerance and accumulation in the noxious weed Zygophyllum fabago. Botany. 2005;83(12):1655-1662

90. Carvalho KM, Martin DF. Removal of aqueous selenium by four aquatic plants. Journal of Aquatic Plant Management. 2001;39:33-36

91. Ebbs S et al. Phytoextraction of cadmium and zinc from a contaminated soil. Journal of Environmental Quality. 1997;26(5):1424-1430

92. Prasad M, Tewari JC. Prosopis juliflora (Sw) DC: Potential for bioremediation and bioeconomy. In: Bioremediation and Bioeconomy. 2016. pp. 49-76

93. Shukla O et al. Growth responses and metal accumulation capabilities of woody plants during the phytoremediation of tannery sludge. Waste Management. 2011;31(1):115-123

94. Nie M et al. Understanding plant-microbe interactions for phytoremediation of petroleum-polluted soil. PLoS One. 2011;6(3):e17961

95. Frick C, Germida J, Farrell R. Assessment of phytoremediation as an in-situ technique for cleaning oil-contaminated sites. In technical seminar on chemical spills. 1999. Environment Canada; 1998

96. Njoku K, Akinola M, Oboh B. Phytoremediation of crude oil contaminated soil: The effect of growth of Glycine max on the physico-chemistry and crude oil contents of soil. Nature and Science. 2009;7(10):79-87

97. Nedjimi B, Daoud Y. Cadmium accumulation in Atriplex halimus subsp. schweinfurthii and its influence on growth, proline, root hydraulic conductivity and nutrient uptake. Flora-Morphology, Distribution, Functional Ecology of Plants. 2009;204(4):316-324

98. Lutts S et al. Heavy metal accumulation by the halophyte species Mediterranean saltbush. Journal of Environmental Quality. 2004;33(4):1271-1279

99. Conesa H et al. Influence of soil properties on trace element availability and plant accumulation in a Mediterranean salt marsh polluted by mining wastes: Implications for phytomanagement. Science of the Total Environment. 2011;409(20):4470-4479

100. Manousaki E, Kalogerakis N. Phytoextraction of Pb and Cd by the Mediterranean saltbush (Atriplex halimus L.): Metal uptake in relation to salinity. Environmental Science and Pollution Research. 2009;16(7):844-854

101. Hegazy A, Abdel-Ghani N, El-Chaghaby G. Phytoremediation of industrial wastewater potentiality by Typha domingensis. International Journal of Environmental Science and Technology. 2011;8(3):639-648

102. Fediuc E, Erdei L. Physiological and biochemical aspects of cadmium toxicity and protective mechanisms induced in Phragmites australis and Typha latifolia. Journal of Plant Physiology. 2002;159(3):265-271

103. Paz-Alberto AM et al. Phytoextraction of lead-contaminated soil using vetivergrass (Vetiveria zizanioides L.), cogongrass (Imperata cylindrica L.) and carabaograss (Paspalum conjugatum L.). Environmental Science and Pollution Research International. 2007;14(7):498-504.

13 Chapter

EVALUATION OF THE PRESENCE AND CONCENTRATION OF HEAVY METALS IN LANDFILL LEACHATE: AN ANALYSIS OF THE THOHOYANDOU LANDFILL IN THE LIMPOPO PROVINCE OF SOUTH AFRICA

1. INTRODUCTION: AN OVERVIEW

1.1. General

The management of solid waste is a matter of worldwide significance in both developing and wealthy nations. Despite extensive efforts to raise awareness and mitigate the trash produced by human activities, there has been a global rise in the output of solid waste [1, 2, 3]. This phenomenon may be attributed, in part, to the rise in population, industrialization, and urbanization. Various initiatives aimed at efficiently managing solid trash in response to the recognized negative health and environmental consequences have been documented [4, 5, 6]. Landfilling is still widely employed as a primary approach for managing solid waste in many regions across the globe. The approach is recommended because to its efficiency, safety, and cost-effectiveness [7, 8, 9]. Various advancements in landfill technology have been documented to improve its appropriateness for the handling of solid waste. Leachate is commonly produced as a result of landfill activities, mostly due to the rising levels of soil moisture and other conducive climatic conditions [1, 5]. Leachate collection and treatment facilities are frequently not included in the construction of landfill sites in many poor nations. Leachate is a harmful consequence of depositing solid waste into landfills, since it leads to the pollution of surface and groundwater. The degree of pollution is determined by the quality of the leachate produced by the landfill [4, 7]. The composition of landfill garbage exhibits regional variations and is contingent upon several factors. Factors that influence the composition of leachate include the site's location, the socioeconomic status of the landfill's users, the technology used in the landfill, and the age of the waste generated. Prevailing environmental factors, such as weather conditions at the landfill site, also play a role [1, 4, 7]. Blight and Fourie [10] documented the fluctuation of the solid waste component that was discarded at landfill sites (Table 1). This encompasses a wide spectrum of substances, including both basic organic and inorganic materials as well as more intricate compounds.

Table 1. Percentage of solid waste composition from different parts of the world [10].

Components	% Composition by undried mass							
	USA	UK	Cape Town	Johannesburg	Gaborone	Lima	Delhi	Abu Dhabi
Ash, soil, building rubble, other	8–14	18	39*	1**	3	16	41	16
Paper and cardboard	34–44	29	19	23	12	14	9	6
Plastic	8–10	7	7*	13	5	7	1+	12
Metals	11–13	8	2+	10	6	4	1+	8
Glass	4–9	10	1+	8	6	3	1+	9
Organic, putrescible	12–22	25	32	45	68	56	47	49

* Building rubble landfilled, in which sand cover is included.

**Building rubble not landfilled, in which cover soil is not included.

+Scavenged before landfilling.

Typically, in the majority of poor nations, the practice of separating hazardous trash from nonhazardous waste prior to disposing it in a landfill is not prevalent. Edokpayi et al. [11] documented the deposition of several types of hazardous waste from households into sanitary landfill sites. If both types of garbage are disposed of in this manner, the quality of the leachate will be worse compared to if the hazardous waste was pre-treated before being placed in a landfill or treated using alternative means.

1.2. Principles of leachate formation

Leachate is generated through the chemical reaction between the waste present in the landfill and water originating from soil moisture, rainfall, snow, and other liquid waste that is deposited in the landfill area. Leachate can also be generated through chemical and biological reactions occurring inside the landfill. The solid waste in a landfill experiences nonuniform and intermittent moisture percolation due to the production of leachate [5, 12, 13]. As a result, the soluble components are extracted from the trash and transferred into the leachate. A diverse range of pollutants, including heavy metals, persistent organic pollutants such as flame retardants, polycyclic aromatic hydrocarbons (PAHs), phenols, polychlorinated biphenyls (PCBs), chlorinated alkyl phosphate, polyfluoroalkyl and other fluorinated chemicals, pharmaceuticals and personal care products, polybrominated diphenyl ethers (PBDEs), polychlorinated dibenzofurans (PCDFs), polychlorinated dibenzodioxins (PCDDs), and polychlorinated diphenyl ethers (PCDEs), are also accumulated in the leachate as it percolates through the waste pile in the landfill [14, 15, 16, 17, 18, 19, 20]. As a result, leachate becomes a highly concentrated effluent due to its elevated quantities of several types of pollutants. Incorporating leachate control is essential when designing solid waste management systems. The volume and flow rate of the leachates are contingent upon the infiltration of water through the layers of waste [5]. The movement and speed of the leachate influence the transmission of pollutants and the breakdown of organic matter. Farquhar [21] detailed the overall process of leachate formation in landfills, as shown in Fig. 1.

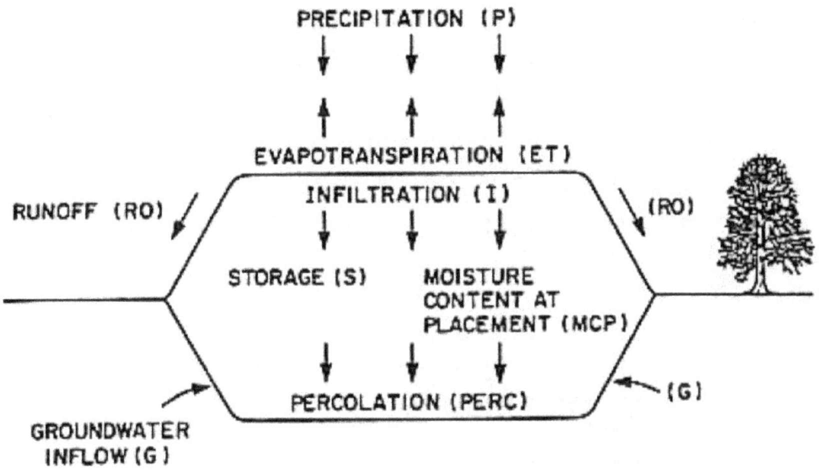

Fig. 1. Generalized pattern of leachate generation [21].

Percolation from the unsaturated zone to the saturated zone allows for the potential of groundwater contamination. Precipitation events might potentially lead to surface water contamination through surface runoff in landfills. If the development of leachate from landfills is not well controlled, it can result in numerous detrimental environmental and health consequences.

1.3. Leachate management and treatment

Effective leachate management is achievable when the landfill's design allows for sufficient liner, collecting, and storage systems. Incorporating landfill liners into the early planning of the waste site is essential. The liners are essential because they serve as a barrier, preventing the leachate produced by the landfill from seeping into the environment or groundwater aquifer [13, 22]. Leachate liners can be made from many materials, such as natural clays, geomembranes (made from plastic materials like polyvinyl chloride (PVC) and high-density polyethylene (HDPE)), geotextiles, geonet, and geosynthetic clay liners [13, 22, 23, 24, 25]. The lining system might be composed of a monolithic material, such as clay, or a composite material consisting of clay and other substances. There are three primary types of lining systems: single, composite, and double lining systems. When choosing a lining system for a landfill, it is important to examine many considerations, as certain lining materials can be susceptible to degradation caused by chemicals present in the leachate.

In certain developing nations, the infrastructure for managing leachate, such as landfill lining and collecting systems, is commonly present in big cities. However, these facilities are typically lacking in other cities, as well as in periurban and rural areas. The collection of leachate in landfills is crucial in order to prevent its accumulation to a level where it can overflow, leading to water pollution in rivers and streams [5, 12]. Insufficient leachate management in the developing countries is mostly hindered by poverty, lack of necessary competence, corruption, and misuse of funding. Various approaches have been documented for the control and treatment of leachate in various regions across the globe. A more straightforward approach is to introduce the leachate as an input into a wastewater treatment facility [26]. This strategy is cost-effective as it eliminates the need for constructing a new treatment facility. The conveyance pipe connecting the landfill to the wastewater treatment facility is adequate, as long as the distance between the two facilities is relatively short. Pretreatment may be necessary prior to merging the leachate with the influent of the municipal wastewater treatment facility, depending on the quality of the leachate. However, in certain circumstances, such as when a significant amount of leachate is produced daily or when the distance between the facilities is considerable, it is preferable to treat the leachate on-site or distribute it back into the landfill. Various techniques have been documented for treating leachate, including physical methods such as evaporation, floatation, air stripping, filtration, and membrane processes. Chemical methods such as coagulation and precipitation, adsorption, ion exchange, and chemical oxidation are also used. Additionally, biological methods involving the use of ponds in aerobic and anaerobic modes have been reported [26, 27, 28, 29, 30]. The choice of treatment method is often determined by factors such as cost, knowledge, the age of the landfill, expected volume of leachate created, climatic conditions, and procedure efficiency. In South Africa, the use of biological techniques for treating leachate is prevalent [26].

1.4. Study area

The landfill site with geographical coordinates 23°0'12.3" S and 30°28'0.16″ E is located in Thulamela Municipality of Limpopo Province, South Africa (Fig. 2a).

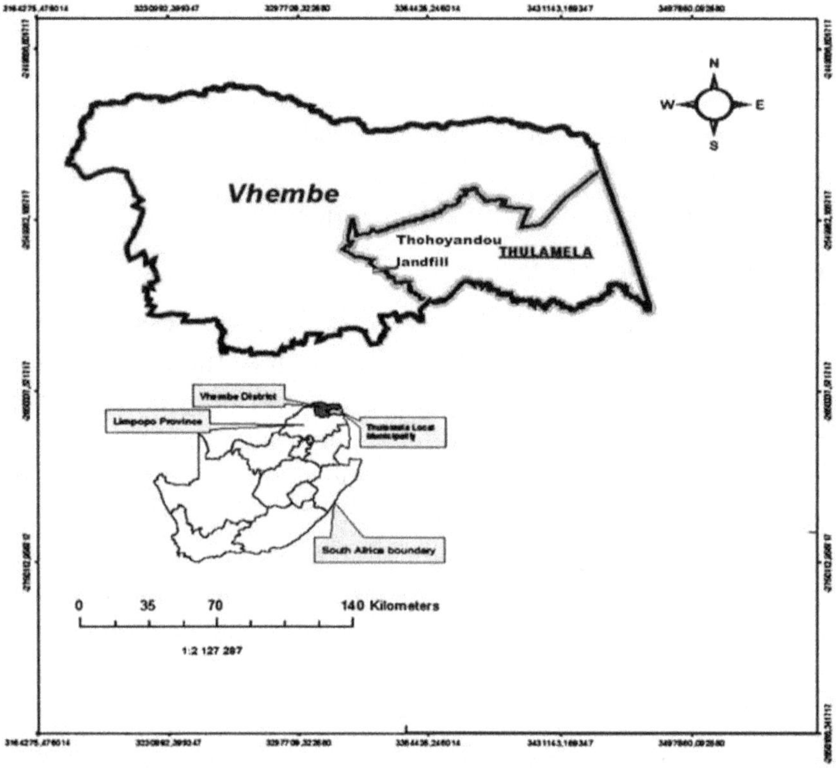

a

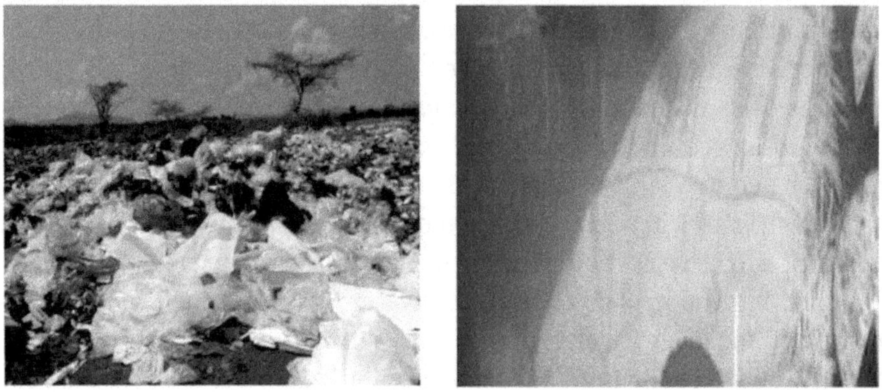

b

Fig. 2. (a) Map of the study area. (b) Some pictures from Thohoyandou landfill site.

The site is located 180 km from the provincial city of Polokwane. In 2014, the facility received 210,000 tons of waste. Thohoyandou is a semiarid region with variations in temperature between 12 and 22°C in the dry season (winter) and between 20 and 40°C in the wet season (summer) [31]. Precipitation also varies between 340 and 2000 mm with an average rainfall of 800 mm. The region is classified as a humid subtropical dry forest biozone [32]. The landfill site is very close to the municipal wastewater treatment plant. Common surface water sources in the region include the Dzindi, Luvhuvhu and Mvudi Rivers. Thohoyandou landfill currently receives all kinds of sanitary wastes from the periurban center, which is Thohoyandou and other villages in the suburb. At the time of this study, there was no leachate collection and treatment system on the site. Residential areas are located around the landfill. Some other major land use around the landfill includes subsistence agriculture, schools and wastewater treatment plant. Some pictures taken from the landfill are presented in Fig. 2b.

2. MATERIALS AND METHODS

2.1. Sampling

Six leachate samples were collected during the wet season of 2014 from various places within the premises of the landfill site. The samples were collected in a precleaned 1-L polyethylene sample container. The containers were washed with livid detergent and soaked with 1% nitric acid. Subsequently, the sample containers were rinsed with deionized water. Physicochemical parameters such as pH, conductivity and turbidity were measured in situ, while samples for the determination of heavy metals were preserved using concentrated nitric acid. The samples were transported on ice to Hydrology and Water Resources Department of the University of Venda. Four major metals (Calcium (Ca), Potassium (K), Magnesium (Mg) and Sodium (Na)) and 10 heavy metals (Aluminum (Al), Cadmium (Cd), Chromium (Cr), Cobalt (Co), Copper (Cu), Iron (Fe), Lead (Pb), Manganese (Mn), Nickel (Ni) and Zinc (Zn)) were selected and analyzed due to their availability in landfill leachate. The samples were preserved at -4°C until further analysis.

2.2. Instrumentation

pH and conductivity multimeter supplied by Fischer Thermoscientific was used for field measurements of pH and conductivity, respectively, while a turbidimeter supplied by HANA was employed to determine the turbidity of the samples. The measuring devices were calibrated following the protocol approved by the manufacturer. Flame atomic absorption spectrometer (FAAS) supplied by Perkin Elmer was used for the analysis of major and heavy metal content in the leachate samples. Calibration standards were prepared from AAS grade reagent for all the metals of interest.

3. RESULTS AND DISCUSSION

The pHs of the leachate from all sites of sampling (Table 2) were found within the acceptable range of wastewater discharge onto the water resource by the South African Department of Water Affairs [33]. An average leachate pH of 7.39 was determined in this study which is suitable for methanogenic bacteria [34]. Various average pH values below and above this value have been reported for leachate in different parts of the developing world (Table 3). pH usually affects the speciation of metals in solutions with most metals easily speciated to their toxic forms in acidic pH below the value of 4, while some precipitate out of the solution at alkaline pH values.

Table 2. Physicochemical parameters of leachate samples.

Samples	pH	Conductivity (μS/m)	Turbidity (NTU)
L1	7.60	486	22.56
L2	7.47	485	28.83
L3	7.68	2288	295.5
L4	7.42	604	25.43
L5	7.20	721	33.15
L6	6.97	426	12.78

Table 3. Comparison of the average physicochemical leachate characteristics in some developing countries.

Parameters	Thohoyandou landfill (South Africa)	Alexandria landfill (Egypt) [34]	Semur landfill (Indian) [35]	El-Jadida landfill (Morocco) [36]
pH	7.39	8.9	6.9	8.8
Conductivity (μS/cm)	835	30,725	NR	26,000
Turbidity (NTU)	69.71	NR	NR	NR
Na (mg/L)	33.88	364	462	NR
K(mg/L)	22.94	5400	1241	NR
Ca (mg/L)	46.27	8.4	NR	190
Mg (mg/L)	16.58	302.5	NR	235
NR indicates values that were not reported.				

Generally, the presence of inorganic charged particles such as carbonates, chlorides, sulfates and bicarbonate ions in water is often estimated in terms of the conductivity of the water. Conductivity oftentimes is also used as a surrogate for total dissolved solids [37]. Table 2 shows the levels of conductivity determined from all the sampling sites. This level complied with the DWA wastewater discharge limit [33] and was much lower than the levels reported in Alexandria landfill leachate in Egypt (Table 3), Columbia (22,000 μS/cm) and Morocco (26,000 μS/cm) [34, 36, 38]. However, Rahim et al. [39] reported a lower conductivity level of 31.68 μS/cm for landfill leachate in Malaysia.

The turbidity of the leachate samples obtained was very high. Liquids with high turbidity often indicate the likelihood of silt, clay, fine organic matter and microscopic organisms that are suspended in the water. Site three recorded a very high turbidity level, much higher than the levels from other sites in the landfill.

Fig. 3 shows the levels of major metals (Na, K, Mg and Ca) in the leachate of Thohoyandou landfill. The levels found are not higher than what can be disposed of into water bodies. The only concern is the continuous migration of salt-rich leachate into surface and groundwater sources which could increase their levels in the river and groundwater aquifers. Various sequences of major metals' occurrence in leachate varied greatly from one location to the other. The sequence for major metals determined in this study follows the trend: Ca > Na > K > Mg (Table 3), while at Alexandria landfill in Egypt, the sequence is K > Na > Mg > Ca [34]. Level of Mg is higher than Ca in the leachate of all the landfills reported except Thohoyandou landfill. Although rock type in groundwater aquifers contributes significantly to its salinity level, continuous percolation of metal-rich leachate may also contribute to increased groundwater salinity.

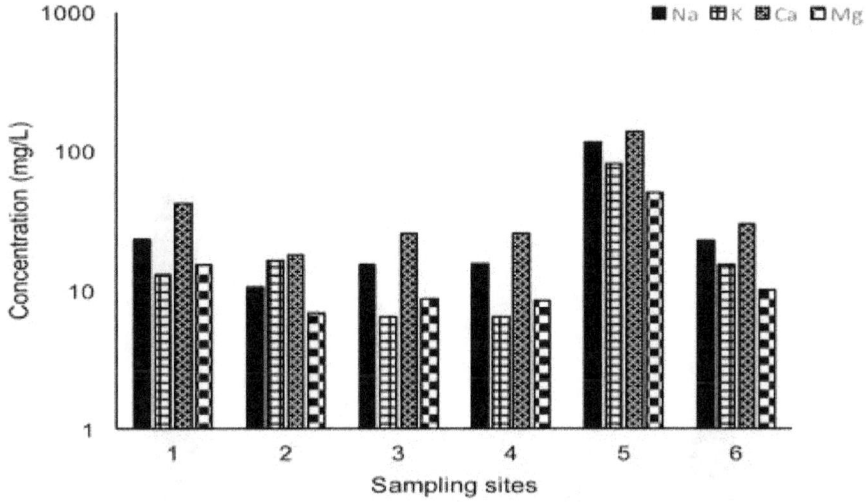

Fig. 3. Concentrations of major metals in leachate samples.

The levels of heavy metals determined in this study are presented in Fig.s 4 and 5. Variable levels of heavy metals were determined in the landfill site. On average, Fe, Mn and Al are present in higher concentrations than the other metals. A possible reason for the high level of these metals could be due to their relative abundance on Earth. Previous studies have shown that Fe and Al are present in higher concentrations in river water, soils and sediments around Thohoyandou when compared to other metals [32, 37, 40]. The relatively low levels of the more hazardous heavy metals could suggest that the waste generated from this region is not largely from a highly industrialized area. It is, however, noteworthy to mention that most of the heavy metals were present in concentrations higher than the normal wastewater discharge guideline values except for Zn and Cr. Ni, Al and Co does not have guideline values for their discharge. Table 4 shows the comparison of average heavy metal concentration in leachate from landfills in some developing countries. Fe recorded the highest level from all the countries except in Jebel Chakir, Landfill of Tunisia, where Fe was not part of the metals determined [34, 36, 37, 41, 42]. Mn and Zn were also part of the metals that generally existed in high concentration.

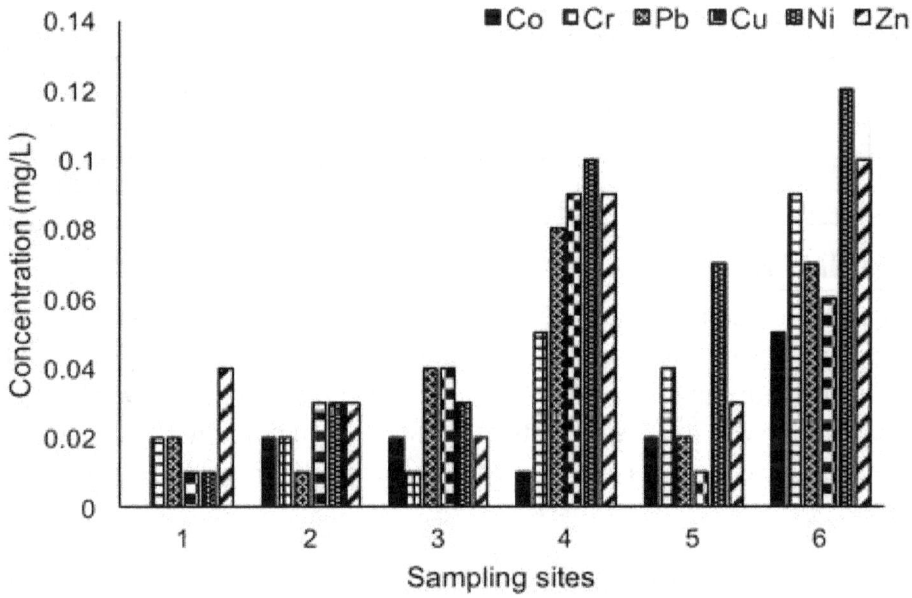

Fig. 4. Levels of Co, Cr, Pb, Cu, Ni and Zn in different sampling sites of Thohoyandou landfill.

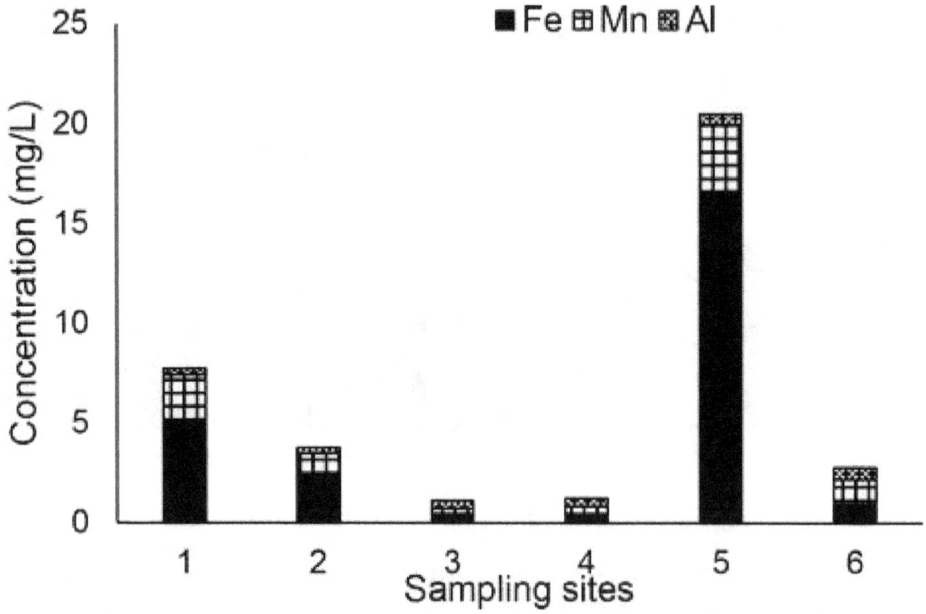

Fig. 5. Levels of Fe, Mn and Al in different sampling sites of Thohoyandou landfill.

Table 4. Comparison of the average heavy metal characteristics in some developing countries.

Heavy metals (mg/L)	Thohoyandou landfill (South Africa)	Alexandria landfill (Egypt) [34]	Semur landfill (Indian) [35]	El-Jadida landfill (Morocco) [36]	Alger landfill (Algeria) [41]	Jebel Chakir landfill (Tunisia) [42]	DWA wastewater discharge limit [33]
Al	0.42	NR	NR	NR	NR	NR	NR
Cd	0.02	0.10	0.02	0.03	NR	0.04	0.005
Cr	0.02	0.06	0.28	0.16	0.50	1.17	0.05
Co	0.02	NR	NR	0.21	NR	NR	NR
Cu	0.07	0.07	0.71	0.16	0.45	0.80	0.01
Fe	4.26	6.31	58.91	24	12.30	NR	0.3
Pb	0.03	0.02	1.20	NR	NR	0.35	0.01
Mn	1.50	0.84	NR	1.25	0.40	0.05	0.1
Ni	0.03	0.10	0.31	0.13	0.25	0.23	NR
Zn	0.08	0.75	1.29	0.75	0.70	0.75	0.1

NR indicates values that were not reported.

4. CONCISE SUMMARY

This chapter has presented the essential foundational statistics about leachate for the Thohoyandou landfill. The concentrations of iron (Fe) and manganese (Mn) measured in the leachate of this study were significantly elevated, surpassing the recommended threshold for wastewater disposal. The concentration of aluminum was likewise elevated. The most intriguing finding revealed that the leachate from Thohoyandou dump has a same level of quality as leachate from other poor nations.

REFERENCES

1. Al-Yaqout AF, Hamoda MF. Evaluation of landfill leachate in arid climate-a case study. Environment International. 2003;29:593-600. DOI: 10.1016/S0160-4120(03)00018-7

2. Scott J, Beydoun D, Amal R, Low G, Cattle J. Landfill management, leachate generation, and leach testing of solid wastes in Australia and overseas. Critical Reviews in Environmental Science and Technology. 2005;35:239-332. DOI: 10.1080/10643380590917969

3. Naveen BP, Mahapatra DM, Sitharam TG, Sivapullaiah PV, Ramachandra TV. Physico-chemical and biological characterization of urban municipal landfill leachate. Environmental Pollution. 2017;220:1-12. DOI: 10.1016/j.envpol.2016.09.002

4. El-Fadel M, Findikakis AN, Leckie JO. Environmental Impacts of solid waste landfilling. Journal of Environmental Management. 1997;50:1-25. DOI: 10.1006/jema.1995.0131

5. Zeiss C, Uguccioni M. Mechanisms and patterns of leachate flow in municipal solid waste landfills. Journal of Environmental Systems. 1994;23:247-270

6. Kassim SM. The importance of recycling in solid waste management. Macromolecular Symposia. 2012;320:43-50. DOI: 10.1002/masy.201251005

7. Capelo J, de Castro MAH. Measuring transient water flow in unsaturated municipal solid waste – A new experimental approach. Waste Management. 2007;27:811-819. DOI: 10.1016/j.wasman.2006.04.012

8. El-Gohary FA, Kamel G. Characterization and biological treatment of pre-treated landfill leachate. Ecological Engineering. 2016;94:268-274. DOI: 10.1016/j.ecoleng.2016.05.074

9. Krčmar D, Tenodi S, Grba N, Kerkez D, Watson M, Rončević S, Dalmacija B. Preremedial assessment of the municipal landfill pollution impact on soil and shallow groundwater in Subotica, Serbia. Science of the Total Environment. 2018;615:1341-1354. DOI: 10.1016/j.scitotenv.2017.09.283

10. Blight GE, Fourie AB. Experimental landfill caps for semi-arid and arid climates. Waste Management & Research–Journal of International Solid Wastes and Public Cleansing Association, ISWA. 2005;23:113-125. DOI: 10.1177/0734242X05052458

11. Edokpayi JN, Odiyo JO, Durowoju OS, Adetoro A. In: Daniel Mmereki, editor. Household Hazardous Waste Management in Sub-Saharan Africa. Rijeka, Croatia: InTech; 2017. DOI:10.5772/66292. Available from: https://www.intechopen.com/books/household-hazardous-waste-management/household-hazardous-waste-management-in-sub-saharan-africa

12. El-Fadel M, Bou-Zeid E, Chahine W, Alayli B. Temporal variation of leachate quality from pre-sorted and baled municipal solid waste with high organic and moisture content. Waste Management. 2002;22:269-282

13. Hughes KL, Christy AD, Heimlich JE. Landfill Types and Liner Systems. United States of America: Ohio State University; 2013

14. Adeniyi A, Dayomi M, Siebe P, Okedeyi O. An assessment of the levels of phthalate esters and metals in the Muledane open dump, Thohoyandou, Limpopo Province, South Africa. Chemistry Central Journal. 2008;2:9. DOI: 10.1186/1752-153X-2-9

15. Chimuka L, Nefale F, Masevhe A. Determination of phenols in water samples using a supported liquid membrane extraction probe and liquid chromatography with photodiode array detection. South African Journal of Chemistry. 2007;60:102-108

16. Huset CA, Barlaz MA, Barofsky DF, Field JA. Quantitative determination of fluorochemicals in municipal landfill leachates. Chemosphere. 2011;82:1380-1386. DOI: 10.1016/j.chemosphere.2010.11.072

17. Busch J, Ahrens L, Sturm R, Ebinghaus R. Polyfluoroalkyl compounds in landfill leachates. Environmental Pollution. 2010;158:1467-1471. DOI: 10.1016/j.envpol.2009.12.031

18. Eggen T, Moeder M, Arukwe A. Municipal landfill leachates: A significant source for new and emerging pollutants. Science of the Total Environment. 2010;408:5147-5157. DOI: 10.1016/j.scitotenv.2010.07.049

19. Clarke BO, Anumol T, Barlaz M, Snyder SA. Investigating landfill leachate as a source of trace organic pollutants. Chemosphere. 2015;127:269-275. DOI: 10.1016/j.chemosphere.2015.02.030

20. Dudzinska MR, Czerwinski J. Persistent Organic Pollutants (POPs) in leachates from municipal landfills. International Journal of Environmental Engineering. 2011;3:253

21. Farquhar GJ. Leachate: Production and characterization. Canadian Journal of Civil Engineering. 1989;16:317-325. DOI: 10.1139/l89-057

22. Sarabian T, Rayhani MT. Hydration of geosynthetic clay liners from clay subsoil under simulated field conditions. Waste Management. 2013;33:67-73. DOI: 10.1016/j.wasman.2012.08.010

23. Benson CH, Kucukkirca IE, Scalia J. Properties of geosynthetics exhumed from a final cover at a solid waste landfill. Geotextiles and Geomembranes. 2010;28:536-546. DOI: 10.1016/j.geotexmem.2010.03.001

24. Divya PV, Viswanadham BVS, Gourc JP. Influence of geomembrane on the deformation behaviour of clay-based landfill covers. Geotextiles and Geomembranes. 2012;34:158-171. DOI: 10.1016/j.geotexmem.2012.06.002

25. Salim IA, Miller CJ, Howard JL. Sorption isotherm-sequential extraction analysis of heavy metal retention in landfill liners. Soil Science Society of America Journal. 1996;60:107-114. DOI: 10.2136/sssaj1996.03615995006000010018x

26. DWA Guidelines for Lechate Control. Pretoria, South Africa: Department of Water Affairs

27. Castrillón L, Fernández-Nava Y, Ulmanu M, Anger I, Marañón E. Physico-chemical and biological treatment of MSW landfill leachate. Waste Management. 2010;30:228-235. DOI: 10.1016/j.wasman.2009.09.013

28. Peng Y. Perspectives on technology for landfill leachate treatment. Arabian Journal of Chemistry. 2017;10:S2567-S2574. DOI: 10.1016/j.arabjc.2013.09.031

29. Foo KY, Hameed BH. An overview of landfill leachate treatment via activated carbon adsorption process. Journal of Hazardous Materials. 2009;171:54-60. DOI: 10.1016/j.jhazmat.2009.06.038

30. Raghab SM, Abd El Meguid AM, Hegazi HA. Treatment of leachate from municipal solid waste landfill. HBRC Journal. 2013;9:187-192. DOI: 10.1016/j.hbrcj.2013.05.007

31. Edokpayi JN, Odiyo JA, Msagati TAM, Potgieter N. Preliminary monitoring of faecal indicator organisms of surface water: A case study of Mvudi River, South Africa. Journal of Science and Technology (Ghana). 2016;36:33-37. DOI: 10.4314/just.v36i1.6

32. Edokpayi JN, Odiyo JO, Popoola OE, Msagati TAM. Assessment of trace metals contamination of surface water and sediment: A case study of Mvudi River, South Africa. Sustainability. 2016;8:135. DOI: 10.3390/su8020135

33. DWA Wastewater limit values applicable to discharge of wastewater into a river resource. National Water Act, Government Gazzette No 20528; 1999; p. Pretoria, South Africa

34. Abd El-Salam MM, Abu-Zuid G. Impact of landfill leachate on the groundwater quality: A case study in Egypt. Journal of Advanced Research. 2015;6:579-586. DOI: 10.1016/j.jare.2014.02.003

35. Nagarajan R, Thirumalaisamy S, Lakshumanan E. Impact of leachate on groundwater pollution due to non-engineered municipal solid waste landfill sites of erode city, Tamil Nadu, India. Journal of Environmental Health Science & Engineering. 2012;9:35. DOI: 10.1186/1735-2746-9-35

36. Chofqi A, Younsi A, Lhadi EK, Mania J, Mudry J, Veron A. Environmental impact of an urban landfill on a coastal aquifer (El Jadida, Morocco). Journal of African Earth Sciences. 2004;39:509-516. DOI: 10.1016/j.jafrearsci.2004.07.013

37. Edokpayi JN, Odiyo JO, Olasoji SO. Assessment of heavy metal contamination of Dzindi River, in Limpopo Province, South Africa. International Journal of Natural Sciences Research. 2014;2:185-194

38. Olivero-Verbel J, Padilla-Bottet C, De la Rosa O. Relationships between physicochemical parameters and the toxicity of leachates from a municipal solid waste landfill. Ecotoxicology and Environmental Safety. 2008;70:294-299. DOI: 10.1016/j.ecoenv.2007.05.016

39. Rahim BEA, Yusoff I, Samsudin AR, Yaacob WZW, Rafek AGM. Deterioration of groundwater quality in the vicinity of an active open-tipping site in West Malaysia. Hydrogeology Journal. 2010;18:997-1006. DOI: 10.1007/s10040-009-0567-3

40. Edokpayi JN, Odiyo JO, Msagati TAM, Popoola EO. Removal efficiency of faecal indicator organisms, nutrients and heavy metals from a peri-urban wastewater treatment plant in Thohoyandou, Limpopo Province, South Africa. International Journal of Environmental Research and Public Health. 2015;12:7300-7320. DOI: 10.3390/ijerph120707300

41. Belkacemi R, Kerbachi M. Characterization and evolution of the landfill leachate of Oued Smar, Alger. Techniques Sciences Methods. 1996;11:615-618

42. Abdelwaheb A, Moncef Z, Hamed BD. Landfill leachate generation and its impact on water at an urban landfill(Jebel Chakir, Tunisia). Hydrology: Current Research. 2012;3:1-6. DOI: 10.4172/2157-7587.1000128.

14 Chapter

THERMODYNAMICS OF THE PROCESS BY WHICH HEAVY METALS ARE ABSORBED ONTO LIGNOCELLULOSIC BIOMASS

1. INTRODUCTION: AN OVERVIEW

Metals can be introduced into the environment through various processes, including the erosion of soils and rocks, volcanic activity, and human activities [1, 2]. Modern industry is a significant contributor to environmental pollution, largely due to anthropogenic sources. These pollutants are regularly discharged into the land and water from a range of sources that cause pollution, including foundries, tanneries, textile, microelectronic, fertilizer and pesticide industries, mining activity, and other industrial activities [3]. Inorganic species exist naturally as ions, compounds, and complexes, and their toxicity and persistent nature can cause health issues and environmental deterioration. The current focus is on the development of sustainable and eco-friendly technology for the extraction of hazardous metals from industrial wastewater. There are various methods available to remove metal ions from water solutions, including both traditional physico-chemical approaches and newer bioremediation methods [3, 4, 5, 6, 7, 8, 9, 10]. Physico-chemical methods, such as chemical precipitation, ion exchange, oxidation/reduction, reverse osmosis, and electrochemical treatment, have been effectively used in industrial environments [2, 11, 12, 13]. However, these approaches have several disadvantages, including expensive acquisition and operation expenses, poor effectiveness at lower but still significant metal concentrations, and the potential for secondary contamination [14, 15, 16]. In order to address the problems outlined above, bioremediation techniques have emerged as promising options for treating heavy metal waste. Bioremediation techniques encompass bioaccumulation, biosorption, and phytoremediation. In certain instances, biosorption has shown remarkable efficacy, similar to the results achieved using ion exchange-based techniques. In conventional sorption studies, the efficacy of a sorbent is assessed by analyzing the rate at which the sorption process occurs and measuring the quantity of sorbate absorbed in relation to the concentration of the sorbate in the solution at equilibrium, resulting in the formation of an isotherm curve. The sorption isotherms themselves do not inherently convey any information regarding the underlying chemistry involved in the sorption phenomenon [17]. Investigating the impact of temperature on the sorption process and analyzing the thermodynamic properties, such as Gibb's free energy, enthalpy, and entropy, can yield valuable insights into the strength of the interactions between the sorbate and sorbent, as well as the energy involved in the sorption process. The ΔG0 (standard free energy) of the reaction is the discrepancy between the starting state (free solute compound) and the final equilibrated state (sorbed compound). This parameter is associated with the spontaneity of the sorption process. Delta G0 readings that are negative suggest that the process occurs spontaneously. The amount of the enthalpy change of the process (ΔH0, $kJ{\cdot}mol^{-1}$) provides insight into the nature of the sorption interactions, whether they are physical or chemical in nature. In physisorption-based processes, the enthalpy range is between 2.1 and 20.9 kJ·mol−1. However, chemisorption often exhibits greater enthalpy values ranging from 20 to 800 kJ·mol−1 [18]. The sign of the change in enthalpy ΔH0, whether it is negative (ΔH0 < 0) or positive (ΔH0 > 0), indicates whether the process is exothermic or endothermic, respectively. The change in entropy (ΔS0) primarily represents the alteration in the level of disorder of a system (at a macroscopic level) during a process. A positive value of this parameter signifies an augmented level of unpredictability at the boundary between the solid and solution, which may also include alterations in the structure of the sorbent and sorbate. Furthermore, if ΔS0 > 0, it indicates that there is an increase in the number of possible configurations for the adsorbed species. The negative change in entropy (ΔS0 < 0) indicates that the adsorption process occurs through an associative mechanism. A negative value of ΔS0 indicates a reduction in disorder at the interface between the sorbent and solution during the sorption process, leading to the escape of sorbate species from the solid phase to the solution phase. This chapter examines the sorption equilibrium of Cu(II), Ni(II), Pb(II), and Cd(II) onto grape stalks (GS), a lignocellulosic material. The experiments were conducted at various temperatures, enabling the collection of significant thermodynamic parameters to more accurately characterize the interactions between the divalent metals and the sorbent.

2. THE IMPACT OF TEMPERATURE ON THE EQUILIBRIUM OF CU(II), NI(II), PB(II), AND CD(II)

The grape stalk wastes from a wine production sector were cleansed with distilled water, fragmented into small pieces, dehydrated, and pulverized to produce a sorbent with particle sizes ranging from 0.25 to 0.50 mm. The adsorption equilibrium of Cu(II), Ni(II), Pb(II), and Cd(II) in individual metal solutions onto GS was investigated at various temperatures ranging from 5 to 60°C. The sorption isotherms were determined by mixing 0.1 g of GS powder with 15 mL of various solutions containing Cu(II), Ni(II), Pb(II), and Cd(II) at concentrations ranging from 5 to 1000 mg·L^{-1}. Sealed glass tubes were used, and the starting pH was set to 5.2. To ensure precise temperature regulation, the samples were placed in an incubator (ICP-500, Memmert). After allowing the sorbent biomaterials to reach equilibrium with the metal solutions for 24 hours, the samples were filtered and acidified by adding 100 μL of high-quality HNO_3. The metal concentration in the solution was then measured using flame atomic absorption spectrophotometry (Varian SpectrAA 220FS). The thermodynamic parameters associated with the affinity and energy of the interaction between the sorbent and sorbate were determined and analyzed based on the results of the experimental sorption equilibrium.

2.1. Sorption isotherms

Equilibrium relationships between adsorbent and adsorbate are described by adsorption isotherms and reflect the relationship between the quantity adsorbed and that remaining in solution at a given temperature [19]. Sorption isotherms provide essential information for optimization of the adsorption mechanism pathways since they are expression of the surface properties and capacities of the sorbents. They become therefore relevant tools in the design of sorption systems, since they help understanding how sorbates interrelate with the sorbent materials [20].

Cu(II), Ni(II), Pb(II) and Cd(II) sorption isotherms onto GS for the different temperatures explored were obtained by plotting the amount of metal adsorbed per GS sorbent mass unit (qe; mol·g^{-1}) as a function of the remaining metal concentration in solution (Ce; mol·L^{-1}). The amount of sorbed metal was computed according to the equation presented below:

$$q_e = \frac{(C_0 - C_f)}{m} * V \tag{1}$$

where Co and Cf are the initial and final metal concentration in solution, respectively ($mol{\cdot}L^{-1}$), m(g) is the sorbent mass (g) and V(L) represents the volume of the solution. The characteristic isotherms are presented in Fig. 1. This Fig. clearly demonstrates that the amount of metal sorbed increases as it does the remaining metal concentration in solution until a maximum value is achieved. In the studied range, the temperature seemed not to have a clear effect on the maximum sorption capacity of GS for the divalent metals, except for Cu(II). In this case, the increase of temperature involved an increase on the maximum sorption capacity at equilibrium from 0.22 $mmol{\cdot}g^{-1}$ (at 5°C) to approximately 0.28 (at 60°C). The experimental Cu(II), Ni(II), Pb(II) and Cd(II) sorption equilibrium results onto GS were submitted to Langmuir, Freundlich and D-R models. The different models and the results obtained are presented and discussed in the next section.

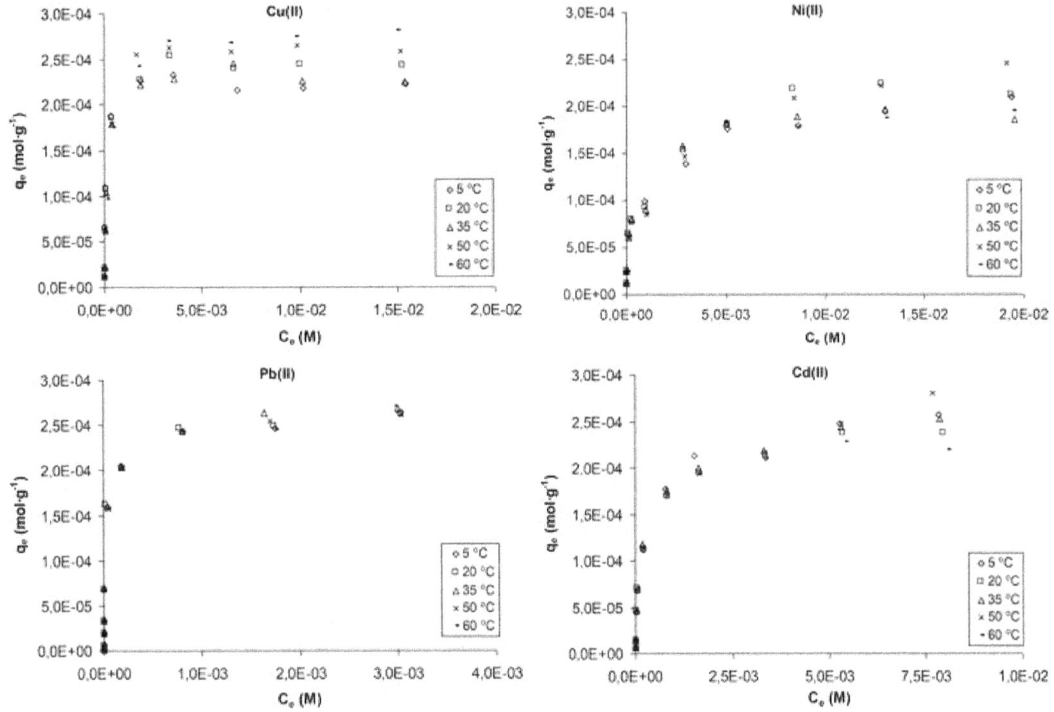

Fig. 1. Sorption isotherms of Cu(II), Ni(II), Pb(II) and Cd(II) onto GS. T: 5–60°C.

2.2. Modelling and computation of sorption equilibrium parameters

The equilibrium adsorption isotherms are one of the most important data that help understanding the sorption mechanism/s and provide fundamental insight for optimization and scale-up of sorption-based processes. Among the different isotherm models available, three of the most representative and largely employed have been chosen for this study: Langmuir, Freundlich and D-R isotherms.

The Langmuir model involves homogenous distribution of sorption sites and is based on the next set of assumptions: (i) the maximum sorption capacity corresponds to a saturated monolayer of solute in the surfaces, (ii) all the active sites are equivalent and the sorption energy remains constant, and (iii) there is no migration of adsorbed species in the plane of the surfaces. From this model, the maximum uptake, qmax ($mol{\cdot}g^{-1}$), and the Langmuir constant, KL ($L{\cdot}mol^{-1}$), can be obtained. While qmax reflects the maximum uptake of the sorbent, the parameter KL is a constant related to the energy of adsorption that quantitatively reflects the affinity between the sorbent and the sorbate. Fitting the experimental dataset to the Langmuir model, the effect of the temperature and of the nature of the metal on the different sorption equilibriums can be ascertained. By means of the Langmuir constant, it can also be discussed whether an adsorption system is favorable or unfavorable. The essential feature of the Langmuir isotherm can be expressed by means of the parameter RL, a dimensionless constant referred to as separation factor or equilibrium parameter. RL is calculated using the following equation [15, 21, 22, 23, 24, 25, 26, 27]:

$$R_L = \frac{1}{1 + K_L \cdot C_0} \quad (2)$$

being KL the Langmuir constant (L·mol−1) and C0 the initial metal concentration (mol·L−1). The RL parameter is considered as a reliable indicator of the adsorption being the next four the possible situations [15, 25] (Table 1).

Table 1. The isotherm characteristics according to the RL value.

R_L value	Type of isotherm
>1	Unfavorable
1	Linear
$0 < R_L < 1$	Favorable
0	Irreversible

The Freundlich model is based in an empirical equation largely employed in the description of sorption processes in heterogeneous systems. On its linear form, the Freundlich equation takes the form:

$$\log q_e = \log K_F + \frac{1}{n} \log C_e \quad (3)$$

where KF and 1/n are empirical constants indicate of the relative sorption capacity and sorption intensity, respectively.

The experimental dataset was also submitted to the D-R isotherm model. The selection of this model was supported by the fact that it is considered as more general than Langmuir and it does not rely necessarily in the formation of a homogenous monolayer surface or a constant adsorption potential. D-R model has been used by several authors to distinguish between physical and chemical adsorption onto different biomaterials [28, 29]. This model can be linearized and described by the next equation:

$$\ln q_e = \ln q_m - \beta \varepsilon^2 \quad (4)$$

being β is a constant related to the mean free energy of adsorption per mole of the adsorbate (mol2·J−2), qm is the theoretical saturation capacity of the monolayer and ε is the Polanyi potential. The expression of this last parameter is RTln(1 + (1/Ce)), being R (8.314 $J \cdot mol^{-1} \cdot K^{-1}$) the gas constant and T (K) the absolute temperature. Hence, by plotting ln(qe) against ε2, it is possible to generate the value of qm (mol·g−1) from the intercept and the value of β from the slope.

The constant β provides information about the mean free energy, E ($J \cdot mol^{-1}$). The parameter E is defined as the free energy change required to transfer 1 mol of sorbate from the solution to the solid surface and can be calculated using the relationship [30, 31]:

$$E = \frac{1}{\sqrt{2\beta}} \quad (5)$$

The fitting of the experimental sorption datasets to the linearized expressions of Langmuir, Freundlich and D-R adsorption isotherms for the different temperatures explored is presented in Fig.s 2–4, respectively. The separation factor (RL) calculated for the different initial metal concentrations has been also plotted and is presented in Fig. 5.

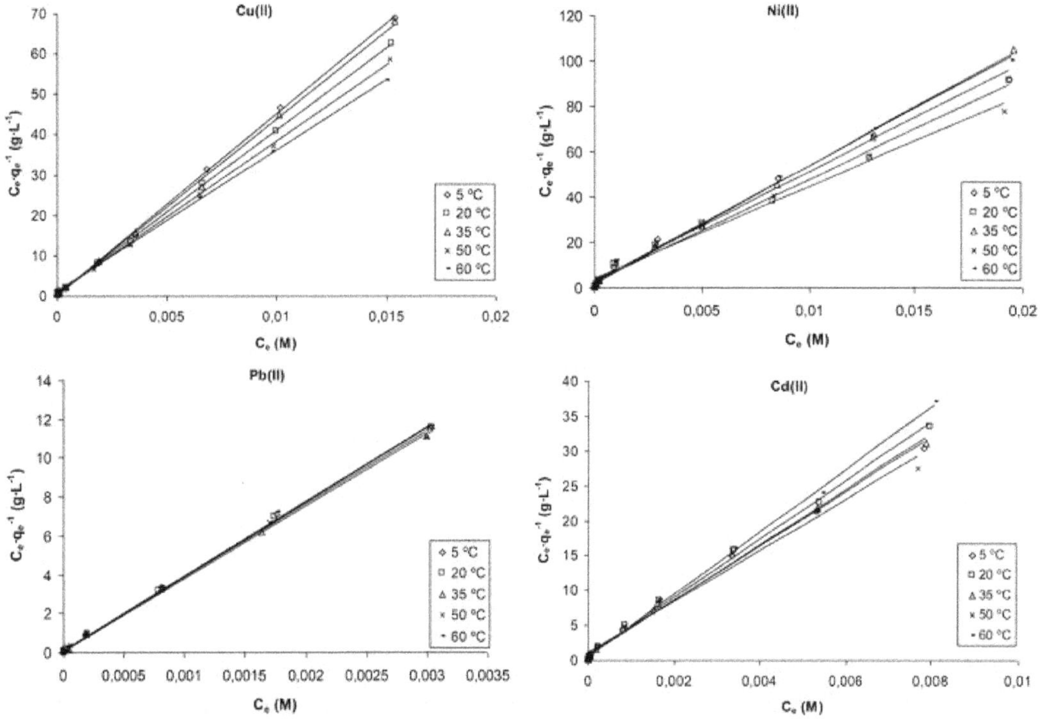

Fig. 2. Langmuir model fitting of Cu(II), Ni(II), Pb(II) and Cd(II) sorption results in the temperature range of 5–60°C.

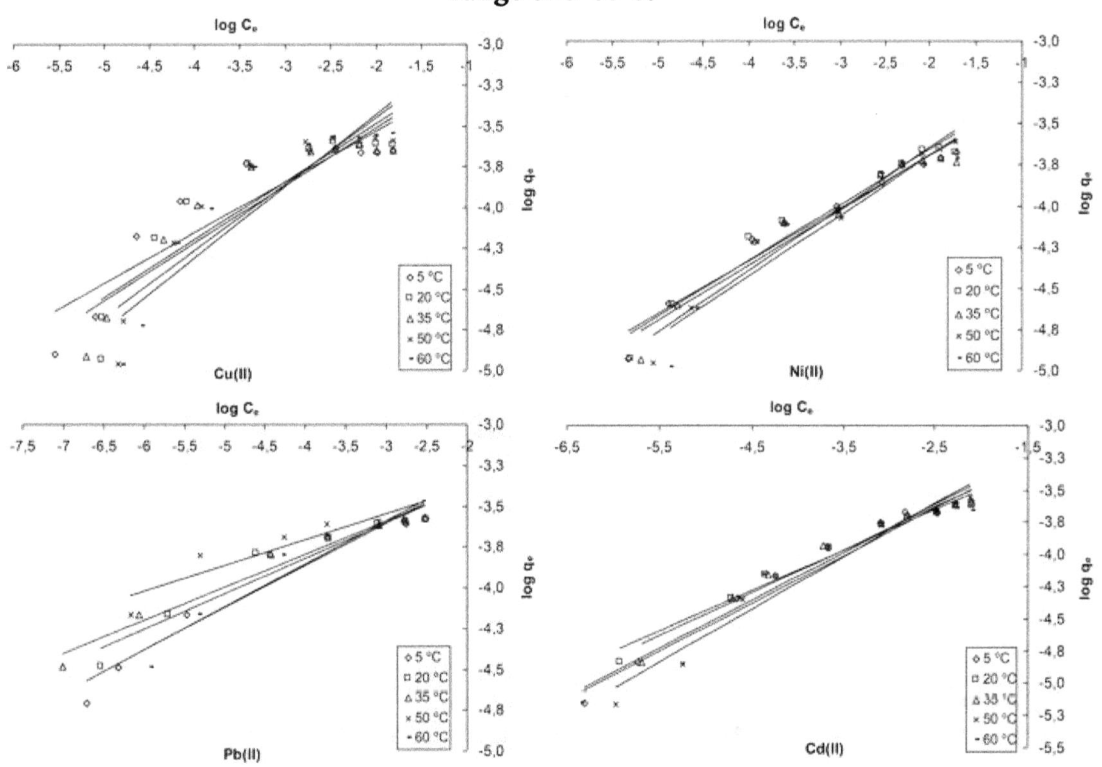

Fig. 3. Freundlich model fitting of Cu(II), Ni(II), Pb(II) and Cd(II) sorption results in the temperature range of 5–60°C.

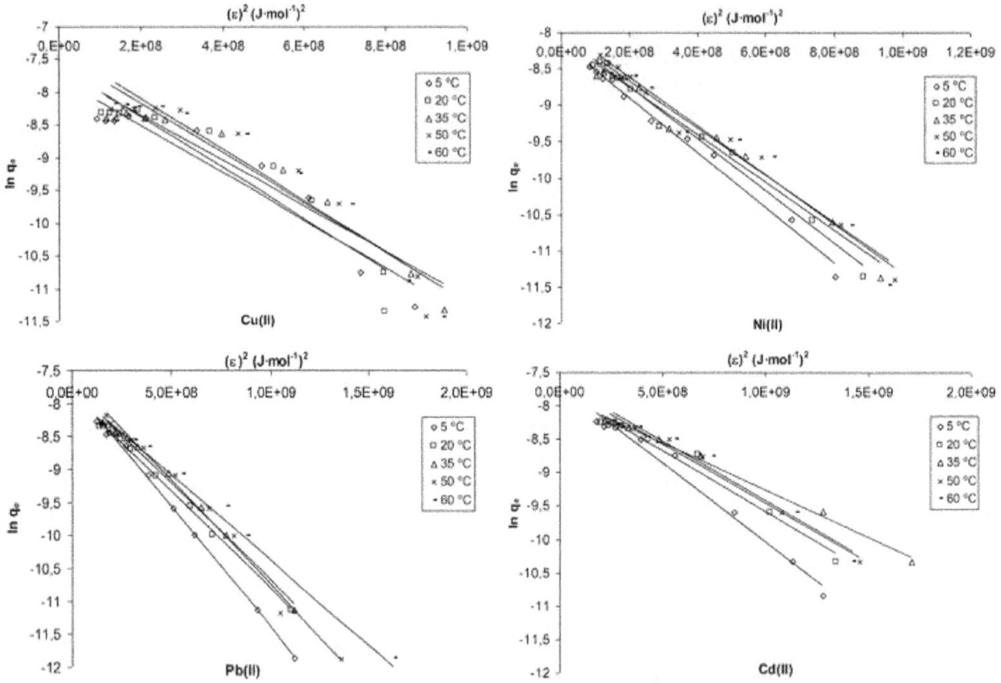

Fig. 4. Dubinin-Radushkevich model fitting of Cu(II), Ni(II), Pb(II) and Cd(II) sorption results in the temperature range of 5–60°C.

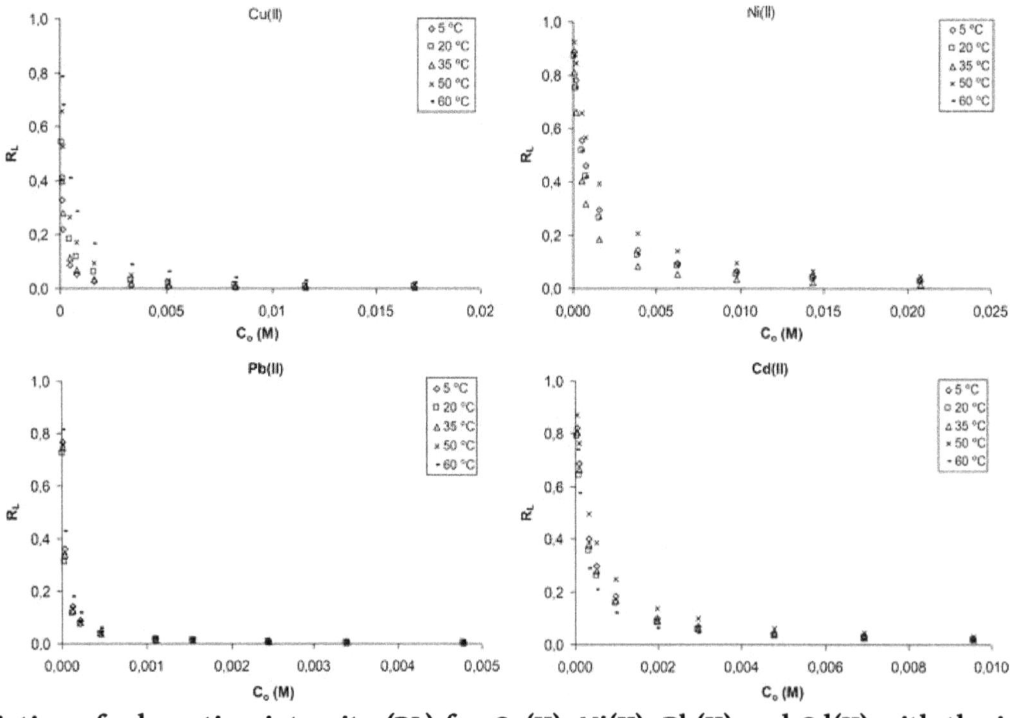

Fig. 5. Variation of adsorption intensity (RL) for Cu(II), Ni(II), Pb(II) and Cd(II) with the initial metal concentration in the temperature range of 5–60°C.

From the linear plots of the different isotherm models, the characteristic sorption parameters of the divalent metals onto GS were calculated and are presented in Table 2. The separation factor (RL) has been also plotted for the different initial metal concentrations and temperatures (Fig. 5). As it may be observed, all the RL values are found in the range 0 < RL < 1, indicating that the sorption of all the metals onto GS is a favorable process regardless on the initial concentration.'

Table 2. Adsorption isotherm constants for the adsorption of cu(II), Ni(II), Pb(II) and cd(II) onto GS as a function of temperature.

		Langmuir			Freundlich			Dubinin-Radushkevich			
Metal	T (°C)	Qmax· 104 (mol·g^{-1})	KL·10^{-4} (L· mol^{-1})	R2	1/n	KF	R2	Qmax·104 (mol· g-1)	β·109 (mol^{2}· kJ^{-2})	E (kJ· mol^{-1})	R2
Cu(II)	5	2.22	2.34	0.999	0.31	2.04	0.836	4.20	3.61	11.76	0.913
	20	2.46	0.97	0.999	0.36	2.26	0.824	5.04	3.89	11.34	0.895
	35	2.29	1.72	0.998	0.31	2.02	0.858	4.67	3.44	12.06	0.931
	50	2.69	0.60	0.999	0.41	2.55	0.840	6.07	3.78	11.50	0.919
	60	2.85	0.32	0.999	0.44	2.75	0.859	6.64	3.88	11.35	0.926
Ni(II)	5	2.09	0.15	0.992	0.32	2.09	0.976	2.93	3.79	11.49	0.913
	20	2.22	0.18	0.993	0.34	2.17	0.958	3.24	3.58	11.82	0.895
	35	1.93	0.28	0.997	0.33	2.16	0.941	3.06	3.28	12.35	0.931
	50	2.45	0.10	0.986	0.37	2.36	0.958	3.55	3.37	12.18	0.919
	60	1.97	0.18	0.997	0.37	2.32	0.914	3.31	3.23	12.44	0.926
Pb(II)	5	2.61	4.71	0.999	0.37	2.35	0.972	4.67	2.36	14.55	0.913
	20	2.60	5.86	0.999	0.31	2.06	0.972	4.22	1.81	16.64	0.895
	35	2.68	5.23	0.999	0.33	2.15	0.957	3.92	1.41	18.81	0.931
	50	2.61	4.95	0.994	0.41	2.55	0.957	4.59	1.76	16.84	0.919
	60	2.65	3.55	0.997	0.37	2.34	0.940	4.81	1.77	16.80	0.926
Cd(II)	5	2.57	0.46	0.993	0.26	1.83	0.938	4.37	3.66	11.69	0.913
	20	2.40	0.56	0.997	0.22	1.65	0.926	3.84	2.95	13.03	0.895
	35	2.52	0.51	0.995	0.20	1.59	0.966	4.33	2.93	13.06	0.931
	50	2.73	0.31	0.985	0.16	1.44	0.850	5.02	3.15	12.60	0.919
	60	2.24	0.76	0.998	0.26	1.81	0.921	4.12	2.58	13.92	0.926

According to the R2 values presented in Table2, the best fitting to the experimental dataset is provided by the Langmuir model. In general, the calculated values of maximum capacity and affinity M(II)-GS obtained through this model indicate that there is not a dramatic effect of the temperature on the sorption process. Cu(II), however, seems exhibiting a slight increase on maximum sorption capacity when the temperature is increased.

When Qmax is compared at a standard temperature of 20°C, it can be observed that a very similar capacity (about $2.5 \cdot 10^{-4}$ mol·g^{-1}) is achieved regardless of the metal. The Qmax values obtained at 20°C through the Langmuir model for the four metals is in agreement with previously reported data [32, 33]. The effect of temperature on the strength of the interaction sorbent-sorbate will be explored later by calculating specifically the thermodynamic parameters of the adsorption process.

The modeling of the experimental dataset according to the D-R equation was also able to provide a good fitting of the experimental trends observed. This model allowed computing the mean free energy of adsorption E (kJ·mol^{-1}) according to Eq. (5). This parameter provides useful information that allows classifying the adsorption mechanism as chemical ion exchange or physical adsorption. If the magnitude of E is between 8 and 16 kJ·mol^{-1}, the adsorption process follows a chemical ion exchange [15, 30]. On the other hand, values of $E < 8$ kJ·mol^{-1} indicate that the adsorption process is of a physical nature [34]. Values higher than 16 kJ·mol^{-1} would be indicative of more energetic interactions than the corresponding to an ion exchange process. As it can be seen in Table 2, the values of adsorption mean free energies are in the range $8 < E$ (kJ·mol^{-1}) < 16 for Cu(II), Ni(II) and Cd(II) sorption at all the temperatures and for Pb(II) at the lowest one, 5°C. These results point out that Cu(II), Ni(II) and Cd(II) sorption onto GS at all the studied temperatures and Pb(II) sorption at 5°C proceeds mainly via ion exchange. For Pb(II) at the temperature of 20°C and higher, the values in the range 16.64 < E < 18.81 indicate that there is an extra contribution to sorption by ion exchange and stronger Pb(II)-GS bonds are being formed.

The thermodynamic parameters of Cu(II), Ni(II), Pb(II), and Cd(II) sorption onto GS were determined using the dataset obtained from equilibrium tests conducted at various temperatures.

2.3. Thermodynamic parameters related to the process of adsorption

The temperature dependence of the sorption process is associated with several thermodynamic parameters that allow concluding whether the process is spontaneous or not. The Gibbs free energy change, ΔG0, is an indicative of the spontaneity of a chemical reaction, and therefore, it is an important criterion when it comes to spontaneity assessment of a sorption process. Both energy and entropy factors must be considered in order to determine the Gibbs free energy of the process. Reactions occur spontaneously at a given temperature if ΔG0 has a negative value, and this parameter can be determined from the following equation:

$$\Delta G^0 = -RT \ln K_L \tag{6}$$

being R the ideal gas constant (8.314 J·mol^{-1}·K^{-1}) and T the absolute temperature (K). Besides, the standard Gibbs free energy can be defined in terms of enthalpy (ΔH0) and entropy (ΔS0) using the equation:

$$\Delta G^0 = \Delta H^0 - T\Delta S^0 \tag{7}$$

Including the Langmuir equilibrium constant (KL) in the Van't Hoff equation, the enthalpy and the entropy of the process can be calculated:

$$\ln K_L = \frac{\Delta S^0}{R} - \frac{\Delta H^0}{RT} \tag{8}$$

From the equation presented above, the values of ΔH0 and ΔS0 can be determined using the slope and the intercept of the plot of ln KL versus 1/T. ΔG0, ΔH0 and ΔS0 were calculated for the different temperatures, and the results are presented in Table 3.

Table 3. Thermodynamic parameters of Cu(II), Ni(II), Pb(II) and Cd(II) sorption onto GS at different temperatures.

Metal	T (°C)	ΔG° (kJ·mol^{-1})	ΔH° (kJ·mol^{-1})	ΔS° (J·mol^{-1}·K^{-1})
Cu(II)	5	-23.27		
	20	-22.38		
	35	-24.98	-17.48	19.99
	50	-23.36		
	60	-22.31		
Ni(II)	5	-16.95		
	20	-18.26		
	35	-20.34	-1.30	57.45
	50	-18.54		
	60	-20.79		
Pb(II)	5	-24.88		
	20	-26.76		
	35	-27.83	-3.45	78.32
	50	-29.04		
	60	-29.02		
Cd(II)	5	-19.52		

Metal	T (°C)	ΔG° ($kJ\cdot mol^{-1}$)	ΔH° ($kJ\cdot mol^{-1}$)	ΔS° ($J\cdot mol^{-1}\cdot K^{-1}$)
	20	-21.05		
	35	-21.87	-5.99	57.34
	50	-21.63		
	60	-24.76		

The negative values of ΔG0 observed for all the M(II)-GS systems indicate that the sorption process is feasible and spontaneous. The negative ΔH0 values obtained for the sorption of all the metals indicate that the sorption process is also exothermic. It is worth noting that, from the four metals, the sorption of Cu(II) is the process that involves a higher exchange of energy, releasing about 17.5 kJ per mol of sorbed metal. The sorption of other three metals involved a much lower energy release, varying from about 6 kJ·mol−1 in the case of Cd(II) to just 1.30 kJ·mol−1 in the case of Ni(II).

On the other hand, the positive values of ΔS0 indicated that the randomness at the solid/liquid interface increases during the adsorption of these divalent metal ions onto GS [22]. A probable explanation for the observed increase of the disorder can be based on the fact that the adsorbed water molecules (which are displaced by the adsorbate species when metals are transferred from the liquid to the solid phase) gain more translational energy than the energy lost by the adsorbate ions [35].

To display the effect of the temperature on the spontaneity of the sorption process, the values of ΔG0 obtained for the different metals have been plotted as a function of temperature in Fig. 6.

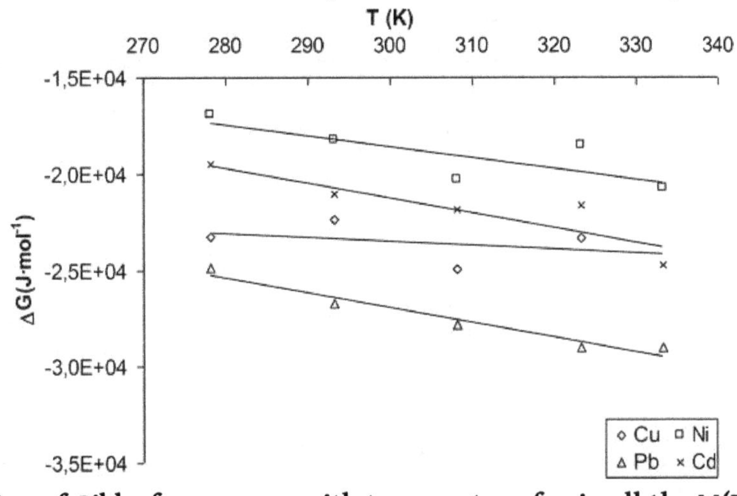

Fig. 6. Variation of Gibbs free energy with temperature for in all the M(II)-GS systems.

As it may be observed in Fig. 6, ΔG exhibits a general decreasing trend when the temperature of the system increased. The thermal effect in the spontaneity of the sorption process can be also assessed through the numerical values of the slopes of the plot for the different metals: Cu(II), -19.26; Ni(II), -56.01; Pb(II), -76.60; and Cd(II), -76.57. The negative values obtained indicate that the sorption process is spontaneous, being Pb(II) and Cd(II) the most temperature-sensitive metals.

The values of variation of enthalpy, entropy and Gibbs free energy presented in Table 3 allow establishing a different set of comparisons between the different metals. So, in basis to the energy released when the metal is adsorbed, the next ranking can be drafted out:

$$\Delta H0{:}Cu > Cd > Pb > Ni$$

In basis to the increase of randomness that metal sorption provokes in the system:

$$\Delta S0{:}Pb > Ni \approx Cd > Cu$$

And lastly, in basis to a more general criterion of spontaneity of the sorption process for temperatures within 5 to 50°C:

ΔG0 (5–50°C): Pb > Cu > Cd > Ni.

It has to be remarked however that for the highest temperature, 60°C, an inversion on the spontaneity of the sorption process takes place between Cd(II) and Cu(II), getting therefore the ranking the next form:

ΔG0 (60°C): Pb > Cd > Cu > Ni.

The thermodynamic results gathered in our study have been compared to those reported by other authors. A summary of the most relevant results found in a bibliographic survey are presented inTable 4.

Table 4. Comparison of the sorption thermodynamic parameters obtained for GS with these observed for other biomass.

Sorbent	Metal	ΔG^0 ($kJ \cdot mol^{-1}$)	ΔH^0 ($kJ \cdot mol^{-1}$)	ΔS^0 ($J \cdot mol^{-1}$)	Reference
Grape stalks	Cu(II)	-22.38	-17.48	19.99	This work
	Ni(II)	-18.26	-1.30	57.45	
	Pb(II)	-27.83	-3.45	78.32	
	Cd(II)	-21.05	-5.99	57.34	
Cellulosic waste orange peel	Cu(II)	-12.48	-19.55	-24.12	[24]
Corn silk (*Zea mays* L)	Cu(II)	-17.10	10.75	95.08	[36]
	Zn(II)	-16.68	7.83	83.68	
Pseudomonas putida	Pb(II)	-21.20	-18.69	8.4	[37]
	Cu(II)	-16.50	23.12	128	
Hazelnut shells	Ni(II)	-15.05	32.49	158.54	[35]
	Pb(II)	-20.94	21.41	142.11	
	Cd(II)	-20.80	12.21	110.73	
Almond shells	Ni(II)	-12.36	47.29	199.00	
	Pb(II)	-21.57	50.55	233.32	
	Cd(II)	-17.45	17.76	118.28	
Capsicum annuum	Cu(II)	-15.93	-7.63	28.35	[22]
Modified spent Chrysanthemum	Cu(II)	-25.19	-11.42	47.0	[25]
Rapeseed biomass	Pb(II)	-35.33	10.05	155.0	[38]
Bacillus pumilus sp. AS1	Pb(II)	-0.56	7.53	0.027	[39]
Loquat (Eriobotrya japonica) leaves	Cd(II)	-8.21	29.73	125.44	[40]
Sargassum filipendula	Cd(II)	-3.82	0.26	0.87	[41]
Penicillium simplicissimum	Cd(II)	-18.27	20.03	130.9	[42]
	Zn(II)	-17.08	25.42	145.5	
	Pb(II)	-20.04	39.13	202.5	
Sporopollenin	Cu(II)	-7.54	17.55	85.58	[43]
	Pb(II)	-13.78	31.97	150.98	
	Cd(II)	-8.85	13.99	76.64	

It is evident that all the writers recorded negative values for ΔG0 and the majority of them also recorded positive values for ΔS0. The data unambiguously demonstrate that sorption is an inherently spontaneous process that mostly occurs in conjunction with an augmentation of the system's disorder. However, the ΔH0 values reported were either positive or negative. The sorption of Cu(II), Ni(II), Pb(II), and Cd(II) upon GS (as described in this study) exhibits negative enthalpy values. Additionally, the sorption of Pb(II) onto *Pseudomonas putida* and the sorption of Cu(II) onto both Capsicum annuum and modified wasted chrysanthemum also demonstrate negative enthalpy values. Conversely, Cu(II) sorption onto Pseudomonas putida, as well as Ni(II), Pb(II), and Cd(II) sorption onto hazelnut and almond shells, and Cu(II), Pb(II), and Cd(II) sorption onto sporopollenin, were found to have positive enthalpy values. Therefore, these findings suggest that the process of metal sorption may occur through the transfer of energy into or out of the system.

3. CONCISE SUMMARY

The adsorption of Cu(II), Ni(II), Pb(II), and Cd(II) onto grape stalk exhibits a Langmuirian sorption pattern across the whole temperature range investigated. The Freundlich and Dubinin-Radushkevich models were able to accurately describe the sorption equilibrium. The Dubinin-Radushkevich model determined that the sorption of Cu(II), Ni(II), and Cd(II) onto grape stalk primarily occurs by ion exchange, as indicated by the computed mean free energy E. An additional effect on the ion exchange process was seen at temperatures over 5°C in the case of Pb(II). This additional contribution would rely on the formation of more robust Pb(II)-GS interactions. The enthalpy and entropy changes observed in the sorption process indicate that the sorption of Cu(II), Ni(II), Pb(II), and Cd(II) onto grape stalks is a spontaneous exothermic process that leads to an increase in the system's disorder. The thermodynamic properties of metal sorption onto GS were used to establish rankings based on different parameters. The ranking based on the energy released during the sorption process, ΔH0, is as follows: Cu > Cd > Pb > Ni. The ranking based on the increase in randomness caused by the sorption process, ΔS0, is as follows: Pb > Ni ≈ Cd > Cu. Finally, the ranking based on the spontaneity of the overall process, ΔG0 (5–50°C), is as follows: Pb > Cu > Cd > Ni.

REFERENCES

1. Deniz F, Karabulut A. Biosorption of heavy metal ions by chemically modified biomass of coastal seaweed community: Studies on phycoremediation system modeling and design. Ecological Engineering. 2017;106:101-108
2. Vijayaraghavan K, Balasubramanian R. Is biosorption suitable for decontamination of metal-bearing wastewaters? A critical review on the state-of-the-art of biosorption processes and future directions. Journal of Environmental Management. 2015;160:283-296
3. Vendruscolo F, da Rocha Ferreira GL, Antoniosi Filho NR. Biosorption of hexavalent chromium by microorganisms, International Biodeterioration & Biodegradation. 2017; 119:87-95
4. Lesmana SO, Febriana N, Soetaredjo FE, Sunarso J, Ismadji S. Studies on potential applications of biomass for the separation of heavy metals from water and wastewater. Biochemical Engineering Journal. 2009;44:19-41
5. Senthilkumaar S, Bharathi S, Nithyanandhi D, Subburam V. Biosorption of toxic heavy metals from aqueous solutions. Bioresource Technology. 2000;75:163-165
6. Ahmad A, Bhat AH, Buang A. Biosorption of transition metals by freely suspended and Ca-alginate immobilised with Chlorella Vulgaris: Kinetic and equilibrium modeling. Journal of Cleaner Production. 2018;171:1361-1375

7. Wu M, Liang J, Tang J, Li G, Shan S, Guo Z, Deng L. Decontamination of multiple heavy metals-containing effluents through microbial biotechnology. Journal of Hazardous Materials. 2017;337:189-197

8. Renu MA, Singh K, Upadhyaya S, Dohare RK. Removal of heavy metals from wastewater using modified agricultural adsorbents. Materials Today: Proceedings. 2017;4:10534-10538

9. Castro L, Blázquez ML, González F, Muñoz JA, Ballester A. Biosorption of Zn(II) from industrial effluents using sugar beet pulp and F. Vesiculosus: From laboratory tests to a pilot approach. Science of the Total Environment. 2017;598:856-866

10. Demey H, Vincent T, Guibal E. A novel algal-based sorbent for heavy metal removal. Chemical Engineering Journal. 2018;332:582-595

11. Sheng PX, Ting Y-P, Chen JP. Biosorption of heavy metal ions (Pb, cu, and cd) from aqueous solutions by the marine alga *Sargassum* sp. in single- and multiple-metal systems. Industrial & Engineering Chemistry Research. 2007;46:2438-2444

12. Jiang L, Zhou W, Liu D, Liu T, Wang Z. Biosorption isotherm study of Cd2+, Pb^{2+} and Zn^{2+} biosorption onto marine bacterium Pseudoalteromonas sp. SCSE709-6 in multiple systems. Journal of Molecular Liquids. 2017;247:230-237

13. Carolin CF, Kumar PS, Saravanan A, Joshiba GJ, Naushad M. Efficient techniques for the removal of toxic heavy metals from aquatic environment: A review. Journal of Environmental Chemical Engineering. 2017;5:2782-2799

14. Ronda A, Martín-Lara MÁ, Blázquez G, Bachs NM, Calero M. Copper biosorption in the presence of lead onto olive stone and pine bark in batch and continuous systems. Environmental Progress & Sustainable Energy. 2014;33:192-204

15. Wei W, Wang Q, Li A, Yang J, Ma F, Pi S, Wu D. Biosorption of Pb (II) from aqueous solution by extracellular polymeric substances extracted from Klebsiella sp. J1: Adsorption behavior and mechanism assessment. Scientific Reports. 2016;6

16. Gupta VK, Rastogi A. Equilibrium and kinetic modelling of cadmium(II) biosorption by nonliving algal biomass Oedogonium sp. from aqueous phase. Journal of Hazardous Materials. 2008;153:759-766

17. Limousin G, Gaudet JP, Charlet L, Szenknect S, Barthès V, Krimissa M. Sorption isotherms: A review on physical bases, modeling and measurement. Applied Geochemistry. 2007;22:249-275

18. Chowdhury PSS. Insight into adsorption thermodynamics. In: Thermodynamics. Rijeka, Croatia: INTECH; 2011, pp. 349-364

19. El-Khaiary MI. Least-squares regression of adsorption equilibrium data: Comparing the options. Journal of Hazardous Materials. 2008;158:73-87

20. Anastopoulos I, Karamesouti M, Mitropoulos AC, Kyzas GZ. A review for coffee adsorbents. Journal of Molecular Liquids. 2017;229:555-565

21. Karunanithi R, Ok YS, Dharmarajan R, Ahmad M, Seshadri B, Bolan N, Naidu R. Sorption, kinetics and thermodynamics of phosphate sorption onto soybean stover derived biochar. Environmental Technology & Innovation. 2017;8:113-125

22. Özcan A, Özcan AS, Tunali S, Akar T, Kiran I. Determination of the equilibrium, kinetic and thermodynamic parameters of adsorption of copper(II) ions onto seeds of Capsicum Annuum. Journal of Hazardous Materials. 2005;124:200-208

23. Saha GC, Hoque MIU, Miah MAM, Holze R, Chowdhury DA, Khandaker S, Chowdhury S. Biosorptive removal of lead from aqueous solutions onto Taro (*Colocasia esculenta*(L.) Schott) as a low cost bioadsorbent: Characterization, equilibria, kinetics and biosorption-mechanism studies. Journal of Environmental Chemical Engineering. 2017;5:2151-2162

24. Guiza S. Biosorption of heavy metal from aqueous solution using cellulosic waste orange peel. Ecological Engineering. 2017;99:134-140

25. Yi Y, Lv J, Zhong N, Wu G. Biosorption of Cu2+ by a novel modified spent chrysanthemum: Kinetics, isotherm and thermodynamics. Journal of Environmental Chemical Engineering. 2017;5:4151-4156

26. Wang N, Jin R-N, Omer AM, Ouyang X-k. Adsorption of Pb(II) from fish sauce using carboxylated cellulose nanocrystal: Isotherm, kinetics, and thermodynamic studies. International Journal of Biological Macromolecules. 2017;102:232-240

27. Khodabandehloo A, Rahbar-Kelishami A, Shayesteh H. Methylene blue removal using Salix Babylonica (weeping willow) leaves powder as a low-cost biosorbent in batch mode: Kinetic, equilibrium, and thermodynamic studies. Journal of Molecular Liquids. 2017;244:540-548

28. Pleşa Chicinaş R, Bedelean H, Stefan R, Măicăneanu A. Ability of a montmorillonitic clay to interact with cationic and anionic dyes in aqueous solutions. Journal of Molecular Structure. 2018;1154:187-195

29. Singh H, Chauhan G, Jain AK, Sharma SK. Adsorptive potential of agricultural wastes for removal of dyes from aqueous solutions. Journal of Environmental Chemical Engineering. 2017;5:122-135

30. Shaker MA. Thermodynamics and kinetics of bivalent cadmium biosorption onto nanoparticles of chitosan-based biopolymers. Journal of the Taiwan Institute of Chemical Engineers. 2015;47:79-90

31. Rangabhashiyam S, Sujata L, Balasubramanian P. Biosorption characteristics of methylene blue and malachite green from simulated wastewater onto Carica papaya wood biosorbent. Surfaces and Interfaces. 2017. https://doi.org/10.1016/j.surfin.2017.09.011

32. Martínez M, Miralles N, Hidalgo S, Fiol N, Villaescusa I, Poch J. Removal of lead(II) and cadmium(II) from aqueous solutions using grape stalk waste. Journal of Hazardous Materials. 2006;133:203-211

33. Villaescusa I, Fiol N, Martínez Ma, Miralles N, Poch J, Serarols J. Removal of copper and nickel ions from aqueous solutions by grape stalks wastes. Water Research. 2004;38:992-1002

34. Onyango MS, Kojima Y, Aoyi O, Bernardo EC, Matsuda H. Adsorption equilibrium modeling and solution chemistry dependence of fluoride removal from water by trivalent-cation-exchanged zeolite F-9. Journal of Colloid and Interface Science. 2004;279:341-350

35. Bulut Y, Tez Z. Adsorption studies on ground shells of hazelnut and almond. Journal of Hazardous Materials. 2007;149:35-41

36. Petrović M, Šoštarić T, Stojanović M, Petrović J, Mihajlović M, Ćosović A, Stanković S. Mechanism of adsorption of Cu2+ and Zn2+ on the corn silk (Zea Mays L.). Ecological Engineering. 2017;99:83-90

37. Uslu G, Tanyol M. Equilibrium and thermodynamic parameters of single and binary mixture biosorption of lead (II) and copper (II) ions onto pseudomonas putida: Effect of temperature. Journal of Hazardous Materials. 2006;135:87-93

38. Morosanu I, Teodosiu C, Paduraru C, Ibanescu D, Tofan L. Biosorption of lead ions from aqueous effluents by rapeseed biomass. New Biotechnology. 2017;39:110-124

39. Sayyadi S, Ahmady-Asbchin S, Kamali K, Tavakoli N. Thermodynamic, equilibrium and kinetic studies on biosorption of Pb+2 from aqueous solution by Bacillus Pumilus sp. AS1 isolated from soil at abandoned lead mine. Journal of the Taiwan Institute of Chemical Engineers. 2017;80:701-708

40. Awwad AM, Salem NM. Kinetics and thermodynamics of cd(II) biosorption onto loquat (Eriobotrya Japonica) leaves. Journal of Saudi Chemical Society. 2014;18:486-493

41. Verma A, Kumar S, Kumar S. Statistical modeling, equilibrium and kinetic studies of cadmium ions biosorption from aqueous solution using S. Filipendula. Journal of Environmental Chemical Engineering. 2017;5:2290-2304

42. Fan T, Liu Y, Feng B, Zeng G, Yang C, Zhou M, Zhou H, Tan Z, Wang X. Biosorption of cadmium(II), zinc(II) and lead(II) by Penicillium simplicissimum: Isotherms, kinetics and thermodynamics. Journal of Hazardous Materials. 2008;160:655-661

43. Ünlü N, Ersoz M. Adsorption characteristics of heavy metal ions onto a low cost biopolymeric sorbent from aqueous solutions. Journal of Hazardous Materials. 2006;136:272-280

15 Chapter

ADSORPTION PROCESSES SUBJECT TO AN EXTERNAL MAGNETIC FIELD FOR THE REMOVAL OF HEAVY METALS

1. INTRODUCTION: AN OVERVIEW

Eliminating dissolved heavy metals from water is a paramount societal problem due to the potential hazards they pose to human and environmental well-being. Heavy metals possess toxicity and carcinogenic properties, making them capable of readily infiltrating the food chain [1, 2]. As per the Environmental Protection Agency (EPA), heavy metals are classified as priority pollutants and must be completely removed or decreased from any water source that may or may not interact with the environment [3, 4]. Several conventional methods, including precipitation, ion-exchange, reverse osmosis, membrane separation, and adsorption, have been proven effective in removing heavy metals from water [5, 6]. Adsorption is a highly promising and commonly employed approach due to its ease of operation, high efficiency, and economic advantages [7, 8]. The magnetization that remains constant and is linked to an adsorption process has garnered significant interest due to its simple implementation, cost-effectiveness, and ecologically beneficial nature. Recent research has shown that the adsorption process can be enhanced by applying a magnetic field [9]. This improvement is believed to be due to many causes and potential alterations that can alter the properties of the solution and material when subjected to a magnetic field [10]. Various characteristics and operating conditions play a crucial role in the adsorption process under a static magnetic field. These include the physicochemical and magnetic properties of the adsorbate and adsorbent, as well as the configuration and strength of the magnetic field. The interaction between the magnetic field and the adsorbate-adsorbent system is intricate. However, by considering factors such as zeta potential, magnetic susceptibility, ionic radius, and ion hydration, it is feasible to gain insight into the potential mechanism involved.

2. MINERAL ADSORBENTS

Table 1 displays the mineral adsorbents employed for eliminating heavy metals from water, including Cd(II), Zn(II), Pb(II), Cu(II), Cr(VI), and Ni(II). The Fe3O4/PANI/MnO2 core-shell hybrid demonstrated the highest efficiency in removing Cd(II), with an adsorption capacity of 154 mg/g [11]. Following closely were hydrotalcite-like compounds (ZnAl-diethylenetriaminepentaacetic acid and ZnAl-meso-2,3-dimercaptosuccinic acid) with adsorption capacities of 112.4 and 49.5 mg/g, respectively [12]. However, the geopolymer made from metakaolin shown a greater ability to adsorb Zn(II) with a capacity of 74.5 mg/g, as compared to natural bentonite and bentonite which had capacities of 52.9 mg/g and 39.5 mg/g, respectively. This capacity was also higher than other adsorbents mentioned in references 11, 13, and 14. In terms of removing Cr(VI), the diatomite/MCM-41 composite that underwent processing had the highest efficiency, with an adsorption capacity of 70.9 mg/g [15]. Following closely behind is the diatomite-supported magnetite nanoparticles, which achieved a value of 69.2 mg/g [16]. In addition, various adsorbents were employed to investigate the elimination of Cu(II), and the most effective one was found to be the hydrotalcite-like complex ZnAl-meso-2,3-dimercaptosuccinic acid, which exhibited an adsorption capacity of 69.9 mg/g [12]. This same material was also utilized to successfully remove Cd(II). The heavy metals Zn(II), Cr(VI), Cu(II), and Ni(II) have been extensively researched using inorganic or mineral adsorbents. Natural bentonite exhibited superior adsorption capacity (50.0 mg/g) for the elimination of Ni(II) [13]. Furthermore, another heavy metal that was examined was Pb(II) with an impressive value of 404.0 mg/g. This was achieved by utilizing the hydrotalcite-like compound ZnAl-meso-2,3-dimercaptosuccinic acid [12].

Table 1. Mineral adsorbents used in the adsorption of heavy metals from water.

Adsorbent	Removed species	Adsorption capacity (mg/g)	Ref
Fe_3O_4/PANI/MnO_2 core-shell hybrids	Cd(II) Zn(II) Pb(II) Cu(II)	154 a88.0 (17.6 mg/g) a99.1 (19.8 mg/g) a99.1 (19.8 mg/g)	[11]
Processed diatomite/MCM-41 composite	Cr(VI)	70.9	[15]
Diatomite-supported magnetite nanoparticles	Cr(VI)	69.2	[16]
Tannin-immobilized activated clay	Cr(VI)	24.1	[17]
Modified natural zeolites	Cr(VI)	3.5–8.8	[18]
Modified rectorite	Cr(VI)	21	[19]
geopolymer/alginate hybrid spheres	Cu(II)	60.8	[20]
Aminopropyltriethoxysilane/sepiolite	Cu(II)	8.9	[21]
Multiwalled carbon nanotubes	Cu(II)	50.3	[22]
Zeolite	Cu(II)	9.0	[23]
Ca-alginate	Cu(II)	42.7	[24]
Bentonite/chitosan	Cu(II)	9.8	[25]
Inorganic-organic clays	Cr(VI)	16.6–17.9	[26]
Hydrotalcite-like compound ZnAl-NO_3			[12]
ZnAl-dtpa (diethylenetriaminepentaacetic acid)	Cu(II)	48.3	
	Pb(II)	145.0	
	Cd(II)	49.5	
ZnAl-dmsa (meso-2,3-dimercaptosuccinic acid)	Cu(II)	69.9	
	Pb(II)	404.0	
	Cd(II)	112.4	
Na-bentonite	Zn(II)	23.1	[13]
Polyphosphate-modified kaolinite clay	Zn(II)	27.8	[13]
Natural zeolite	Zn(II)	13.4	[13]
Kaolinite	Zn(II)	7.2	[13]
Metakaolinite	Zn(II)	12.4	[13]
Vermiculite	Zn(II)	31.7	[13]
Bentonite	Zn(II)	39.5	[13]
Zeolite	Zn(II)	18.7	[13]
Natural bentonite	Zn(II)	52.9	[13]
Metakaolin-based geopolymer	Zn(II)	74.5	[13]
Vermiculite	Ni(II)	25.3	[13]
Natural bentonite	Ni(II)	50.0	[13]
Zeolite	Ni(II)	1.98	[13]

Adsorbent	Removed species	Adsorption capacity (mg/g)	Ref
Na-bentonite	Ni(II)	24.2	[13]
Ca-bentonite	Ni(II)	6.3	[13]
Natural clinoptilolite	Ni(II)	8.7	[13]
Montmorillonite	Ni(II)	21.1	[13]
Metakaolin-based geopolymer	Ni(II)	42.6	[13]
Mixed silica-alumina oxide	Cu(II) Ni(II) Zn(II)	0.1 0.2 4.2	[14]
[a]Removal percentage.			

3. MINERAL CARBONS

The minerals utilized for investigating the adsorption of heavy metals from water include peat, lignite, anthracite, graphite, and primarily bituminous carbon, which is the most commonly employed. The bituminous carbon underwent physical activation using steam and was then changed through impregnation with calcium, phosphoric acid, sulfur, and nitric acid. Following this, a thermal treatment was applied [27, 28, 29, 30, 31, 32].

Table 2 compiles the mineral carbons that have been documented in the literature as being employed for the adsorption of heavy metals from water. There is limited information available regarding the utilization of these types of adsorbents, with the bituminous carbon being the most extensively researched. The bituminous carbon treated with calcium and H3PO4 is the most effective adsorbent for removing Zn(II), Cd(II), and Ni(II), with respective capacities of 3.4, 8.0, and 5.0 mg/g [27]. However, the untreated bituminous carbon demonstrates a greater ability to adsorb Cu(II) and Cr(VI) with values of 13.3 and 10.5 mg/g, respectively [28]. Furthermore, it is worth noting that bituminous charcoal and lignite charcoal were employed in adsorption experiments involving a metallurgical effluent comprising Cu(II), Cr(VI), and Zn(II). However, the adsorption capabilities were found to be minimal, as reported in reference [30]. The study investigated the effects of sulfur and HNO_3 modified bituminous coal on Hg0 and Hg(II). The adsorption capacities obtained were 1.9 mg/g and 0.7 mg/g, respectively [29].

Table 2. Mineral carbons used in the adsorption of heavy metals from water.

Mineral carbon	Modification method used	Modifying agent	Removed species	Adsorption capacity (mg/g)	Ref
Bituminous carbon	Impregnation-thermal treatment	Calcium-H_3PO_4	Zn(II) Cd(II) Ni(II)	3.4 8.0 5.0	[27]
Bituminous carbon	—	—	Cd(II) Cu(II) Cr(VI)	5.1 13.3 10.5	[28]
Bituminous coal (AC-109)	Impregnation-thermal treatment	Sulfur	Hg^0 Hg(II)	1.6 0.7	[29]
Bituminous coal (AC-107)	Impregnation-thermal treatment	Sulfur	Hg^0 Hg(II)	1.9 0.7	[29]
Bituminous coal (AC-C2)	Impregnation-thermal treatment	HNO_3	Hg^0 Hg(II)	1.8 0.7	[29]

Mineral carbon	Modification method used	Modifying agent	Removed species	Adsorption capacity (mg/g)	Ref
Bituminous charcoal	Physical activation	Steam	Cu(II) Cr(VI) Zn(II)	a0.1 a0.3 a0.3	[30]
Lignite charcoal	Physical activation	Steam	Cu(II) Cr(VI) Zn(II)	a0.1 a0.3 a0.3	[30]
Bituminous coal	Physical activation	Water vapor	Cd(II) Zn(II)	1.7 2.0	[31]
Bituminous carbon	Impregnation-thermal treatment	Calcium-H_3PO_4	Zn(II) Cd(II) Ni(II)	1.6 2.7 2.3	[32]
[a] Industrial effluent.					

4. MAGNETIC ADSORBENTS

The term "magnetic adsorbent" is commonly used to describe the materials derived from iron (particles of γ-Fe_2O_3, Fe_3O_4 and $CoFe_2O_3$) or composites prepared from different types of matrix modified with iron compounds, which are used in the adsorption of water contaminants [33]. Particularly, for the adsorption of heavy metals from water, the principal magnetic adsorbents used are the composites, which are prepared with different methodologies using iron compounds and a specific matrix such as activated carbon, biochar, carbon nanotubes, α-cellulose, chitosan, cotton, resins, silica, starch, polymers, different type of biomass, and so on (see Table 3) [34, 35, 36, 37, 38, 39, 40, 41, 42, 43, 44, 45, 46, 47, 48, 49, 50, 51, 52, 53, 54, 55]. The methods used in the preparation of magnetic composites can be grouped in the following categories according to the data reported in the literature [56]: (1) chemical co-precipitation, (2) thermal decomposition, (3) hydrothermal, (4) polyol process, (5) sol-gel and (6) chemical reduction. In this context, it is relevant to mention that the chemical co-precipitation is the most used method for the preparation of magnetic composites and this can be carried out in different routes, which include the following: (1) mixture of iron solutions ($FeSO_4$·$7H_2O$ and $FeCl_3$·$6H_2O$ or $FeCl_3$·$6H_2O$ and $FeCl_2$·$4H_2O$ or $Fe(NO_3)_3$·$9H_2O$) with the matrix, and then, the addition of ammonium or sodium hydroxide or NH3·H2O to precipitate the iron nanoparticles on the matrix surface [33, 34, 36, 38, 39, 41, 45, 49, 55]. (2) The preparation of nanoparticles of iron (Fe_3O_4) by precipitation using $FeCl_3$·$6H_2O$ and $FeCl_2$·$_4H_2O$ with additives such as $CaCl_2$·$2H_2O$, Na_3PO_4·$12H_2O$, EDTA [57, 58]. (3) First, the precipitation of nanoparticles of iron (Fe_3O_4), second, the mixture of Fe_3O_4 with materials such as Cyclosorus interruptus, silica and p-phenylenediamine-thiourea-formaldehyde polymer; and finally, the addition of 3-aminopropyltriethoxysilane or tetraethyl orthosilicate or potassium persulfate, depending on the final application of composite [33, 51, 53].

Table 3. Magnetic composites used in the adsorption of heavy metals and metalloids from water.

Matrix	Modified method	Iron compounds	Removed species	Adsorption capacity (mg/g)	Ref
Activated carbon	Co-precipitation	$FeSO_4$·$7H_2O$, $FeCl_3$·$6H_2O$	Cu(II)	a99.8	[34]
Activated carbon	Co-precipitation	$Fe(NO_3)_3$·$9H_2O$	As(III) As(V)	38.8 51.3	[35]

Matrix	Modified method	Iron compounds	Removed species	Adsorption capacity (mg/g)	Ref
Activated carbon	Co-precipitation	$FeCl_2$	As(V)	1.95	[36]
Activated carbon	Reduction reaction	FeO_4S	As(III) As(V)	18.19 12.02	[37]
Carbon nanotubes	Co-precipitation	$FeSO_4{\cdot}4H_2O$, $FeCl_3{\cdot}6H_2O$	Ni(II)	51.62	[38]
Biochar from *Astragalus mongholicus*	Co-precipitation	$FeSO_4{\cdot}4H_2O$, $FeCl_3{\cdot}6H_2O$	Cr(VI)	44.74	[39]
α-Cellulose	Extrusion dropping technology	Fe_3O_4	Cu(II) Pb(II) Zn(II)	47.573 37.994 20.800	[40]
Chitosan	Co-precipitation	$FeCl_3{\cdot}6H_2O$, $FeSO_4$	Cu(II) Cd(II) Zn(II) Cr(VI)	188 159 72 46	[41]
Chitosan	Hydrothermal	$FeCl_3{\cdot}6H_2O$	Pb(II) Cu(II)	97.97 83.65	[42]
Chitosan	Crosslinking method	Fe_3O_4	Cu(II)	143.27	[43]
Cotton	Impregnation-thermal treatment	$Fe(NO_3)_3{\cdot}9H_2O$	Cr(VI)	3.74	[44]
Cyclosorus interruptus	Co-precipitation	$FeSO_4{\cdot}7H_2O$, $FeCl_3{\cdot}6H_2O$	Pb(II)	90.1	[33]
Dowex 50 WX4 resin	Co-precipitation	$FeCl_3{\cdot}6H_2O$, $FeCl_2{\cdot}4H_2O$	Cr(VI) Ni(II) Cu(II) Cd(II) Pb(II)	400 384 416 398 380	[45]
Graphene oxide	Co-precipitation	$FeSO_4{\cdot}7H_2O$, $FeCl_3{\cdot}6H_2O$	Hg(II)	400	[46]
Macroalgal biomass (Kelp)	Impregnation-thermal treatment	$FeCl_3{\cdot}6H_2O$	Cd(II) Cu(II) Zn(II)	34.89 46.74 95.49	[47]
Mesoporous silica	Thermal decomposition	$FeCl_3{\cdot}6H_2O$	Fe(III)	a79.9	[48]
Palygorskite	Co-precipitation	$FeCl_3{\cdot}6H_2O$, $FeCl_2{\cdot}4H_2O$	Pb(II)	26.7	[49]
Pine bark	Impregnation-thermal treatment	$Fe(NO_3)_3{\cdot}9H_2O$	Pb(II) Cd(II)	25.294 14.960	[50]
Poly p-phenylenediamine	Co-precipitation	$FeCl_3{\cdot}6H_2O$, $FeSO_4{\cdot}7H_2O$	As(V)	35.14	[51]
Poly(vinylidene fluoride)	Impregnation-thermal treatment	$Fe(NO_3)_3{\cdot}9H_2O$	Cr(VI)	1423.4	[52]

Matrix	Modified method	Iron compounds	Removed species	Adsorption capacity (mg/g)	Ref
Silica	Co-precipitation	$FeCl_2 \cdot 4H_2O$, $FeCl_3 \cdot 6H_2O$	Pb(II) Hg(II) Pb(II)	270 303 256.4	[53]
Starch	Crosslinking method	Fe_3O_4	Pb(II) Cu(II)	a83.1 a66.5	[54]
Waste orange peel	Co-precipitation	$FeSO_4 \cdot 7H_2O$, $FeCl_3 \cdot 6H_2O$	As(III)	11.12	[55]
[a] Removal percentage.					

The second most popular method used in the preparation of magnetic composites includes the impregnation of the matrix with iron compounds such as $Fe(NO_3)_3 \cdot 9H_2O$ and $FeCl_3 \cdot 6H_2O$ and a thermal treatment of impregned materials using drying on a regular oven overnight [44], pyrolysis in conventional systems [47, 50, 52, 57, 58, 59] or pyrolysis in microwave system [60].

Table 3 summarizes the different types of magnetic composites used in the removal of some heavy metals and metalloids from water such as Pb(II), Cu(II), Ni(II), Cd(II), Zn(II), Cr(VI), Fe(III), Hg(II), As(II) and As(V). In general, the composites prepared with the resin Dowex 50 WX4 were the most efficient in the removal of Ni(II), Cu(II), Cd(II) and Pb(II), showing an adsorption capacity of 384, 416, 398 and 380 mg/g, respectively [45]. These values are higher than the data registered for other composites obtained from carbon nanotubes [34], α-cellulose [40], chitosan [41, 42, 43], silica [53], starch [54] and different types of biomass such as waste orange peel, pine bark and Cyclosorus interruptus (see Table 3) [33, 50, 55]. Additionally, it is relevant to mention that for the removal of Cr(VI), the composite obtained with poly(vinylidene fluoride) shows the higher adsorption capacity (1423.4 mg/g) in comparison with the samples prepared from chitosan (46 mg/g), cotton (3.74 mg/g) and Dowex 50 WX4 resin (400 mg/g) [41, 44, 45]. In this context, the adsorption of As(III) and As(V) has also been studied employing magnetic composites prepared with activated carbon principally [35, 36, 37]. For example, the composite prepared with bituminous coal carbon is showing a high adsorption capacity of As(III) and As(V) (38.8 and 51.3 mg/g, respectively), in comparison with the materials prepared with an inorganic matrix as the poly p-phenylenediamine (34.15 mg of As(V)/g) [35, 51]. Finally, it is important to mention that one of the most toxic heavy metals, Hg(II), was studied employing magnetic composites obtained with silica and with graphene oxide, highlighting the latter material with an adsorption capacity of 400 mg/g [46].

To summarize, the adsorption of heavy metals has been investigated using magnetic composites primarily synthesized through the co-precipitation method. However, it is not feasible to make a direct comparison due to variations in adsorption conditions and physicochemical properties of the magnetic adsorbents, such as porosity and chemical functionality. However, the research conducted in this field has been growing in recent years, and the information presented in this chapter will serve as a foundation for forthcoming publications.

5. ADSORBATE PROPERTIES AND ADSORPTION CONDITIONS

Heavy metals are adsorbed onto various adsorbents through mechanisms such as ion-exchange, complexation, coordination, chelation, microprecipitation, electrostatic contact, or a combination of these processes [61]. Therefore, the adsorption mechanisms are determined by the physicochemical features of adsorbents (such as texture and chemical functionality), the conditions under which adsorption occurs (such as pH, temperature, mass of adsorbent, treated volume, type of adsorption system, etc.), and the characteristics of the substances being adsorbed. The key characteristics of heavy metals that are taken into account in the adsorption mechanism include: (1) ionic radius, (2) electronegativity, (3) atomic weight, (4) hydration radius, (5) oxidation number, and (6) electrostatic potential.

Based on the information presented in the literature, the ionic radius and electronegativity are the main parameters that are utilized to understand the mechanisms of heavy metal adsorption. Fig. 1 illustrates the adsorption outcomes of Pb(II), Cd(II), Cu(II), Ni(II), and Zn(II) in relation to their ionic radius. The adsorbents utilized in the study were Kappaphycus alvarezii, a red marine alga, waste tire rubber carbon, mesoporous silica, natural zeolite, and activated carbon cloths [61, 62, 63, 64, 65]. Typically, a distinct behavior is not specified, except for the adsorption of Pb(II), Cd(II), Cu(II), and Ni(II) on red marine alga, where a direct relationship was observed between the amount of metal ions adsorbed and their ionic radius. For this scenario, the quantity of adsorbed material was greater for the heavy metal with a larger ionic radius (Pb: 1.19 Å), demonstrating a maximum adsorption capacity of 0.51 mmol/g by the process of electrostatic contact [61]. The adsorption of Pb(II), Cd(II), Ni(II), and Zn(II) using natural zeolite and carbon derived from waste tire rubber as an adsorbent exhibited contrasting behavior. The amount of adsorbed ions reduced as the ionic radius increased, as found in previous studies [62, 68]. Furthermore, the use of the hydration radius has been employed to downplay the adsorption of heavy metals, and the outcomes also lack a discernible pattern [66].

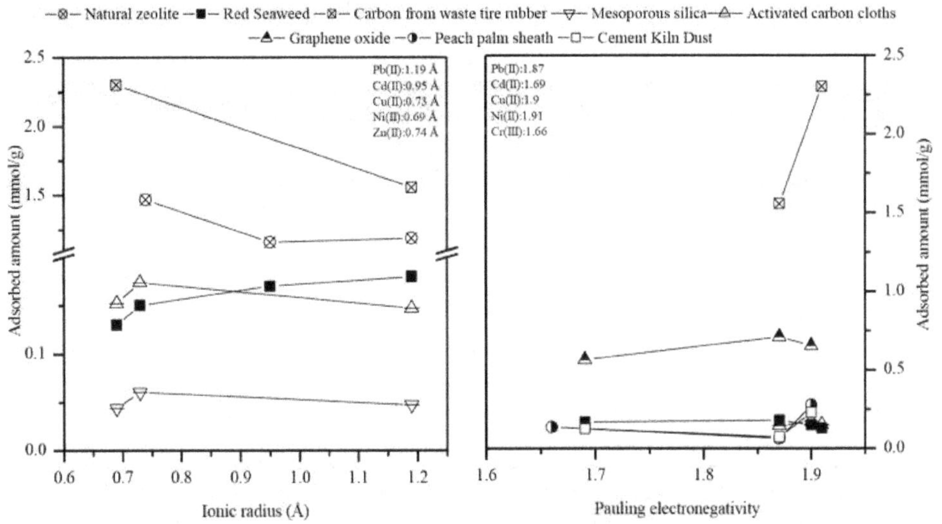

Fig. 1. Adsorption results of heavy metals from water of different adsorbents in function of ionic radius and Pauling electronegativity.

On the other hand, the Pauling electronegativity of heavy metals has also been correlated with the adsorption of heavy metals. Fig. 1 shows the adsorption results of Pb(II), Cd(II), Cu(II), Ni(II) and Cr(III) employing different adsorbents in function of Pauling electronegativity [61, 62, 65, 67, 68, 69]. Particularly, it is interesting to note that a defined behavior was observed in the adsorption of Pb(II) and Ni(II) using the carbon from waste tire rubber as an adsorbent [62]. Specifically, this adsorbent has the higher adsorption capacity of Ni(II) and Pb(II), and the adsorption was higher for the more electronegative metal. This behavior can be associated with the chemical functionality of carbon because this material was activated with hydrogen peroxide and consequently, the presence of O—H group of phenols and C—O group of esters favored the adsorption of heavy metals [62]. Additionally, for the adsorption of Pb(II), Cd(II), Cu(II) and Cr(III) on cement kiln dust and peach palm sheath, the electronegativity does not keep a relation with the absorbed quantity of metals [68, 69]. However, it is important to mention that the Cr has a different oxidation number (III) in comparison with the other heavy metals (II). In this context, there are few reports that have studied the effect of oxidation number of heavy metals in the adsorption process and particularly, the works related with the adsorption by complexation mechanism have considered this property. For example, in the removal of Ni^{2+}, Fe^{2+} and Fe^{3+} (with external hydroxyl functional groups) was demonstrated that high-valence species complex more quickly than the low-valence due to the ability to accept more of the ligand's electron donation, where the adsorption followed the trend $Ni^{2+} > Fe^{3+} > Fe^{2+}$ [70].

It is crucial to acknowledge that there are multiple factors that affect the adsorption of heavy metals, such as the initial concentration, temperature, adsorbent dosage, pH, contact time, and stirring speed. The pH level has a substantial impact on the maximum adsorption capacity of different adsorbents, such as activated carbons, carbon nanotubes, graphene, biosorbents, low-cost adsorbents (such as vegetal and industrial wastes), silica, chitosan, zeolites, alumina, clay, algalbiomass, red mud, magnetic composites, and others. The pH of a solution affects the surface charge of the adsorbent material, as well as the degree of ionization and speciation of the surface functional groups and metal ions. This topic has been examined in multiple studies [71, 72, 73, 74]. Generally, the adsorption of metallic species is intensified when the pH rises within a particular range. This occurs due to the competition between the metallic species and the H+ ions in the aqueous solution at low pH levels.

6. MAGNETIC ORDERING AND ZETA POTENTIAL

The response of materials to a magnetic field varies depending on their physicochemical and magnetic properties. This variation is documented in Table 4 [75].

Magnetic behavior	Description
Diamagnetic	A magnetic field acting on any atom induces a magnetic dipole for the entire atom by influencing the magnetic moment caused by the orbiting electrons. These dipoles can oppose the magnetic field, causing the magnetization to be less than zero. A diamagnetic behavior gives a negative susceptibility. Materials such as copper, silver, gold and alumina are diamagnetic at room temperature
Paramagnetic	A net magnetic moment due to electronic spin is associated with each atom when materials have unpaired electrons. When a magnetic field is applied, the dipoles line up with the field, causing a positive magnetization. Because the dipoles do not interact, extremely large magnetic fields are required to align all of the dipoles; so, the effect is lost as soon as the magnetic field is removed. This behavior can be observed in metals such as aluminum and titanium
Ferromagnetic	This type of behavior is promoted by the unfilled energy levels in the *3d* level of iron, nickel and cobalt. In this kind of materials, the permanent unpaired dipoles easily line up with the imposed magnetic field due to the exchange interaction. Large magnetizations are obtained even for small magnetic fields
Antiferromagnetic	In this behavior, the magnetic moments produced in neighboring dipoles line up in opposition to one another in the magnetic field, even though the strength of each dipole is very high. Examples of these materials are manganese, chromium, MnO and NiO
Ferrimagnetic	Different ions have different magnetic moments, like in ceramic materials. In a magnetic field, the dipoles of ion *A* may line up with the field, while dipoles of ion *B* oppose the field. But because the strengths of the dipoles are not equal, a net magnetization results. Ceramics called ferrites perfectly represent this behavior

On the other hand, superparamagnetism is observed in very small particles of transition metals and their compounds, particularly their oxides. It can be used to characterize fine dispersions of metal, alloy and their oxides and has applications in several areas. Parameters such as the magnetization per cubic centimeter or magnetization per gram are better parameters than the susceptibility for describing superparamagnetism and magnetically ordered ferromagnetic and ferromagnetic materials. The applied magnetic field is expressed in oersteds (Oe). Furthermore, the magnetic moment in ferromagnetic and ferromagnetic materials refers to the saturation moment and not the effective moment. Ferromagnetic metals, such as Fe, Co, Ni and insulators such as γ-Fe_2O_3 and Fe_3O_4, show the well-known hysteresis curve; in their unmagnetized state, the unpaired electrons associated with each atom have a net magnetic moment or magnetization, which is the vector sum of all unpaired electrons in that domain. In this sense, it is important to explain that when ferromagnetic and ferromagnetic materials are heated above a critical Curie temperature, they change over to paramagnetic behavior; thus the hysteresis disappears. In contrast, ideal superparamagnetic systems, when sufficiently cooled below a critical blocking temperature will experience a very slow relaxation time. Their net magnetic moment will align parallel to the applied field and appear to behave as if they had an apparent ferromagnetic behavior. This aspect will result in hysteresis of "apparent" ferromagnetic behavior. Conversely, above the critical blocking temperature, the hysteresis will disappear and the clusters will show a unique curve with no hysteresis [76].

It is possible to classify the materials from the magnetic point of view as a function of the number of Bohr magnetons, whose parameter can be calculated from the Langevin function, first derived for the noninteracting paramagnetic spins and to the noninteracting:

$$\sigma/\sigma_s = coth(\mu_c H/kT) - (kT/\mu_c H) \tag{1}$$

where σ is the magnetization per gram; σs is the magnetization of saturation; μc is the magnetic moment of the cluster; H is the applied magnetic field; k is the Boltzmann constant; T is the temperature (K). A parameter used to corroborate the effect of the magnetic field on an adsorbate-adsorbent system is the zeta potential. Some studies have reported this parameter, such as Zhang et al. in 2004 where the zeta potential was measured in magnetized and non-magnetized Ca-rectorite suspensions observing that in the magnetized solutions the zeta potential was larger than in the nonmagnetized ones when the Cu concentration was zero. In contrast, the magnetic treatment reduced the zeta potential in Ca-rectorite suspensions containing Cu [77]. This same behavior was observed in a study made with Na-rectorite dispersions magnetized and not magnetized in the presence and absence of Zn [78]. There is limited information in the literature regarding the mechanism by which magnetic influences influence adsorbate-adsorbent systems. One of the most evident indications is that exposure to magnets changes the organization of water molecules, ions, and hydrated ions that are attached to the surface of the particle. This causes the layer of attached particles to become thicker, which in turn causes the slipping plane to move away from the particle surface. As a result, the zeta potential decreases and the particles appear larger, leading to a decrease in their ability to move around [79].

7. THE SIGNIFICANCE OF THE PHYSICOCHEMICAL AND MAGNETIC CHARACTERISTICS OF THE ADSORBATES IN THE ADSORPTION PROCESS FACILITATED BY AN EXTERNAL MAGNETIC FIELD

The physicochemical characteristics of adsorbates are crucial in determining the process of adsorption. Parameters such as ionic radius, electronegativity, valence, charge, and hydration number are crucial in selecting the adsorbent material and comprehending the adsorption mechanism. These properties aid in determining forces and mobilities [80]. Adsorbates, such as heavy metals, exist in a positively charged state. This creates an imbalance in charges. However, when these adsorbates are dissolved in water, they maintain stability by balancing their charge with the highly negative electronic cloud of oxygen in water molecules. As a result, the ions are surrounded by water molecules and are quasi-stabilized. This phenomenon is known as ion hydration. The hydration of ions plays a crucial role in the adsorption process as it determines the strength of adsorption for each ion on the surface of an adsorbent material. It also determines the competitive advantage of certain ions in occupying active sites, which is explained by the Gouy-Chapman theory.

The Gouy-Chapman theory posits that ions with the same charge are adsorbed with equal force on the surface of the adsorbent. However, the strength of adsorption is determined by the hydration radius, with ions having a smaller hydrated radius being more strongly retained. This leads to an increase in the size of the ion and a decrease in its mobility [81]. According to this, the hydration of each ion will vary as it is surrounded by a specific number of water molecules, which depends on the ion's size and charge. This quantity of water molecules surrounding the ion is referred to as the number of hydration water molecules.To know the hydration number of an ion, it is necessary to study the mobility of the ion within the solution, this mobility is mainly affected by two forces: the electric force FeFe (Eq. (2)) and the viscous force FvFv (Eq. (3)):

$$Fe = z_i e_0 x \tag{2}$$

where zie0 is the electric charge; x is the electric field.

$$Fv = 6\pi r \eta v \tag{3}$$

where r is the ionic radius; η is the viscosity; v is the ion speed.

The phenomenon of hydration is explained when the ion reaches a constant speed, which implies that the electric and viscous forces are equalized (Eq. (4)).

$$r = \frac{z_i e_0 x}{6\pi \eta v} \tag{4}$$

Bearing in mind that the electric mobility u is the result of the velocity ratio on the electric field (Eq. (5)), we find the hydrated radius rh (Eq. (6)):

$$u = \frac{v}{x} \tag{5}$$

$$r_h = \frac{z_i e_0}{6\pi \eta u} \tag{6}$$

Finally, the hydration number nh is found by differentiating the hydrated radius rh with the crystallographic radius of the rCris ion divided by the molecular radius of the water rH2O (Eq. (7)):

$$n_h = \frac{r_h^3 - r_{Cris}^3}{r_{H_2O}^3} \tag{7}$$

The phenomenon of ionic hydration is mostly caused by thermodynamic processes [82]. However, there have been studies that have detected changes in ion hydration due to exposure to an external magnetic field, including alterations in ion displacement and the formation of complexes [83]. The impact of a magnetic field on ionic hydration varies depending on the thickness of the hydration layer of each ion and the thermodynamic conditions of hydration. This suggests that magnetic field treatment can alter the structure of the water that hydrates the ions [84]. Theoretical studies suggest that the presence of a magnetic field promotes the hydration of ionic species by generating normal Lorentz forces. These forces aid in penetrating the boundary layer of natural hydration, resulting in a molecular configuration that facilitates enhanced hydration [85]. Furthermore, the transportation of the adsorbates to the active sites is facilitated by the momentum generated by the Lorentz forces and the electrostatic forces of Van der Waals, providing an advantage over viscous forces [86]. However, research has demonstrated that divalent cations, including heavy metals, are highly influenced by magnetic fields. These cations appear to enhance their polarization and exhibit a "magnetic memory" effect, which can be disrupted by increasing temperatures [87, 88]. Additionally, it has been observed that the magnetic memory effect can diminish non-linearly over time, depending on the specific configuration of the magnetic field and the duration of exposure to it [89]. The shift in the polarity of the hydrated ions caused by the magnetic field exposure also leads to modifications in the zeta potential (ζ) of the solutions, which in turn affects the pH values and promotes the adsorption of metallic species [77, 78, 79, 90]. Fig. 2 illustrates how a magnetic field enhances the process of ionic hydration by allowing water molecules to more effectively surround the ion. This effect is observed not only in the first layer of hydration, but also in the second layer, albeit with less intensity than in the first layer. The enhanced ion mobility in the presence of a magnetic field is attributed to the magnetic susceptibility of both the ion and the water molecules, which generates a magnetic force (Fmag). This force, combined with the electrostatic force (Fe), overcomes the viscous force (Fv), thereby increasing the likelihood of ion collisions with the adsorbent surface.

The interactions that occur between the ions and the magnetic field can be explained by understanding the magnetic susceptibility; this is an intrinsic property that has all the matter. Magnetic susceptibility (χMag) is a magnitude that represents the sensitivity to the magnetization of a material influenced by a magnetic field, determined by the sum of a diamagnetic component or diamagnetic susceptibility (χDia) and a paramagnetic component or paramagnetic susceptibility (χParaχPara) [91] (Eq. (8)), understanding that the diamagnetic matter opposes the flow of the magnetic field and the paramagnetic matter is oriented in the same direction of the magnetic field.

$$\chi_{Mag} = \chi_{Para} + \chi_{Dia} \tag{8}$$

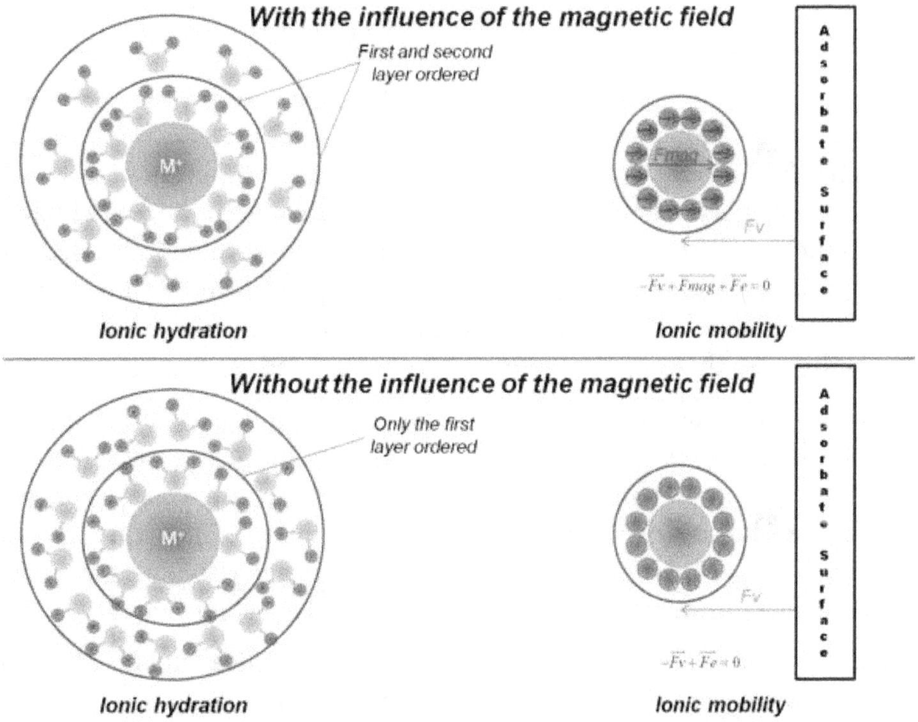

Fig. 2. Mechanism of adsorption of heavy metals with and without the presence of a magnetic field.

Substances that are magnetized in the same direction of the magnetic or paramagnetic field have an odd electronic orbital and their magnetic susceptibility is positive, while those that are magnetized against the magnetic field or diamagnetic have their electronic orbital even and have negative magnetic susceptibility [92, 93], and each substance has a different value of magnetic susceptibility, inclusive, the same substance with different oxidation state can vary in its magnitude of magnetic susceptibility. In the case of ionic species such as heavy metals, they have distinct values of magnetic susceptibility between them, coupled with this, the water molecules that surround a hydrated ion also have a magnetic susceptibility, so as a whole, generates a new magnitude of magnetic susceptibility, being the sum of the susceptibility of the solitary ion and that of each of the water molecules that surround it in the main layer. Based on this, the net susceptibility of the hydrated ion will depend on the magnetic susceptibility value of the ion χMagand the multiplication of the magnetic susceptibility of water χH_2O by the hydration number nh (Eq. (9)).

$$X_h = X_{mag\ ion} + [(n_h)(X_{H_2O})] \tag{9}$$

As an example, the susceptibilities of three heavy metals are shown in Table 5: Pb, Cd and Zn, where the Pb has a magnetic susceptibility lower than that of the Cd and Zn; contrary to the values of ionic radio and electronegativity; however, we found that the hydration index nh of Pb is lower than that of Cd and Zn, which is consistent with the Gouy-Chapman theory. On the other hand, the value of the magnetic susceptibility of water, which, like that of heavy metals is negative, which implies its diamagnetism and with which it will be possible to determine the hydrated magnetic susceptibility, resulting to be now larger than the hydrated magnetic susceptibilities of both Cd and Zn, compared to that of Pb.

Table 5. Physicochemical and magnetic parameters of different ions and hydrated ions.

	χmag (1 × 10^{-6} emu/mol)	rion (pm)	nh (min–max) [2]	*χh* (1 × 10^{-6} emu/mol)
Pb	-23	1.2	4–7.5	-120.275
Cd	-19.7	0.97	10–12.5	-181.825
Zn	-9.5	0.74	10–12.5	-171.625
H_2O	-12.97	—	—	—

As can be seen in the specific case of Pb, Cd and Zn are diamagnetic species despite being heavy metals; however, many transition metal complexes exhibit both diamagnetic and paramagnetic complexes. For example, the compounds of the complex ion $[Co(NH_3)_6]^{3+}$ do not have unpaired electrons, but the compounds of $[CoF_6]_3^-$ ion have four per metal ion. Both complexes contain Co(III) with 3d6 electronic configuration [94], although both Zn and Cd are elements of electronic configuration d, they exhibit diamagnetic behavior because they are at the limit of groups IIB and tend to behave similar to group IIIA.

Alternatively, the magnetic ordering of ionic species in an aqueous solution can be determined by observing the color of the salt they originate from. This is because heavy metal compounds often have orbitals that are divided into two distinct sets by division energies, which correspond to wavelengths of light in the visible region. The absorption of visible light induces electronic transitions between orbitals within these sets, establishing a relationship between the paramagnetism and the coloration of the salt. Conversely, certain metallic nitrates exhibit diamagnetism and lack color [94].

In general, there is a clear relationship between the physicochemical and magnetic characteristics of ionic adsorbates, and how they are affected by exposure to a magnetic field.

8. GENERATION AND SETUP OF A MAGNETIC FIELD FOR USE IN AN ADSORPTION PROCESS

Knowing the characteristics of the magnetic field that will interact with both the adsorbate and the adsorbent is overriding, three design parameters can be mentioned in the application of a magnetic field for assistance in the adsorption process: (1) type of source, (2) arrangement or trajectory of field lines and (3) magnetic field strength.

The source of the magnetic field can be classified by its generation in three different ways: alternating current (AC) electromagnets, direct current (DC) electromagnets and permanent magnets.

In general, the electromagnets consist of a solenoid made from cable made of an electrically conductive material, these solenoids generate a magnetic field in their cavity to the passage of electric current. The design of the solenoid goes from linear, annular and cylindrical forms according to the application, where the magnetic field strength that could generate depends on the number of turns, intensity and current potential [95].

The AC electromagnets are distinguished by passing electric current in which the magnitude and direction of the current vary in cyclical periods. This type of electromagnets have been used to generate magnetic fields applied in multiple ways ranging from heating by induction micellar particles to remove dyes from water [96], the acceleration in the germination of seeds under the application of a magnetic pulse [97], to be used as an incentive for the adsorption of volatile organic compounds [98]. They present some advantages in the generation of temperatures by induction and work with magnetic pulses of high frequency; however, it has some other disadvantages such as the creation of a heterogeneous and discontinuous magnetic field.

DC electromagnets are usually the most used for assistance in processes to eliminate water pollutants, and they are completely similar to AC electromagnets, with the difference that the field they generate is homogeneous and continuous due to the nature of the electric current, although they can also generate pulsations [99, 100, 101]. Studies have been carried out on the application of an electromagnet in adsorption processes, where it has been found that the field generated by them decreases the equilibrium time [101], increases the efficiency of dye removal by orienting the molecules and creating micromagnets from magnetic adsorbent materials [102]. It was also found that there is an effect of the pulsed magnetic field on the morphology of the surface of the adsorbents and the intensification of the adsorption activity [103]. However, the heating of the solenoid generates a disadvantage because the temperature control in adsorption studies is fundamental, so it would be necessary to use a cooling system for temperature control [9], raising the equipment costs.

Within the electromagnets, there is a new modality that is also used for water purification processes; it is the superconducting magnets that generate the high-gradient magnetic field. The superconducting magnets are capable of generating a magnetic field from 2 to 10 T [95, 104].

Finally, permanent magnets, which are manufactured from ceramic and nonceramic ferromagnetic materials that have iron, nickel, cobalt, rare earth and even silica as the main elements [105, 106, 107, 108].

The magnetic force that permanent magnets could present depends on two factors: the coercive field that presents the material and the field strength with which it is loaded in its manufacture where it is possible to find a wide variety of intensities that can be higher than 2 T per unit of length. In addition to this, they have the advantage of offering a reduction in the operational cost related to the consumption of electric power for the generation of magnetic field [95]. In addition, both the polarity of the magnet and the versatility of its shape can be an advantage in the design of devices for assisting the adsorption process. On the other hand, its fragility, oxidation in environments with humidity and limitation in size are rejected for industrial application. Like electromagnets, permanent magnets have been used for their application in adsorption processes, showing changes in kinetics [108]; modification in the magnetic ordering of materials [109], modification in the magnetic ordering in addition to the removal of metals from groundwater without modifying components of Ca and Mg materials [110].

Like the source, the configuration of the magnetic field is significant in the design of an adsorption system assisted by magnetic fields, since the influence, gradient of intensity and direction will determine the type of effect in some of the adsorption phases. The configuration of the magnetic field will influence in various ways ranging from orienting, hydrating and redirecting the adsorbate, to magnetize, activate and prevent the degeneration of the adsorbent [111, 112, 113, 114, 115].

In the case of adsorption in continuous systems, the application of the magnetic field could be based on the orientation of the magnetic field, which could be parallel or perpendicular to the directions of the fluid flow [95]. Particularly in the use of permanent magnets, they have been used in the treatment of electrolytes where they were arranged around the treatment circuit, so that the applied field was orthogonal to the direction of fluid flow [116].

There are several setups available for conducting adsorption experiments in batch systems. An experiment was conducted using a parallel arrangement of permanent magnets, with an adsorption system placed between them. The findings of the experiment were found to be favorable [117]. This phenomenon can be explained by the summation of magnetic intensities. It occurs when two or more magnets are placed in parallel and their polarities attract each other. However, this effect is only effective at short distances, and the distance at which it occurs depends on the specific intensities of each permanent magnet involved. Alternatively, the study examined the impact of various magnetic arrangements on the adsorption outcomes. The removal of the adsorbate was used as the output variable, and three distinct configurations of permanent magnets were employed and assessed. Specifically, the study focused on evaluating the behavior of the fluidized adsorbent particles. This work elucidates that the Lorentz forces alter the paths of the adsorbent particles within the container, counteracting the turbulent forces generated by mechanical agitation. As a result, varying degrees of dye removal are achieved in the polluted solution [112]. However, research has shown that there is a correlation between the magnetic configuration and the gradient of magnetic intensities. This, together with the arrangement of magnetic fields, can have an impact on the solution, the adsorbent material, and both factors [113].

Computer simulations can be used to evaluate various magnetic systems by quantifying and visually analyzing the trajectories of field lines, the gradient of magnetic intensities, and the spatial distribution profiles of magnetic flux density [84, 118, 119, 120]. This ensures that the magnetic field will indeed impact the adsorption system.

Based on the information provided in this chapter, it can be concluded that the magnetic field is a practical tool for efficiently and eco-friendly removal of heavy metals through adsorption processes. However, it is crucial to consider various parameters to achieve the desired outcomes. The interaction between the magnetic field and the adsorbate-adsorbent system is intricate and relies on various variables inside the system.

REFERENCES

1. Alijani H, Beyki MH, Shariatinia Z, Bayat M, Shemirani F. A new approach for one step synthesis of magnetic carbon nanotubes/diatomite earth composite by chemical vapor deposition method: Application for removal of lead ions. Chemical Engineering Journal. 2014;253:456-463. DOI: 10.1016/j.cej.2014.05.021

2. Chen G, Qiao C, Wang Y, Yao J. Synthesis of magnetic gelatin and its adsorption property for Cr(VI). Industrial and Engineering Chemistry Research. 2014;53:15576-15581. DOI: 10.1021/ie502709u

3. Agency for Environmental Protection (EPA) Federal Law on Control of Water Pollution "Clean Water Act". United States of America. 1977

4. Hernández-Montoya V, Pérez-Cruz MA, Mendoza-Castillo DI, Moreno-Virgen MR, Bonilla-Petriciolet A. Competitive adsorption of dyes and heavy metals on zeolitic structures. Journal of Environmental Management. 2013;116:213-221. DOI: 10.1016/j.jenvman.2012.12.010

5. Tang WW, Zeng GM, Gong JL, Liang J, Xu P, Zhang C, Huang BB. Impact of humic/fulvic acid on the removal of heavy metals from aqueous solutions using nanomaterials: A review. Science of the Total Environment. 2014;468:1014-1027. DOI: 10.1016/j.scitotenv.2013.09.044

6. Lesmana SO, Febriana N, Soetaredjo FE, Sunarso J, Ismadji S. Studies on potential applications of biomass for the separation of heavy metals from water and wastewater. Biochemical Engineering Journal. 2009;44:19-41. DOI: 10.1016/j.bej.2008.12.009

7. Zeng G, Liu Y, Tang L, Yang G, Pang Y, Zhang Y, Zhou Y, Li Z, Li M, Lai M, He X, He Y. Enhancement of Cd(II) adsorption by polyacrylic acid modified magnetic mesoporous carbon. Chemical Engineering Journal. 2015;259:153-160. DOI: 10.1016/j.cej.2014.07.115

8. Masoumi A, Ghaemy M, Bakht AN. Removal of metal ions from water using poly(MMA-co-MA)/modified-Fe3O4 magnetic nanocomposite: Isotherm and kinetic study. Industrial and Engineering Chemistry Research. 2014;53:8188-8197. DOI: 10.1021/ie5000906

9. Ferreira J, Oliveira L, Ragozoni M, Paulo da Silva J, Ramalho T. Adsorption of aromatic compounds under magnetic field influence. Water, Air, and Soil Pollution. 2012;223:3545-3551. DOI: 10.1007/s11270-012-1132-6

10. Tireli A, Firmino M, Ferreira L, Guimaraes I, Guerreiro M, Paulo J. Influence of magnetic field on the adsorption of organic by clays modified with iron. Applied Clay Science. 2014;97:1-7. DOI: 10.1016/j.clay.2014.05.014

11. Zhang J, Han J, Wang M, Guo R. Fe3O4/PANI/MnO2 core–shell hybrids as advanced adsorbents for heavy metal ions. Journal of Materials Chemistry A. 2017;5:4058-4066. DOI: 10.1039/c6ta10499a

12. Pavlovic I, Pérez MR, Barriga C, Ulibarri MA. Adsorption of Cu2+, Cd2+ and Pb2+ ions by layered double hydroxides intercalated with the chelating agents diethylenetriaminepentaacetate and meso-2,3-dimercaptosuccinate. Applied Clay Science. 2009;43:125-129. DOI: 10.1016/j.clay.2008.07.020

13. Kara I, Yilmazer D, Akar ST. Metakaolin based geopolymer as an effective adsorbent for adsorption of zinc(II) and nickel(II) ions from aqueous solutions. Applied Clay Science. 2017;139:54-63. DOI: 10.1016/j.clay.2017.01.008

14. Wawrzkiewicz M, Wisniewska M, Wołowicz A, Gun'ko VM, Zarko VI. Mixed silica-alumina oxide as sorbent for dyes and metal ions removal from aqueous solutions and wastewaters. Microporous and Mesoporous Materials. 2017;250:128-147. DOI: 10.1016/j.micromeso.2017.05.016

15. Selim AQ, Mohamed EA, Mobarak M, Zayed AM, Seliem MK, Komarneni S. Cr(VI) uptake by a composite of processed diatomite with MCM-41: Isotherm, kinetic and thermodynamic studies. Microporous and Mesoporous Materials. 2018;260:84-92. DOI: 10.1016/j.micromeso.2017.10.041

16. Yuan P, Liu D, Fan M, Yang D, Zhu R, Ge F, Zhu J, He H. Removal of hexavalent chromium [Cr(VI)] from aqueous solutions by the diatomite-supported/unsupported magnetite nanoparticles. Journal of Hazardous Materials. 2010;173:614-621. DOI: 10.1016/j.jhazmat.2009.08.129

17. Li W, Tang Y, Zeng Y, Tong Z, Liang D, Cui W. Adsorption behavior of Cr(VI) ions on tannin-immobilized activated clay. Chemical Engineering Journal. 2012;193-194:88-95. DOI: 10.1016/j.cej.2012.03.084

18. Zeng Y, Woo H, Lee G, Park J. Removal of chromate from water using surfactant modified Pohang clinoptilolite and Haruna chabazite. Desalination. 2010;257:102-109. DOI: 10.1016/j.desal.2010.02.039

19. Hong H, Jiang W, Zhang X, Tie L, Li Z. Adsorption of Cr(VI) on STAC-modified rectorite. Applied Clay Science. 2008;42:292-299. DOI: 10.1016/j.clay.2008.01.015

20. Ge Y, Cui X, Liao C, Li Z. Facile fabrication of green geopolymer/alginate hybrid spheres for efficient removal of Cu(II) in water: Batch and column studies. Chemical Engineering Journal. 2017;311:126-134. DOI: 10.1016/j.cej.2016.11.079

21. Catauro M, Bollino F, Papale F, Lamanna G. Investigation of the sample preparation and curing treatment effects on mechanical properties and bioactivity of silica rich metakaolin geopolymer. Materials Science and Engineering C: Materials for Biological Applications. 2014;36:20-24. DOI: 10.1016/j.msec.2013.11.026

22. Tofighy MA, Mohammadi T. Adsorption of divalent heavy metal ions from water using carbon nanotube sheets. Journal of Hazardous Materials. 2011;185:140-147. DOI: 10.1016/j.jhazmat.2010.09.008

23. Erdem E, Karapinar N, Donat R. The removal of heavy metal cations by natural zeolites. Journal of Colloid and Interface Science. 2004;280:309-314. DOI: 10.1016/j.jcis.2004.08.028

24. Algothmi WM, Bandaru NM, Yu Y, Shapter JG, Ellis AV. Alginate-graphene oxide hybrid gel beads: An efficient copper adsorbent material. Journal of Colloid and Interface Science. 2013;397:32-38. DOI: 10.1016/j.jcis.2013.01.051

25. Grisdanurak N, Akewaranugulsiri S, Futalan CM, Tsai WC, Kan CC, Hsu CW, Wan MW. The study of copper adsorption from aqueous solution using crosslinked chitosan immobilized on bentonite. Journal of Applied Polymer Science. 2012;125:E132-E142. DOI: 10.1002/app.35541

26. Rathnayake SI, Martens WN, Xi Y, Frost RL, Ayoko GA. Remediation of Cr (VI) by inorganic-organic clay. Journal of Colloid and Interface Science. 2017;490:163-173. DOI: 10.1016/j.jcis.2016.11.070

27. Tovar-Gómez R, Rivera-Ramírez DA, Hernández-Montoya V, Bonilla-Petriciolet A, Durán-Valle CJ, Montes-Morán MA. Synergic adsorption in the simultaneous removal of acid blue 25 and heavy metals from water using a $Ca(PO_3)_2$-modified carbon. Journal of Hazardous Materials. 2012;199-200:290-300. DOI: 10.1016/j.jhazmat.2011.11.015

28. Schier de Lima L, Machado Araujo MD, Pércio Quináia S, Migliorine DW, Garcia JR. Adsorption modeling of Cr, Cd and Cu on activated carbon of different origins by using fractional factorial design. Chemical Engineering Journal. 2011;166:881-889. DOI: 10.1016/j.cej.2010.11.062

29. Hsi HC, Rood MJ, Massoud RA, Chang YM. Effects of sulfur, nitric acid, and thermal treatments on the properties and mercury adsorption of activated carbons from bituminous coals. Aerosol and Air Quality Research. 2013;13:730-738. DOI: 10.4209/aaqr.2012.07.0177

30. Schier de Lima L, Pércio Quináia S, Melquiades FL, de Biasi GEV, Garcia JR. Characterization of activated carbons from different sources and the simultaneous adsorption of Cu, Cr, and Zn from metallurgic effluent. Separation and Purification Technology. 2014;122:421-430. DOI: 10.1016/j.seppur.2013.11.034

31. Aguayo-Villarreal IA, Hernández-Montoya V, Bonilla-Petriciolet A, Tovar-Gómez R, Ramírez-López EM, Montes-Morán MA. Role of acid blue 25 dye as active site for the adsorption of Cd2+ and Zn2+ using activated carbons. Dyes and Pigments. 2013;96:459-466. DOI: 10.1016/j.dyepig.2012.08.027

32. Tovar-Gómez R, Moreno-Virgen MR, Moreno-Pérez J, Bonilla-Petriciolet A, Hernández-Montoya V, Durán-Valle CJ. Analysis of synergistic and antagonistic adsorptionof heavy metals and acid blue 25 on activated carbon from ternary systems. Chemical Engineering Research and Design. 2015;93:755-772. DOI: 10.1016/j.cherd.2014.07.012

33. Zhou J, Liu Y, Zhou X, Ren J, Zhong C. Magnetic multi-porous bio-adsorbent modified with amino siloxane for fast removal of Pb(II) from aqueous solution. Applied Surface Science. 2018;427:976-985. DOI: 10.1016/j.apsusc.2017.08.110

34. Ebrahimi Zarandi MJ, Reza Sohrabi M, Khosravi M, Mansouriieh N, Davallo M, Khosravan A. Optimizing Cu(II) removal from aqueous solution by magnetic nanoparticles immobilized on activated carbon using Taguchi method. Water Science and Technology. 2016;74:38-47. DOI: 10.2166/wst.2016.152

35. Chen W, Parette R, Zou J, Cannon FS, Dempsey BA. Arsenic removal by iron-modified activated carbon. Water Research. 2007;41:1851-1858. DOI: 10.1016/j.watres.2007.01.052

36. Chang Q, Wei L, W-C Y. Preparation of iron-impregnated granular activated carbon for arsenic removal from drinking water. Journal of Hazardous Materials. 2010;184:515-522. DOI: 10.1016/j.jhazmat.2010.08.066

37. Zhu H, Jia Y, Wu X, Wang H. Removal of arsenic from water by supported nano zero-valent iron on activated carbon. Journal of Hazardous Materials. 2009;172:1591-1596. DOI: 10.1016/j.jhazmat.2009.08.031

38. He J, Shang H, Zhang X, Sun X. Synthesis and application of ion imprinting polymer coated magnetic multi-walled carbon nanotubes for selective adsorption of nickel ion. Applied Surface Science. 2018;428:110-117. DOI: 10.1016/j.apsusc.2017.09.123

39. Shang J, Pi J, Zong M, Wang Y, Li W, Liao Q. Chromium removal using magnetic biochar derived from herb residue. Journal of the Taiwan Institute of Chemical Engineers. 2016;68:289-294. DOI: 10.1016/j.jtice.2016.09.012

40. Luo X, Lei X, Cai N, Xie X, Xue Y, Yu F. Removal of heavy metal ions from water by magnetic cellulose-based beads with embedded chemically modified magnetite nanoparticles and activated carbon. Sustainable Chemistry & Engineering. 2016;4:3960-3969. DOI: 10.1021/acssuschemeng.6b00790

41. Horst MF, Álvarez M, Lassalle VL. Removal of heavy metals from wastewater using magnetic nanocomposites: Analysis of the experimental conditions. Separation Science and Technology. 2016;51:550-563. DOI: 10.1080/01496395.2015.1086801

42. Liu T, Han X, Wang Y, Yan L, Du B, Wei Q, Wei D. Magnetic chitosan/anaerobic granular sludge composite: Synthesis, characterization and application in heavy metal ions removal. Journal of Colloid and Interface Science. 2017;508:405-414. DOI: 10.1016/j.jcis.2017.08.067

43. Yi X, He J, Guo Y, Han Z, Yang M, Jin J, Gu J, Ou M, Xu X. Encapsulating Fe3O4 into calcium alginate coated chitosan hydrochloride hydrogel beads for removal of Cu (II) and U (VI) from aqueous solutions. Ecotoxicology and Environmental Safety. 2018;147:699-707. DOI: 10.1016/j.ecoenv.2017.09.036

44. Zhu J, Gu H, Guo J, Chen M, Wei H, Luo Z, Colorado HA, Yerra N, Ding D, Ho TC, Haldolaarachchige N, Hopper J, Young DP, Guo Z, Wei S. Mesoporous magnetic carbon nanocomposite fabrics for highly efficient Cr(VI) removal. Journal of Materials Chemistry A. 2014;2:2256-2265. DOI: 10.1039/C3TA13957C

45. Lasheen MR, El-Sherif IY, El-Wakeel ST, Sabry DY, El-Shahat MF. Heavy metals removal from aqueous solution using magnetite Dowex 50WX4 resin nanocomposite. Journal of Materials and Environmental Science. 2017;8:503-511

46. Zhou C, Zhu H, Wang Q, Wang J, Cheng J, Guo Y, Zhou X, Bai R. Adsorption of mercury(II) with an Fe3O4 magnetic polypyrrole-graphene oxide nanocomposite. RSC Advances. 2017;7:18466-18479. DOI: 10.1039/c7ra01147d

47. Son E-B, Poo K-M, Chang J-S, Chae K-J. Heavy metal removal from aqueous solutions using engineered magnetic biochars derived from waste marine macro-algal biomass. Science of the Total Environment. 2018;615:161-168. DOI: 10.1016/j.scitotenv.2017.09.171

48. Meng C, Zhikun W, Qiang LV, Li C, Shuangqing S, Songqing H. Preparation of amino-functionalized Fe3O4@mSiO2 core-shell magnetic nanoparticles and their application for aqueous Fe3+ removal. Journal of Hazardous Materials. 2018;341:198-206. DOI: 10.1016/j.jhazmat.2017.07.062

49. Rusmin R, Sarkar B, Tsuzuki T, Kawashima N, Naidu R. Removal of lead from aqueous solution using superparamagnetic palygorskite nanocomposite: Material characterization and regeneration studies. Chemosphere. 2017;186:1006-1015. DOI: 10.1016/j.chemosphere.2017.08.036

50. Kumar Reddy DH, Lee SM. Magnetic biochar composite: Facile synthesis, characterization and application for heavy metal removal. Colloid Surface A. 2014;20:96-103. DOI: 10.1016/j.colsurfa.2014.03.105

51. Elwakeel KZ, Al-Bogami AS. Influence of Mo(VI) immobilization and temperature on As(V)sorption onto magnetic separable polyp-phenylenediamine-thiourea-formaldehyde polymer. Journal of Hazardous Materials. 2018;342:335-346. DOI: 10.1016/j.jhazmat.2017.08.007

52. Cao Y, Huang J, Peng X, Cao D, Galaska A, Qiu S, Liu J, Khan AM, Young DP, Ryu JE, Feng H, Yerra N, Guo Z. Poly(vinylidene fluoride) derived fluorine-doped magnetic carbon nanoadsorbents for enhanced chromium removal. Carbon. 2017;115:503-514. DOI: 10.1016/j.carbon.2017.01.033

53. Vojoudi H, Badiei A, Bahar S, Ziarani GM, Faridbod F, Ganjali MR. A new nano-sorbent for fast and efficient removal of heavy metals from aqueous solutions based on modification of magnetic mesoporous silica nanospheres. Journal of Magnetism and Magnetic Materials. 2017;441:193-203. DOI: 10.1016/j.jmmm.2017.05.065

54. Wang S, Zhang C, Chang Q. Synthesis of magnetic crosslinked starch-graftpoly(acrylamide)-co-sodium xanthate and its application in removing heavy metal ions. Journal of Experimental Nanoscience. 2017;12:270-284. DOI: 10.1080/17458080.2017.1321793

55. Shehzad K, Xie C, He J, Cai X, Xu W, Liu J. Facile synthesis of novel calcined magnetic orange peel composites for efficient removal of arsenite through simultaneous oxidation and adsorption. Journal of Colloid and Interface Science. 2018;511:155-164. DOI: 10.1016/j.jcis.2017.09.110

56. Noor NM, Othman R, Mubarak NM, Abdullah EC. Agricultural biomass-derived magnetic adsorbents: Preparation and application for heavy metals removal. Journal of the Taiwan Institute of Chemical Engineers. 2017;78:168-177. DOI: 10.1016/j.jtice.2017.05.023

57. Keochaiyom B, Wan J, Zeng G, Huang D, Xue W, Hu L, Huang C, Zhang C, Cheng M. Synthesis and application of magnetic chlorapatite nanoparticles for zinc (II), cadmium (II) and lead (II) removal from water solutions. Journal of Colloid and Interface Science. 2017;505:824-835. DOI: 10.1016/j.jcis.2017.06.056

58. Ghasemi E, Heydari A, Sillanpää M. Superparamagnetic Fe3O4@EDTA nanoparticles as an efficient adsorbent for simultaneous removal of Ag(I), Hg(II), Mn(II), Zn(II), Pb(II) and Cd(II) from water and soil environmental samples. Microchemical Journal. 2017;131:51-56. DOI: 10.1016/j.microc.2016.11.011

59. Rahman Khattak MMU, Zahoor M, Muhammad B, Khan FA, Ullah R, AbdEI-Salam NM. Removal of heavy metals from drinking water by magnetic carbon nanostructures prepared from biomass. Journal of Nanomaterials. 2017;1:1-10. DOI: 10.1155/2017/5670371

60. Yap MW, Mubarak NM, Sahu JN, Abdullah EC. Microwave induced synthesis of magnetic biochar from agricultural biomass for removal of lead and cadmium from wastewater. Journal of Industrial and Engineering Chemistry. 2017;45:287-295. DOI: 10.1016/j.jiec.2016.09.036

61. Praveen RS, Vijayaraghavan K. Optimization of Cu(II), Ni(II), Cd(II) and Pb(II) biosorption by red marine alga Kappaphycus alvarezii. Desalination and Water Treatment. 2014;55:1-9. DOI: 10.1080/19443994.2014.927334

62. Gupta VK, Ganjali MR, Nayak A, Bhushan B, Agarwal S. Enhanced heavy metals removal and recovery by mesoporous adsorbent prepared from waste rubber tire. Chemical Engineering Journal. 2012;197:330-342. DOI: 10.1016/j.cej.2012.04.104

63. Lin L-C, Thirumavalavan M, Wang Y-T, Lee J-F. Surface area and pore size tailoring of mesoporous silica materials by different hydrothermal treatments and adsorption of heavy metal ions. Colloid Surface A. 2010;369:223-231. DOI: 10.1016/j.colsurfa.2010.08.032

64. Castaldi P, Santona L, Enzo S, Melis P. Sorption processes and XRD analysis of a natural zeolite exchanged with Pb2+, Cd2+ and Zn2+ cations. Journal of Hazardous Materials. 2008;156:428-434. DOI: 10.1016/j.jhazmat.2007.12.040

65. Faur-Brasquet C, Reddad Z, Kadirvelu K, Le Cloirec P. Modeling the adsorption of metal ions (Cu2+, Ni2+, Pb+2) onto ACCs using surface complexation models. Applied Surface Science. 2002;196:356-365. DOI: 10.1016/S0169-4332(02)00073-9

66. Chen SB, Ma YB, Chen L, Xian K. Adsorption of aqueous Cd2+, Pb2+, Cu2+ ions by nano-hydroxiapatite: Single-and multi-metal competitive adsorption study. Geochemical Journal. 2010;44:233-239. DOI: 10.2343/geochemj.1.0065

67. Yang K, Chen B, Zhu X, Xing B. Aggregation, adsorption and morphological transformation of graphene oxide in aqueous solutions containing different metal cations. Environmental Science & Technology. 2016;50:11066-11075. DOI: 10.1021/acs.est.6b04235

68. Massocatto CL, de Andrade M, Honorato AC, Caetano J, Tarley CRT, Gonçalves-Júnior A, Colauto NB, Linde Colauto GA, Cardoso Dragunski D. Biosorption of Pb^{2+}, Cr^{3+}, and Cu^{2+} by peach palm sheath modified colonized by Agaricus Blazei. Desalination and Water Treatment. 2016;57:19927-19938. DOI: 10.1080/19443994.2015.1107503

69. El Zayat M, Elagroudy S, El Haggar S. Removal of some heavy metals in selected wastewater using cement kiln dust. In: World Environmental and Water Resources Congress. 2015. pp. 2459-2469

70. Castillo VA, Barakat MA, Ramadan MH, Woodcock HL, Kuhn JN. Metal ion remediation by polyamidoamine dendrimers: A comparison of metal ion, oxidation state, and titania immobilization. International Journal of Environmental Science and Technology. 2014;11:1497-1502. DOI: 10.1007/s13762-013-0346-5

71. Fu F, Wang Q. Removal of heavy metal ions from wastewaters: A review. Journal of Environmental Management. 2011;92:407-418. DOI: 10.1016/j.jenvman.2010.11.011

72. Bisht R, Agarwal M, Singh K. Heavy metal removal from wastewater using various adsorbents: A review. Journal of Water Reuse and Desalination. 2017;07(4):387-419. DOI: 10.2166/wrd.2016.104

73. Deliyanni EA, Kyzas GZ, Triantafyllidis KS, Matis KA. Activated carbons for the removal of heavy metal ions: A systematic review of recent literature focused on lead and arsenic ions. Open Chemistry. 2015;13:699-708. DOI: 10.1515/chem-2015-0087

74. Aeisyah Abas SN, Shah Ismail MH, Kamal ML, Izhar S. Adsorption process of heavy metals by low-cost adsorbent: A review. World Applied Sciences Journal. 2013;28:1518-1530. DOI: idosi.wasj.2013.28.11.1874

75. Askeland DR. Ciencia e ingeniería de los materiales. 3rd ed. Mexico: International Thompson; 1998

76. Drago RS. Magnetism. Physical Methods for Chemists. 2nd ed. United States: Saunders College Publishing; 1992

77. Zhang G, Yang X, Liu Y, Jia Y, Yu G, Ouyang S. Copper(II) adsorption on Ca-rectorite, and effect of static magnetic field on the adsorption. Journal of Colloid and Interface Science. 2004;278:265-269. DOI: 10.1016/j.jcis.2004.05.046

78. Zhang G, Liu Y, Xie Y, Yang X, Hu B, Ouyang S, Liu H, Wang H. Zinc adsorption on Na-rectorite and effect of static magnetic field on the adsorption. Applied Clay Science. 2005;29:15-21. DOI: 10.1016/j.clay.2004.09.001

79. Higashitani K, Iseri H, Okuhara K, Kage A, Hatade S. Magnetic effects on zeta potential and diffusivity of nonmagnetic colloidal particles. Journal of Colloid and Interface Science. 1995;172:383-388. DOI: 10.1006/jcis.1995.1268

80. Elwakeel KZ. Removal of Cr(VI) from alkaline aqueous solutions using chemically modified magnetic chitosan resins. Desalination. 2010;250:105-112. DOI: 10.1016/j.desal.2009.02.063

81. Bockris JO, Reddy AKN. Modern Electrochemistry 1. 2nd ed. Boston: Kluwer Academic Publishers; 2002

82. Higashitani K, Oshitani J. Magnetic effects on thickness of adsorbed layer in aqueous solutions evaluated directly by atomic force microscope. Journal of Colloid and Interface Science. 1998;204:363-378. DOI: 10.1006/jcis.1998.5590

83. Gehr R, Zhai ZA, Finch JA, Rao SR. Reduction of soluble mineral concentrations in CaSO4 saturated water using a magnetic field. Water Research. 1995;29:933-940. DOI: 10.1016/0043-1354(94)00214-R

84. Holysz L, Szczes A, Chibowski E. Effects of a static magnetic field on water and electrolyte solutions. Journal of Colloid and Interface Science. 2007;316:996-1002. DOI: 10.1016/j.jcis.2007.08.026

85. Lipus LC, Krope J, Crepinsek L. Dispersion destabilization in magnetic water treatment. Journal of Colloid and Interface Science. 2001;236:60-66. DOI: 10.1006/jcis.2000.7392

86. Jiang X, Qiao J, Lo IMC, Wang L, Guan X, Lu Z, Zhou G, Xu C. Enhanced paramagnetic Cu^{2+} ions removal by coupling a weak magnetic field with zero valent iron. Journal of Hazardous Materials. 2015;283:880-887. DOI: 10.1016/j.jhazmat.2014.10.044

87. Silva IB, Queiroz Neto JC, Petri DFS. The effect of magnetic field on ion hydration and sulfate scale formation. Colloids and Surfaces A: Physicochemical and Engineering Aspects. 2015;465:175-183. DOI: 10.1016/j.colsurfa.2014.10.054

88. Oshitani J, Yamada D, Miyahara M, Higashitani K. Magnetic effect on ion-exchange kinetics. Journal of Colloid and Interface Science. 1999;210:1-7. DOI: 10.1006/jcis.1998.5952

89. Holysz L, Chibowski M, Chibowski E. Time-dependent changes of zeta potential and other parameters of in situ calcium carbonate due to magnetic field treatment. Colloids and Surfaces A: Physicochemical and Engineering Aspects. 2002;208:231-240. DOI: 10.1016/S0927-7757(02)00149-8

90. Higashitani K, Oshitani J. Measurements of magnetic effects on electrolyte solutions by atomic force microscope. Institution of Chemical Engineers. 1997;75:115-119. DOI: 10.1205/095758297528887

91. Bain A, Berry JF. Diamagnetic corrections and Pascal's constants. Journal of Chemical Education. 2008;85:532

92. Lide DR. Magnetic susceptibility of the elements and inorganic compounds. Handbook of Chemistry and Physics. 87th ed. Boca Raton: Taylor and Francis; 2005. pp. 130-135

93. Hollens WRA, Spencer JF. The magnetic susceptibility of cadmium compounds. Journal of the Chemical Society. 1935;0:495

94. Whitten KW, Davis RE, Stanley GG, Peck ML. Chemistry. 9th ed. Belmont, CA. USA: Mary Finch; 2010

95. Ambashta RD, Sillanpää M. Water purification using magnetic assistance: A review. Journal of Hazardous Materials. 2010;180:38-49. DOI: 10.1016/j.jhazmat.2010.04.105

96. Zhang Q, Lu T, Bai DM, Lin DQ, Yao SJ. Self-immobilization of a magnetic biosorbent and magnetic induction heated dye adsorption processes. Chemical Engineering Journal. 2016;284:972-978. DOI: 10.1016/j.cej.2015.09.047

97. Moon JD, Chung HS. Acceleration of germination of tomato seed by applying AC electric and magnetic fields. Journal of Electrostatics. 2000;48:103-114. DOI: 10.1016/S0304-3886(99)00054-6

98. Samonin VV, Podvyaznikov ML, Chentsov MS, Spiridonova EA, Kiseleva VL. Effect of AC magnetic field on adsorption of benzene and ethanol vapors by activated carbons. Russian Journal of Applied Chemistry. 2012;85:1176-1181. DOI: 10.1134/S107042721208006X

99. Yavuz H, Çelebi SS. Effects of magnetic field on activity of activated sludge in wastewater treatment. Enzyme and Microbial Technology. 2000;26:22-27. DOI: 10.1016/S0141-0229(99)00121-0

100. Novickij V, Grainys A, Švediene J, Markovskaja S, Paškevičius A, Novickij J. Microsecond pulsed magnetic field improves efficacy of antifungal agents on pathogenic microorganisms. Bioelectromagnetics. 2014;35:347-353. DOI: 10.1002/bem.21848

101. Duan L, Guo S, Yang J. Study on the effect of a magnetic field on Pb (II) removal using modified chitosan. Advances in Chemical Engineering and Science. 2012;2:101-107. DOI: 10.4236/aces.2012.21011

102. Tireli AA, Marcos FCF, Oliveira LF, Guimarães IR, Guerreiro MC, Silva JP. Influence of magnetic field on the adsorption of organic compound by clays modified with iron. Applied Clay Science. 2014;97-98:1-7. DOI: 10.1016/j.clay.2014.05.014

103. Bel'chinskaya LI, Khodosova NA, Bityutskaya LA. Adsorption of formaldehyde at mineral nanoporous sorbents exposed to a pulse magnetic field. Protection of Metals and Physical Chemistry of Surfaces. 2009;45:203-206. DOI: 10.1134/S2070205109020130

104. Li XL, Yao KL, Liu HR, Liu ZL. The investigation of capture behaviors of different shape magnetic sources in the high-gradient magnetic field. Journal of Magnetism and Magnetic Materials. 2007;311:481-488. DOI: 10.1016/j.jmmm.2006.07.040

105. Szymura S, Sojka L. The effect of silicon on the structure and properties of Fe-Cr-Co permanent magnet alloys. Journal of Magnetism and Magnetic Materials. 1986;53:379-389. DOI: 10.1016/0304-8853(86)90184-8

106. Herbst JF, Croat JJ. Neodymium-iron-boron permanent magnets. Journal of Magnetism and Magnetic Materials. 1991;100:57-78. DOI: 10.1016/0304-8853(91)90812-O

107. Sagawa M, Hirosawa S, Yamamoto H, Fujimura S, Matsuura Y. Nd–Fe–B permanent magnet materials. Japanese Journal of Applied Physics. 1987;26:785-800. DOI: 10.1143/JJAP.26.785

108. Jia YY, Zhang GK, Hu B, Dong W. Adsorption capacity of kaolinite for copper (II) under magnetic field. Journal of Wuhan University of Technology-Materials Science Edition. 2004;19:52-54. DOI: 10.1007/BF03000168

109. Patkowski J, Myśliwiec D, Chibowski S. Adsorption of polyethyleneimine (PEI) on hematite. Influence of magnetic field on adsorption of PEI on hematite. Materials Chemistry and Physics. 2014;144:451-461. DOI: 10.1016/j.matchemphys.2014.01.019

110. Cotton GB, Navratil JD, Bradley H. Magnetic adsorption method for the treatment of metal contaminated aqueous waste. Waste Management Symposia. 1999;2:1-11

111. Brown P, Bushmelev A, Butts CP, Eloi J, Grillo I, Baker PJ, Schmidt AM, Eastoe J. Properties of new magnetic surfactants. Langmuir. 2013;29:3246-3251. DOI: 10.1021/la400113r

112. Hao X, Liu H, Zhang G, Zou H, Zhang Y, Zhou M, Gu Y. Magnetic field assisted adsorption of methylene blue onto organo-bentonite. Applied Clay Science. 2012;55:177-180. DOI: 10.1016/j.clay.2011.11.019

113. González Vázquez OF, Moreno Virgen MR, Hernández Montoya V, Tovar Gómez R, Alcántara Flores JL, Pérez Cruz MA, Montes Morán MA. Adsorption of heavy metals in the presence of a magnetic field on adsorbents with different magnetic properties. Industrial and Engineering Chemistry Research. 2016;55:9323-9331. DOI: 10.1021/acs.iecr.6b01990

114. Zubov VE, Kudakov AD, Levshin NL, Fedulova TS. The influence of adsorption of water molecules on magnetic susceptibility of amorphous ferromagnets. Sensors and Actuators A. 2001;91:214-217. DOI: 10.1016/S0924-4247(01)00494-0

115. Ozeki S, Miyamoto J, Ono S, Wakai C, Watanabe T. Water-solid interactions under steady magnetic fields: Magnetic-field-induced adsorption and desorption of water. The Journal of Physical Chemistry. 1996;100:4205-4212. DOI: 10.1021/jp9528774

116. Tombácz E, Ma C, Busch KW, Busch MA. Effect of a weak magnetic field on hematite sol in stationary and flowing systems. Colloid & Polymer Science. 1991;269:278-289. DOI: 10.1007/BF00665502

117. Mohammed RR, Ketabachi MR, McKay G. Combined magnetic field and adsorption process for treatment of biologically treated palm oil mill effluent (POME). Chemical Engineering Journal. 2014;243:31-42. DOI: 10.1016/j.cej.2013.12.084

118. Petkovska L, Ieee SM, Cvetkovski G. FEM Based Simulation of a Permanent Magnet Synchronous Motor Performance Characteristics. In: IEEE 5th International Power Electronics and Motion Control Conference. 2006. DOI: 10.1109/IPEMC.2006.4777986

119. Zakaria Z, Saiful M, Mansor B, Jahidin AH, Shuhanaz M, Azlan Z, Rahim RA. Simulation of magnetic flux leakage (MFL) analysis using FEMM software. In: IEEE Symposium on Industrial Electronics and Applications. 2010. pp. 481-486. DOI: 10.1109/ISIEA.2010.5679417

120. Vergallo C, Ahmadi M, Mobasheri H, Dini L. Impact of inhomogeneous static magnetic field (31.7-232.0T) exposure on human neuroblastoma SH-SY5Y cells during cisplatin administration. PLoS One. 2014;9:1-19. DOI: 10.1371/journal.pone.0113530

16 Chapter

UTILIZING BIOSORBENTS FOR THE DETERMINATION OF METALLIC IONS

1. INTRODUCTION: AN OVERVIEW

Metallic ions are present in a wide variety of samples at quantities ranging from milligrams per liter to levels believed to be trace amounts. Every day, we come into touch with different metallic ions without intending to, through sources such as the air, water, industrial products (such as cosmetics, makeup, remedies,deodorants,andtattootinctures),andeveninourfood(fromfertilizers,packaging,pesticides,etc.). These ions are essential for sustaining crucial biological processes in organisms, as their absence can lead to numerous disorders. Additionally, they can be highly poisonous even at low doses if swallowed. Mercury, cadmium, arsenic, chromium, and lead are the metallic ions that primarily contribute to health issues [1]. Given the potential harm or essential role of these species for organisms, it is crucial to develop analytical approaches that can accurately determine and facilitate their monitoring in many types of samples. For many years, flame atomization (FAAS) has been widely employed as a technique in atomic absorption spectrometry to determine inorganic ions in various materials. Nevertheless, it poses several challenges especially concerning the detection threshold and specificity. To overcome these challenges, extraction and preconcentration methods, such as liquid-liquid extraction, co-precipitation, and solid phase extraction (SPE), can be employed [2]. The technique of solid-phase extraction (SPE) was first introduced in 1976 as a solution to the drawbacks associated with liquid-liquid extraction. Since then, it has been the most often used method for sample preparation. This technique involves the transfer of a liquid phase (sample) to a solid phase, known as the adsorbent. The majority of adsorbents employed in these techniques are synthetic. The primary goals of solid-phase extraction (SPE) are to eliminate interference from the surrounding matrix, concentrate the desired analytes, and isolate them [3]. Current advancements in solid phase extraction (SPE) suggest a growing trend towards the development of smaller-scale techniques, such as flow-injection analysis systems (FIA), sequential injection analysis (SIA), and lab-on-a-valve (Lov) [4]. The adsorbents employed in solid phase extraction, namely as ion exchangers, prior to mid-1934, were natural substances known as lignites. Following this timeframe, other researchers focused their efforts on the advancement of novel ionic exchangers, including polymerization resins. In the study conducted by Reis and colleagues in 1996, almost 90% of the investigations on the extraction or preconcentration of metallic ions utilized polymeric resins, namely adsorbers [5]. Currently, over 90% of scientific articles utilize commercial absorbents in solid phase extraction systems [6]. Given the growing demand for green chemistry approaches, there has been a notable focus on studying alternate adsorbents in solid-phase extraction (SPE) procedures. Alternative adsorbent materials have been effectively utilized in metal adsorption processes. The majority of these materials are easily obtained and, in certain circumstances, are considered by-products of industrial operations, such as sugarcane bagasse, onion skin, peanut shells, corn cobs, and rice husks [7]. Hence, the selection of an alternative adsorbent is determined not only by its adsorption capacity, but also by its affordability, accessibility, and prevalence in the natural environment. This chapter provides a comprehensive review of the primary absorbers utilized in solid phase extraction techniques that have been introduced in recent years.

2. ADSORBENTS USED IN SOLID PHASE EXTRACTION SYSTEMS

2.1. Lignocellulosic adsorbents

Lignocellulosic materials are abundant renewable resources that are typically not fully used. The agricultural sector produces a significant amount of biomass that is not well utilized, resulting in its transformation into industrial waste [8]. Therefore, there has been a significant focus on the advancement of techniques that utilize these waste materials in recent years. Among these ways, the utilization of lignocellulosic materials has gained prominence due to its exceptional ability to adsorb metallic ions and organic compounds. Lignocellulosic biomass refers to the dry matter of plants, mostly consisting of a combination of lignin and holocellulose (the entire carbohydrate content in a plant). Lignocellulosic materials typically consist of 48% carbon, 6% hydrogen, and 45% oxygen in terms of their chemical composition. Proteins, organic acids, and other non-structural elements are present in small quantities [10]. These materials can be classified as lignocellulosic materials, including rice straw, wheat straw, cotton stalk, rice hull, sugarcane bagasse, corncob, sawdust, and others [9]. The precise mechanism by which these materials interact with metallic ions and organic compounds remains incompletely understood. However, it can be asserted that the adsorption capacity is correlated with the presence of functional groups such as hydroxyl and amines. The pulp, hemicellulose, and lignin structures contain these functional groups. It is crucial to acknowledge that this process does not rely on a single mechanism. The sequestration of the metal ion might take place by various complex mechanisms, such as ion exchange and complexation. It is likely that some of these systems operate simultaneously to different extents, depending on the biomass and the metal ion [11]. Davis et al. [12] stated that ion-exchange is a crucial concept in biosorption as it elucidates numerous observations reported during research on the absorption of heavy metals. Within this particular framework, the phrase ion-exchange does not specifically indicate how heavy metals attach to biomass. Instead, the mechanisms of chemical binding, such as ionic and covalent interactions, should be attributed to electrostatic or London-van der Waals forces. Hence, due to their nature, availability, and low cost, lignocellulosic materials are increasingly being studied for the development of novel analytical techniques to detect metallic ions in trace amounts across different sample types. There are numerous publications in the literature that examine the utilization of lignocellulosic materials as natural adsorbents. Table 1 illustrates the utilization of lignocellulosic adsorbents for the advancement of analytical techniques.

Table 1. Lignocellulosic adsorbents used in development of preconcentration and metal ion determination.

Material	Treatment	Sample	Analyte	L.O.D (μg L-1)	Preconcent ration factor	References
Moringa oleifera seeds	—	Beer	Pb	7.5	93	[13]
Rice husks	—	Water and Red Wine	Cd and Pb	1.14 and 14.1	72.4 and 46	[14]
Coffee	—	Water	Cu	0.22	—	[15]
Chinese herb Pang Da Hai	—	Water	Pb and Cd	0.0096 and 0.0032		[16]
Peel of *Citrus reticulata* (mandarin orange)	—	Alcoholic beverages	Ni	3.1	12	[17]

Moringa oleifera is a highly utilized lignocellulosic adsorbent in the advancement of analytical methods. Originally indigenous to northern India, this plant now thrives in several areas such as Africa, Southeast Asia, the Pacific and Caribbean Islands, and South America [18, 19]. The most well-known application of M. oleifera seeds and fruits is in the research of water purification and treatment of turbid water [11]. It is regarded as an effective natural substance for adsorbing small amounts of metals, making it useful for creating new analytical techniques. The seeds were utilized for the analysis of silver in water [20], inorganic arsenic in natural waterways [21], cadmium and zinc in alcohol fuel [22, 23], and lead in beer samples [13]. The husks of this substance were utilized in the targeted extraction of chromium from water [24], the analysis of copper in gasoline samples [25], the detection of zinc in cachaça, and the assessment of lead in chicken feed [27]. Araújo and colleagues [20] outline a method for quantifying Ag(I) in water solutions using solid phase extraction in a preconcentration system using flow injection, and detection through flame atomic absorption spectrometry (FAAS). A mini-column was constructed using 35 mg of *Moringa oleifera* seeds in this method. In order to identify the optimal conditions for the determination of silver, the preconcentration system of analysis by flow injection was optimized. After optimizing the system, the analytical properties were assessed. These included a preconcentration factor of 35, a detection limit of 0.22 µg L^{-1}, a quantification limit of 0.73 µg L^{-1}, a linear range from 5 to 20 µg L^{-1}, and a precision of 3.8% (at a concentration of 5 µg L-1, with a sample size of 7). Therefore, demonstrating exceptional consistency. The technique was utilized to analyze mineral and tap water samples. The accuracy of the approach was evaluated by analyzing a certified reference material for water (APS-1071 NIST) and conducting recovery tests, which yielded values ranging from 94% to 101%. Therefore, the authors concluded that the approach had acceptable detection limits and precision. Alves and colleagues [22] devised a technology for the on-line preconcentration and determination of Cd(II) in alcohol fuel samples using *Moringa oleifera* seeds as a bioadsorbent. The detection was done using FAAS. The flow and chemical parameters were fine-tuned to identify the optimal values. It was determined that using 10 mL of an enriched solution containing 50 µg g^{-1} of Cd(II) at a pH of 9.1, with 0.5 mol L^{-1} of HNO_3 as the eluent, 200 mg of adsorbent to fill the mini column, a sample flow rate of 6.0 mL min-1, and a buffering solution concentration of 0.05 mol L^{-1} yielded the best results. By optimizing the settings, we achieved analytical characteristics that included a detection limit of 5.50 µg L^{-1} and a precision below 2.3% (35.0 µg L^{-1}, n = 9). The linear range of the analysis was between 5 and 150 µg L^{-1}. The approach was effectively utilized on alcohol fuel samples and its precision was assessed by the recuperation test, which exhibited a variation of 97.5–100%. Carmo and colleagues [25] outline a comparable approach [22, 24, 26] for solid phase extraction utilizing *Moringa oleifera* husks as a natural adsorbent to directly determine Cu(II) in gasoline samples, with detection by FAAS. Gasoline is very insoluble in water, whereas ethanol is fully miscible in both gasoline and water at any concentration. The authors chose to dilute the gasoline samples with ethanol in a 50% (v/v) ratio. This facilitates the sorption of copper ions in the sample. The present study aimed to assess the factors that affected the adsorption and desorption of Cu(II) by *Moringa oleifera* husks in gasoline samples. A 15.0 mL solution containing 10 µg L^{-1} of Cu(II) was used for the evaluation. The optimal conditions were seen at a pH of 4.0, using an eluent of 0.5 mol L^{-1} HNO_3, a sample flow rate of 6.0 mL min^{-1}, and an adsorbent mass of 96.0 mg. Once the optimal parameters were determined, the method's analytical features were evaluated. The acquired results include a preconcentration factor of 14, a detection limit of 0.75 µg L^{-1}, a linear range from 0.75 to 500 µg L^{-1}, and a precision of 1.86% (50.0 µg L^{-1}, n = 9). Oliveira et al. [27] described a technique that is comparable to the one employed by Araújo et al. [20] for analyzing lead levels in chicken feed samples. Chicken feed contains metallic ions such as chromium (Cr), copper (Cu), iron (Fe), zinc (Zn), silver (Ag), calcium (Ca), lead (Pb), and molybdenum (Mo) [28]. After being consumed, these ions can persist in the chicken meat and be indirectly transmitted to individuals.

Various analytical approaches can be employed to ascertain metallic ions, while the intricacy of the matrix can provide some challenges. In this chapter, the authors utilized a method of extraction/ preconcentration to determine the presence of lead in chicken feed. They employed *Moringa oleifera* husks as a bioadsorbent and detected the lead using FAAS. The offline preconcentration system was optimized, and the optimal conditions were achieved at pH 9, with 30 mg of adsorbent mass. Elution was performed using 100 µL of HNO_3 0.5 mol L^{-1} as the eluent, with an elution and preconcentration flow rate of 2.0 mL min^{-1}. The analytical performance of the developed method was assessed by implementing the optimized system. Consequently, a detection limit of 1.66 μL^{-1}, a linear range spanning from 5.0 to 1000 µg L^{-1}, and a precision of 1.63% were determined. A preconcentration factor of 550 was determined, with a consumption rate of 0.02 mL. Bioadsorbents have been effectively utilized in the development of speciation analysis methodologies, in addition to their ability to detect metallic ions at low concentrations. Speciation analysis is challenging, mostly because of the intricate nature of the matrix and the low concentration of these elements. In order to reduce interference and improve the limit of detection (LOD) of the technique, certain matrices necessitate a significant dilution factor [29]. Alves and colleagues [21] devised a method for extracting inorganic arsenic from natural streams that is highly specific. The seeds of *Moringa oleifera* were employed as a biosorbent for solid phase extraction and subsequently analyzed using graphite furnace atomic absorption spectroscopy (GFAAS). The behavior of the As(III) and As(V) species was assessed throughout a wide range of pH values. The findings indicated that As(III) exhibited superior adsorption at pH 7, while As(V) had lower adsorption values and remained predominantly unbound in the solution. This characteristic enabled direct determination of As(V) by GFAAS. In order to achieve efficient adsorption of As(III), the method was optimized. It was then determined that a quantity of 1.0 g of adsorbent was adequate to maintain agitation for a duration of 1 hour. The limit of detection for As(V) was 6.3 µg L^{-1}, with a precision of less than 1.23%. The recovery outcomes (92%) were obtained by employing four distinct water samples, which were combined with the specified concentration of As(III) and As(V). The accuracy was demonstrated by utilizing the Drinking Water Standard Solution (APS 1071). Consequently, the authors were able to deduce that the determination method is a highly effective and appealing choice for determining inorganic arsenic in water samples. The study conducted by Alves and Coelho [24] examined an analytical methodology for selectively extracting inorganic chromium from water. The study used a flow injection analysis apparatus, similar to the one used by the previous group [23]. The *Moringa oleifera* husks were utilized as a biosorbent for the purpose of selectively extracting and preconcentrating chromium species. The detection of these species was achieved using the FAAS approach. The *Moringa oleifera* husks demonstrated efficacy in separating the Cr(III) and Cr(VI) species.

The trivalent species is absorbed within the pH range of 7 to 9, while the Cr(VI) is absorbed within the pH range of 1 to 2. The process was subsequently refined to determine the essential parameters for achieving an efficient extraction of Cr(III). The best conditions were identified as a pH of 7, an adsorbent mass of 58.0 mg, an eluent concentration of 0.01 mol L^{-1} of HCL, and a sample flow rate of 4.5 mL min^{-1}. The same parameters were utilized to determine the concentration of Cr(VI), with the exception that the pH range of the solution was limited to 1 and 2. The detection limits obtained for Cr(III) and Cr(VI) using this approach were 1.92 and 2.45 mg L^{-1}, respectively. The precision for Cr(III) and Cr(VI) was determined to be 1.63% and 0.08%, respectively [24]. The utilization of lignocellulosic adsorbent has numerous benefits, one of which is the cost-effectiveness due to the material's capacity to maintain stability during 100 preconcentration/elution cycles. In a related study, Sousa Neto [30] utilized Luffa cylindrica fibers to separate inorganic chromium. The material exhibits a positive charge on its surface in solutions with a pH below 6.0. Conversely, when the pH exceeds this value, the surface becomes negatively charged. A study was conducted to assess the impact of pH on the adsorption of chromium inorganic species. The results indicated that at a pH of 4.0, it is feasible to separate the chromium species (III and VI) (Fig. 1).

The analytical performance of the method was assessed using many parameters, including the detection limit (LOD), quantification limit (LOQ), and accuracy (D.P.R). The values obtained for these parameters were 19.2 μg L^{-1}, 63.99 μg L^{-1}, and 0.19%, respectively. The linear range of the method was determined to be 63.90–2000 μg L^{-1}. The accuracy of the procedure was assessed by analyzing water certificate reference material (APS-1071) and sediment (APS-1066). Recovery tests were conducted on water samples, revealing energy and isotonic values ranging from 87.60.

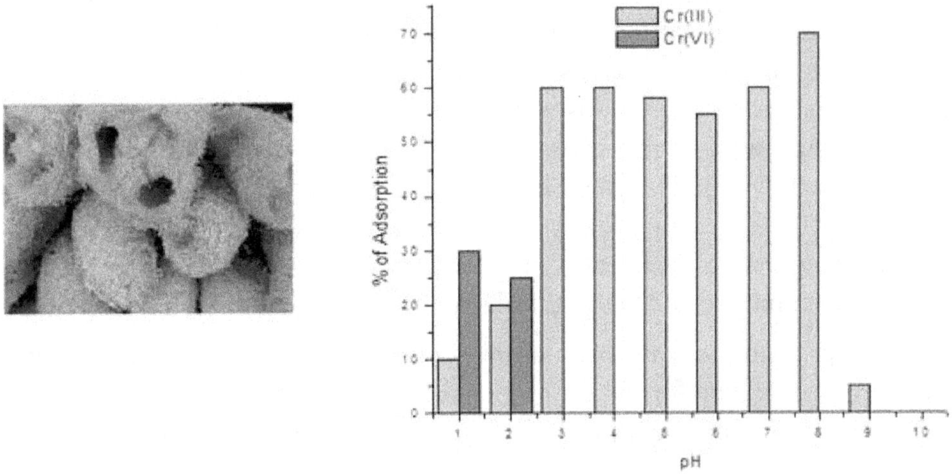

Fig. 1. Inorganic chromium adsorption on Luffa cylindrica fibers.

Ribeiro et al. [17] employed Mandarin peel (*Citrus reticulata*) as a natural adsorbent to establish an analytical technique using solid phase extraction for the direct quantification of nickel in alcoholic beverages, specifically whiskies and cachaça. This study utilized an online flow injection analysis method, similar to the one discussed in prior research, with detection by flame atomic absorption spectroscopy (FAAS) [17]. A volume of 10 mL of a solution containing 30.0 μg L^{-1} of Ni(II) was utilized. The optimal conditions were determined to be a mass of 50 mg for the adsorbent, a pH of 6, a sample flow rate of 4.3 mL min^{-1}, and an eluent consisting of HCI 1.0 mol L^{-1} with a flow rate of 2.0 mL min^{-1}. The approach exhibited the following analytical characteristics: a linear range of 10 to 75 mg L^{-1}, a detection limit of 3.1 μg L^{-1}, a preconcentration factor of 12, and a standard deviation of 0.9%. Liu and colleagues [16] evaluated the utilization of Sterculia lychnophora hance seeds, a natural adsorbent, in analyzing lead and cadmium in solid waste samples derived from the Chinese herb Pang Da Hai (PDH). The identification of these elements was achieved using flame atomic absorption spectroscopy (FAAS). The study examined the parameters that affected the development of the preconcentration method, and determined that the optimal conditions for both assessed ions were a concentration of 0.1 mol L^{-1} of NHO3 as the eluent and a pH of 6.5. The PDH material exhibited adsorption capabilities of 27.1 mg g^{-1} for Pb(II) and 17.5 mg g^{-1} for Cd(II). The standard deviation associated with the suggested method was 10% for both metals. A solution with a concentration of 100 μg L^{-1} and 20 mg of adsorbent were utilized. The method was verified through analysis using the recognized standard (GBW 08301), and the values obtained from the samples studied meet the values specified by the reference material. The relative error falls within the range of 2.5% to 11%. The detection limits for both Pb and Cd were 0.096 and 0.032 micrograms per liter (μg L^{-1}), respectively. The procedure was implemented on actual samples of tap and river water, and the recovery outcomes were satisfactory for all examined samples.

2.2. Mineral adsorbents and humic acid

Minerals are naturally occurring inorganic compounds that typically possess a well-defined chemical structure and are responsible for various physical properties. While these minerals occur naturally, they can also be artificially created in a laboratory to achieve a greater level of purity and consistency in their crystalline structure. These natural resources are highly valuable for industrial production and serve as raw materials in various sectors, including the manufacturing of domestic appliances, electrical wiring, jewelry, construction materials, and as an energy source. Minerals, whether natural or manmade, possess chemical-physical features that make them effective ionic exchangers. This enables them to efficiently adsorb metallic ions and organic molecules. The minerals often employed in adsorption processes include montmorillonites, vermiculites, zeolites, carbon nanotubes, and certain oxides such as graphene oxides, iron oxides, and titanium oxides [31, 32, 33]. These minerals are commonly used in adsorption procedures for environmental remediation, namely for treating and recovering water contaminated with heavy metals, contaminants, and both organic and inorganic chemicals. Only a limited number of studies have been conducted on the development of analytical techniques for the determination and pre-concentration of metallic ions. Nevertheless, these compounds have recently garnered significant interest due to their affordability and accessibility, as they are abundantly available in nature and can be easily synthesized.

Synthetic zeolites, for example, have been used as adsorbent material on copper ions determination and preconcentration through the formation of a chelate of copper in minicolunas of synthetic zeolite lynde type (Na-LTA) and faujasite Type X (Na-Fau). To evaluate the performance of these materials, a flow injection system was used for water samples from different sources. The preconcentration factors ranged from 35 to 125 for the LTA and 30–65 to Na-Fau. The accuracy was evaluated by recovery experiments and analysis of certified reference materials. The limits of detection, the standard deviation and relative linear calibration charts were 0.1 ng mL^{-1}, 2.6% and 2.5–30 ng mL^{-1} to Na-LTA and 0.4 ng mL^{-1}, 2.8% and 2.5–40 ng mL^{-1} for In-Fau respectively. The results indicate that the zeolites are promising materials for application in copper extraction preconcentration in trace levels with detection by flame atomic absorption spectrometry (FAAS) [34].

The adsorption capacity of natural zeolites was evaluated through the development of an analytical methodology for cobalt ions determination in natural waters through the formation of a neutral chelate with pyrrolidine acid ammonium diocarbamato (APDC). The neutral chelate formed was retained in a column packed with natural zeolites and then eluted with isomethyl-butyl ketone (MIBK) and detected by flame atomic absorption spectrometry (FAAS). The detection limit and the relative standard deviation (%RSD) was 1.9–2.3 μg L^{-1} and 2.3–4.5%, respectively. The sampling frequency of 27 h^{-1} and preconcentration factor of 130–160 have been achieved, to a 6 mL sample volume, indicating a high analyte retention on natural zeolites [33].

The use of these adsorbents increases more and more, because efficient synthesis procedures, purification and characterization ensures the availability of pure materials and well characterized. Soylak and Unsal [35], have developed a separation and preconcentration procedure based on solid phase extraction for Fe(III) and Cr(III) ions, using single-walled carbon nanotubes (SWCNTs). The ions were recovered quantitatively analyzed at pH 8.0 using 30 mg of carbon nanotube. The detection limits were 2.12 and 4.08 μg L^{-1}, for iron and chromium, respectively. The method was validated by analyzing of water and food samples certificates.

For chromium speciation Tuzen and Soylak [36] used carbon nanotubes (MWNTs) multilayer in a solid phase extraction procedure and the formation of a chelate by using of ammonium pyrrolidine dithiocarbamate (APDC). The procedure had some validated analytical parameters including pH, solvent type and effects of array. The detection limit was 0.90 g L^{-1} for Cr(VI). The adsorption capacity of the multilayer carbon nanotube was 9.50 mg g^{-1} Cr(VI). The presented method was applied to the speciation of chromium in natural water showing a 95% recovery factor indicating that the proposed methodology using multilayer carbon nanotubes is more efficient in terms of selectivity, detection limit and preconcentration factor when compared to other methods of preconcentration of chrome described in the literature.

Iron oxides are found in nature in the hematite form (αFe_2O_3), goethite (αFeOOH) and magnetite (Fe_3O_4). In recent times, the processes involving adsorption using these oxides as adsorbent materials have attracted great interest, mainly due to the high adsorptive capacity and magnetic property of these materials. These solids can be used in preconcentration methods, magnetic separation and molecular identification of biomolecules. Separation techniques using magnetic solids also offer some advantages over centrifugation, filtration and solid phase extraction once these steps are simplified [37]. Parham and Rahbar [38], used magnetic iron oxide nanoparticles (MIONs) as adsorbent to develop a methodology for extracting and determination of fluoride ion in water samples. The method is based on the discoloration of the Fe(III)-SCN formed with fluoride ions extracted and sequentially, monitoring by spectrophotometry at λmax = 458 nm. The detection and quantification limit were 0.042 and 0.015 µg mL^{-1}, respectively. In this method the preconcentration factor achieved was 50 and the method was applied to the determination of fluoride in water samples and concentration values obtained were compared with the default values that are presenting SPADNS according to the method of the comparison. The no use of toxic compounds, the short procedure (about 8 min per sample) and a low LD are the main advantages of the proposed method.

Metal oxides such as iron oxide cited previously, zirconia oxide and titanium oxide obtained in the nanoscale have become important for various lines of research due to its special properties such as high chemical activity and mainly the good capacity to adsorb a wide variety of metallic ions. In addition, the particle size is another factor that contributes to the use of those materials in adsorption processes, because fine particles and nanoscale enable high adsorption speeds when compared the classic adsorbents [39].

Most of the nano adsorbents are used with their surfaces modified by complexing agents or microorganisms to increase the capacity to adsorb analytes. However, this modification limits the repeated use of adsorbent in adsorption processes. There are synthesized materials in the form of mixed oxides that enable their use without requiring any modification by another reagent. The nano compound of boron trioxide and titanium dioxide ($B2O_3/TiO_2$) is an example of these mixed oxides. Kalfa and contributors [40] developed a methodology for cadmium separation and preconcentration in tap water and tea samples using this nano compound. Under the best experimental conditions, preconcentration factor and the detection limit were determined being 50 and 1.44 µg L^{-1}, respectively. The precision of the method was confirmed by analysis of certified reference materials (tea leaves GBW-07605). The results showed good agreement with the certified values (< 10%) relative error. Cadmium recovery in optimum conditions was 96 ± 3% at a 95% confidence level. The use of solid phase extraction enables the development of a simple method, selective, precise, fast and economical for preconcentration and determination of cadmium.

Some hybrid materials based on silica (SiO_2) have been synthesized by mixed metal oxides form as the titanium dioxide (TiO_2) and used to preconcentrate a large quantity of metallic ions the trace level. Lima and contributors [41] describe the analytical performance of a hybrid composite material of SiO_2, Al_2O_3 and TiO_2. The material was prepared by a sol–gel process to act as adsorbent in copper preconcentration system. The retention capacity was 1.4 mg g^{-1} in dynamic conditions. The detection and quantification limits were of 0.50 and 1.4 µg L^{-1}, respectively, and the calibration curve is linear in the range of 5.0–245.0 μg L^{-1} (r = -0999). The relative standard deviation is 3.20 (for n = 7 to a concentration of Cu(II) of 10 μg L^{-1}). The method was applied to the determination of copper ions in water, vegetables and alcohol fuel using flame atomic absorption spectrometry (FAAS).

Among the wide variety of materials that can be used as adsorbents, humic acid, although little reported in the literature has a good application in the area of analytical chemistry. Humic substances are natural materials and called major components from the decomposition of organic matter. This decomposition is subdivided into three parts: humic acid, fulvic acid and humina [42]. These substances have a high capacity of cationic exchange for different species due to the presence of carbonyl groups, phenolic hydroxyl, and alcoholics in their structure as well, metals to interact with the humic substances, are retained by active sites of adsorption.

Pereira & Arruda [43] employed vermicompost, humic substances generated by the metabolic processes of organic residues in the soil by earthworms, and commercially available humic acid to preconcentrate Cd(II) and Pb(II) utilizing flow systems coupled with flame atomic absorption spectrometry (FAAS). The Cd(II) ion was concentrated for a duration of 4 minutes at a flow rate of 4.0 mL min^{-1} and subsequently extracted using 220 mL of nitric acid at a concentration of 3.0 mol L^{-1}. The preconcentration factor achieved under these circumstances was 46 for vermicompost and 27 for humic acid. With the exception of using nitric acid 1.0 mol L^{-1} for humic acid, all the conditions for preconcentration of Pb(II) were the same as those for Cd(II). The obtained preconcentration factors were 62 and 83 for vermicompost and humic acid, respectively. This technique was utilized in samples of both mineral and tap water. Bianchin et al. [44] assessed the quantification of Cd(II) using vermicompost in a solid phase extraction technique coupled with flow injection analysis and detected by flame atomic absorption spectrometry. The ideal conditions for extracting the sample were determined to be a pH range of 7.3–8.3, using tris (hydroxymethyl) aminomethane buffer at a concentration of 50 mmol L^{-1}. The sample was extracted at a flow rate of 4 mL min^{-1}, using 160 mg of adsorbent material. Under the optimum conditions, the preconcentration factor was determined to be 32, the detection limit was calculated to be 1.7 μg L^{-1}, and the analytical frequency was found to be 20 samples per hour. The linear portion of the analytic curve extended to a concentration of at least 50 gL^{-1}. The correlation coefficient was determined to be 0.998, indicating a strong relationship between the variables. The relative standard deviation was found to be 2.4% at a concentration of 35 μgL^{-1}, based on a sample size of 7. The approach was utilized on ethanol fuel samples to evaluate its accuracy by assessing the recovery rate. The results showed a 94% recovery rate, with some variation from the expected 100%. This indicates that worm compost is an effective and reliable material for preconcentrating cadmium in complex samples.

2.3. Fungi and bacteria

Biosorption is the utilization of microbial biomass to accumulate organic and inorganic compounds from diluted solutions. Originally, biosorption research focused on utilizing biological materials to eliminate metals [45]. Currently, algae, fungi, bacteria, and yeast are recognized as crucial components of the natural absorbers employed in the advancement of analytical techniques. Bioadsorbents can be classified into two categories based on their metabolic activity: one involves a quick interaction between the metallic species and the material's surface, while the other involves an initial rapid superficial interaction between the metallic species and the living bioadsorbent. Afterwards, a more gradual and intricate process is developed, triggered by the bonding of the metallic species with the cellular membrane, which is then followed by the passage through it, ultimately leading to intracellular processes such as methylation, reduction, and oxidation [46]. Industrial activities involved in the production of antibiotics, enzymes, and fermentation generate a significant quantity of biological material as by-products. Typically, these by-products are wasted, but they can be utilized in the adsorption process [47]. The microbial biomass contains various chemical groups, including polysaccharides, proteins, and lipids, which are linked together through functional groups such as hydroxyl, sulfhydryl, carboxyl, amine, and phosphate. These groups enable the biomass to attract and retain metals, and the biomass possesses mechanisms for this purpose, such as ion exchange, complexation, coordination, adsorption, electrostatic interaction, chelation, and micro precipitation [48]. Various microorganisms have been examined to determine the presence of metal ions. Fungi are microorganisms that are highly efficient bioadsorbents due to their ability to rapidly multiply and generate large amounts of biomass. Additionally, fungi are capable of thriving in harsh settings, making them highly adaptable microorganisms [48].

Mendil et al. [49] developed a preconcentration system using Penicillium italicum to determine Cd(II), Mn(II), Pb(II), Fe(II), Ni(II) and Co(II) in trace levels. The adsorption occurs in pH range 8.5–9.5. After optimizing the method, the detection limit was 0.41 and 1.60 μg L−1, for cadmium and iron, respectively. The method was successfully applied for the determinations of analytes in natural water, cultivated mushroom, lichen (Bryum capillare Hedw), moss (Homalothecium sericeum) and refined table salt samples. The proposed method is safe, fast and handy.

The fungus Penicillium was used by Baytak and collaborators [50] to determine copper, zinc and lead in the dam water, waste water, spring water, parsley and carrots. The fungi were immobilized in pumice stone, and the ideal conditions for adsorption, such as solid phase mass, pH, solution flow rate and elution solution was studied. The analyte was determined by flame atomic absorption spectrometry (FAAS). The preconcentration factor obtained was 50 and the detection limits of detection were 1.8; 1.3 and 5.8 for copper, zinc and lead, respectively. Already the accuracy of the method was evaluated in samples of tea standard reference (GBW-07605).

To determine arsenic in drinking water samples, Shahlaei and collaborators [51] used dead fungal biomass (Aspergillus niger) in a solid phase extraction (SPE) system. We evaluated the effects of parameters such as time of biosorption, sample volume, pH, type and concentration of the eluent and effect of interfering ions. The recovery was evaluated by precision between 99 and 102%. The detection limit found for the method was of 1 mg mL−1.

Alpodogan and collaborators [52] also used dead fungal biomass from Aspergillus niger loaded into TiO_2 nanoparticles for copper, nickel, cadmium, manganese, and chromium preconcentration in milk samples, with determination by flame atomic absorption spectrometry (FAAS). The preconcentration factor was 600, detection limits for copper, nickel, cadmium, manganese, chromium and lead were 0.21; 0.34; 018; 0.08; 0.53; 0.61 g L−1, respectively.

Among the fungi, yeasts are the most exploited scientifically, due to the fact that eukaryotic organisms are easier to manipulate, serving as an excellent model for studies. Ohta and collaborators [53] evaluated a preconcentration method for cooper determination in trace level in river water, using yeast Saccharomyces cerevisiae.

Bacteria is another microorganism commonly employed for the adsorption of metallic ions. Bacteria possess adsorptive capacity as a result of the nitrogen (N) and oxygen (O) ligands found in their cell walls. This allows for the mechanism of micro precipitation between metals and bacteria, as well as electrostatic attraction and ion exchange with groups on the cell wall [54]. Pagnanelli et al. [55] investigate the adsorption capacity of Sphaerotilus natans biomass for lead, copper, zinc, and cadmium in aqueous solutions. The biosorption process was evaluated at various pH levels. The experimental findings demonstrated a beneficial correlation between higher pH values and the absorption of the analyzed metals. This indicates that the biomass possesses a noteworthy ability to adsorb the researched ions.

2.4. Functionalized biosorbent

The application of bio-organic species with high sorption capacity for metallic ions to modify supporting materials has garnered significant interest in the field of analytical chemistry. This approach aims to enhance the preconcentration of metallic ions that are frequently present in food, vegetables, drinking water, and river samples at trace and ultra-trace quantities. This functionalization can be achieved through structural alterations of biomass surfaces or by immobilizing bio-organic species onto the surfaces of biosynthetic, inorganic, polymeric materials, and biomass itself. Moreover, the bio-organic compounds can undergo treatment with acid and base solutions or organic and inorganic solvents, which often enhance the adsorption, selectivity, and recovery capability of the metal ions. This discussion will focus on several bio-organic systems that have been modified and attached to solid supports. These systems are utilized for the adsorption of metal ions [56].

Yang and contributors [57] described in their article that various efforts have been made to generate selectivity and adsorbent capacity through chemical modification and genetic engineering, due to the low selectivity of biological cells towards certain metal species and a low capacity for anionic metal species in neutral pH. In this section, we will describe some methods that have been used with great success in metal ion analyzers after functionalization or immobilization of the biological cells and other biosorbents on a support material.

The main biological organisms used as functionalizing agent on supporting material in analytical chemistry for trace metal determination are shown in Table 2. They offer several advantages, such as ease in immobilization on the surface of supporting material, their biodegradability, natural abundance, low cost, and simple production.

Table 2. Types of biological organisms used as functionalizing agent on surface of supporting material.

Biological organisms	Types
Bacteria	*Bacillus* sp., *Streptococcus* sp., *Streptomyces* sp., *Escherichia* sp., *Enterobacter* sp., *Klebsiella* sp.
Fungi	Filamentous fungi (*Aspergillus niger*, Alternaria sp., *Rhizopus oryzae*), unicellular yeasts (*Saccharomyces* sp., *Yamadazyma* sp.), *Pleurotus* sp., *Helvella* sp., *Coriolus* sp., *Agaricus* sp.,
Plant-derivatives	*Musa paradisiaca*, *Citrus reticulata*, *Glycine max*, *Zea mays*, *Sorghum bicolor*, *Agave sisalana perrine*, *Bambusa vulgaris*
Other biological organisms	Algae (*Chlorella* sp.), resting eggs, sea sponge, eggs shell membrane
Source: Adapted from Escudero and collaborators [58].	

Bathocuproine (BCP) or 2,9-dimethyl-4,7-diphenyl-1,10-phenanthroline is well-known reagent that respond selectively to Cu by forming a colored 2:1 complex with copper. It was successfully used to determine the Pb(II), Fe(III), Co(II), Cr(III) and Zn(II) in water and food samples after preconcentration [59]. Teodoro and collaborators [60] in their study, functionalized bamboo fiber with bathocuproine for determination of copper in foods using FAAS. Copper preconcentration was performed in an off-line system consisting of a glass column containing bamboo fiber functionalized with bathocuproine and peristaltic pump. This method made it possible to separate, preconcentrate and determine the copper ion in wine, cassava flour, corn flour, oregano and coriander with concentrations much lower than mg L−1, that is, in the μg L−1 for liquid samples and μg g−1 for solid samples.

In the study developed by Baytak and collaborators [61] the fungus Rhizopus oryzae functionalized on the almond bark (natural cellulose) was used to preconcentrate and determine the Cu(II), Fe(III), Mn(II) and Zn(II) in water, fish and vegetable tissues samples. The glass column (150 mm × 8.0 mm i.d) containing the Rhizopus oryzae-loaded almond shells was used during the preconcentration of the analytes. Analyte detection was based on high resolution continuous source atomic absorption spectrometry. According to the author, the proposed method presented several advantages, such as precision, accuracy, preconcentration factor, reuse, environmental friendly material and did not require chelating or complexing agents during ion analyzes.

Boas and collaborators [62] used macadamia nuts waste to investigate the adsorption of copper ions. They proved that by chemical modifications of the residues of macadamia nuts was treated by NaOH solution and citric acid, and other residues treated by NaOH solution showed a better adsorption of Cu2+ ions compared to "in natura" residues. The adsorption process consisted of the addition of 0.5 g of the "in natura" and modified residues in a solution of Cu2+ ions at 100 mg L−1 at optimized pHs and after 24 hours the supernatants were analyzed with flame atomic absorption spectrophotometer (FAAS).

Studies have shown that sisal fiber, obtained from the plant Agave sisalana perrine of Mexican origin, can be used for preconcentration of metals. In a recent study by Dias and collaborators [63] first used the sisal fiber functionalized with 3-aminomethylalizarin-N,N-diacetic acid or Alizarine Fluorine Blue (complexing reagent) to preconcentrate and determine the copper ion in tobacco leaves. According to these results, the concentration of copper in tobacco leaves was very low. Chen and colleagues [64] have found that the eggshell surface contains carboxylic groups that decrease the adsorption of arsenic ions on the shell's surface. The researchers conducted a chemical modification of the eggshell using methyl esterification in order to enhance the effectiveness of bioadsorption for arsenic ions. They also confirmed a strong preference for these ions. By utilizing the keywords "sorghum" and "bioadsorption" in various journal search engines, numerous studies may be found that employ sorghum as an adsorbent material for the preconcentration, determination, and removal of naturally occurring heavy metals. In a recent study, researchers utilized sorghum crop residue in its natural form and chemically changed it with phosphoric acid. They employed flame atomic absorption spectrometry (FAAS) as a detection method to analyze samples of black tea and river water for the presence of cadmium, copper, manganese, and lead ions [65]. The authors of this study confirmed that chemically modified sorghum agricultural waste has an adsorption capacity that is 4.5 times higher than that of the natural trash. Following this observation, the researchers utilized the altered residue as a solid phase to concentrate and analyze the Cd, Cu, Mn, and Pb ions found in samples of tea and river water. Based on their findings, the chemically altered sorghum showed significant promise in effectively separating and concentrating metallic ions in actual samples. Passive sampler disks were created using a concentration gradient-based diffusion method in thin films. These disks were generated by functionalizing Saccharomyces cerevisiae yeast in agarose gel, which is a polysaccharide isolated from red seaweed. The purpose of these disks was to preconcentrate and measure the concentration of Cd(II) using ICP-OES [66]. The disks were fabricated using varying ratios of gel and yeast in order to achieve an ideal composition for detecting the Cd(II) ion in river and sea water. The yeast *Saccharomyces cerevisiae* was selected for the investigation based on prior research that shown its exceptional capacity to separate and preconcentrate metal ions.

3. CONCISE SUMMARY

Given the ubiquitous occurrence of metallic ions in individuals' everyday experiences, it is imperative to devise techniques that can accurately detect them, even at minimal levels of concentration. There has been a growing interest in recent years to develop ways that align with the principles of chemical "green". These approaches entail the utilization of materials that contribute to a cleaner, healthier, and more sustainable environment. In recent years, there has been a focus on the advancement of techniques utilizing alternative adsorbents due to their exceptional ability to adsorb metallic ions and organic molecules. The study and advancement of novel materials that are inexpensive and can be reused has emerged as a fascinating field in analytical chemistry.

REFERENCES

1. Soares ARS. Desenvolvimento de Métodos para Determinação de Chumbo e Níquel em Produtos Cosméticos e Cabelo por GF AAS [dissertation]. Belo Horizonte: Universidade Federal de Minas Gerais; 2012
2. Ferreira SLC, Lemos VA, Moreira BC, Costa ACS, Santelli RE. An online continuous flow system for copper enrichment and determination by flame atomic absorption spectrometry. Analytica Chimica Acta. 2000;403:259-264. DOI: 10.1016/S0003-2670(99)00632-7
3. Jardim ICSF. Extração em Fase Sólida: Fundamentos Teóricos e Novas Estratégias para Preparação de Fases Sólidas. Scientia Chromatographica. 2010;2:13-25
4. Pacheco PH, Gil RA, Cerutti SE, Smichowski P, Martinez L. Biosorption: A new rise for elemental solid phase extraction methods. Talanta. 2011;85:2290-2300. DOI: 10.1016/j.talanta.2011.08.043
5. Reis BF, Miranda CES, Baccan N. Pré-concenttação empregando extração fase líquida-fase sólida em sistemas de análise em fluxo com detecção espectrométrica. Química Nova. 1996;19:623-635
6. Science direct [Internet]. 2017. Available from: http://www.sciencedirect.com. [Accessed: Dec 1, 2017]
7. Tarley CRT, Silviera G, Santos WNL, Matos GD, Silva EGP, Bezerra MA, Miró M, Ferreira SLC. Chemometric tools in electroanalytical chemistry: Methods for optimization based on factorial design and response surface methodology. Microchemical Journal. 2009;92:58-67. DOI: 10.1016/j.microc.2009.02.002
8. Demirbas A, Demirbas MF. Importance of algae oil as a source of biodiesel. Energy Convers. Manage. 2011;52:163-170. DOI: 10.1016/j.enconman.2010.06.055
9. Cai J, He Y, Yu X, Banks SW, Yang Y, Zhang X, Yu Y, Liu R, Bridgwater AV. Review of physicochemical properties and analytical characterization of lignocellulosic biomass. Renewable and Sustainable Energy Reviews. 2017;76:309-322. DOI: 10.1016/j.rser.2017.03.072
10. Correia DS, Santos FHR, Correia MEF. Enzimas oxidativas microbianas envolvidas na biodegradação da lignocelulose: produção, características bioquímicas e importância biotecnológica. Embrapa Agrobiologia. 2011;1:284
11. Araújo CST, Carvalho DC, Rezende HC, Almeida ILS, Coelho LM, Coelho, NMM, Marques TL, Alves, VN. Chapter 10: Bioremediation of waters contaminated with heavy metals using *Moringa oleifera* seeds as biosorbent. In: Applied Bioremediation - Active and Passive Approache. Intech; 2013. pp. 227-255

12. Davis TA, Mucci A, Volesky B. A review of the biochemistry of heavy metal biosorption by brown algae. Water Research. 2003;37:4311-4330. DOI: 10.1016/S0043-1354(03)00293-8

13. Alves VN, Borges SSO, Neto WB, Coelho NMM. Determination of low levels of lead in beer using solid-phase extraction and detection by flame atomic absorption spectrometry. Journal of Automated Methods and Management in Chemistry. 2011;1:1-6. DOI: 10.1155/2011/464102

14. Tarley CRT, Ferreira SLC, Arruda MAZ. Use of modified rice husks as a natural solid adsorbent of trace metals: Characterization and development of an on-line preconcentration system for cadmium and lead determination by FAAS. Microchemical Journal. 2004;77:163-175. DOI: 10.1016/j.microc.2004.02.019

15. Dávila-Guzmán NE, Cerino FJ, Rangel-Mendez JR, Flores PED. Copper biosorption by spent coffee ground: Equilibrium, kinetics, and mechanism. Clean Soil, Air, Water. 2013;41:557-564

16. Liu Y, Chang X, Guo Y, Meng S. Biosorption and preconcentration of lead and cádmium on waste Chinese herb pang Da Hai. Journal of Hazardous Materials. 2006;B135:389-394. DOI: 10.1016/j.jhazmat.2005.11.078

17. Ribeiro GC, Coelho LM, Coelho NMM. Determination of nickel in alcoholic beverages by FAAS after online preconcentration using mandarin peel (Citrus reticulada) as biosorbent. Journal of the Brazilian Chemical Society. 2013;24(6):1072-1078. DOI: 10.5935/0103-5053.20130138

18. Bhatti HN, Mumtaz B, Hanif MA, Nadeem R. Removal of Zn(II) ions from aqueous solution using *Moringa oleifera* Lam. (horseradish tree) biomass. Process Biochemistry. 2007;42:547-553. DOI: 10.1016/j.procbio.2006.10.009

19. Reddy DHK, Seshaiah K, Reddy AVR, Rao MM, Wang MC. Biosorption of Pb2+ from aqueous solutions by *Moringa oleifera* bark: Equilibrium and kinetic studies. Journal of Hazardous Materials. 2010;174:831-838. DOI: 10.1016/j.jhazmat.2009.09.128

20. Araújo CST, Alves VN, Rezende HC, Coelho NMM. Development of a flow system for the determination of low concentrations of silver using Moringa oleífera seeds as biosorbent and flame atomic absorption spectrometry. Microchemical Journal. 2010;96:82-85. DOI: 10.1016/j.microc.2010.02.002

21. Alves VN, Neri TS, Borges SSO, Carvalho DC, Coelho NMM. Determination of inorganic arsenic in natural waters after selective extraction using Moringa oleífera seeds. Ecological Engineering. 2017;106:431-435. DOI: 10.1016/j.ecoleng.2017.06.007

22. Alves VN, Mosqueta R, Coelho NMM, Bianchin JN, Roux KCDP, Martendal E, Carasek E. Determination of cádmium in alcohol fuel using Moringa oleífera seeds as a biosorbent in an on-line system coupled to FAAS. Talanta. 2010;80:1133-1138. DOI: 10.1016/j.talanta.2009.08.040

23. Alves VN, Mosquetta R, Carasek E, Coelho NMM. Determination of Zn(II) in alcohol fuel by flame atomic absorption spectrometry after on-line preconcentration using a solid phase extraction system. Journal of Analytical Chemistry. 2012;67(5):448-454. DOI: 10.1134/S1061934812050139

24. Alves VN, Coelho NMM. Selective extraction and preconcentration of chromium using Moringa oleífera husks as biosorbent and flame atomic absorption spectrometry. Microchemical Journal. 2013;109:16-22. DOI: 10.1016/j.microc.2012.05.030

25. Carmo SN, Damásio FQ, Alves VN, Marques TL, Coelho NMM. Direct determination of copper in gasoline by flame atomic absorption spectrometry after sorption and preconcentration os Moringa oleífera husks. Microchemical Journal. 2013;110:320-325. DOI: 10.1016/j.microc.2013.04.010

26. Alves VN, Borges SSO, Coelho NMM. Direct zinc determination in Brazilian sugar cane spirit by solid phase extraction using *Moringa oleifera* husks in a flow system with detection by FAAS. International Journal of Analytical Chemistry. 2011;1:1-8

27. Oliveira JAN, Siqueira LMC, Sousa Neto JA, Coelho NMM, Alves VN. Preconcentration system for determination of lead in chicken feed using Moringa oleífera husks as a biosorbent. Microchemical Journal. 2017;133:327-332. DOI: 10.1016/j.microc.2017.04.001

28. Gattás G, Barbosa FF. Cobre na nutrição de aves e suínos. Revista Eletrônica Nutrine. 2004;1:133-177

29. Coelho LM, Arruda MAZ. Preconcentration procedure using cloud point extraction in the presence of electrolyte for cadmium determination by flame atomic absorption spectrometry. Spectrochimica Acta, Part B: Atomic Spectroscopy. 2005;60:743-748. DOI: 10.1016/j.sab.2005.02.016

30. Sousa Neto JA. Avaliação e desenvolvimento de um método para extração seletiva de cromo utilizando a luffa cylindrica como bioadsorvente [dissertation]. Catalão: Universidade Federal de Goiás; 2016

31. Anjos VE, Rohwedder JR, Cadore S, Abate G, Grassi MT. Montmorillonite and vermiculite as solid phases for the preconcentration of trace elements in natural waters: Adsorption and desorption studies of As, Ba, Cu, Cd, Co, Cr, Mn, Ni, Pb, Sr, V, and Zn. Applied Clay Science. 2014;289:296-299. DOI: 10.1016/j.clay.2014.07.013

32. Ortega EP, Ramos RL, Barron JM. Role of electrostatic interactions in the adsorption of cadmium(II) from aqueous solution onto vermiculite. Applied Clay Science. 2014;10(17-88):89. DOI: 10.1016/j.clay.2013.12.012

33. Peña PY, Rondón W. Mini-columns packed with natural zeolites of Cuba for determination of cobalt using the ligand APDC in a flow injection system coupled to a FAAS. Rev. Téc. Ing. Univ. Zulia. 2014;215:224

34. Peña PY, López W, Burguera JL, Burguera M, Gallignani M, Brunetto R, Carrero P, Rondon C, Imbert F. Synthetic zeolites as sorbent material for on-line preconcentration of copper traces and its determination using flame atomic absorption spectrometry. Analytica Chimica Acta. 2000;249:258-403. DOI: 10.1016/S0003-2670(99)00566-8

35. Soylak M, Unsal YE. Chromium and iron determinations in food and herbal plant samples by atomic absorption spectrometry after solid phase extraction on single-walled carbon nanotubes (SWCNTs) disk. Food and Chemical Toxicology. 2010;1511:1515-1548. DOI: 10.1016/j.fct.2010.03.017

36. Tuzen M, Soylak M. Multiwalled carbon nanotubes for speciation of chromium in environmental samples. Journal of Hazardous Materials. 2007;219:225-147. DOI: 10.1016/j.jhazmat.2006.12.069

37. Arteaga KA, Rodriguez JA, Barrado E. Magnetic solids in analytical chemistry: A review. Analytica Chimica Acta. 2010;157:165-674. DOI: 10.1016/j.aca.2010.06.043

38. Parham H, Rahbar N. Solid phase extraction–spectrophotometric determination of fluoride in water samples using magnetic iron oxide nanoparticles. Talanta. 2009;664:669-680. DOI: 10.1016/j.talanta.2009.07.045

39. Türker AR. New sorbents for solid-phase extraction for metal enrichment. Clean. 2007;548:557-535. DOI: 10.1002/clen.200700130

40. Kalfa OM, Yalçinkaya Ö, Türker AR. Synthesis of nano B2O3/TiO2 composite material as a new solid phase extractor and its application to preconcentration and separation of cadmium. Journal of Hazardous Materials. 2009;455:461-166. DOI: 10.1016/j.jhazmat.2008.11.112

41. Lima FG, Ohara MO, Clausen DN, Nascimento RD, Ribeiro ES, Segatelli MG, Bezerra MA, Tarley CRT. Flow injection on-line minicolumn preconcentration and determination of trace copper ions using an alumina/ titanium oxide grafted silica matrix and FAAS. Microchimica Acta. 2012;61:70-178. DOI: 10.1007/s00604-012-0807-4

42. Landgraf MD, Alves MR, Silva SC, Rezende MOO. Caracterização de ácidos húmicos de vermicomposto de esterco bovino compostado durante 3 e 6 meses. Química Nova. 1999;22:484-486. DOI: 10.1590/S0100-40421999000400003

43. Pereira MG, Arruda MAZ. Preconcentration of Cd(II) and Pb(II) using humic substances and flow systems coupled to flame atomic absorption spectrometry. Microchimica Acta. 2004;215:222-146. DOI: 10.1007/s00604-004-0231-5

44. Bianchin JN, Martendal E, Mior R, Alves VN, Araújo CST, Coelho NMM, Carasek E. Development of a flow system for the determination of cadmium in fuel alcohol using vermicompost as biosorbent and flame atomic absorption spectrometry. Talanta. 2009;333:336-378. DOI: 10.1016/j.talanta.2008.11.012

45. Aksu Z. Equilibrium and kinetic modelling of biosorption of Remazol Black B by Rhizopus arrhizus in a batch system: Effect of temperature. Process Biochemistry. 2000;36:431-439

46. Madrid Y, Camara C. Biological substrates for metal preconcentration and speciation. Trends in Analytical Chemistry. 1997;16:36-44

47. Crini G. Non-conventional low-cost adsorbents for dye removal: A review. Bioresource Technology. 2006;97:1061-1085. DOI: 10.1016/j.biortech.2005.05.001

48. Veglio F, Beolchini F. Removal of metals by biosorption : A review. Hydrometallurgy. 1997;44:301-316. DOI: 10.1016/S0304-386X(96)00059-X

49. Mendil D, Tuzen M, Soylak MA. Biosorption system for metal ions on Penicillium italicum – Loaded on Sepabeads SP 70 prior to flame atomic absorption spectrometric determinations. Journal of Hazardous Materials. 2008;152:1171-1178. DOI: 10.1016/j.jhazmat.2007.07.097

50. Baytak S, Kenduzler E, Turker AR, Gok N. Penicillium digitatum immobilized on pumice stone as a new solid phase extractor for preconcentration and/or separation of trace metals in environmental samples. Journal of Hazardous Materials. 2008;153:975-983. DOI: 10.1016/j.jhazmat.2007.09.049

51. Shahlaei M, Pourhossein A. Determination of arsenic in drinking water samples by electrothermal atomic absorption spectrometry after preconcentration using the biomass of Aspergillus niger loaded on activated charcoal. Journal of Chemistry. 2014;1:1-6

52. Alpdogan G, Zor SD, Peksel A. Analysis of some trace metals in milk samples by FAAS after Preconcentration by Aspergillus niger loaded on TiO2 nanoparticles. Current Analytical Chemistry. 2016;12:642-650

53. Ohta K, Tanahasi H, Suzuki T, Kanec S. Preconcentration of trace copper with yeast for river water analysis. Talanta. 2001;53:715-720. DOI: 10.1016/S0039-9140(00)00502-6

54. Buratto AP, Costa RD, Ferreira ES. Aplicação de biomassa fúngica de Pleurotus ostreatus em processo de biossorção de íons cobre(II). Eng. Sanit. Ambiente. 2012;17:413-420

55. Pagnanelli F, Esposito A, Toro L, Vegliò F. Metal speciation and pH effect on Pb, Cu, Zn and Cd biosorption onto Sphaerotilus natans: Langmuir-type empirical model. Water Research. 2003;37:627-633. DOI: 10.1016/S0043-1354(02)00358-5

56. Bag H, Turker AR, Lale M. Determination of Cu, Zn, Fe, Ni and Cd by flame atomic absorption spectrophotometry after preconcentration by Escherichia coli immobilized on sepiolite. Talanta. 2000;51:1035-1043. DOI: 10.1016/S0039-9140(00)00289-7

57. Yang T, Chen M-L, Wang J-H. Genetic and chemical modification of cells for selective separation and analysis of heavy metals of biological or environmental significance. TrAC Trends in Analytical Chemistry. 2015;66:90-102. DOI: 10.1016/j.trac.2014.11.016

58. Escudero LB, Avila Maniero M, Agostini E, Smichowski PN. Biological substrates: Green alternatives in trace elemental preconcentration and speciation analysis. Trac-Trends in Analytical Chemistry. 2016;80:531-546. DOI: 10.1016/j.trac.2016.04.002

59. Mendil D, Karatas M, Tuzen M. Separation and preconcentration of Cu(II), Pb(II), Zn(II), Fe(III) and Cr(III) ions with coprecipitation method without carrier element and their determination in food and water samples. Food Chemistry. 2015;177:320-324. DOI: 10.1016/j.foodchem.2015.01.008

60. Teodoro MTF, de S, Dias F, da Silva DG, Bezerra MA, Dantas AF, Teixeira LSG, et al. Determination of copper total and speciation in food samples by flame atomic absorption spectrometry in association with solid-phase extraction with bamboo (Bambusa vulgaris) fiber loaded with bathocuproine. Microchemical Journal. 2017;132:351-357. DOI: 10.1016/j.microc.2017.01.033

61. Baytak S, Mert R, Turker AR. Determination of Cu(II), Fe(III), Mn(II) and Zn(II) in various samples after preconcentration with Rhizopus oryzae loaded natural cellulose (almond bark). International Journal of Environmental Analytical Chemistry. 2014;94:975-987. DOI: 10.1080/03067319.2014.900679

62. Boas NV, Casarin J, Caetano J, Gonçalves Junior AC, Tarley CRT, Dragunski DC. Biossorção de cobre utilizando-se o mesocarpo e o endocarpo da macadâmia natural e quimicamente tratados. Revista Brasileira de Engenharia Agrícola e Ambiental. 2012;16:1359-1366. DOI: 10.1590/S1415-43662012001200014

63. Dias Fde S, Bonsucesso JS, Oliveira LC, dos Santos WN. Preconcentration and determination of copper in tobacco leaves samples by using a minicolumn of sisal fiber (Agave sisalana) loaded with Alizarin fluorine blue by FAAS. Talanta. 2012;89:276-279. DOI: 10.1016/j.talanta.2011.12.027

64. Chen ML, Gu CB, Yang T, Sun Y, Wang JH. A green sorbent of esterified egg-shell membrane for highly selective uptake of arsenate and speciation of inorganic arsenic. Talanta. 2013;116:688-694. DOI: 10.1016/j.talanta.2013.07.061

65. Uçar G, Bakircioglu D, Kurtulus YB. Determination of metal ions in water and tea samples by flame-AAS after preconcentration using sorghum in nature form and chemically activated. Journal of Analytical Chemistry. 2014;69:420-425. DOI: 10.1134/S1061934814050098

66. Menegario AA, Tonello PS, Durrant SF. Use of Saccharomyces cerevisiae immobilized in agarose gel as a binding agent for diffusive gradients in thin films. Analytica Chimica Acta. 2010;683:107-112. DOI: 10.1016/j.aca

17 Chapter

URBAN DUST CONTAINS HIGH CONCENTRATIONS OF HEAVY METALS

1. INTRODUCTION: AN OVERVIEW

In recent years, several countries affiliated with the United Nations have made commitments to address and enhance air pollution and air quality, biodiversity, chemical and waste management, energy, environmental governance, forests, freshwater resources, land, and oceans and seas, as part of the Global Environmental Goals (GEGs). One of the main concerns in environmental issues is the rising concentration of metals [1]. Several heavy metals of concern include arsenic (As), copper (Cu), zinc (Zn), lead (Pb), cadmium (Cd), and mercury (Hg). Heavy metals typically exist in soil at low quantities resulting from natural processes such as rock and mineral weathering, volcanic eruptions, and so on [2]. Over the past few decades, human intervention in the natural geochemical cycle has accelerated the introduction of heavy metals into our environment [3]. Heavy metals become harmful and are considered pollutants in the environment when they are generated at a faster rate through human activities compared to natural processes. The elevated metal levels seen in both urban and rural contexts, exceeding the natural background concentrations, can have detrimental effects on ecosystems, including plants and animals. Additionally, these elevated metal levels pose a significant risk to human health. Notable instances of metal toxicity include Minamata illness, which is caused by methylmercury poisoning, and itai-itai disease, which is caused by cadmium poisoning. This chapter specifically examines the presence of heavy metals in urban dust. Urban dust contamination arises from multiple dispersed sources and is frequently challenging to manage. While the contamination level caused by industrial and mining pollutants is typically higher, it nevertheless contributes to an increase in the overall exposure of heavy metals to individuals and ecosystems in urban areas. This contamination can have both obvious and subtle effects. This chapter provides a summary of heavy metals found in urban dust, the atmosphere, and on road surfaces. Next, the road runoff process and the subsequent infiltration into the soil are examined in terms of the movement of heavy metals. The current advancements in studies about the ecotoxicity of road dust are discussed.

2. PRESENCE OF HIGH CONCENTRATIONS OF HEAVY METALS IN THE ATMOSPHERE AND ON ROAD SURFACES

2.1. Atmospheric particulate matter

The urban environment is exposed to significant amounts of human-made pollutants originating from stationary sources (such as power plants, industries, incinerators, and residential heating) and mobile sources (including activities related to automobiles and transportation on roads) [2]. Additionally, there are natural sources in the surrounding area, such as windblown dust, the weathering of rocks and minerals, sea salt, and volcanic eruptions. These contaminants are linked to particle matter, liquid form, and gaseous form. They are then carried to the Earth's surface through both dry and moist deposition. Air pollution can consist of inorganic, organic, and/or microbiological substances. Inorganic heavy metals, including copper, zinc, cadmium, lead, arsenic, and chromium, are of significant concern as pollutants due to their toxicity and ability to cause cancer [6]. Aryal et al. conducted a study on the presence of heavy metals (namely Ni, Cu, Zn, Pb, and Cd) in particles smaller than 10 μm in the urban environment of Ulsan, an industrial city in South Korea [7]. Fig. 1 displays the levels of heavy metals in air particulate matter over the summer and fall seasons at two different sample locations. The outcome revealed that the Zn content was the greatest among the five elements, and heavy metal concentrations were generally greater in summer, except for Ni in the nonferrous metal industry region, which may be attributed to input from the surrounding area. Their comprehensive examination of the size-fractionated dust particles revealed that the particles had a bimodal distribution, with one peak representing particles ranging in size from 1.1 to 4.7 μm, and the other peak representing particles ranging from 4.7 to 10 μm. Out of the metals that were analyzed, it was shown that the distribution of Pb varied based on size, but Zn was uniformly distributed across all sizes.

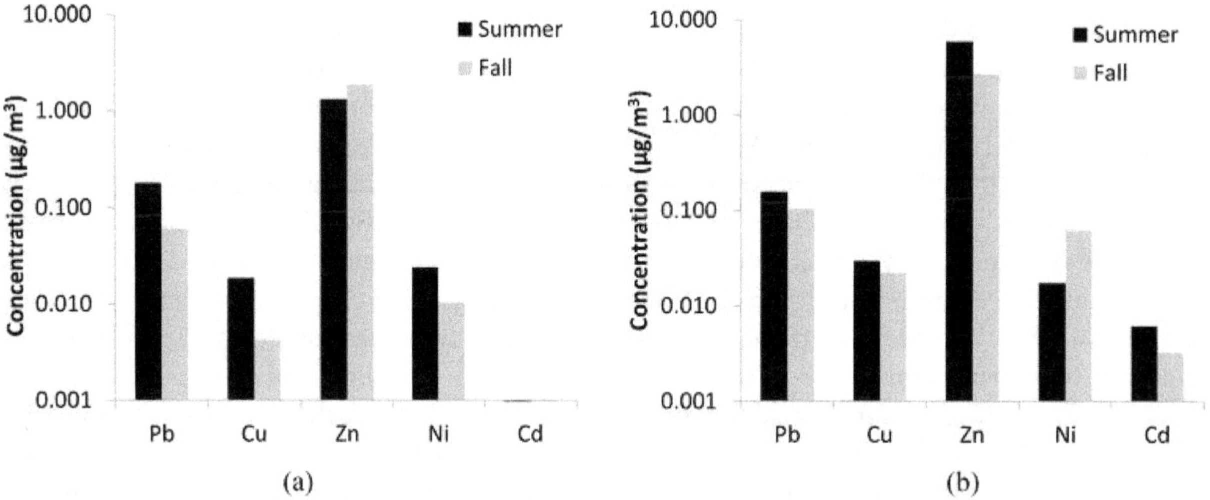

Fig. 1. Metal concentrations (μg/m³) in the atmospheric particulate matter (<10 μm) in (a) petrochemical area and (b) nonferrous metal industry area [6].

Several urban areas are regularly affected by dust storms originating from distant locations. Examples encompass phenomena such as yellow sand-dust storms occurring in Asia and Sahara dust storms observed in Northern Europe, among others. Due to its extensive trip, the dust storm has the potential to accumulate numerous harmful contaminants throughout its trajectory. The makeup of these dust particles is a significant matter of concern. Aryal et al. examined the physical and chemical characteristics of particles in the dust storm that affected towns on the east coast of Australia, including Sydney, on September 23, 2009. The storm covered the area for several hours [8]. Sydney residents experienced the most severe dust storm since 1942. The dust particle examination, conducted using wide-angle X-ray scattering (WAXS) and electron diffraction X-ray techniques, revealed that it predominantly consisted of crustal elemental oxides, including SiO_2, Al_2O_3, and Fe_2O_3. Fig. 2 displays the elemental composition acquired from electron diffraction X-ray spectra and the shape of the dust particle using scanning imaging microscopy (Fig. 2). The Al/Si ratio in the Sydney dust storm was approximately 0.39, suggesting that the dust originated from a desert. In addition, the researchers examined the organic matter present in the dust and determined that it was improbable for the dust to include any biohazardous substances that could pose a risk to human health.

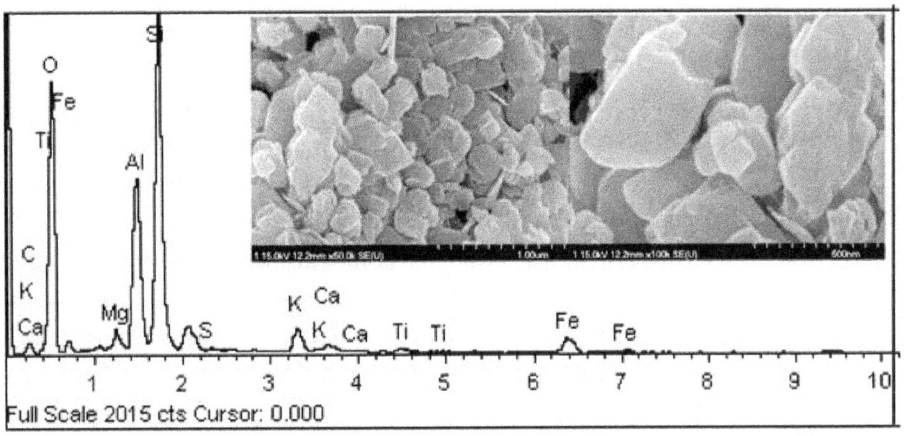

Fig. 2. X-ray spectrum of the Sydney dust storm particles [7].

2.2. Road dust

Accumulation of heavy metal on both urban and rural road surfaces can lead to substantial and severe pollution. The sources of hazardous metals include motor fuel exhaust, abrasion of vehicle parts such as tires and body, and industrial pollutants [9, 10]. Particles of road dust disperse into the atmosphere, leading to the contamination of the air and their subsequent transportation to more extensive regions [11]. The resuspended road dust has the potential to enter the human body through direct eating, inhalation of dust particles through the mouth and nose, and skin absorption. This poses a threat to people's health [12]. Atmospheric deposition and surface runoff of precipitation transfer road dust to aquatic environments. The presence of polluted dust may have a negative impact on the ecological well-being of the waterbodies where it is deposited. Hence, in order to safeguard human and ecological well-being, it is imperative to address the presence of heavy metals in road dust within major urban areas. Several research have been conducted on the topic of road dust and heavy metals. A research conducted in Winterthur, Switzerland, found that heavy metals are deposited in different diameters of road dust [13]. A separate investigation documented the presence of high levels of heavy metals in road dust in Ulsan, South Korea, across several particle sizes (<75, 75–180, 180–800, and 800–2000 μm) [1]. The heavy metal concentrations varied from 0.4 to 21 μg/g for Cd, 25 to 204 μg/g for Cu, 6 to 58 μg/g for Ni, 43 to 343 μg/g for Pb, and 79 to 326 μg/g for Zn. The heavy metal concentration was ranked in the following order: industrial complex > industrial vicinity > heavy traffic > residential > backdrop.

3. HEAVY METALS IN ROAD RUNOFF AND INFILTRATION

3.1. Road runoff

Urban runoff, especially from the roadways (road runoff), often contains significant loads of dissolved and particulate metals along with other pollutants such as hydrocarbons and nutrients [14, 15, 16, 17, 18, 19]. Unlike organic compounds, metal elements are not degraded in the environment and constitute an important class of constituents generated by traffic activities due to their non-degradability.

During wet weather period, the surface-deposited metals are washed off to the pipe network and finally to the receiving waterbodies including rivers, lakes, ponds, reservoirs, etc. The deposited pollutants are washed off mostly at the early stage of the runoff, called first flush [20]. Under heavy traffic conditions, the depositions of toxic pollutants are usually more concerned rather than those of other nonpoint pollutants such as nitrogen and phosphorus.

A monitoring on highway runoff in Winterthur, Switzerland, was conducted from September to December 2000 [13, 21]. The study covered nine runoff events, and samples were collected first 3 mm rainfall in order to capture the first flush. Furumai et al. reported that the runoff samples with higher SS concentrations had higher heavy metal concentrations [13]. The particle-bound Zn, Cu, and Pb concentrations were 155–524, 29–69, and 13–46 μg/L, respectively. The heavy metal concentrations were close to other studies [22, 23, 24, 25, 26, 27]. Aryal et al. reported the fractionated samples in this monitoring campaign [21]. The collected serial total suspended solid (TSS) samples were fractionated into five size fractions (<20, 20–45, 45–106, 106–250, and >250 μm) to understand the runoff behavior of the particles. The wash-off behavior of fine and coarse fractions was quite different. With respect to the TSS, a coarse fraction (>106 μm) showed power growth relationship, whereas the finer fractions (<106 μm) showed clustering at lower range TSS and almost saturated at higher TSS region (Fig. 3). This result indicated that particle sizes were very important in metal transport in runoff.

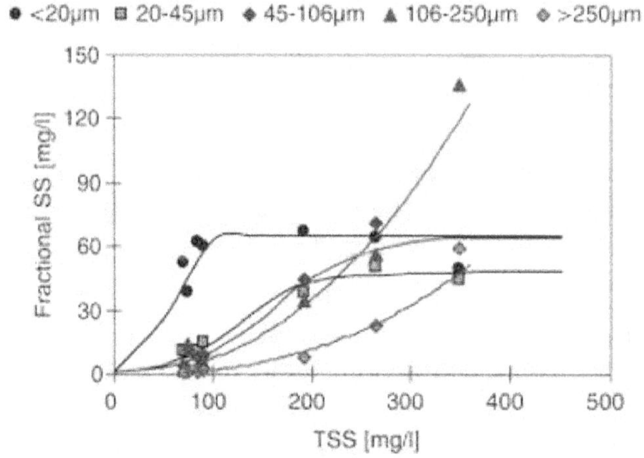

Fig. 3. Relationship between total suspended solid (TSS) concentrations and size-fractionated SS concentrations in highway runoff [20].

Aryal and Lee also studied the relationship between heavy metal content in particles and SS concentrations in urban runoff [28]. They found that all the runoff samples had low heavy metal content in particles at the beginning of the rainfall with higher SS concentration. As the time elapsed, the heavy metal content increased with lower SS concentration. The content of heavy metals reached the lowest level when SS concentration reached the highest value. The high contents of heavy metals at low SS concentration were believed to be due to higher contribution of fine particles to the total suspended solids (Fig. 4).

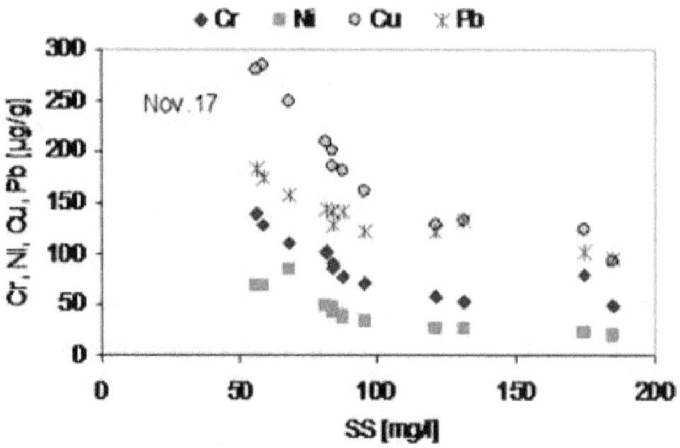

Fig. 4. Relationship between SS concentrations in the runoff samples and heavy metal content in SS (adopted from [28]).

3.2. Stormwater infiltration

The management of urban runoff in the city is accomplished through many methods. Within the stormwater management system, stormwater infiltration facilities are included as a means of flood control. These facilities are specifically designed to decrease the maximum flow rate by allowing water to seep into the ground [29]. Nevertheless, there is a significant apprehension regarding the potential contamination of groundwater by heavy metals. This is due to the fact that infiltration system designs are typically not designed with any regard for the retention of pollutants [30]. The mobility of deposited metals is determined by the environmental conditions within the facility, such as pH, oxidation reduction potential (oxygen availability), minerals, and organic matter [31, 32, 33, 34, 35]. Various heavy metals exhibit distinct patterns of accumulation within sediment, depending on the size of the particles [36]. Aryal et al. conducted a survey of over 200 infiltration inlets for road runoff in Tokyo [37]. The researchers conducted measurements of sediment depth, collected samples of sediment, and analyzed the presence of heavy metals in both total amounts and various particle sizes. Additionally, they examined the speciation of heavy metals. The depth of sediment accumulation ranged from a few millimeters to about 14 cm. The sediment's particle size examination revealed a varied distribution of sizes at different depths, indicating heterogeneity. The concentrations of heavy metals in the sediment at the inflow varied as follows: chromium (Cr) ranged from 7 to 143 µg/g, manganese (Mn) ranged from 383 to 1009 µg/g, cobalt (Co) ranged from 9 to 31 µg/g, nickel (Ni) ranged from 10 to 84 µg/g, copper (Cu) ranged from 51 to 316 µg/g, zinc (Zn) ranged from 288 to 1974 µg/g, and lead (Pb) ranged from 24 to 230 µg/g. The sediment's heavy metal concentrations were predominantly similar to those found in road dust, although in few instances they were lower than the levels observed in the surrounding soils. Table 1 presents the concentrations of heavy metals in the sediments at two different inlets. The study found that the bottom sediment in the inlet had lower levels of heavy metals compared to the top sediment. This indicates that there is a possibility that heavy metals have seeped into the soil below the surface [37, 38].

Table 1. Heavy metal content (µg/g) in the top and bottom sediment of two infiltration inlets of road runoff in Tokyo (#4 and #172) [36].

	Cr		Mn		Co		Ni		Cu		Zn		Pb	
Inlet	T	B	T	B	T	B	T	B	T	B	T	B	T	B
#4	31	94	842	765	31	25	80	62	259	214	1827	1350	271	220
#172	74	67	965	699	33	28	67	66	221	201	1071	791	96	85
Note: T, top (0–3 cm) and B, bottom (12–15 cm).														

Organic matter is one of the most important factors in heavy metal mobility in soil and sediment. Organic matter associated with the particulate matter may contain wide ranges of organics with potentially different sources. Humic substances are often the dominant organic fraction in soils and sediments [39]. They consist of heterogeneous mixtures of molecules having molecular weights ranging from a few hundred to several hundred thousand Daltons and are polyfunctional (each molecule may have a large number of different complexing sites, e.g., carboxylic, phenolic, phthalic, salicylic, and amine functional groups) [40, 41, 42]. Due to the large variation in structural properties, different organic matters can possess different strengths in terms of metal bonding. Also, the competitiveness among metals to bind to organic matter may vary [32, 43, 44]. It is reported that the general order of affinity of metal cations complexed by organic matter is Cu > Cd > Fe > Pb > Ni > Co > Mn > Zn [45]. Organic matter with a different functional group shows different affinities to the same heavy metals. This means that a single metal may show different affinities to different organic substances and may attach to the site where it is attracted more.

Aryal et al. studied road dust organic matter in different particle sizes and the organics' possible role in binding heavy metals [46]. The road dust was collected from industrial, business, and commercial districts and residential areas in the city of Ulsan, South Korea. Their study used fluorescent spectroscopy in combination with parallel factor analysis (PARAFAC). The study found that the finest fraction contained more microbial humic-like substances, whereas the coarse fraction was enriched with fulvic acid. The organic matter in two fractions (75–180 and 180–800 μm) showed dual characteristics. It was also observed that Zn had high affinity with aromatic protein, humic substances, and microbial by-products, whereas Cd showed a negative correlation with fulvic acids.

The fate of heavy metals in infiltration systems is highly dependent upon the sorptive and desorptive behavior of organic matter received from various sources such as road surfaces, parks, and agricultural areas. The organic matter found in the infiltration inlet sediment seemed changed with time due to prolonged deposition. Aryal et al. investigated the potential role of organic matter in binding and mobilizing heavy metals in the deposited sediment of infiltration facilities in Tokyo by using fluorescent spectroscopy combined with parallel factor analysis [47]. The surface sediment (few cm deep) showed a variation in organic matter and heavy metal content place by place. The sediment samples in lower layers had been relatively similar in different infiltration facilities. The result suggested that the sediment surface was used to be disturbed by scouring and mixing while receiving the runoff. The heavy metal content was found lower at the higher depth indicating possible release to the subsurface soil. The study also found that Cu and Zn had a positive linear relationship in the sediment samples which reflected that both of them had a common source. The PARAFAC analysis identified three major components with different spectral shapes and volumes in different excitation emission locations. The components reflected UV A humic-like substances (defined in [48]) and fresh organic matter [49], UV C-like organic matter having higher molecular weight humic substances [50, 51], and commercial humic acid [52]. While correlating organic components with heavy metals, they observed that UV A-type substances had a negative relationship with all metals, whereas commercial humic-type substances had a positive relationship with heavy metals.

4. ECOTOXICITY OF URBAN DUST

As described above, urban road dust contains toxic heavy metals as well as other various toxic chemicals such as polycyclic aromatic hydrocarbons (PAHs), perfluorinated surfactants (PFSs), etc. [53, 54, 55]. Stormwater runoff brings the road dust to the receiving waterbody and poses the harmful effect to the aquatic lives. In this section, recent studies on ecotoxicity of urban road dust are reviewed, and the heavy metal contribution to the effect is discussed.

Watanabe et al. and Khanal et al. reported the toxicity of the dilution series of road dust in Tokyo to calculate LC20 and LC50 by direct contact test with a benthic crustacean (ostracod, Heterocypris incongruens) (ISO 14371) [56, 57]. Those results are summarized in Table 2. Watanabe et al. showed the apparent lethal toxicity in six out of seven road dust samples tested although the dose-response relationship was shown only for the St. 6 in Table 2 [56]. Six out of 10 samples in [57] were toxic as shown in Table 2, and the rest of the four samples had <50% mortality (so that LC20 and LC50 were not calculated).

Table 2. Toxicity of urban road dust to benthic ostracod Heterocypris incongruens.

Road dust sample name in reference	Toxicity to benthic ostracod *H. incongruens* LC20 (% (v/v))*	LC50 (% (v/v))*	Ref.
St. 6	20	30	[56]
St. 1	13	31	[57]
St. 2	9	41	
St. 3	10	13	
St. 4	49	67	
St. 7	13	24	
St. 8	1.6	3.8	

*The unit is a volumetric percentage of road dust in the tested sediment. The sediment was composed of road dust and nontoxic reference sediment (quartz sand [56] and commercially available reference soil [57]).

Niyommaneerat et al. recently developed a new method of sediment chronic toxicity evaluation using ostracod reproduction and applied the method to urban road dust [58]. The road dust sample in [58] caused delayed egg production in 6.25% (v/v) dilution although LC20 for 6-day mortality test was 14% (v/v). Acute chronic ratio (ACR) was calculated as 6.8, and the authors estimated that 179 times of dilution was necessary to make the most toxic road dust (St.8 in [57]; see Table 2) nontoxic in the reproduction endpoint. Such dilution is not always achieved in the cities as the road dust contaminated the sediment in 3.7% in a river receiving road runoff in Tokyo [59].

Urban road dust, originally in dry form, is likely to change its toxic nature by wetting process. Watanabe et al. showed the toxicity of water-soluble (easily desorbable) components and remaining solids (particle-bound components), separately [56]. As a result, the primary toxic compounds of road dust existed mainly in the water-soluble form rather than the particle-bound form. Interestingly, after 7 days of incubation in water, the solid part also showed significant lethal toxicity. The similar phenomena of toxicity change in solid phase were also confirmed in other road dust samples [60]. The causal substances of the toxicity observed after several days are still unknown.

Toxicity change was also observed in seawater [61]. Road dust toxicity was tested in different salinities with the estuarine amphipod Grandidierella japonica. For three road dust samples, the toxicity in 3.5% salinity was significantly higher than in 0.5 and 2.0% salinity. The toxicity change may be caused by the change of bioavailability and/or toxicity of chemicals in the road dust.

In most of the cases, it is generally difficult to identify the causal substances of toxic environmental samples. Watanabe et al. and Khanal et al. applied toxicity identification evaluation (TIE) approach to the road dust toxicity assessment by benthic ostracod H. incongruens [62, 63]. In both of the studies, the two different types of adsorbents were used to manipulate the toxic substances in road dust, namely, Ambersorb/XAD for hydrophobic organic matter and Chelex/SIR-300 for cationic metals. Watanabe et al. showed that the adsorbent for hydrophobic organic matter was effective to reduce the toxicity of all the three road dust tested and concluded that the heavy metals were not considered to be major toxicants in the dust sample [62]. In contrast, Khanal et al. indicated that the adsorbent effectiveness depended on the dust samples (Fig. 5) and one of the dust samples (St. 7 in Fig. 5) became less toxic only by the adsorbent for cationic metals [63].

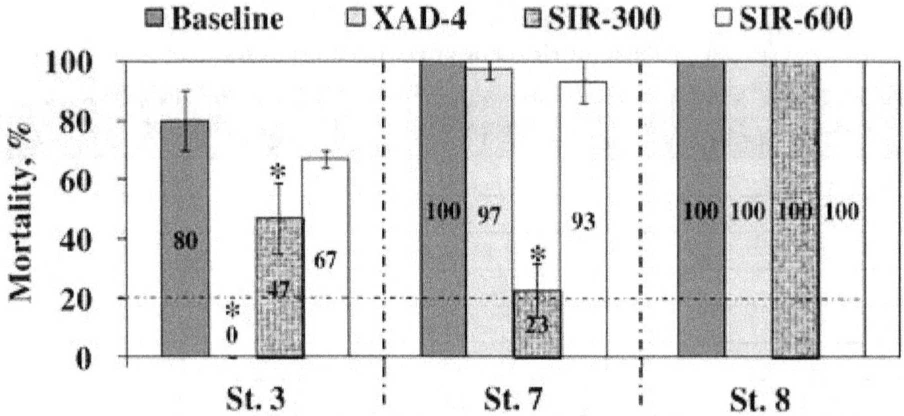

Fig. 5. Toxicity reduction by adsorbent addition to urban road dust [63]. Baseline indicates toxicity of urban road dust without adsorbent addition. XAD-4, SIR-300, and SIR-600 are adsorbents for hydrophobic organic matter, cationic metals, and ammonia, respectively. The toxicity was evaluated in a 6-day ostracod test.

One of the limitations in the TIE approach is that the adsorbent mainly removes the dissolved toxicant and may not reduce the dietary exposure through the ingestion of contaminated particles in the sediment tests. To assess the dietary exposure pathway, we should have dose-response relationship for the contaminated solids. However, such information is very limited even for the sediment toxicity test species. Sevilla et al. showed such dose-response relationship for a benthic ostracod to heavy metal-contaminated food algae (Fig. 6) [64, 65]. This will be useful to interpret the toxicity test result. For example, Watanabe et al. showed the overall toxicity of urban road dust but could not discuss the dietary pathway of heavy metal exposure [56]. The toxic urban road dust contained heavy metals in 1300–2980 μg Zn/g, 279–1200 μg Cu/g, and 0.64–2.0 μg Cd/g. Assuming that the heavy metal bioavailability of the dust and algae is similar (probably it is a safer side assessment), zinc and copper may cause the mortality of the ostracod species.

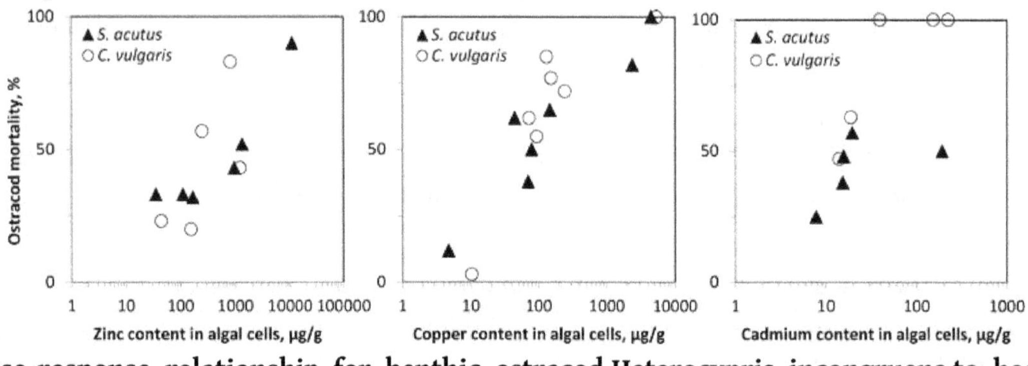

Fig. 6. Dose-response relationship for benthic ostracod Heterocypris incongruens to heavy metal-contaminated food algae (*Scenedesmus acutus and Chlorella vulgaris*) (adopted from [64]).

Biomarker is a potential method for detecting heavy metal exposure from road dust. However, when it comes to small benthic organisms utilized in sediment toxicity assessments, it poses a technological challenge to gather a significant amount of tissues for accurately measuring conventional biomarkers like metallothionein. Advancements in molecular techniques now allow for a more complete detection of biological responses to hazardous chemicals, specifically through gene expression. Hiki et al. documented the difficulty in investigating the biomarker genes of the benthic amphipod *G. japonica* in response to zinc, copper, and nicotine [66]. The researchers utilized the cDNA-amplified fragment length polymorphism (AFLP) approach on a non-model species that lacked genomic sequence information. The researchers discovered distinct loci for each exposure treatment and regarded them as potentially valuable biomarkers for each toxicant. The technology was utilized to analyze a sample of road dust, revealing that there were no shared gene expressions between the zinc/copper therapy and road dust exposure (Fig. 7). The presence of zinc and copper in the road dust was observed, although it appeared that these metals had little bioavailability to the species being tested.

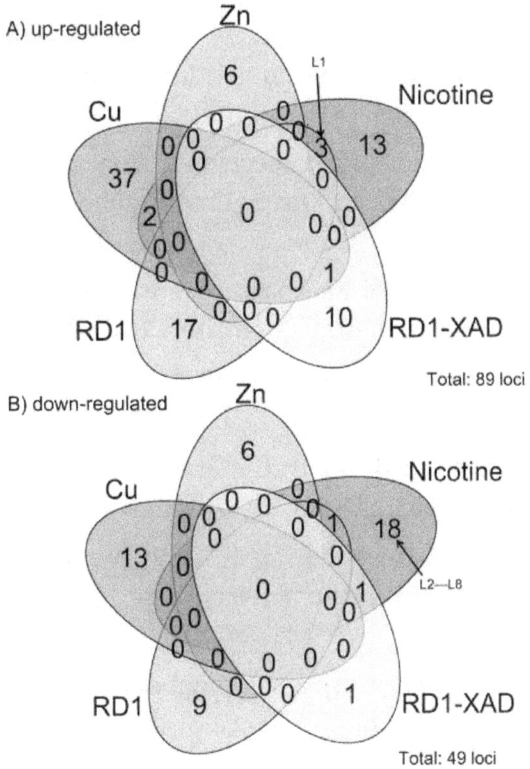

Fig. 7. Venn diagram showing the number of upregulated loci compared with control (A) and downregulated ones (B) [66].

5. CONCISE SUMMARY

This chapter provides an overview of the latest scientific advancements on heavy metals found in urban dust, with a particular focus on urban road dust. The discussion revolves around the mobility and toxicity of these heavy metals. The composition of urban dust varies locally, resulting in differences in its mobility and toxicity. The ecotoxicity of road dust has been demonstrated in numerous samples, but, additional investigation is required to elucidate the specific toxicant responsible and its long-term presence in the environment.

REFERENCES

1. GEGs. Global Environmental Goals. United Nations Environment Program [Internet]. 2010. Available from: http://geodata.grid.unep.ch/gegslive/ [Accessed 2017-12-12]

2. Bradl HB. Chapter 1. Sources and origins of heavy metals. In: Bradl HB, editor. Heavy Metals in the Environment: Origin, Interaction and Remediation. Interface Science and Technology. Vol. 6. The Netherlands: Elsevier; 2005. pp. 1-27. DOI: 10.1016/S1573-4285(05)80020-1

3. Vidal O, Rostom F, François C, Giraud G. Global trends in metal consumption and supply: The raw material-energy nexus. Elements. 2002;13:319-324. DOI: 10.2138/gselements.13.5.319

4. McAlpine D, Araki S. Minamata disease an unusual neurological disorder caused by contaminated fish. The Lancet. 1958;272(7047):629-631. DOI: 10.1016/S0140-6736(58)90348-9

5. Kaji M. Itai-itai disease: Lessons for the way to environmental regeneration. In: Fujigaki Y, editor. Lessons from Fukushima: Japanese Case Studies on Science, Technology and Society. Cham: Springer International Publishing; 2015. pp. 141-165. DOI: 10.1007/978-3-319-15353-7_7

6. Tchounwou PB, Yedjou CG, Patlolla AK, Sutton DJ. Heavy metal toxicity and the environment. In: Luch A, editor. Molecular, Clinical and Environmental Toxicology. Experientia Supplementum. Vol. 101. Basel: Springer; 2012. pp. 133-164. DOI: 10.1007/978-3-7643-8340-4_6

7. Aryal R, Kim A, Lee BK, Kamruzzaman M, Beecham S. Characteristics of atmospheric particulate matter and metals in industrial sites in Korea. Environment and Pollution. 2013;2(4):10-21. DOI: 10.5539/ep.v2n4p10

8. Aryal R, Kandel D, Acharya D, Chong MN, Beecham S. Unusual Sydney dust storm and its mineralogical and organic characteristics. Environmental Chemistry. 2012;9(6):537-546. DOI: 10.1071/EN12131

9. Li X, Poon CS, Liu PS. Heavy metal contamination of urban soils and street dusts in Hong Kong. Applied Geochemistry. 2001;16(11):1361-1368. DOI: 10.1016/S0883-2927(01)00045-2

10. Johansson C, Norman M, Burman L. Road traffic emission factors for heavy metals. Atmospheric Environment. 2009;43(31):4681-4688. DOI: 10.1016/j.atmosenv.2008.10.024

11. Apeagyei E, Bank MS, Spengler JD. Distribution of heavy metals in road dust along an urban-rural gradient in Massachusetts. Atmospheric Environment. 2011;45(13):2310-2330. DOI: 10.1016/j.atmosenv.2010.11.015

12. Zheng N, Liu J, Wang Q, Liang Z. Health risk assessment of heavy metal exposure to street dust in the zinc smelting district, northeast of China. Science of the Total Environment. 2010;408(4):726-733. DOI: 10.1016/j.scitotenv.2009.10.075

13. Furumai H, Aryal RK, Nakajima F. Profile analysis of polycyclic aromatic hydrocarbons and heavy metals in size fractionated highway dust and runoff samples. In: Proceedings of 9th International Conference on Urban Drainage; 8-13 September 2002; Portland. Reston: ASCE; 2012. pp. 1-13. DOI: 10.1061/40644(2002)205

14. Launay MA, Dittmer U, Steinmetz H. Organic micropollutants discharged by combined sewer overflows—Characterisation of pollutant sources and stormwater-related processes. Water Research. 2016;104:82-92. DOI: 10.1016/j.watres.2016.07.068

15. Krein A, Schorer M. Road runoff pollution by polycyclic aromatic hydrocarbons and its contribution to river sediments. Water Research. 2000;34(16):4110-4115. DOI: 10.1016/S0043-1354(00)00156-1

16. Han Y, Lau SL, Kayhanian M, Stenstrom MK. Characteristics of highway stormwater runoff. Water Environment Research. 2006;78(12):2377-2388. DOI: 10.2175/106143006X95447

17. Zhao H, Li X, Wang X. Heavy metal contents of road-deposited sediment along the urban–rural gradient around Beijing and its potential contribution to runoff pollution. Environmental Science & Technology. 2011;45(17):7120-7127. DOI: 10.1021/es2003233

18. Birch H, Mikkelsen PS, Jensen JK, Holten Lützhøft HC. Micropollutants in stormwater runoff and combined sewer overflow in the Copenhagen area, Denmark. Water Science and Technology. 2011;64:485-493. DOI: 10.2166/wst.2011.687

19. Wang S, He Q, Ai H, Wang Z, Zhang Q. Pollutant concentrations and pollution loads in stormwater runoff from different land uses in Chongqing. Journal of Environmental Sciences. 2013;25(3):502-510. DOI: 10.1016/S1001-0742(11)61032-2

20. Herngren L, Goonetilleke A, Ayoko GA. Understanding heavy metal and suspended solids relationships in urban stormwater using simulated rainfall. Journal of Environmental Management. 2005;76(2):149-158. DOI: 10.1016/j.jenvman.2005.01.013

21. Aryal RK, Furumai H, Nakajima F, Boller M. Dynamic behavior of fractional suspended solids and particle-bound polycyclic aromatic hydrocarbons in highway runoff. Water Research. 2005;39(20):5126-5134. DOI: 10.1016/j.watres.2005.09.045

22. Legret M, Pagotto C. Evaluation of pollutant loadings in the runoff waters from a major rural highway. Science of the Total Environment. 1999;235:143-150. DOI: 10.1016/S0048-9697(99)00207-7

23. Drapper D, Tomlinson R, Williams P. Pollutant concentrations in road runoff: Southeast Queensland case study. Journal of Environmental Engineering. 2000;126:313-320. DOI: 10.1061/(ASCE)0733-9372

24. Gromaire MC, Garnaud S, Saad M, Chebbo G. Contribution of different sources to the pollution of wet weather flows in combined sewers. Water Research. 2001;35:521-533. DOI: 10.1016/S0043-1354(00)00261-X

25. Zhao H, Li X, Wang X, Tian D. Grain size distribution of road-deposited sediment and its contribution to heavy metal pollution in urban runoff in Beijing, China. Journal of Hazardous Materials. 2010;183(1-3):203-210. DOI: 10.1016/j.jhazmat.2010.07.012

26. Göbel P, Dierkes C, Coldewey WG. Storm water runoff concentration matrix for urban areas. Journal of Contaminant Hydrology. 2007;91:26-42. DOI: 10.1016/j.jconhyd.2006.08.008

27. Helmreich B, Hilliges R, Schriewer A, Horn H. Runoff pollutants of a highly trafficked urban road—Correlation analysis and seasonal influences. Chemosphere. 2010;80:991-997. DOI: 10.1016/j.chemosphere.2010.05.037

28. Aryal RK, Lee BK. Characteristics of suspended solids and micropollutants in first-flush highway runoff. Water, Air, and Soil Pollution: Focus. 2009;9(5-6):339-346. DOI: 10.1007/s11267-009-9243-9

29. Furumai H, Jinadasa HKPK, Murakami M, Nakajima F, Aryal RK. Model description of storage and infiltration functions of infiltration facilities for urban runoff analysis by a distributed model. Water Science and Technology. 2005;52(5):53-60

30. Mikkelsen PS, Häfliger M, Ochs M, Jacobsen P, Tjell JC, Boller M. Pollution of soil and groundwater from infiltration of highly contaminated stormwater—A case study. Water Science and Technology. 1997;36(8-9):325-330. DOI: 10.1016/S0273-1223(97)00578-7

31. Tyler LD, McBride MB. Mobility and extractability of cadmium, copper, nickel, and zinc in organic and mineral soil columns. Soil Science. 1982;134(3):198-205. DOI: 10.1097/00010694-198209000-00009

32. Elliott HA, Liberati MR, Huang CP. Competitive adsorption of heavy metals by soils. Journal of Environmental Quality. 1986;15(3):214-219. DOI: 10.2134/jeq1986.00472425001500030002x

33. Lo KSL, Yang WF, Lin YC. Effects of organic matter on the specific adsorption of heavy metals by soil. Toxicological & Environmental Chemistry. 1992;34(2-4):139-153. DOI: 10.1080/02772249209357787

34. Yong RN, Phadungchewit Y. pH influence on selectivity and retention of heavy metals in some clay soils. Canadian Geotechnical Journal. 1993;30(5):821-833. DOI: 10.1139/t93-073

35. Mesquita ME, Carranca C, Menino MR. Influence of pH on copper-zinc competitive adsorption by a sandy soil. Environmental Technology. 2002;23(9):1043-1050. DOI: 10.1080/09593332308618352

36. Marsalek J, Marsalek PM. Characteristics of sediments from a stormwater management pond. Water Science and Technology. 1997;36(8-9):117-122. DOI: 10.1016/S0273-1223(97)00610-0

37. Aryal RK, Murakami M, Furumai H, Nakajima F, Jinadasa HKPK. Prolonged deposition of heavy metals in infiltration facilities and its possible threat to groundwater contamination. Water Science and Technology. 2006;54(6-7):205-212. DOI: 10.2166/wst.2006.584

38. Hossain MA, Furumai H, Nakajima F, Aryal RK. Heavy metals speciation in soakaways sediment and evaluation of metal retention properties of surrounding soil. Water Science and Technology. 2007;56(11):81-89. DOI: 10.2166/wst.2007.746

39. MacCarthy P, Malcolm RL, Clapp CE, Bloom PR. An introduction to soil humic substances. In: MacCarthy P, Clapp CE, Malcolm RL, Bloom PR, editors. Humic Substances in Soil and Crop Sciences: Selected Readings. Madison: SSSA; 1990. pp. 1-12. DOI: 10.2136/1990.humicsubstances.c1

40. Kramer RW, Kuawinski EB, Zang X, Green-Church KB, Jones RB, Freitas MA, Hatche PG. Studies of the structure of humic substances by electrospray ionization coupled to a quadrupole-time of flight (QQ-TOF) mass spectrometer. In: Ghabbour EA, Davis G, editors. Humic Substances: Structures, Models and Function. Cambridge: Royal Society of Chemistry; 2001. pp. 95-107

41. Chin YP, Aiken G, O'Loughlin E. Molecular weight, polydispersity, and spectroscopic properties of aquatic humic substances. Environmental Science & Technology. 1994;28(11):1853-1858. DOI: 10.1021/es00060a015

42. Buffle J, Filella M. Physico-chemical heterogeneity of natural complexants: Clarification. Analytica Chimica Acta. 1995;313(1-2):144-150. DOI: 10.1016/0003-2670(95)00204-D

43. Atanassova I. Competitive effect of copper, zinc, cadmium and nickel on ion adsorption and desorption by soil clays. Water, Air, and Soil Pollution. 1999;113(1-4):115-125. DOI: 10.1023/A:1005076821325

44. Jalali M, Moradi F. Competitive sorption of Cd, Cu, Mn, Ni, Pb and Zn in polluted and unpolluted calcareous soils. Environmental Monitoring and Assessment. 2013;185(11):8831-8846. DOI: 10.1007/s10661-013-3216-1

45. Adriano D, Vangronsveld J, Bolan N, Wenzel W. Heavy metals. In: Encyclopedia of Soils in the Environment. New York, USA: Springer; 2005. pp. 175-182. DOI: 10.1016/B0-12-348530-4/00196-X

46. Aryal RK, Duong TTT, Lee BK, Hossain MA, Kandel D, Kamruzzaman M, Beecham S, Chong MN. Organic matter composition variability in road sediment and its role in binding heavy metals. Sustainable Environment Research. 2014;24(2):81-91

47. Aryal R, Furumai H, Nakajima F, Beecham S, Kandasamy J. Characterisation of prolonged deposits of organic matter in infiltration system inlets and their binding with heavy metals: A PARAFAC approach. Water, Air, and Soil Pollution. 2015;226(6). DOI: 10.1007/s11270-015-2426-2

48. Cory RM, McKnight DM. Fluorescence spectroscopy reveals ubiquitous presence of oxidized and reduced quinones in dissolved organic matter. Environmental Science & Technology. 2005;39(21):8142-8149. DOI: 10.1021/es0506962

49. Klapper L, McKnight DM, Fulton JR, Blunt-Harris EL, Nevin KP, Lovley DR, Hatcher PG. Fulvic acid oxidation state detection using fluorescence spectroscopy. Environmental Science & Technology. 2002;36(14):3170-3175. DOI: 10.1021/es0109702

50. Stedmon CA, Markager S, Bro R. Tracing dissolved organic matter in aquatic environments using a new approach to fluorescence spectroscopy. Marine Chemistry. 2003;82(3-4):239-254. DOI: 10.1016/S0304-4203(03)00072-0

51. Stedmon CA, Markager S. Resolving the variability in dissolved organic matter fluorescence in a temperate estuary and its catchment using PARAFAC analysis. Limnology and Oceanography. 2005;50(2):686-697. DOI: 10.4319/lo.2005.50.2.0686

52. Pingqing F, Fengchang W, Congqiang L. Fluorescence excitation-emission matrix characterization of a commercial humic acid. Chinese Journal of Geochemistry. 2004;23(4):309-318. DOI: 10.1007/BF02871302

53. Takada H, Onda T, Harada M, Ogura N. Distribution and sources of polycyclic aromatic hydrocarbons (PAHs) in street dust from the Tokyo metropolitan area. Science of the Total Environment. 1991;107:45-69. DOI: 10.1016/0048-9697(91)90249-E

54. Pengchai P, Furumai H, Nakajima F. Source apportionment of polycyclic aromatic hydrocarbons in road dust in Tokyo. Polycyclic Aromatic Compounds. 2004;24(4-5):713-789. DOI: 10.1080/10406630490487828

55. Murakami M, Takada H. Perfluorinated surfactants (PFSs) in size-fractionated street dust in Tokyo. Chemosphere. 2008;73:1172-1177. DOI: 10.1016/j.chemosphere.2008.07.063

56. Watanabe H, Nakajima F, Kasuga I, Furumai H. Toxicity evaluation of road dust in the runoff process using a benthic ostracod Heterocypris incongruens. Science of the Total Environment. 2011;409:2366-2372. DOI: 10.1016/j.scitotenv.2011.03.001

57. Khanal R, Furumai H, Nakajima F. Toxicity assessment of size-fractionated urban road dust using ostracod Heterocypris incongruens direct contact test. Journal of Hazardous Materials. 2014;264:53-64. DOI: 10.1016/j.jhazmat.2013.10.058

58. Niyommaneerat W, Nakajima F, Tobino T, Yamamoto K. Development of a chronic sediment toxicity test using the benthic ostracod Heterocypris incongruens and their application to toxicity assessments of urban road dust. Ecotoxicology and Environmental Safety. 2017;143:266-274. DOI: 10.1016/j.ecoenv.2017.05.011

59. Kumata H, Yamada J, Masuda K, Takada H, Sato Y, Sakurai T, Fujiwara K. Benzothiazolamines as tire-derived molecular markers: Sorptive behavior in street runoff and application to source apportioning. Environmental Science & Technology. 2002;36(4):702-708. DOI: 10.1021/es0155229

60. Khanal R. Characterization of Toxicants in Size-Fractionated Urban Road Dust Using Ostracod Heterocypris incongruens Direct Contact Test and Toxicity Identification Evaluation Procedure [Thesis]. Tokyo: The University of Tokyo; 2013

61. Hiki K, Nakajima F. Effect of salinity on the toxicity of road dust in an estuarine amphipod Grandidierella japonica. Water Science and Technology. 2015;72(6):1022-1028. DOI: 10.2166/wst.2015.304

62. Watanabe H, Nakajima F, Kasuga I, Furumai H. Application of whole sediment toxicity identification evaluation procedures to road dust using a benthic ostracod Heterocypris incongruens. Ecotoxicology and Environmental Safety. 2013;89:245-251. DOI: 10.1016/j.ecoenv.2012.12.003

63. Khanal R, Furumai H, Nakajima F. Characterization of toxicants in urban road dust by toxicity identification evaluation using ostracod Heterocypris incongruens direct contact test. Science of the Total Environment. 2015;530-531:96-102. DOI: 10.1016/j.scitotenv.2015.05.090

64. Nakajima F, Sevilla JB, Kasuga I. Dose-response relationship of dietary heavy metals to mortality of ISO sediment toxicity test species Heterocypris incongruens. In: Proceedings of 50th Environmental Engineering Research Forum; 19-21 November 2013; Sapporo. Tokyo: Japan Society of Civil Engineers; 2013. pp. 160-162 (in Japanese)

65. Sevilla JB, Nakajima F, Kasuga I. Comparison of aquatic and dietary exposure of heavy metals Cd, Cu and Zn to benthic ostracod *Heterocypris incongruens*. Environmental Toxicology and Chemistry. 2014;33(7):1624-1630. DOI: 10.1002/etc.2596

66. Hiki K, Nakajima F, Tobino T. Application of cDNA-AFLP to biomarker exploration in a non-model species Grandidierella japonica. Ecotoxicology and Environmental Safety. 2017;140:206-213. DOI: 10.1016/j.ecoenv.2017.02.037

18

Chapter

AN EVALUATION OF THE POTENTIAL HEALTH HAZARDS POSED BY HEAVY METALS FOUND IN DUST AND SOIL WITHIN PRIMARY SCHOOL PREMISES IN LAGOS STATE, NIGERIA

1. INTRODUCTION: AN OVERVIEW

Lagos State is the most densely populated city in Nigeria, the second most rapidly developing city in Africa, and the seventh worldwide [1]. In recent decades, the fast growth of industrialization and urbanization has led to significant emissions of both metal and organic pollutants. As a result, the environment has become particularly vulnerable to ecological degradation and pollution [2, 3, 4]. Heavy metals are persistent and accumulate in the environment without any recognized mechanism for their elimination [5]. Elevated concentrations of heavy metals can have detrimental effects on human health by impeding the proper functioning of organs and systems, including the liver, kidney, central nervous system, and bones. Additionally, heavy metals can act as cofactors in the development of various diseases [5, 6]. Several metals are essential for life and play crucial roles as sources of vitamins and minerals in the functioning of bodily organs. Every living organism has certain requirements for metals, but when these metals are present in larger concentrations, they become harmful [7]. After being absorbed by the body, heavy metals continue to accumulate in crucial organs such as the brain, liver, bones, and kidneys over extended periods of time, leading to severe health implications [8]. Essential metals such as copper (Cu), cobalt (Co), and zinc (Zn) are necessary for normal bodily development and functioning, while elevated levels of other metals like cadmium (Cd), chromium (Cr), manganese (Mn), and lead (Pb) are considered highly toxic to both humans and aquatic life [9]. Typical physiological capabilities necessitate a measurement of Cr, but its elevated levels can lead to toxicity, including liver and kidney problems and genotoxic carcinogenesis [10, 11, 12]. Elevated levels of manganese (Mn) and copper (Cu) in drinking water can lead to the development of mental disorders such as Alzheimer's disease and Manganism. These conditions mostly impact the cognitive abilities of children who are 10 years old. Pb is a very toxic and carcinogenic metal that can lead to chronic health risks, such as headaches, irritability, abdominal pain, nerve damage, kidney damage, high blood pressure, lung cancer, stomach cancer, and gliomas [14, 15, 16]. Children are highly vulnerable to lead toxicity. When exposed to high levels of lead, they have significant health complications such as behavioral disturbances, memory impairment, reduced cognitive abilities, and long-term exposure can also lead to anemia [16]. Lead (Pb) also causes renal tumors and disrupts the proper operation of the kidneys, joints, reproductive system, and brain system [17]. Soils and dust have emerged as a highly effective diagnostic tool for assessing environmental factors that impact human health [18]. Several research undertaken during the past decade have examined the chemical makeup of soil and dust. Considerable focus has been given to research on urban parking lots and the accumulation of dust on roads [19, 20, 21, 22, 23, 24, 25]. Soil particles undergo direct or indirect transformation into house dust, which can be consumed by both adults and children through several methods [26]. The primary means by which hazardous metals are encountered in urban environments are through dermal contact, ingestion, and inhalation [26, 27]. The likelihood of being exposed to negative effects from ingesting soil is greater in children compared to adults [26]. Children mostly engage with soil in schools and play spaces. Children have the potential to consume significant amounts of toxic metals from soil, dust, and air [27]. Due to their limited resistance to toxins and the tendency to ingest substances by hand-to-mouth contact, the health risk is elevated in this population [28, 29]. Therefore, it is crucial to maintain low levels of potentially harmful compounds in soil in locations where children frequently visit [30]. School playgrounds and classrooms are purposefully created environments that facilitate children's play, promote their well-being, and enhance their cognitive development [31]. These activities are specifically created to improve the connections between children, offer amusement and leisure, and promote physical health and adaptability. In modern times, playgrounds are increasingly equipped with metal structures such as sandboxes, jungle gyms, carousels, slides, and other materials that enhance recreational activities [31]. Safety, specifically in the context of playgrounds, pertains to the prevention of injuries, however there is insufficient focus on the potential health hazards posed by heavy metals. The health hazard is linked to the exposure to harmful substances. The Human Exposure to Soil Pollutants (HESP) model is commonly utilized to assess the risks associated with exposure to both carcinogenic and non-carcinogenic sources. This model relies solely on data related to soil pollution [32]. The risk assessment for heavy metals is determined by comparing the daily oral intake of heavy metals to the standard daily oral intake. The target hazard quotient (HQ), as defined by the USEPA [32], is the ratio of the expected exposure to a pollutant to the level at which no adverse effects are observed. It is now used as the approach for assessing non-cancer risks. If the anticipated proportion exceeds 1, then it is possible to conceive potential repercussions on well-being. While direct exposure is a fundamental aspect of human health risk assessments, it is not as extensively studied in the literature compared to the first pathway, which is indirect exposure [33, 34, 35]. Research has consistently shown that engaging in outdoor activities that include contact with soil polluted with heavy metals can be a significant way for humans, especially children, to be exposed to and consume these harmful substances [33, 34, 35]. The objective of this chapter was to analyze the risk to children's health by evaluating the presence of hazardous metals in soil from playgrounds and classroom dust in certain elementary schools in Lagos State. This study involved the examination of various hazardous metals, including Pb, Mn, and Cr, in soil and dust samples taken from playgrounds and classrooms in selected primary schools in Lagos State. Consequently, this chapter examined the levels of heavy metals and conducted a risk assessment to determine the potential health hazards that children may face in playgrounds and classrooms at selected schools in Lagos State.

1.1. Study area

Lagos State (Fig. 1) represents the most urbanized state in Nigeria with about 300 industries on 12 industrial estates, representing more than 60% of all industries activities in the country. It is the smallest state in Nigeria, yet it has the second biggest populace of more than 9 million [1]. The rate of populace development is around 275,000 people for every annum with a populace thickness of 2594 people for each km2. Lagos State primary schools list was acquired from the Lagos State Primary School Education Board (SPEB), Maryland. Two schools were selected in each of the 20 local government areas (namely Agege, Ajeromi-Ifelodun, Apapa, Alimosho, Amuwo-Odofin, Badagry, Eti-Osa, Epe, Ifako-Ijaye, Ikeja, Ojo, Oshodi-Isolo, Surulere, Mushin, Shomolu, Lagos Mainland, Lagos Island, Kosofe, Ikorodu, and Ibeju-Lekki), one in highly populated area and the other in the low-density area. One of the two primary schools selected was from the planned areas (low traffic density areas, government reserved areas or residential estates) and the other from the unplanned area (high traffic density, commercial, or industrial areas).

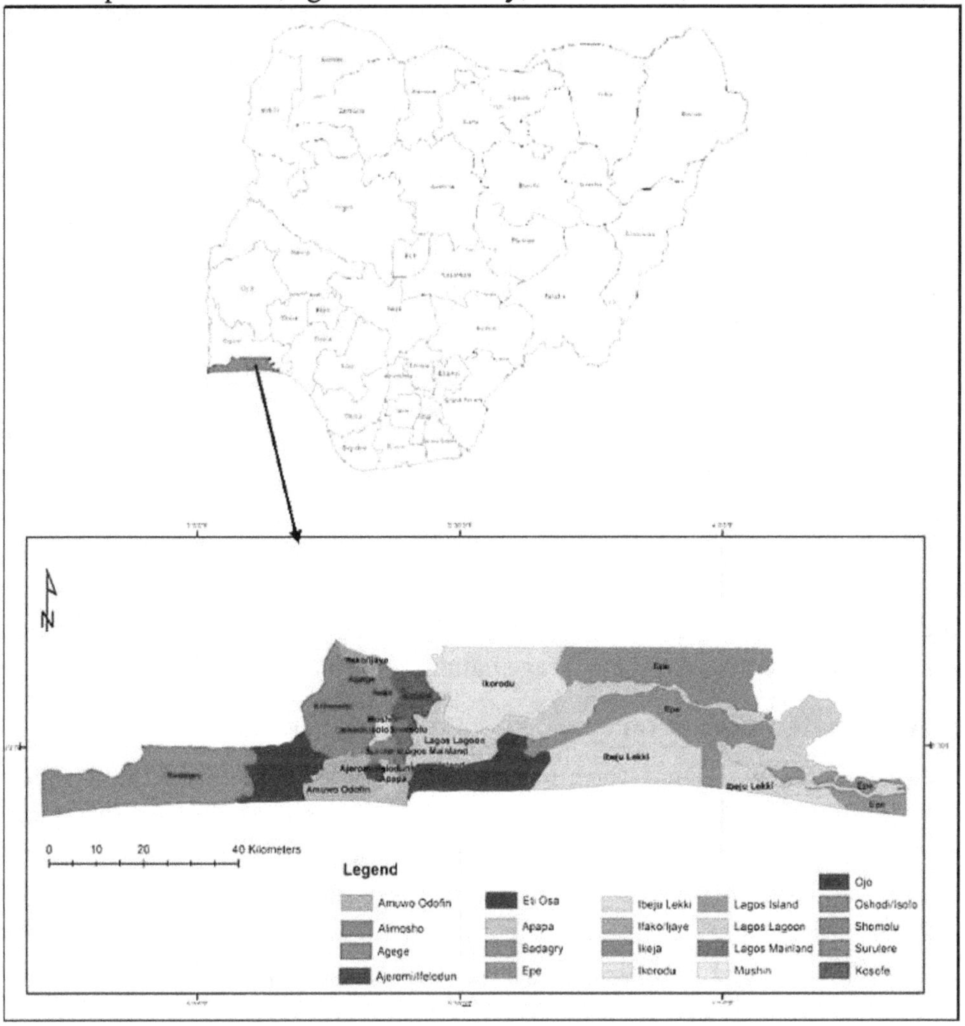

Fig. 1. Lagos State map showing local government areas.

2. METHODOLOGY

2.1. Sampling and analysis

Samples were collected in two primary schools in each of the 20 local government areas of Lagos State. The schools were selected based on the proximity of the school to major roads and where possible, nearness to industrial activities. A point was picked aimlessly in the play area and entrance and a squared framework of around 2 km side was drawn from the spot. In the vicinity of four and five discrete soil samples were gathered from each of the grid. The number of samples from each site was determined based on the circumstances existing at that site, i.e., high foot traffic (entrance), spots where the children frequently play during recess (playgrounds). A total of 80 topsoil samples were obtained. Plastic parker and brush were used to collect the surface soil. Composite samples were obtained by mixing all the subsamples to form one single sample (about 2 kg of soil), which was used for chemical analysis. Particulates such as grass, leaves, polythene bags, and papers present in soil samples were removed after gentle shaking to remove the soil attached around them. All soil samples were air-dried and sieved to 2 mm using a stainless-steel sieve and stored in plastic vials until analyzed.

In the classrooms, 40 dusts samples were gathered in the nursery classes (kids in the age scope of 2–4 years of age are more inclined to pica propensities). Dusts were gathered from window ledges, bookshelves, corners in the classrooms, and so forth. Around 3–5 tests were gathered and were blended to frame one composite sample. The soil and dust samples were air-dried and sieved to <2 mm using a stainless-steel sieve and stored in plastic vials until analyzed.

2.2. Sample preparation

The specimens were allowed to desiccate in the laboratory for a period of 24 hours, with precautions taken to prevent contamination from laboratory dust. Prior to homogenization using a mortar and pestle, the dried dust samples were first purified by removing larger particles of grit and debris. Prior to examination, the obtained samples were packed and stored in a sealed polythene.

2.3. Glassware, reagents, and standards

The glassware and plastic items were initially cleaned using a high-quality laboratory cleanser, followed by rinsing with deionized water. Subsequently, a 10% nitric acid solution was added and left overnight, after which it was rinsed again with deionized water. All solutions and dilutions were prepared using double deionized water (18.2 M cm) acquired from the Simplicity water purification system. All compounds used in the analysis were of analytical grade. No additional cleaning was performed to purify all reagents, and Merck Pty Ltd. South Africa provided the necessary adjustment standards. The analysis was conducted three times to achieve the average value.

2.4. Microwave-assisted digestion

Davidson et al. [36] method for determination of pseudototal metal concentration was adopted. This digestion method is enhanced by extracting most of the potentially mobile fractions but leaves the more resistant silicates undissolved. Approximately, 1 g of the sample and 20 mL of acid were heated at temperature of 180°C and pressure of 120 psi for 10 min. Then, the digested samples were filtered through Whatman No. 125 filter paper into 100 mL volumetric flasks. The digestion vessels were then rinsed with distilled water and filtered into the flasks. Each filtrate was made up to the mark with further distilled water, to give a final sample solution containing 20% (v/v) aqua regia.

2.5. Instrumentation calibration and measurement

1000 µg/mL standard stock solutions of Pb, Cr, Cd, and Mn were prepared from 1000 mg/L certified standard solutions (Merck Ltd., South Africa). The stock solutions were acidified with nitric acid and used for calibration. Calibration solutions concentrations of 5, 10, 20, 30, 40, 50, and 60 ppb (µg/mL) were prepared, respectively, for Pb, Cr, and Cd, while 1, 1.5, 2, 2.5, 3, 3.5, and 4 µg/mL (ppb) were prepared for Mn. A Shimadzu AA-6300 Graphite Furnace Atomic Absorption Spectrophotometer (GFA-EX7) fitted with an Autosampler (ASC)-6100 was used for determination of the heavy metals. The GFAAS was equipped with Cr, Cd, Pb, and Mn hollow cathode lamps (Varian cathode lamps and photon cathode lamps) and were employed for the measurement of the absorbance.

2.6. Risk assessment

The measured total heavy metals from the analysis were used to assess the health risk of the metal on children. The mean heavy metals concentrations were used to estimate intake at different pathways using standard USEPA's exposure equations [37, 38]. Children could be exposed to contaminants from soil via three different pathways that include oral intake (Iingestion), inhalation intake (Iinhalation), and through skin exposure (Idermal) [38]. Based on this fact, cancer/noncancer risk assessment in this study was estimated. For intake estimation via each exposure pathways, the following equations were used.

$$Intake_{ingestion} = \frac{C \; X \; IngR \; X \; EF \; X \; ED}{BW \; X \; AT} X\, 10^{-6} \quad (1)$$

where C—concentration of a contaminant in soil (mg/kg), IngR—ingestion rate of soil (mg/day), EF—exposure frequency (days/year), ED—exposure duration (years), BW—average body weight (kg), and AT—average time (days) = ED*365

$$Intake_{inhalation} = \frac{C \; X \; InhR \; X \; EF \; X \; ED}{PEF \; X \; BW \; X \; AT} \quad (2)$$

where InhR—inhalation rate (m^3/day) and PEF—particle emission factor (m^3/kg).

$$Intake_{dermal} = \frac{C \; X \; SA \; X \; SAF \; X \; ABS \; X \; EF \; X \; ED}{BW \; X \; AT} X\, 10^{-6} \quad (3)$$

where SA—surface area of the skin that contacts the soil (cm^2), SAF—skin adherence factor for soil (mg/cm2), and ABS—dermal absorption factor (chemical specific) = 0.001 (for all metals).

After the three exposure pathways were calculated, hazard quotient (HQ) and HI based on cancer/non-cancer toxic risk were calculated as follows [38]:

$$HQ = \frac{Intake}{RfD} \quad (4)$$

$$HI_{ex}P = \sum HQexP \quad (5)$$

where exP are different exposure pathways, respectively. Reference dose (RfD) (mg/kg/day) is an estimated value of the daily exposure, maximum permissible risk, to the human population, including sensitive subgroups (children) during a lifetime. Tables 1 and 2 show the exposure parameters, reference doses, and cancer slope factors used for the health risk assessment for standard residential exposure scenario through different exposure pathways.

Table 1. Exposure parameters used for the health risk assessment through different exposure pathways for soil.

Parameter	Unit	Child	References
Body weight (*BW*)	kg	15	[39]
Exposure frequency (*EF*)	days/year	360	[38]
Exposure duration (*ED*)	years	6	[40]
Ingestion rate (*IR*)	mg/day	200	[41, 42]
Inhalation rate (*IRair*)	m^3/day	7.6	[43]
Skin surface area (*SA*)	cm^2	2800	[40]
Soil adherence factor (*SAF*)	mg/cm^2	0.2	[40]
Dermal Absorption factor (*ABS*)	none	0.001	[38, 39, 40, 43, 44]
Particulate emission factor (*PEF*)	m^3/kg	1.36×10^9	[38, 39]
Average time (*AT*)	days		[39]
For carcinogens		365 × 70	[39]
For non-carcinogens		365 × ED	[39]

Table 2. Reference doses (RfD) in (mg/kg-day) and cancer slope factors (exP) for the different heavy metals.

Heavy metal	$Rf_{Dingestion}$	$Rf_{Ddermal}$	$RfD_{Inhalation}$	$ex_{Pingestion}$	$ex_{Pdermal}$	$exP_{Inhalation}$	Ref.
Pb	3.50E-03	5.25E-04	3.50E-03	8.50E-03	—	4.20E-02	[39, 45]
Cr	3.00E-03	3.00E-03	3.00E-05	5.00E-01	—	4.10E+01	[39, 46]
Cd	1.00E-03	1.00E-03	5.70E-05	6.30E+00	—	6.30E+00	[39, 45]
Mn	1.40E-1	1.40E-1	—	—	—	—	[39, 46]

3. RESULTS AND DISCUSSION

3.1. Concentrations of heavy metals in soil from classrooms and playgrounds

Descriptive statistics of the four groups of schools are shown in Table 3. Average concentrations of heavy metals in mg/kg from playgrounds and classrooms in selected primary schools in Lagos State were used to calculate average daily intakes for carcinogenic/non-carcinogenic risk assessment. The results presented showed the average concentrations of heavy metals in classrooms and playgrounds varied significantly and decreased in the order from Group 1 > Group 3 > Group 2 > Group 4 for Pb; Group 4 > Group 3 > Group 1 > Group 2 for Cr; Group 2 > Group 1 > Group 3 = Group 4 for Cd and Group 2 > Group 3 > Group 1 = Group 4 for Mn in the samples. This implies that the heavy metals are distributed irrespective of the locality either in high or low-density playgrounds/classrooms. The results show the magnitude of the heavy metals in the classrooms and playgrounds decreases in the following order: Pb > Cr > Cd > Mn. The high concentrations of Pb and Cr in the samples could be attributed to the paint chippings peeling off to settle on the classrooms floors as dust [30]. A study from Wright et al. [47] show that chipping house paint is a critical determinant of high blood lead levels more than 10 µg/L found in 70% of children aged 6–35 months. It has been accounted for that emulsion and gloss sorts of paints, which are produced and sold in Nigeria contained considerable levels of lead [48].

Table 3. Statistical parameter for the distribution of Pb, Cr, Cd, and Mn in classroom dusts and playgrounds from selected schools in Lagos State.

		GROUP 1	GROUP 2	GROUP 3	GROUP 4
Pb (µg/g)	Min	23.04	24.30	24.96	8.33
	Max	79.01	37.05	41.91	41.91
	Mean + SD	35.17 ± 16.54	30.08 ± 5.35	30.75 ± 6.72	24.21 ± 12.05
	Range	23.04–79.01	24.30–37.05	24.96–41.91	8.33–41.91
Cr (µg/g)	Min	3.25	0.05	6.66	11.42
	Max	16.93	14.52	19.15	20.24
	Mean + SD	11.88 ± 4.50	6.08 ± 4.63	14.58 ± 4.33	15.73 ± 2.87
	Range	3.25–16.93	0.05–14.52	6.66–19.15	11.42–20.24
Cd (µg/g)	Min	1.05	0.90	ND	ND
	Max	1.63	3.10	ND	ND
	Mean + SD	1.28 ± 0.31	1.90 ± 1.07	ND	ND
	Range	1.05–1.63	0.90–3.10	ND	ND

		GROUP 1	GROUP 2	GROUP 3	GROUP 4
Mn (μg/g)	Min	ND	1.18	1.18	ND
	Max	ND	1.87	1.18	ND
	Mean + SD	ND	1.47 ± 0.36	1.18 ± .00	ND
	Range	ND	1.18–1.87	1.18	ND
Group 1—High density playground; Group 2—Low-density playground; Group 3—High density classroom, and Group 4—Low-density classroom, ND—Not detected.					

Higher concentrations of investigated metals could be due to proximity of the schools to traffic roads and some industrial locations. Lagos State has been under high urbanization in the past few decades. In this chapter, there were no specific pollution sources of toxic metals [30]; hence, the toxic metal contamination of the soils from the classrooms and playgrounds was most likely from continuous urbanization and development, which can influence human health in the contaminated area. It is important to emphasize that Lagos State with her increasing population is known to have very intensive and heavy traffic [30]. The high Pb concentration of the soil could be ascribed to vehicular discharges and metal plating and greasing up oils. It could likewise be because of harsh surfaces of the streets which increment the wearing of tires and run-offs from the roadsides [48]. Lead contamination in urban soil has been ascribed to ignition of fuel that contains tetraethyl lead as anti-knock agent [48, 49]. Compared with recommended maximum allowable limits for Nigeria and WHO as shown in Table 4, Pb, Cr, and Cd concentrations were found to be higher than the permissible limits, and hence the classrooms and playgrounds soils were contaminated, and the learners were at high risk. These highest levels of Pb and Cr could possibly be the cause of alleged sickness in children suffering from diarrheal diseases and chest pains. The high concentrations of heavy metals could be because of the nearness of the schools and playgrounds to bus stops, auto workshops, nearby industries, gas stations in addition to paints chippings as reported in the city of Ibadan [53].

Table 4. DPR [50], FEPA [51], and WHO [52] permissible limits.

	Pb	Cr	Cd	Mn
DPR(μg/g)	0.05	0.03	0.01	—
FEPA(μg/g)	0.05	0.03	0.01	—
WHO (μg/g)	0.05	0.02	0.005	5

3.2. Health risk of heavy metals for children

The mean values of the heavy metals were higher than permissible limits by DPR, FEPA, and WHO, which make the soils/dusts contaminated except for Mn (Table 4). The obtained results of non-carcinogenic health risk on children, based on metal concentrations in playgrounds and classrooms soils and exposure by three different pathways (ingestion, inhalation, and dermal) are shown in Table 5. When HQ and HI values are less than one, there is no obvious risk to the population, but if these values exceed one, there may be concern for potential non-carcinogenic effects [38]. The results showed that the major path of children exposure to playgrounds soils and classroom dust that may have adverse effect on their health by Pb, Cr, Cd, and Mn is ingestion, followed by dermal exposure (Fig. 2). Contribution of inhalation exposure to the HQ and HI is the smallest (Fig. 2). This finding supports other studies to establish the ingestion pathway as the major contributor to HI, particularly for children [6, 17, 28, 31].

Table 5. Noncarcinogenic risk for children in soil/dust from the classrooms and playgrounds.

		GROUP 1	GROUP 2	GROUP 3	GROUP 4
Pb	Intake(ingestion)	4.63E-04	3.96E-04	4.04E-04	3.18E-04
	Intake(inhalation)	1.29E-08	1.11E-08	1.13E-08	8.90E-09
	Intake(dermal)	1.30E-06	1.11E-06	1.13E-06	8.91E-07
	Total Intake	4.64E-04	3.97E-04	4.06E-04	3.19E-04
	HQ(ingestion)	1.32E-01	1.13E-01	1.16E-01	9.10E-02
	HQ(inhalation)	2.46E-05	2.11E-05	2.15E-05	1.69E-05
	HQ(dermal)	3.70E-04	3.16E-04	3.24E-04	2.55E-04
	Total HQ	1.33E-01	1.13E-01	1.16E-01	9.12E-02
	HI(ingestion)	1.12E-03	9.61E-04	9.82E-04	7.73E-04
	HI(inhalation)	1.03E-06	8.84E-07	9.04E-07	7.12E-07
	HI(dermal)	—	—	—	—
	Total HI	1.12E-03	9.62E-04	9.83E-04	7.74E-04
Cr	Intake(ingestion)	1.56E-04	8.00E-05	1.92E-04	2.07E-04
	Intake(inhalation)	4.37E-09	2.23E-09	5.36E-09	5.78E-09
	Intake(dermal)	4.37E-07	2.24E-07	5.37E-07	5.79E-07
	Total Intake	1.57E-04	8.02E-05	1.92E-04	2.07E-04
	HQ(ingestion)	5.21E-02	2.67E-02	6.39E-02	6.90E-02
	HQ(inhalation)	1.46E-06	7.45E-07	1.79E-06	1.93E-06
	HQ(dermal)	1.46E-02	7.46E-03	1.79E-02	1.93E-02
	Total HQ	6.67E-02	3.41E-02	8.18E-02	8.83E-02
	HI(ingestion)	2.60E-02	1.33E-02	3.20E-02	3.45E-02
	HI(inhalation)	5.97E-05	3.05E-05	7.32E-05	7.90E-05
	HI(dermal)	—	—	—	—
	Total HI	2.61E-02	1.34E-02	3.20E-02	3.46E-02
Cd	Intake(ingestion)	1.68E-05	2.50E-05	—	—
	Intake(inhalation)	4.70E-10	6.98E-10	—	—
	Intake(dermal)	4.71E-08	7.00E-08	—	—
	Total Intake	1.69E-05	2.51E-05	—	—
	HQ(ingestion)	1.68E-02	2.50E-02	—	—
	HQ(inhalation)	8.25E-06	1.22E-05	—	—
	HQ(dermal)	4.71E-05	7.00E-05	—	—
	Total HQ	1.69E-02	2.51E-02	—	—
	HI(ingestion)	1.06E-01	1.57E-01	—	—
	HI(inhalation)	5.20E-05	7.72E-05	—	—
	HI(dermal)	—	—	—	—
	Total HI	1.06E-01	1.57E-01	—	—

		GROUP 1	GROUP 2	GROUP 3	GROUP 4
Mn	Intake(ingestion)	—	1.93E-05	1.55E-05	—
	Intake(inhalation)	—	5.40E-10	4.34E-10	—
	Intake(dermal)	—	5.41E-08	4.34E-08	—
	Total Intake	—	1.94E-05	1.56E-05	—
	HQ(ingestion)	—	1.38E-04	1.11E-04	—
	HQ(inhalation)	—	—	—	—
	HQ(dermal)	—	3.87E-07	3.10E-07	—
	Total HQ	—	1.38E-04	1.11E-04	—

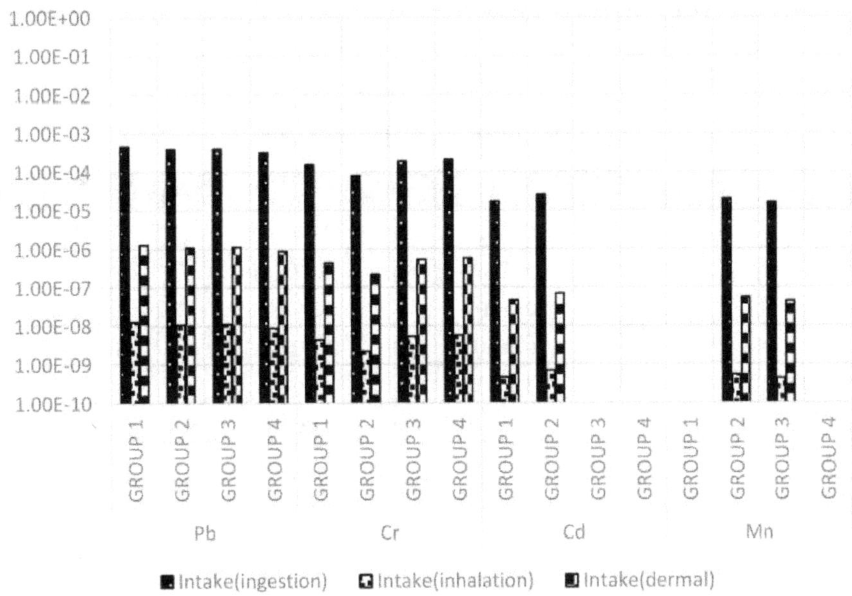

Fig. 2. Heavy metal intake (three pathways) values for children in selected school in Lagos State.

In this study, the ingestion had HI values less than 1 for all the observed heavy metals in all the groups. The summation of the total HI-value is approximately 0.4, which is less than one and ingestion pathway is a major contributor though there is no health effect (Table 5). Although, samples (classroom dusts and playground soils) were contaminated with Pb, Cd, and Cr but there is no obvious non-cancer health risk to the children. The high concentrations of indicated heavy metal pollution may pose a non-cancer health risk to children learning in such schools. This could be attributed to the short-term exposure duration (ED) of 6 years as recommended by USEPA health assessment model deployed in the study. The results also indicate that children via ingestion pathway could be exposed to potential possibility of non-carcinogenic risk followed by the inhalation pathway if considered at long-term exposure duration.

Literature shows that the daily ingestion rates of soil by children were calculated to vary between 39 and 270 mg/day [54]. Monitoring of Pb content in soil is of great necessity owing to the adverse effect on the children's central nervous system [55]. Numerous neurological and formative issue might be seen in children' populace because of a long time of exposure and ingestion of specific amount of Pb from contaminated soil; the potential wellbeing dangers incorporate iron deficiency, kidney harm, colic, muscle shortcoming, and cerebrum harm [56]. Ingestion of small Pb from dust may be harmful for blood, development, behavior, and intellectual functioning, as well [56]. Hence, children are more at risk particularly in both high- and low-density classrooms/playgrounds.

The excess lifetime cancer risks for children are calculated from the mean concentrations of individual heavy metals in classrooms and playgrounds for the pathways using Eqs. (4) and (5). Table 6 presented the calculated carcinogenic risk values for Pb, Cr, Cd, and Mn. Cr and Cd was found to be highest contributor to the cancer risk. The US Environmental Protection Agency considers adequate for administrative purposes a malignancy hazard point of confinement of 1×10^{-6} to 10^{-4} [32]. The summation of cancer risk for learners was found to be 3.2×10^{-2} (1 in 31 individuals), which is higher than the acceptable values. Be that as it may, ingestion pathway is by all accounts the significant contributor to overabundance lifetime cancer risk, then inhalation pathway. Tables 5 and 6 show the difference in the exposure duration, which results to the no obvious non-cancer risk (exposure duration of 6 years) and potential cancer risk (exposure duration of 70 years) of the learners in the selected schools. Therefore, there is a need to take necessary action to decontaminate the sites in order to protect children's health. This approach may be helpful in decision-making process in every growing city in the world like Lagos State.

Table 6. Carcinogenic (three exposure pathways) risk for children in soil and dust from the classrooms and playgrounds.

		GROUP 1	GROUP 2	GROUP 3	GROUP 4
Pb	Intake(ingestion)	3.96E-05	3.39E-05	3.47E-05	2.73E-05
	Intake(inhalation)	1.11E-09	9.47E-10	9.68E-10	7.63E-10
	Intake(dermal)	1.11E-07	9.49E-08	9.71E-08	7.64E-08
	Total Intake	**3.98E-05**	**3.40E-05**	**3.48E-05**	**2.74E-05**
	HQ(ingestion)	1.13E-02	9.69E-03	9.90E-03	7.80E-03
	HQ(inhalation)	2.11E-06	1.80E-06	1.84E-06	1.45E-06
	HQ(dermal)	3.17E-05	2.71E-05	2.77E-05	2.18E-05
	Total HQ	**1.14E-02**	**9.72E-03**	**9.93E-03**	**7.82E-03**
	HI(ingestion)	9.63E-05	8.23E-05	8.42E-05	6.63E-05
	HI(inhalation)	8.86E-08	7.58E-08	7.75E-08	6.10E-08
	HI(dermal)	—	—	—	—
	Total HI	**9.64E-05**	**8.24E-05**	**8.43E-05**	**6.63E-05**
Cr	Intake(ingestion)	1.34E-05	6.85E-06	1.64E-05	1.77E-05
	Intake(inhalation)	3.74E-10	1.91E-10	4.59E-10	4.95E-10
	Intake(dermal)	3.75E-08	1.92E-08	4.60E-08	4.96E-08
	Total intake	**1.34E-05**	**6.87E-06**	**1.65E-05**	**1.78E-05**
	HQ(ingestion)	4.46E-03	2.28E-03	5.48E-03	5.91E-03
	HQ(inhalation)	1.25E-07	6.38E-08	1.53E-07	1.65E-07
	HQ(dermal)	1.25E-03	6.40E-04	1.53E-03	1.65E-03
	Total HQ	**5.71E-03**	**2.92E-03**	**7.01E-03**	**7.57E-03**
	HI(ingestion)	2.23E-03	1.14E-03	2.74E-03	2.96E-03
	HI(inhalation)	5.11E-06	2.62E-06	6.28E-06	6.77E-06
	HI(dermal)	—	—	—	—
	Total HI	**2.24E-03**	**1.14E-03**	**2.75E-03**	**2.96E-03**

Cd	Intake(ingestion)	1.44E-06	2.14E-06	—	—
	Intake(inhalation)	4.03E-11	5.98E-11	—	—
	Intake(dermal)	4.04E-09	6.00E-09	—	—
	Total intake	**1.45E-06**	**2.15E-06**	**—**	**—**
	HQ(ingestion)	1.44E-03	2.14E-03	—	—
	HQ(inhalation)	7.07E-07	1.05E-06	—	—
	HQ(dermal)	4.04E-06	6.00E-06	—	—
	Total HQ	**1.45E-03**	**2.15E-03**	**—**	**—**
	HI(ingestion)	9.09E-03	1.35E-02	—	—
	HI(inhalation)	4.46E-06	6.61E-06	—	—
	HI(dermal)	—	—	—	—
	Total HI	**9.09E-03**	**1.35E-02**	**—**	**—**
Mn	Intake(ingestion)	—	1.66E-06	1.33E-06	—
	Intake(inhalation)	—	4.63E-11	3.72E-11	—
	Intake(dermal)	—	4.64E-09	3.72E-09	—
	Total intake	**—**	**1.66E-06**	**1.33E-06**	**—**
	HQ(ingestion)	—	1.18E-05	9.50E-06	—
	HQ(inhalation)	—	—	—	—
	HQ(dermal)	—	3.31E-08	2.66E-08	—
	Total HQ	**—**	**1.19E-05**	**9.53E-06**	**—**

4. CONCISE SUMMARY

An analysis was conducted to evaluate the concentrations of heavy metals (Pb, Cr, Cd, and Mn) in the playgrounds and classrooms of specific primary schools in Lagos State. The concentrations of metals found in the dust of classrooms and the dirt of playgrounds above the recommended allowable limits set by DPR, FEPA, and WHO, with the exception of Mn. The elevated levels of heavy metals may be attributed to the close proximity of schools to bus stops, auto workshops, local industry, gas stations, as well as paint chippings. The findings indicated that the primary route of children's exposure to soil in playgrounds and dust in classrooms is through ingestion, with cutaneous exposure being the subsequent route. The HI value suggests significant levels of heavy metal contamination, which could potentially provide no apparent non-cancerous health hazards to youngsters studying in these schools. This may be attributed to the relatively little duration of exposure, which spans only 6 years. The findings also suggest that the primary source of carcinogenic risk is children who are exposed through ingestion, followed by the breathing pathway due to a longer length of exposure spanning 70 years. The incidence of cancer among learners was determined to be 3.2 × 10-2 (1 in 31 individuals), beyond the permissible range of 1 × 10-6 to 10-4. The results will offer conclusive evidence required by local environmental authorities and school administrators to alert students about the potential health hazards posed by high levels of heavy metals found in dust in playgrounds and classrooms.

REFERENCES

1. Population-Lagos State. Lagos State Government. Archived from the original on 18 October 2015. Retrieved 21 December 2017
2. Nriagu JO. A silent epidemic of environmental metal poisoning. Environmental Pollution. 1988;50:139-161
3. Kreimer A. Environmental management and urban vulnerability. World Bank Discussion Paper 168. World Bank, Washington, DC; 1992

4. Thornton I. Environmental geochemistry and health in the (1990): A global perspective. Applied Geochemistry (Supplement). 1993;2:203-210

5. Tong ST, Lam KC. Home sweet home? A case study of household dust contamination in Hong Kong. Science of the Total Environment. 2000;256(2-3):115-123

6. Lane TW, Morel FM. A biological function for cadmium in marine diatoms. Proceedings of the National Academy of Sciences of the United States of America. 2009;9:462-431

7. Kabata-Pendias A. Trace Elements in Soil and Plants. 4th ed. Boca Raton, FL, USA: Taylor & Francis; 2011

8. Ouyang Y, Higman J, Thompson J, Toole OT, Campbell D. Characterization and spatial distribution of heavy metals in sediment from Cedar and Ortega Rivers sub-basin. Journal of Contaminant Hydrology. 2002;54:19-35

9. Knight C, Kaiser GC, Lailor H, Robothum JV. Witter, heavy metals in surface water and stream sediments in Jamaica. Environmental Geochemistry and Health. 1997;19:63-66

10. Loubieres Y, Lassence AD, Bernier M, Baron AV, Schmitt JM, Page B, Jardin F. Acute, fatal, oral chromic acid poisoning. Journal of Toxicology. Clinical Toxicology. 1999;37:333-336

11. Strachan S. Heavy metal. Current Anaesthesia and Critical Care. 2010;21:44-48

12. Dieter HH, Bayer TA, Multhaup G. Environmental copper and manganese in the pathophysiology of neurologic diseases (Alzheimer's disease and Manganism). Acta Hydrochimica et Hydrobiologica. 2005;33:72-78

13. Steenland K, Boffetta P. Lead and cancer in humans: Where are we now? American Journal of Industrial Medicine. 2000;38:295-299

14. Mortada WI, Sobh MA, El-Defrawy MM, Farahat SE. Study of lead exposure from automobile exhaust as a risk for nephrotoxicity among traffic policemen. American Journal of Nephrology. 2001;21:274-279

15. Jarup L. Hazards of heavy metal contamination. British Medical Bulletin. 2003;68:167-182

16. Ogwuegbu MOC, Muhanga W. Investigation of lead concentration in the blood of people in the copper belt province of Zambia. Journal of Environment. 2005;1:66-75

17. Banerjee ADK. Heavy metal levels and solid phase speciation in street dusts of Delhi, India. Environmental Pollution. 2003;123:95-105

18. Davydova S. Heavy metals as toxicants in big cities. Microchemical Journal. 2005;79(1-2):133-316

19. Bhargava AK, Gupta R, Bhargava S, Paridhi. Effect of automobile exhaust on total N, P and heavy metals of road side sugarcane at district Saharanpur. Advances in Plant Sciences. 2003;16:557-560

20. Ahmed F, Ishiga H. Trace metal concentration in street dusts of Dhaka City, Bangladesh. Atmospheric Environment. 2006;40:3835-3844

21. Lu X, Wang L, Lei K, Huaing J, Zhai Y. Contamination assessment of copper, lead, zinc, manganese and nickel in street dust of Boaji, N.W china. Journal of Hazardous Materials. 2009;161:1058-1062

22. De Miguel E, Llamas JF, Chacon E, Berg T, Larssen S, R›yset O, Vadset M. Origin and patterns of distribution of trace elements in street dust: Unloaded petrol and urban lead. Atmospheric Environment. 1997;31:2733-2740

23. Turer D. Effect of non-vehicular sources on heavy metal concentrations of roadside soils. Water, Air, and Soil Pollution. 2005;166:251-264

24. Shinggu DY, Ogugbuaja VO, Toma I, Barminas JT. Determination of heavy metals in street duct in Yola, Nigeria. African Journal of Pure and Applied Chemistry. 2010;4(1):17-21

25. Abrahams PW. Soils: Their implications to human health. Science of the Total Environment. 2002;291(1-3):1-32

26. Poggio L, Vrscaj B, Schulin R, Hepperle E, Ajmone Marsan F. Metals pollution and human bioaccessibility of topsoils in Grugliasco (Italy). Environmental Pollution. 2009;157(2):680-689

27. Ljung K, Selinus O, Otabbong E. Metals in soils of children's urban environments in the small northern European city of Uppsala. Science of the Total Environment. 2006;366(2-3):749-759

28. Acosta JA, Cano AF, Arocena JM, Debela F, Martínez-Martínez S. Distribution of metals in soil particle size fractions and its implication to risk assessment of playground in Murcia City (Spain). Geoderma. 2009;149(1-2):101-109

29. Popoola OE, Bamgbose O, Okonkwo OJ, Arowolo TA, Odukoya AT, Popoola AO. Heavy metals content in playground topsoil of some public primary schools in metropolitan Lagos, Nigeria. Research Journal of Environmental and Earth Sciences. 2012;4(4):434-439

30. Okereke CJ, Amadi PU. Accumulation and risk assessment of heavy metal contents in school playgrounds in Port Harcourt Metropolis, Rivers State, Nigeria. Journal of Chemical Health and Safety. 2017. http://dx.doi.org/10.1016/j.jchas.2017.01.002

31. USEPA. Risk-Based Concentration Table. Philadelphia PA. Washington DC, USA: United States Environmental Protection Agency; 2000

32. Dooyema CA, Neri A, Lo YC, Durant J, Dargan PI, Swarthout T, Biya O, Gidado SO, Haladu S, Sani-Gwarzo N, Nguku PM, Akpan H, Idris S, Bashir AM, Brown MJ. Outbreak of fatal childhood lead poisoning related to artisanal gold mining in Northwestern Nigeria, 2010. Environmental Health Perspectives. 2012;120(4):601-607

33. Su C, Jiang L, Zhang W. A review of heavy metal contamination in the soil worldwide: Situation impact and remediation techniques. Environmental Skeptics and Critics. 2014;3(2):24-38

34. Wang Z, Chai L, Yang Z, Wang Y, Wang H. Identifying sources and assessing potential risk of heavy metals in soils from direct exposure to children in a mine-impacted city, Changsha, China. Journal of Environmental Quality. 2010;39(5):1616-1623

35. Davidson CM, Duncan AL, Littlejohn D, Ure AM, Garden LM. Critical evaluation of the three-stage BCR sequential extraction procedure to assess the potential mobility and toxicity of heavy metals in industrially-contaminated land. Analytica Chimica Acta. 1998;363:45-55

36. US EPA (United States Environmental Protection Agency). Risk Assessment Guidance for Superfund. Human Health Evaluation Manual. EPA/540/1-89/002, Vol. I. Office of Solid Waste and Emergency Response. 1989. Available from: http://www.epa.gov/superfund/programs/risk/ragsa/index.htm

37. US DoE. The Risk Assessment Information System (RAIS). Argonne, IL: U.S. Department of Energy's Oak Ridge Operations Office (ORO); 2011

38. Department of Environmental Affairs. The Framework for the Management of Contaminated Land, South Africa. 2010. Available from: http://sawic.environment.gov.za/documents/562.pdf [Accessed: 22 December 2017]

39. Zheng N, Liu J, Wang Q, Liang Z. Health risk assessment of heavy metal exposure to street dust in the zinc smelting district, northeast of China. Sci Total Environ 2010;408(4):726-33

40. World Health Organization. Growth reference data for 5-19 years. Geneva: World Health Organization; 2000. Available from: http://www.who.int/growthref/en

41. US EPA (United States Environmental Protection Agency). Supplemental Guidance for Developing Soil Screening Levels for Super Fund Sites. OSWER 9355.4-24.2001. Office of Solid Waste and Emergency Response. 2001. Available from: http://www.epa.gov/superfund/resources/soil/ssgmarch 01.pdf

42. Calabrese EJ, Kostecki PT, Gilbert CE. How much dirt do children eat? An emerging environmental health question. Comments Toxicol. 1987;1:229-41

43. US EPA (United States Environmental Protection Agency). Exposure Factors Handbook. EPA/600/P-95/002F. Washington, DC: Environmental Protection Agency, Office of Research and Development; 1997

44. Luo XS, Ding J, Xu B. Incorporating bioaccessibility into human health risk assessments of heavy metals in urban park soils. Science of the Total Environment. 2012;424:88-96

45. U.S. Environmental Protection Agency. Framework for Determining a Mutagenic Mode of Action for Carcinogenicity: Review Draft. 2007. Available online: http: //www.epa.gov/osa/mmoaframework/pdfs/MMOA-ERD-FINAL-83007.pdf (accessed on 20 December 2017)

46. Wright NJ, Thacher TD, Pfitzner MA, Fischer PR, Pettifor JM. Causes of lead toxicity in a Nigerian city. Archives of Disease in Childhood. 2005;90:262-626

47. Adebamowo EO, Agbede OA, Sridhar MKC, Adebamowo CA. An evaluation of lead levels in residential paints sold in Nigeria markets. Indoor and Built Environment. 2006;15:551-554

48. Yusuf AA, Arowolo TA, Bamgbose O. Cadmium, copper and nickel levels in vegetables from industrial and residential areas of Lagos. City, Nigeria. Food and Chemical Toxicology. 2003;41:375-378

49. Tuzen M. Determination of heavy metals in soil, mushroom and plant samples by atomic absorption spectrometry. Microchemical Journal. 2003;74:289-297

50. DPR. Environmental Guidelines and Standards of the Petroleum Industry in Nigeria. UK: Ministry of Petroleum Resources Lagos; 1991. pp. 35-75

51. FEPA. Guideline and Standard for Environmental Pollution Control in Nigeria. Federal Republic of Nigeria; 1991. pp. 61-63

52. WHO/WMO. Air Monitoring Programme Designed for Urban and Industrial Area Published for Global Environmental Monitoring System by UNEP. Geneva: WHO and WMO; 1971

53. Onianwa PC, Jaiyeola OM, Egekenze RN. Heavy metals contamination of topsoil in the vicinities of auto-repair workshops, gas stations and motor-parks in a Nigerian City. Geneva, Switzerland: Toxicological and Environmental Chemistry. 2003;84(1):33-39

54. Ljung K, Oomen A, Duits M, Selinus O, Berglund M. Bioaccessibility of metals in urban playground soils. Journal of Environmental Science and Health. 2007;42(9):1241-1250

55. Cicchella D, de Vivo B, Lima A, Albanese S, Mcgill RA, Parrish RR. Heavy metal pollution and Pb isotopes in urban soils of Napoli, Italy. Geochemistry: Exploration, Environment, Analysis. 2008;8(1):103-112

56. Osman K. Health effects of environmental lead exposure in children [Dissertation]. Stockholm: Karolinska Institute, Institute Environmental Medicine; 1998. (Swedish)

19 Chapter

ASSESSMENT AND MANAGEMENT OF HEALTH RISKS ASSOCIATED WITH HEAVY METAL CONTAMINATION IN URBAN SOIL

1. INTRODUCTION: AN OVERVIEW

World Urbanization Prospects 2014 Revision published by United Nations Population Division showed that from 1950 to 2014, the global urban population increased from 746 million to 3.9 billion, accounting for 54% of the world's total population [1]. Despite the low urbanization rate in Asia, it is still the most populous city in the world due to its large population base, accounting for 53% of the total urban population in the world [2]. Urban is the place where human activities are highly concentrated. The population in urban is dense and unevenly distributed, which consumes a large amount of natural resources and produces a large amount of pollutants [3]. When the pollutants exceed the load of the urban environment, the urban environment would be polluted or even destroyed. With rapid urbanization and industrialization in developing country, relatively short-term and intensive human activities will bring a large number of organic pollutants (such as polycyclic aromatic hydrocarbons and phenols) and inorganic pollutants (such as heavy metals, arsenic compounds, and so on) into the urban environment. These pollutants will be accumulated in the urban environment, and may migrate or transform directly through the soil, water, atmosphere, and other environmental multi-media, directly or indirectly causing risk to human health and ecological security in urban [4, 5, 6]. Urban soil is an important part of urban ecological environment, and it is closely related to urban water, atmosphere, groundwater, and other environmental factors [7]. Urban soil is located in the central position of urban ecological environment, which is not only the sink of various pollutants, but also the source of various pollutants; so it is often used as an important indicator of comprehensive environmental quality of urban [8, 9]. Heavy metals are a kind of toxic pollutants, which are difficult to degrade in the environment. Heavy metal pollution in urban soils is difficult to control because of its wide sources, many forms, and complex migration [10]. The three wastes, traffic exhaust, sewage irrigation, improper fertilizer, and pesticide application have become the main sources of heavy metals in urban soils [4, 8, 11]. The land use type, the disturbance time, and degree of human activities are the important factors affecting heavy metal pollution in urban soil [8, 12]. In recent years, with the acceleration of urbanization and the occurrence of environmental public hazards related to heavy metal pollution in urban, the research, and practice of the characteristics, mechanism, assessment, and management of heavy metal pollution in urban soil have become major issues that all countries must be faced with. The research results will be of great significance to the sustainable development of urban economy and society, the maintenance of regional environmental equity, and the construction of ecological civilization in urban.

The Chinese National "12th five-year plan" for Environmental Protection and National Environment and Health Action Plans (2007–2015) pointed out that "With the rapid development of industrialization and urbanization in China, problems of environmental pollution affecting people›s health and safety have become prominent. Protecting environment and ensuring human health have become the most pressing need of the people". Urban residents may directly or indirectly be exposed to heavy metals in urban soil through the ingestion of soil, skin contact, and inhalation, causing short-term harm or long-term risks [13]. By 2050, the global urban population will increase by 2.5 billion, and the proportion of global urban population will reach 66%, nearly 90% of which is concentrated in Asia and Africa. Countries with the largest increase in urban population will be India, China, and Nigeria in the future. From 2014 to 2050, these three countries will increase by 404, 292, and 212 million of the urban population, totally accounting for 37% of the global new urban population. In developing countries such as China, its urbanization rate has reached 53.73% and the total exceeding standard rate of soil was 16.1% [14]. The research on the relationship between heavy metals in urban soil and urban population health is in line with the actual needs of the country and people. According to the soil census data of 2005–2013 in China, the spatial distribution of the total noncarcinogenic risk factors of heavy metals (including cadmium, mercury, lead, chromium, arsenic, copper, zinc, and nickel) in three pathways of oral intake, skin contact, and respiratory inhalation was obtained, which can be seen from Fig. 1. For children, 45% of 29 provinces, autonomous regions and municipalities in the mainland of China have a moderate potential noncarcinogenic risk, which did not exceed the acceptable noncarcinogenic risk limits [15]. Moreover, research and practice have shown that the current soil environmental management and control system based on Soil Environmental Quality Standard (GB15618-1995) has been unable to adapt to changes of the current complex environment and requirements of "green" socio-economic development. In 2016, the State Council promulgated Action Plan for Soil Pollution Prevention and Control. It indicates that the health risk of urban people exposure to heavy metal pollution in urban soil cannot be ignored.

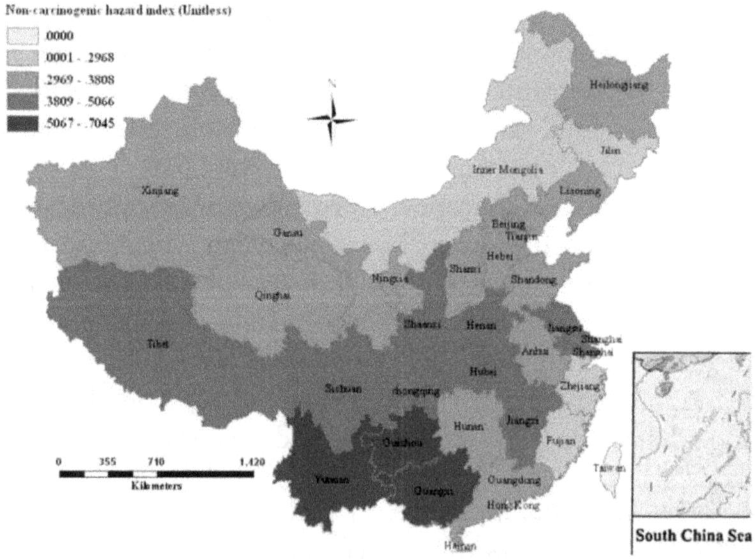

Fig. 1. Spatial distribution of total noncarcinogenic risk factors of heavy metals [15].

Therefore, hopes are placed on the health risk assessment and management technology, which has been vigorously promoted and developed by the United States Environmental Protection Agency (USEPA) and the International Program on Chemical Safety (IPCS) in the past 20 years. A mature framework for health risk assessment and management of contaminated sites has been developed both at home and abroad. However, there is a big difference in land area, function, and characteristics between soil in contaminated sites and urban soil; mechanism of urban soil heavy metal pollution is more complex and involved more factors, so there is still no formation of a scientific and efficient urban soil environmental health risk assessment and management system in China. The World Bank Overview of the Current Situation on Brownfield Remediation and Redevelopment in China points out that at present, China has a large area of Brownfield and a limited amount of restoration funds, and a highly efficient and prioritized Brownfield risk rating system should be established according to the actual situation on Brownfield in China. Considering the latest researches on soil science and the significant differences in land use status, receptor spatial distribution, exposure characteristics, risk assessment and management, and economic input degree between different countries, the key technologies of health risk assessment, risk source analysis and risk quantification management of heavy metal in soil should be further improved, integrated, and developed.

Overall, the scientific and rational assessment and management of heavy metal contamination in urban soil have significant scientific, practical, and strategic importance in safeguarding the health of urban populations. It aids in the process of making decisions regarding regional soil remediation, setting regional soil standard values and action values, and developing and enhancing applicable laws and regulations. Starting from the scientific health risk assessment and China's national context, it is necessary to explore the development of a health risk assessment and management system for environmental pollutants in urban soil that is tailored to Chinese characteristics and easy to implement. In 2016, the State Council issued the Action Plan for Soil Pollution Prevention and Control. As a result, the market size for assessing health risks and remediation of pollutants in urban soil in China is expected to exceed 1000 billion Chinese yuan. Undoubtedly, this study has enormous social and economic prospects, as it serves as a significant catalyst for promoting the growth of linked research projects.

2. HEAVY METAL POLLUTION IN URBAN SOILS

2.1. Introduction of soil pollution

2.1.1. Characteristics of soil pollution

Soil is an ancient, naturally occurring entity. The loose and heterogeneous aggregate layer with vitality and production is found on the surface of the Earth and at the bottom of the phreatic area. It is an integral part of the earth system and a crucial element in regulating environmental quality [12]. The soil environment is an intricate system that consists of solid, liquid, and gas phases, with its components undergoing significant changes. Furthermore, the interplay between the soil environment system and the plant system that thrives in soil forms the soil-plant ecosystem [12, 16]. Soil pollution has become prevalent as a result of industrial development, primarily caused by human activities. Toxic substances or preparations are deliberately or accidentally introduced into the soil, leading to an increase in the concentration of a new component or a significant increase in the concentration of an existing component. This ultimately leads to the degradation of the soil environment, both currently and potentially.

According to soil science, soil pollution has three types of characteristics [12, 16, 17]:

- Concealed and latent. Water or air pollution is intuitive. It could be easily found by people when water or air pollution is heavy. On the contrary, soil pollution has characters of concealment and latency. It commonly reflected the growth status of crops including grain, vegetables, fruits, and grass and health status of related human or animals, and there is a gradual cumulative process from happening pollution to causing harm.

- Irreversible and long-term effects. Once the soil is contaminated, it is difficult to recover. The process of soil pollution by heavy metals is irreversible.
- Consequences are serious. Once the soil is polluted, biodiversity, biological cycles, and water cycles (including water quality and water cycle processes) are bound to be affected accordingly. Besides, because of the concealment and irreversibility of soil pollution, health of animals and human beings often suffer from soil pollution via food chain.

2.1.2. Sources of soil pollution

- Soil pollution sources can be divided into natural sources and anthropogenic sources. Natural sources of soil pollution relate to natural disasters such as volcano eruption that will produce harmful substances. These harmful substances will be discharged into the natural environment and result in soil pollution. Anthropogenic sources refer to the pollution produced by anthropogenic activities such as sewage irrigation, utilization of solid waste, use of pesticides and chemical fertilizers, atmospheric deposition and so on, which are the main objects of scientific research.
- Sewage irrigation. Domestic sewage and industrial wastewater contain nutrient elements that plants growth needed, such as nitrogen, phosphorus and potassium. Therefore, using sewage probably to irrigate farmland can gradually increase production. However, the sewage also contains heavy metals, phenols, cyanide, and many other toxic and harmful substances. If the sewage is used for agricultural irrigation directly without pretreatment, toxic and harmful substances in the sewage will enter into the farmland and pollute the soil.
- Utilization of solid waste. Industrial waste and municipal solid waste are solid pollutants in soil. For example, a variety of agricultural plastic film is widely used as in greenhouse and plastic mulching. If the management and recovery are not good, a large number of pieces of plastic film will scatter in the field and cause white pollution. Such solid pollutants are not easily evaporated, volatilized, or decomposed by microorganisms, which will remain in the soil for a long time.
- Use of pesticides and chemical fertilizers. The application of chemical fertilizer is an important measure to increase agricultural yields, but unreasonable use of chemical fertilizer will result in soil pollution. Long-term use of nitrogen fertilizer will destroy soil structure, result in soil compaction, deteriorate biological properties, and finally affect the yields and quality of crops. Part of the pesticide sprayed on the object absorbed by plants escapes into the atmosphere. About half of pesticide for object spraying are scattered on farmland, and this part of pesticide and pesticide directly applied to the farmland constitute the basic sources of pesticides in soil.
- Atmospheric deposition. The harmful gases in the atmosphere are mainly poisonous waste gases discharged from industry. Its pollution surface is big, which will cause serious pollution to the soil. Industrial waste gas pollutions include gas pollution and aerosol pollution. Gas pollution is usually caused by sulfur dioxide, fluoride, ozone, nitrogen oxides, and hydrocarbons. The reason for causing aerosol pollution is that solid particles (such as dust, soot, and smoke) and liquid particles (such as mist) enter into soil via deposition and precipitation.

2.1.3. Types of soil pollution

- There is no strict classification of soil pollution. Considering from the material properties, soil pollution can be divided into organic pollution, inorganic pollution, microbial pollution, and radioactive pollution.

- **Organic pollution.** Natural organic pollutants and synthetic organic pollutants can cause organic pollution. Synthetic organic pollutants include organic wastes and pesticides. Organic pollutants in soil have adverse effects on the growth of crops and the survival of soil organisms. After people contacted with polluted soil, red rash appears on their hands with the feeling of nausea and dizziness. The application of pesticide in agricultural production has received good results, but its residues have contaminated the soil and corresponding food chain.
- **Inorganic pollution.** Some inorganic pollutants come into the soil with the natural processes such as crustal change, volcanic eruption and rock weathering, while some enter into the soil with the human production and consumption activities. Human production activities such as mining, smelting, machinery manufacturing, building, and chemical producing produce a great number of inorganic pollutants including harmful heavy metals and their compounds, acids, alkalis, and salts.
- **Microbial pollution.** Soil biological pollution means that one or more harmful microbial populations invade the soil from the outside, and multiply in a sharply speed, destroy the original ecological balance, and cause adverse effects on soil ecosystem and human health. The main sources of soil biological pollution are untreated feces, garbage, municipal sewage, and waste of feed yard and slaughterhouse. Soil organisms may do harm to human health, and some plant pathogens that survive in the soil for a long time will seriously harm plants, and resulting in agricultural yield reduction.
- **Radioactive pollution.** Radioactive pollution refers to that radioactive materials emitted by human activities entering into soil makes the radioactivity of soil become higher than its background level. Radioactive pollutants refer to various radionuclides, radioactivity of which has nothing with their chemical states. Radionuclides can pollute soils in many ways. Radioactive wastewater to the ground, radioactive solid waste buried in the underground, and radioactive emission accident in nuclear power plant will cause serious soil pollution in some areas. Deposition of radioactive substances in the atmosphere, application of phosphate fertilizer containing uranium, radium and other radionuclides, and irrigation of water with radioactive river water will also cause soil radioactive pollution. Although the pollution is generally mild, its scope is relatively large. Polluted by radioactive materials, soil can produce alpha, beta, and gamma rays by radioactive decay. These rays can penetrate human tissues, damage cells, cause external radiation damage, and enter the human body through respiratory system or food chain, causing internal radiation damage.

2.2. Overview of urban soil

2.2.1. Definition of urban soil

Bockheim, the originator of the term "urban soil," posited that it refers to soil found in urban or suburban areas that is approximately 50 cm thick and is a result of human activities such as mixing, landfilling, or land pollution [18]. De Kimpe and Morel [19] define urban soil as the soil found in urban and suburban areas that has been significantly disrupted by human activities. According to Chinese researchers, the prevailing understanding of urban soil is that it forms as a result of prolonged disruption or direct intervention by human activities, within the specific environmental conditions of urban areas [20]. Urban dirt possesses both the inherent qualities of natural soil and distinct soil environment and formation process, distinguishing it from both natural and agricultural soil. Urban soil possesses distinct physical and chemical qualities, nutrient cycling, and soil biological traits. Urban soil is prevalent in residential neighborhoods, urban waterways, roadways, parks, suburban regions, stadiums, dump sites, abandoned factories, and other similar locations. Considering that heavy metals in urban soil across all types of land use are expected to come into contact with adjacent populations and pose health hazards, this study defines urban soil as encompassing urban soil, forest soil, agricultural soil, and soil that may be contaminated [21].

2.2.2. Current situation of urban soil pollution

The earliest cases of heavy metal pollution in foreign was the copper poisoning caused by drainage from the Ashio Copper Mine in Tochigi prefecture, beginning as early as 1878. It was found that the extremely painful itai-itai disease and Minamata disease were caused by mercury and cadmium pollution. Heavy metal pollution caused by the food chain has begun to be much concerned. In the early 1970s, British have carried out studies on heavy metals in soil and dust in London and other big city. It was found that heavy metal pollution in urban soil is closely related to industrial activities and emission of automobile exhaust, and heavy metals in urban surface soil and road dust can be used as indicators of urban air pollution. In 1990s, urban soil research has gradually become new hot topic in the field of soil research all over the world. Heavy metal pollution in urban soil in cities such as major cities in British, major cities in America, Naples and Sicily in Italy, Seville in Spain, Munich in Germany, Hanoi in Vietnam, and Bangkok in Thailand have been studied by many a scholar. Pb, Zn, and Hg were the main pollution factors in Sicily City of Italy, and the enrichment coefficient of Pb was 5–10, and Hg was 35. The correlation between soil heavy metal content and soil properties was determined, and the heavy metal pollution's relationships with urban scale, population density, traffic flow, land use types, and history were discussed. Therefore, the International Union of Soil Sciences established the Soils in Urban and Industrial, Traffic and Mining Areas (SUITMA) working group. Up to now, three international academic conferences had been held, of which heavy metal pollution in urban soils was the main topic.

The National Soil Survey Report was produced in 2014 by the Ministry of Land and Resources and the Ministry of Environmental Protection in a collaborative effort. The findings indicated that the soil conditions across the entire nation were unfavorable, with several regions experiencing severe soil pollution. The soil quality of agricultural land and the soil problems resulting from abandoned mining activity were significant and deserving of careful attention [14]. In China, the total rate of soil sample sites that exceeded standard values was 16.1%. The percentage of slight pollution, mild pollution, moderate pollution, and severe pollution were 11.2%, 2.3%, 1.5%, and 1.1% correspondingly, when compared to the standard values. Among the sampling sites that exceeded the pollution threshold, nonorganic pollution was the most prevalent type (82.8%), followed by organic pollution, and the least common was compound pollution. The issue of heavy metal contamination in urban soil in China is prominent. Liao et al. [22] conducted a study on the distribution features and pollution sources of heavy metals in soils in Jiangsu Province. An analysis revealed that the levels of Cd and Hg in soil in the southern region of Jiangsu were generally elevated, but the soil in the northern region of Jiangsu had a high concentration of As. The levels of heavy metal pollution in the soil in Suzhou and Wuxi were relatively severe, primarily as a result of human activities throughout the industrialization and urbanization process. Guo et al. [23] conducted a study on the features of heavy metal pollution in soil in several functional zones of Hohhot. The results revealed that the average concentrations of Cu and Zn were 2.33 times and 1.85 times higher than their respective background values. The soil heavy metal pollution near the commercial district and urban roadways is very severe, primarily resulting from traffic pollution and the disposal of household waste. The urban population experiences a higher level of intensity, with heavy metals in the soil readily entering the human body through multiple exposure paths, either directly or indirectly. Heavy metals have the ability to form strong interactions with proteins and enzymes, leading to a loss of activity in these biomolecules. Heavy metals can also accumulate in human organs. When the amount of heavy metals in the human body above the tolerable limit, it can result in significant injury or danger. This can manifest as acute poisoning, subacute poisoning, chronic poisoning, and other related conditions [24, 25].

2.3. Assessment methods of heavy metal pollution

2.3.1. Geo-accumulation index

To assess the enrichment degree of heavy metals in the soils, the geo-accumulation index (Igeo) was used. Igeo can be calculated by the following formula [26, 27, 28, 29]:

$$I_{geo} = log_2(k\, C_i / B_i) \qquad (1)$$

where Ci is the actually measured concentration of the heavy metal in the samples. k is corrected coefficient, which take account of variation of background value caused by anthropogenic influences or litho logic variations in the soil (in general k = 1.5). Bi is the reference value of heavy metal concentration in soil. Igeo is classified into seven levels and its corresponding contamination degrees of heavy metal are shown in Table 1.

Table 1. Seven levels and its corresponding contamination degrees.

Level	Value	Extent of pollution
I	$I_{geo} \leq 0$	Uncontaminated
II	$0 < I_{geo} \leq 1$	Uncontaminated to moderately contaminated
III	$1 < I_{geo} \leq 2$	Moderately contaminated
IV	$2 < I_{geo} \leq 3$	Moderately contaminated to heavily contaminated
V	$3 < I_{geo} \leq 4$	Heavily contaminated
VI	$4 < I_{geo} \leq 5$	Heavily to extremely contaminated
VII	$I_{geo} > 5$	Extremely contaminated

2.3.2. Potential ecological index

In order to evaluate the potential ecological risk caused by heavy metals in soil, the potential ecological risk index (PEI) was utilized. The PEI was established by Hanson to assess the characteristics of heavy metal contaminants on the basis of sedimentary theory [30, 31].

$$RI = \sum_{i=1}^{n} E_r^i,\ E_r^i = T_r^i \cdot C_r^i = C_D^i \cdot C_B^i \qquad (2)$$

where RI is the sum of the potential risk of individual heavy metal, Eir is the potential risk of individual heavy metal, Tir is the toxic-response factor for a given heavy metal, value of Tir for Cd, Ni, Zn, Cu, and Cr are 30, 5, 1, 5, and 2, respectively [30, 32, 33]. Cir is the contamination factor, CiD is the present concentration of heavy metals in soil, and CiB is the reference value of heavy metal concentration in soil. Hanson defined five categories of Eir, and four categories of RI, as listed in Table 2.

Table 2. Indices and corresponding grades of potential eco-risk of heavy metals.

I	$I^{geo} \leq 0$	Uncontaminated
II	$0 < ^{Igeo} \leq 1$	Uncontaminated to moderately contaminated
III	$1 < ^{Igeo} \leq 2$	Moderately contaminated
IV	$2 < ^{Igeo} \leq 3$	Moderately contaminated to heavily contaminated
V	$3 < ^{Igeo} \leq 4$	Heavily contaminated
VI	$4 < ^{Igeo} \leq 5$	Heavily to extremely contaminated
VII	$I^{geo} > 5$	Extremely contaminated

3. HEALTH RISK ASSESSMENT OF HEAVY METALS IN URBAN SOILS

3.1. Health risk assessment

The current soil environmental management system has some ability to mitigate soil heavy metal pollution. However, it is evident that there is a lack of attention for the connection between soil heavy metal contamination and the health of urban populations. The absence of a comprehensive collection of interconnected laws and regulations results in a low level of effectiveness in terms of quantitative control. Over the past two decades, there has been rapid development in health risk assessment management, primarily due to the efforts of influential organizations such as the Organization for Economic Co-operation and Development (OECD), World Health Organization (WHO), United States Environmental Protection Agency (USEPA), International Program on Chemical Safety (IPCS), and European Center for Ecotoxicology and Toxicology of Chemicals (ECETOC) [13]. The European Union, the United States, Canada, and Japan have enacted pertinent legislation and regulations, making health risk assessment a crucial foundation and scientific benchmark at both national and international levels for addressing the issue of chemical pollution risk management [34]. It offers crucial sources for the development of a health risk assessment and management system for urban soil. Health risk pertains to the potential for individuals to be exposed to environmental contaminants that can result in harm, illness, or fatality. In 1983, the United States National Academy of Sciences (NAS) released a publication called "Risk Assessment in the Federal Government: Managing the Process," also referred to as the "Red Book." This publication principally introduced the definition of health risk assessment. It is a precise delineation of the possible harmful health consequences resulting from human exposure to environmental dangers. Health risk assessment comprises various components [13, 35, 36]: an evaluation of potential detrimental health effects using epidemiological, clinical, toxicological, and environmental studies, extrapolation of assessment findings to anticipate and estimate the nature and magnitude of human health effects under specific exposure conditions, identification of the number and characteristics of individuals exposed to varying intensities and durations, and a summary of the presence and overall magnitude of public health concerns. Consequently, the process of risk assessment consists of four distinct steps: identifying hazards, assessing the response to different doses, evaluating exposure levels, and finally, characterizing the overall risk. Afterwards, risk management is conducted in accordance with the findings of the risk assessment. Fig. 2 illustrates the fundamental components of health risk assessment and management.

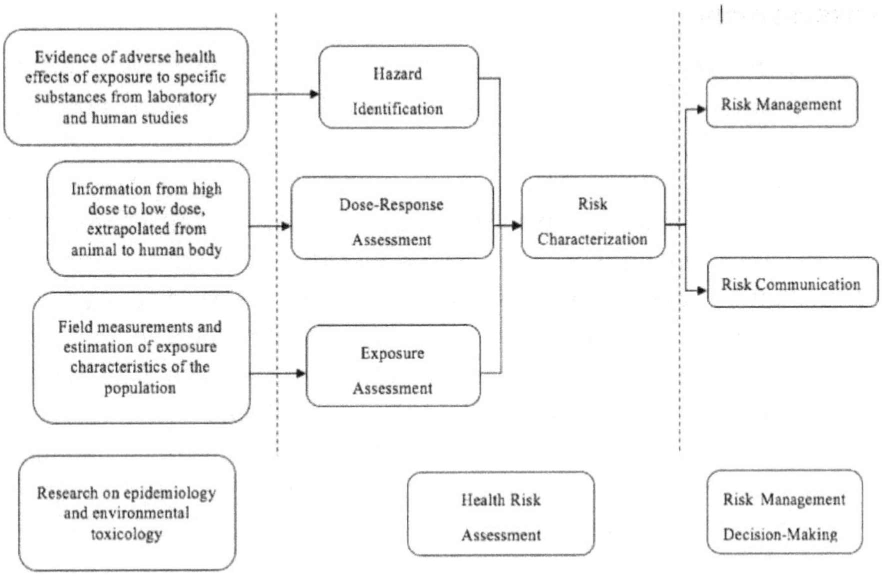

Fig. 2. Risk assessment and management of environmental health.

3.1.1. Hazard identification

Hazard identification is the initial phase of a human health risk assessment, with the objective of determining the nature and magnitude of risk factors. According to the NAS, hazard identification is the procedure of assessing whether being exposed to a stressor can lead to a higher occurrence of particular negative health impacts, such as cancer or birth defects. Hazards can be classified into three categories: physical, chemical, and biological. When a person is exposed to a sufficient amount of these hazards, it can result in injury, disease, or death. Scientists evaluate the impact of detrimental compounds on the health of humans and laboratory animals. They aim to comprehend the properties of the drug, determine its possible health risks, and provide clear evidence of its considerable toxicity, either through statistical or biological means [13, 37]. Conducting research on humans who are directly exposed to hazardous substances can offer the most reliable evidence. However, due to limited availability of human clinical trial data, scientists often rely on animal experiments, in vitro experiments, and analysis of chemical structure relationships to estimate the potential health risks associated with human exposure to chemicals.

3.1.2. Dose-response assessment

Dose-response assessment is a quantitative estimating procedure that examines the link between the quantity of exposure to harmful variables and the occurrence of health hazards in people. It is a crucial step in quantifying health risk assessment [13, 37]. Dose-response evaluation is a fundamental aspect of toxicology research, guided by the principle that "Dose makes the poison". For instance, the dosage level in clinical settings plays a crucial role in determining whether a medicine is therapeutic or possibly fatal. The dose-response relationship can be broadly classified into two distinct groups. The first factor to consider is the correlation between the dosage of exposure to harmful contaminants and the biological intensity response of an individual. The second one refers to the correlation between the dosage of harmful pollutants and the percentage of individuals who exhibit a specific response within a group, such as death and tumor occurrence. The toxicity of each chemical agent is characterized by a unique dose-response relationship.

3.1.3. Exposure assessment

Exposure assessment involves the use of quantitative or qualitative techniques to determine or compute the amount of exposure, the frequency of exposure, the duration of exposure, and the route of exposure [38]. The primary routes of exposure include ingestion, dermal contact, and inhalation. Nevertheless, it is important to emphasize that regardless of the level of danger posed by the pollutant, there is no health risk if there is no exposure to it. Exposure assessment typically necessitates the consideration of several exposure elements, such as body height, weight, breathing rate, exposure frequency, and other relevant characteristics. Additionally, it needs the consideration of factors associated to the transportation of pollutants in the environmental media, taking into account the specific local environmental circumstances. A national exposure factors database has been produced by the United States, Japan, South Korea, and other countries. Given China's expansive land area, substantial population, and significant variations in lifestyle, it is imperative to gradually develop our own database of exposure factors. In 2013, the Ministry of Environmental Protection released the Exposure Factors Handbook of Chinese Population (adults), which addressed the deficiencies in China's national exposure factor manual. Nevertheless, it is evident that the comprehensive national exposure factor database is still in the phase of gathering data, which introduces some degree of ambiguity to the health risk assessment in China [39].

3.1.4. Risk characterization

As the last step of risk assessment, risk characterization based on the data and analysis results of the three steps mentioned above. Finally, the probability of the occurrence of harmful results, the acceptable risk level and the uncertainty of the evaluation results are quantified [13]. Risk characterization is a bridge linking risk assessment and risk management, and risk managers can use risk characterization results to formulate strategies of risk control and pollution remediation. Risk communicators can use risk characterization conclusions to inform the types, sizes, and possibilities of adverse health effects among stakeholders, and disseminate knowledge about risk prevention.

3.1.5. Risk management

Health risk assessment requires collaboration among multiple scientific teams. Similarly, risk management is also a multidisciplinary process. Risk management is a decision-making process base on health risk assessment. It is necessary to balance the political, social, economic and technical information and other related information, to develop, analyze and compare management methods, and ultimately choose appropriate management methods for health risks. That is, to reduce the risk to acceptable levels with acceptable costs [13].

3.2. Researches and issues on health risk assessment and management

3.2.1. Hazard identification

With the development of industrial technology and social economy, the types of materials that people produce and discharge into the environment are increasing rapidly, and the amount of harmful substances that can be analyzed and detected is also increasing. Although people pay more and more attention to the environmental pollution and its health damage, it is impossible to carry out comprehensive management of the pollutants in the environment because of the limit of manpower, material resources, financial resources, and science and technology. At the same time, researches on chemical toxicology have shown that not all of the pollutants have the same harm to human health. Therefore, targeted research and treatment of pollutants with high health hazard effect have gradually become an effective environmental management strategy. Since the mid-twentieth century, many countries, regions and international organizations have developed their own screening methods for pollutants, and compiled their own list of environmental priority pollutants, and applied them to the management of environmental pollutants in practice [13, 40]. The European community, the United States, Japan, Germany, Holland, and other countries have published list of environmental hazardous substances one after another, which plays an important role in promoting environmental protection and governance.

With a short period of about 30 years, China has achieved development, which can catch up with the development of developed countries for 100 years, and achieved brilliant economic achievements. However, due to the extensive development mode and many other reasons, the ecological environment of our country has suffered serious damage, which leads to the environmental problems that should occur at different stages are reflected and erupted in a short time. In view of this, the State Environmental Protection Administration of China presided over research and proposed "Black list of pollution substance for preferential control in water" in the period of seventh five-year. However, over time, the characteristics of China's environmental problems have undergone profound changes, the existing list obviously not able to fully reflect the current situation of environmental risk and research level, and failed to cover the comprehensive environment system of water, soil, and atmosphere [40, 41]. In view of the new changes in the national and world environmental issues, with the reference of the developed countries' catalog making methods, the State Environmental Protection Administration (SEPA) published the National catalog of Pollutants' Environmental Health Risk in 2007. It represents the significant progress in the field of hazard identification in China, and provides the basis and method for the decision-making of government, monitoring of environment, formulation of environmental emergency plans, and emergency treatment of environmental pollution accidents.

However, after more than half a century of development, directory management has gradually exposed its limitations. Most of the early catalogs are static, which are balancing results of environmental pollution status, understanding level and management objectives at the beginning of the catalog editing. It cannot reflect the long-term dynamic changes of various factors, making many catalogs cannot effectively support the current environmental management work. Besides, the early catalogs commonly compiled by a single country or region. With the deepening understanding of transboundary migration of pollutants, people have realized that it is necessary to take the joint action of river basin and region to effectively cope with the common environmental problems. Therefore, the hot research spots in the field of hazard identification are mainly [13, 40, 42]: (a) new screening methods of priority pollutants considering pollutants migration and transformation characteristics, (b) construction of dynamic risk list updating mechanism in regional environment, (c) and risk identification of non-occupational long-term exposure with low dose.

3.2.2. Dose-response assessment

No threshold effect (carcinogenic effect) [13]. In view of humanitarian considerations, most carcinogenic chemicals are based on low dose extrapolation models to assess the risk probability of human exposure. Varieties of evidences have been applied to dose-response assessment of carcinogens, including human epidemiological data, which should be regarded as primary basis. In the absence of appropriate human clinical studies, data on animal species that close to humans should be used, and in the long-term animal studies, the most sensitive and biologically acceptable data should be given to the greatest extent of attention.

Threshold effect (noncarcinogenic effects) [13]. The reference dose (RfD) is an important basic factor, which means that when the dose is below this level, the risk of harmful effects will not be expected. At present, the following methods are used to evaluate the dose-response effect of threshold dose: firstly, determine the key toxic effects (the initial deleterious effects at this dose) and the highest dose that does not occur harmful effects (no observed adverse effect level, NOAEL) through reading literatures. Then, the NOAEL is divided by the uncertainty factor to obtain the safety limit value, and the uncertainty factor range from 10 to 1000. The uncertainty factor expresses a variety of internal uncertainties related to the existing data.

The focus of toxicity testing of environmental chemicals is to determine the safety level of human exposure to toxic chemicals. At present, the test has been extended from simple acute and subacute tests to full consideration of various toxicity data, including acute, subacute and chronic toxicity, and some specific toxicities, such as carcinogenicity, mutagenicity, reproductive toxicity, immune toxicity, neurotoxicity, skin toxicity, and other organ tests. In addition to these studies, the toxic dynamics of chemicals and their related mechanisms in the tissues, cells, subcellular, and receptor levels are also expected to further elucidate the toxic mechanisms of environmental chemicals and provide scientific references for potential hazard assessment of human bodies. In recent years, the main research focuses of dose-response assessment are [42, 43]: (a) dose-response relationship of chemical mixture interaction, (b) study on biological dynamics characteristics and establishment of bio dynamic model, (c) study on toxicity mechanism and toxicity kinetics, (d) and study on toxicity mechanism of cellular and subcellular.

3.2.3. Exposure assessment

Humans may expose to a variety of pollutants through different kinds of exposure pathways, which can be broadly divided into external exposure and internal exposure [39]. External exposure refers to the concentration of a substance in contact with the receptor, which can be understood as the gastrointestinal epithelium, the lung epithelium during respiration, and the epidermis in contact with the skin [37]. Internal exposure refers to the amount that a substance has been absorbed, that is, the amount that has entered the systemic circulation of the receptor. Bioavailability is defined as the proportion of absorbed in external dose.

Exposure assessment research mainly includes the following aspects [39, 44, 45, 46, 47]: (a) analysis and characterization of exposure environment. That is, to describe characteristics of common environmental physics and population. The climate, vegetation, groundwater literature, and surface water in regional need to be determined, and the target receptor population also should be determined, and characteristics of exposure population like the location of the source of pollution and the activity characteristics of the population should be described. Moreover, it is necessary to take into account characteristics of both the current exposure population and future exposure population. (b) Analysis of exposure pathways. According to the location and release of pollution sources, the migration and transformation process of possible chemical substances in multi-media environment and the potential exposure position and activities of people, to analyze and determine exposure points and approaches via each exposure pathway (such as direct skin absorption, ingestion, and so on). (c) Quantification of exposure. To quantify the exposure size, exposure frequency, and exposure duration of each pathway, step of exposure quantification including estimating exposure concentration and calculating intake. (d) Estimation of exposure concentration. To determine concentrations of chemical pollutants in the exposure period, and the exposure concentration is generally estimated by monitoring data or chemical environmental fate model. (e) Calculation of intake. To determine specific chemical exposure dose via each exposure route, and exposure dose is expressed in terms of unit weight per unit time, and the mass of body exposure to chemicals. In recent years, hot research spots in field of exposure assessment mainly include: (a) investigation on regional epidemiology or occupational exposure factors, (b) study on the mechanism and mass model of chemical migration in atmosphere, (c) study on migration mechanism and model of chemicals in water and sediment, (d) study on migration mechanism and model of chemicals in soil, (e) and study on chemo taxis mechanism and corresponding integrated exposure model of chemicals in multi-media environment.

3.2.4. Risk characterization

Risk characterization includes qualitative risk characterization and quantitative risk characterization [13]. The qualitative risk characterization use semi quantitative words such as "negligible", "slight", "medium", or "serious" to describe the degree of risk. Quantitative risk characterization can quantify the degree of risk by digital expression, which can more intuitively and effectively describe the degree of risk, and facilitate the screening and ranking of the risk of pollution factors, providing scientific reference for decision makers. The risk characterization methods both at home and abroad are established based on the framework of risk assessment system of USEPA. The quantitative risk characterization can be divided into the carcinogenic risk characterization and noncarcinogenic risk characterization. The conclusion of environmental risk characterization will provide a scientific basis for the formulation of environmental standards or environmental management strategies at the national, local, and organizational levels. And the formulation and implementation of environmental standards is the starting point of environmental administration and an important basis for environmental management.

3.2.5. Risk management

The risk management process is triggered by the concerns of risk of specific uses or special scenarios of chemicals. Environmental risk assessment and risk management are closely related, but their process is different. The characteristics of the risk management decision affect the breadth and depth of risk assessment, while risk assessment provides a scientific basis for risk management. The last game between risk assessment and legal, political, social, and economic and technology based on the present situation and put forward reasonable control measures. Risk management is based on the relationship of risk assessment with law, politics, society, economy, and technology, and puts forward reasonable control measures. Risk management includes the following tasks: classification of risks, determination of risk reduction measures and benefit analysis of risk management, risk reduction, and monitoring and review. In recent years, the new development of risk management research mainly includes [13, 48, 49]: (a) focus on risk reduction and responsibility concerns, that is, the establishment of appropriate laws and regulations, and the responsibility of the original competent authorities transferred to the manufacturing risk industry (manufacturers and importers and its users), (b) risk communication and stakeholder participation. Risk communication is the link between environmental risk assessment and risk management, and the active participation of stakeholders will help to ensure that the assessment results and management practices are better understood and implemented, (c) the integrated framework of information platform for environmental risk assessment and management system. To analyze human health comprehensively, environment, ecological security and socio-economic factors has been one of the main trends in this field, (d) risk cognition. Because the risk (or interest) cognitions of individuals, public, enterprises or employees are different, which will change with time as well, it is of great significance to explore different groups for their own risk assessment or cognitive inertia.

3.2.6. Uncertainty analysis of system

The disputes surrounding risk assessment often center on discrepancies in the evaluation of inadequate and uncertain data, as well as the nature, explanation, and presentation of assessment methodologies and models. When science is applied to management, decision makers must possess not just the current scientific knowledge, but also a comprehension of the uncertain and incomplete aspects of this knowledge, as well as the capacity to differentiate between uncertainty and variability. Uncertainty can arise from risk assessment due to factors such as faulty and incorrect parameters, variability in parameters, flawed or unreasonable simplified model theory, and the design of receptor exposure scenarios that are not realistic [13, 34]. In 1999, Cullen and Frey categorized the uncertainty in risk assessment into three distinct types: parameter uncertainty, model uncertainty, and variability. Due to its quantifiability, the majority of uncertainty research is centered around parameter uncertainty. Zhang Yinghua, Liang Jie, and Babendreier utilized different methods, namely the Monte-Carlo algorithm, Bayesian Monte-Carlo neural network method, Monte-Carlo method, and fuzzy mathematics method, to effectively manage and measure the uncertainty of parameters in the assessment. However, they overlooked the uncertainty and variability of the model during the evaluation process. Wang Yongjie, Moschander, and Karuchit conducted a study and discovered that the impact of model uncertainty and variability on the reliability of assessment outcomes is significantly more than that of parameter uncertainty. They performed a qualitative/semi-quantitative analysis. However, due to the difficulty in quantifying model uncertainty and variability, related research tends to go slowly. In summary, based on the risk assessment systems in the United States, the European Union, Australia, Canada, Japan, and other countries, and after reviewing a significant amount of literature, the main areas of deficiency and difficulty in the research of health risks posed by soil environmental pollutants can be identified as follows:

- The urban is the center of the crowd gathering, which is easy to occur environmental public hazards. Present researches on the urban soil environment and population health is not deep enough at home and abroad.
- In the practice of risk assessment and management, how to obtain the regional characteristics of risk assessment and management more efficiently under limited economic budget still needs further exploration.
- In the field sampling analysis of health risk assessment, it is often used to monitor the distribution of sites, which requires a large amount of human and material input. In view of the significant temporal and spatial variability of soil environment, it is necessary to find out more efficient sampling methods.
- A large number of studies show that the chemical fractions of heavy metals in soil can more accurately characterize soil metal bioavailability and mobility, so how to embed heavy metal fraction parameters into the classical model of heavy metal content evaluation needs further study and verification in practice.
- For health risk assessment, the scientific identification of exposure pathways of different receptor populations is the key to determine the credibility of risk assessment. Most of the existing studies base on questionnaire surveys or theoretical analysis hypotheses. However, these two methods have high cost and high uncertainty, so it is a new research direction to find a low cost and high credibility exposure path analysis and determination method.
- In the risk assessment systems implemented in China and abroad, most of them only make the relevant regulations on the uncertainty control of the parameters in the assessment process ignoring model uncertainty and variability. Researches show that influences of model uncertainty and variability on the credibility of assessment results is much higher than that of parameter uncertainty. Because model uncertainty and variability are not so easy to quantify, related researches tend to move along at a slow.

- In most of the current studies, the pollutant source analysis is only multivariate statistical analysis of regional pollutant content data. But considering the complexity of urban pollution, single pollution data statistical analysis is limited by the amount and quality of pollution data, which cannot fully reflect the problem obviously. Therefore, it is necessary to take into account ecological environment, human health and social economy and further study the correlation of risk sources, so as to help decision makers make more reasonable and accurate risk management decisions.
- In the process of current health risk assessment, the participation rights and discourse rights of stakeholders are obviously unequal. How to embed the concept of environmental equity into environmental risk assessment system and further develop feasible technical methods has become a new direction of research.

REFERENCES

1. Shi JR, Chen KL. Urban environmental safety. Beijing: Chemical Industry Press; 2010
2. Li QL. Characteristics of Heavy Metals in Soil and Crops in Regional Ecosystems: A Case Study of Chongqing. Beijing: China Environmental Science Press; 2010
3. Wang Y, Chen YC, Li ZP. Contamination pattern of heavy metals in Chinese urban soils. Environmental Chemistry. 2012;31(6):763-770
4. Cheng HG, Li M, Zhao CD, Li K, Peng M, Qin AH, Cheng XM. Overview of trace metals in the urban soil of 31 metropolises in China. Journal of Geochemical Exploration. 2014;139:31-52
5. Li F, Huang J, Zeng G, Huang XL, Liu WC, Wu HP, Yuan YJ, He XX, Lai MY. Spatial distributions and health risk assessment of heavy metals associated with receptor population density in street dust: A case study of Xiandao District, Middle China. Environmental Science and Pollution Research. 2015;22(9):6732-6742
6. Li F, Huang J, Zeng G, Yuan XZ, Li XD, Liang J, Wang XY, Tang XJ, Bai B. Spatial risk assessment and sources identification of heavy metals in surface sediments from the Dongting Lake, Middle China. Journal of Geochemical Exploration. 2013;132:75-83
7. Qishlaqi A, Moore F, Forghani G. Characterization of metal pollution in soils under two landuse patterns in the Angouran region, NW Iran: A study based on multivariate data analysis. Journal of Hazardous Materials. 2009;172(1):374-384
8. Mihailović A, Budinski-Petković L, Popov S, Ninkov J, Vasin J, Ralevića NM, Vučinić Vasića M. Spatial distribution of metals in urban soil of Novi Sad, Serbia: GIS based approach. Journal of Geochemical Exploration. 2015;150:104-114
9. Wang XT, Chen L, Wang XK, Lei BL, Sun YF, Zhou J, Wua MH. Occurrence, sources and health risk assessment of polycyclic aromatic hydrocarbons in urban (Pudong) and suburban soils from Shanghai in China. Chemosphere. 2015;119:1224-1232
10. Singh R, Gautam N, Mishra A, Gupta R. Heavy metals and living systems: An overview. Indian Journal of Pharmacology. 2011;43(3):246-253
11. Xue JL, Zhi YY, Yang LP, Shi JC, Zeng LZ, Wu LS. Positive matrix factorization as source apportionment of soil lead and cadmium around a battery plant (Changxing County, China). Environmental Science and Pollution Research. 2014;21(12):7698-7707

12. Chen HM. Environmental Soil Science. 2nd ed. Beijing: Science Press; 2010

13. Leeuwen CJV, Vermeire TG. Risk Assessment of Chemicals: An Introduction. 2nd ed. Heidelberg: Springer Press; 2007

14. Ministry of Environmental Protection of China (MEPC), Ministry of Land and Resources of China (MLRC). Report on the National General Survey of Soil Contamination. Beijing: MEPC and MLRC; 2014

15. Chen HY, Teng YG, Lu SJ, Wang YY, Wang JS. Contamination features and health risk of soil heavy metals in China. Science of the Total Environment. 2015;512-513:143-153

16. Wu QT. Environmental Soil Science. Beijing: China Agriculture Press; 2011

17. Huang CY. Soil Science. Beijing: China Agriculture Press; 2004

18. Bockheim JG. Nature and properties of highly disturbed urban soils, Philadelphia, Pennsylvania. In: Paper Presented Before Division S-5, Soil Genesis, Morphology and Classification, Annual Meeting of the Soil Science Society of America; 1974. Chicago, America: Soil Science Society; 1974

19. De Kimpe CR, Morel JL. Urban soil management: A growing concern. Soil Science. 2000;28(4):31-40

20. Zhang GL, Wu YJ, Gong ZT. Urban soils: An ecological protector of urban environment. Chinese Journal of Nature. 2006;28(4):205-209

21. Chu CJ, Ma JH, Zhu YT. Comparison between heavy metal contents in urban soils and their potential ecological hazards in city and town: A case study of Zhengzhou, Zhongmu and Hansi, Henan, China. Chinese Journal of Soil Science. 2010;41(2):467-472

22. Liao QL, Hua M, Jin Y, Huang SS, Zhu BW, Weng ZH, Pan YM. A preliminary study of the distribution and pollution source of heavy metals in soils of Jiangsu Province. Geology in China. 2009;36(5):1163-1171

23. Guo W, Sun WH, Zhao RX, Zhao WJ, Fu RY, Zhang J. Characteristic and evaluation of soil pollution by heavy metal in different functional zones of Hohhot. Environmental Science. 2013;34(4):1561-1567

24. Wu S, Peng S, Zhang X, Wu S, Peng SQ, Zhang XX, Wu DL, Luo W, Zhang TB. Levels and health risk assessments of heavy metals in urban soils in Dongguan, China. Journal of Geochemical Exploration. 2015;148:71-78

25. Xi DL, Sun YS, Liu XY. Environmental Monitoring. 3rd ed. Beijing: China Higher Education Press; 2004

26. He DM, Wang XF, Chen LJ, Su R. Assessment on heavy metals contaminations of sugarcane soil in Guangxi Province by the geo-accumulation index and potential ecological risk index. Journal of Agricultural Resources and Environment. 2014;31(2):126-131

27. Muller G. Index of geoaccumlation in sediments of the Rhine river. GeoJournal. 1969;2(3):108-118

28. Nezhad MTK, Tabatabaii SM, Gholami A. Geochemical assessment of steel smelter–impacted urban soils, Ahvaz, Iran. Journal of Geochemical Exploration. 2015;152:91-109

29. Karim Z, Qureshi BA, Mumtaz M. Geochemical baseline determination and pollution assessment of heavy metals in urban soils of Karachi, Pakistan. Ecological Indicators. 2015;48:358-364

30. Hakanson L. An ecology risk index for squatic pollution control: A sedimentological approach. Water Research. 1980;14(8):975-1001

31. Islam S, Ahmed K, Mamun HA, Masunaga S. Potential ecological risk of hazardous elements in different land-use urban soils of Bangladesh. Science of the Total Environment. 2015;512-513:94-102
32. Zhang L, Liao Q, Shao S, Zhang N, Shen Q, Liu C. Heavy metal pollution, fractionation, and potential ecological risks in sediments from lake chaohu (eastern china) and the surrounding rivers. International Journal of Environmental Research and Public Health. 2015;12(11):14115-14131
33. Xu ZQ, Ni SJ, Tuo XG, Zhang CJ. Calculation of heavy metals' toxicity coefficient in the evaluation of potential ecological risk index. Environmental Science and Technology. 2008;31(2):112-115
34. Cullen AC, Frey HC. Probabilistic Techniques in Exposure Assessment. New York and London: Plenum Press; 1999
35. National Research Council. Risk Assessment in the Federal Government: Managing the Process. Washington DC: National Academy Press; 1983
36. Chen HH, Chen HW, He JT, Li F, Shen ZL, Han B, Sun J. Health-based risk assessment of contaminated sites: Principles and methods. Earth Science Frontiers. 2006;13(1):216-223
37. Zhou ZC. Evaluation of toxic effect threshold and dose response relationship. Journal of Toxicology. 2013;27(6):407-408
38. Qiu FG, Gao ST, Chen Q. Review on research advances in health risk exposure assessment. Journal of Safety and Environment. 2012;12(1):126-129
39. Duan XL. Research Methods of Exposure Factors and Its Application in Environmental Health Risk Assessment. Beijing: Science Press; 2012
40. Cui XY, editor. Comparative Study on Environmental Health Risk of Chemical Pollutants at Home and Abroad. Beijing: Science Press; 2010
41. Pei SW, Zhou JL, Liu ZT. Research progress on screening of environment priority pollutants. Journal of Environmental Engineering Technology. 2013;3(4):363-368
42. Mackay D. Multimedia Environmental Models: The Fugacity Approach 2nd ed. UK: Taylor and Francis Group; 2001
43. Dong WY, Ding ZB, Sun SQ. Algorithm of radiological risk estimation for exposures to depleted uranium based on icrp biokinetic models. Environmental Science and Management. 2008;33(2):21-24
44. Leggett RW. A biokinetic model for zinc for use in radiation protection. Science of the Total Environment. 2012;420:1-12
45. Wethasinghe C, Yuen STS, Kaluarachchi JJ, Hughes R. Uncertainty in biokinetic parameters on bioremediation: Health risks and economic implications. Environment International. 2006;32(2):312-323
46. Duan XL, Zhang K, Qian Y, Li Q, Wang ZS, Zhang JL. The latest developments of human exposure assessment. Academic Symposium and Conference Papers of Chinese Toxicology Society Management Toxicology Specialized Committee; Beijing. 2009
47. Jin YL, Zhang YF, Min J, Sun JF, Liu P. The application of individual margin of exposure in the health risk assessment of dietary exposure to lead. Chinese Journal of Health Statistics. 2014;31(6):943-945

48. Wang JN, Cao GZ, Cao D, Yu F, Bi J. A framework analysis on national environmental risk management system for China. China Environmental Science. 2013;33(1):186-191

49. Zhang HY, Ge Y, Li FY, Yan J, Bi J. A review of psychometric paradigm in environmental risk perception. Journal of Natural Disasters. 2010;19(1):78-82

20 Chapter

INDUSTRIALIZATION AND URBANIZATI1ON IN THE REPUBLIC OF AZERBAIJAN HAVE LED TO SIGNIFICANT POLLUTION OF THE ECOSYSTEM WITH HEAVY METALS

1. INTRODUCTION: AN OVERVIEW

Azerbaijan is situated between 38°25’ and 41°55′ North Latitude and 44°50’ and 50°51’ East Longitude. It holds the distinction of being the largest country in the South Caucasus region in terms of its land area, population size, abundant fuel energy, and other resources. It covers a total area of 866,000 square kilometers. The country’s overall border length is 2646 km, which includes 390 km with the Russian Federation, 480 km with Georgia, 1007 km with Armenia, 756 km with the Islamic Republic of Iran, and 13 km with Turkey. The Caspian Sea constitutes the eastern boundary. Azerbaijan is distinguished by its unique and captivating natural environment, climate conditions, and physiogeographical characteristics. The region possesses abundant and important natural resources, such as oil and gas reserves, mineral deposits, and mineral springs. The industry focuses on the exploration and extraction of hydrocarbon and other mineral reserves, as well as the refining of oil and the production of petrochemicals and chemicals. The primary agricultural pursuits are cattle husbandry, cultivation of grains, gardening, as well as the cultivation of cotton, tea, and vegetables [1]. Azerbaijan has had substantial economic growth in recent years, primarily driven by its huge oil and gas reserves. An enduring oil-industrial infrastructure has been developed in the republic. In addition to the improvement of the social and economic conditions in the country, these accomplishments also foster the growth of the non-oil industry. Azerbaijan also functions as a crucial conduit for the transit of oil and gas. The construction of the Baku-Tbilisi-Ceyhan (BTC) oil pipeline in 2005 somewhat helped to the economic expansion of the country. There has been a substantial rise in the gross domestic production during the past few decades. The majority of the country’s income is derived from the oil sector. Simultaneously, Azerbaijan faces significant environmental challenges as a result of the extensive exploitation of the area’s hydrocarbon reserves, leading to a rise in fossil fuel use and the emission of greenhouse gases. The primary environmental challenges are as follows [1, 2]:

- Pollution of water resources with wastewater, including transboundary pollution
- Inadequate supply of quality water to human settlements, wastage in delivery, and shortage of sewer lines (insufficient quantity of sewerage systems)
- Air pollution by industrial plants and vehicles
- Degradation of fertile soil lands (contamination, erosion, salinity, etc.)
- Improper disposal of solid industrial and municipal wastes including hazardous wastes
- Biodiversity decline including the depletion of forests, flora, and fauna

It is well known that pollution of the environment with toxic compounds causes serious problems due to their negative impact on all the components of ecosystem especially on human health. Unfavorable consequences of ecosystem pollution are appearing more acutely in urbanized territories characterized with a high density of population and industrial enterprises. The Absheron peninsula is the most urbanized territory of the republic of Azerbaijan, where two large industrial cities—Baku and Sumgait—and about 60% of onshore oil production are located.

The Absheron peninsula has about 200,000-ha territory on the coastal line of the Caspian Sea in the southwestern edge of the Great Caucasus Mountain. The region is characterized by a dry semi-desert climate. Strong northern winds are the main distinguishing feature of the peninsula. The average annual air temperature is +14.2°C and the average annual rainfall is about 200–300 mm. The soil cover of the territory mainly consists of gray-brown and coastal sandy soils. The Absheron peninsula is also the location of a great number of oil and gas fields. The region is specialized on the production, initial processing and transportation of crude oil and gas, oil machine building, and fishery. Much of the peninsula is crossed by oil and gas pipelines. Despite the fact that the history of oil production in Absheron dates to ancient times, the industrial development of oil-gas fields has started from the middle of the nineteenth century. About 70% of the republic's industrial potential is concentrated in this territory. Like other industrialized and urbanized regions of the world, the Absheron peninsula could not avoid environmental pollution. There are a number of technogenic-polluted aerials. Oil industry is the main pollution source of the environment. Since the end of the nineteenth century, the peninsula has been subject to intense industrialization and urbanization. Accelerated development of hydrocarbon producing and refining industries has radically changed the historically formed geochemical balance of the environment due to the removal of different chemical elements to the Earth's surface. The territory has over 200 basins of natural and artificial origin. A huge amount of toxic hydrocarbons and heavy metals accumulate in soil and deep sediments, thus leading to the change of the qualities of natural ecosystem. According to its ecological status and climatic landscape character, the Absheron region is categorized to be a territory with the most acute environmental situation in the republic [3, 4].

Presently, global environmental contamination is a significant and pressing issue. Oil and its derivatives have a substantial impact on environmental pollution in areas where oil is found. The investigation of hydrocarbon reservoirs, along with the retrieval, conveyance, retention, and utilization of oil and gas, leads to the contamination of the environment with perilous compounds. Crude oil consists of hydrocarbons, including alkanes, cycloalkanes, unsaturated aliphatic hydrocarbons, and aromatic hydrocarbons. It also contains heterocyclic compounds that contain nitrogen, oxygen, and sulfur, as well as metals and natural radionuclides. Prior research has demonstrated that the soil and water systems of the Absheron peninsula contain significant levels of pollutants [5, 6, 7]. The level of oil contamination in soil might range from 20 to 30% or higher. A significant portion of the peninsula is comprised of devastated oil-contaminated and bituminous terrain. Contaminants that have built up in soils move into lakes, reservoirs, surface waters, and groundwater across the entire area. The migration of contaminants into deep layers of soil and groundwater poses a significant threat from both local and regional perspectives. Currently, the environmental strain on the peninsula is steadily intensifying as a result of the continuously growing population, as well as the expansion of industrial and urban areas.

1.1. Heavy metals and their environmental effects

Heavy metals pose the greatest risk among environmental pollutants because of their high toxicity and potential to accumulate. Heavy metals, as classified, have a density exceeding 5 g/cm^3 (about 6.0 g/cm3 or more), which is significantly greater than the average density of soil particles (2.65 g/cm3).

They are naturally present in rocks but often exist in larger concentrations due to contamination. Exotic and precious metals are not classified as heavy metals. Based on the biological classification of chemical elements, heavy metals are categorized as micro- and ultra-microelements. According to the majority of studies, Pb, Cu, Zn, Ni, Cd, Co, Sb, Sn, Bi, and Hg are widely recognized as heavy metals [8, 9, 10]. Heavy metals possess two primary characteristics: they have hazardous effects on living beings even at low concentrations, and they have the ability to accumulate in the bodies of organisms over time. Due to the high concentration of heavy metals in the soil, plants and bottom sediments gradually accumulate toxicity, posing a significant threat to all forms of life. The global scientific community extensively investigates both the deposition of heavy metals in the environment and their detrimental effects on the ecosystem. Several scientists conducted extensive research on heavy metals in the biosphere [11, 12, 13, 14]. Soil has the highest concentration of heavy metals compared to other components of the ecosystem. Soil plays a crucial role in the introduction of heavy metals into the human body by its interaction with other components of the ecosystem, such as the atmosphere, hydrosphere, and plants. Upon entering the soil, these substances accumulate in various components of agricultural produce via the root system, as well as in aquatic organisms and silt at the bottom when exposed to surface water. Acid rain, a significant global issue in modern times, is contributing to the elevated levels of heavy metals in ecosystems. Exposing soil rocks to acid rain enhances the concentration of metals in lakes and other water basins. The movement of metals within the ecosystem facilitates their buildup in the human body (Fig. 1).

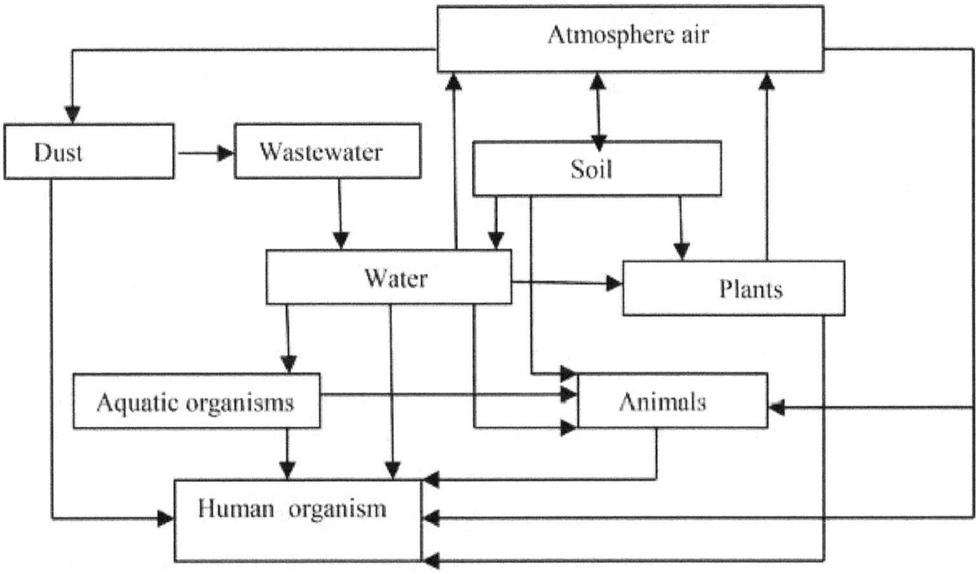

Fig. 1. Scheme of heavy metals penetrating into the human organism.

Heavy metals can originate from both natural sources and human activities. The primary origins of these substances are rocks and soils, primarily in forms that are not easily accessible to wildlife, such as components of rock and soil minerals. Anthropogenic causes of metal contamination can have a significant impact on all components of the environment. Anthropogenic sources of heavy metals include waste disposal, transport emissions, oil-gas and mining enterprises, air depositions, and land application of fertilizers [10, 15, 16]. The majority of oils contain a specific amount of heavy metals. All fractions of oil, including gasoline, contain metals. The crude oil obtained from the hydrocarbon reserves of Azerbaijan is primarily characterized by the presence of iron (up to 0.74%), chrome, and nickel (often exceeding $n \times 10^{-2}$ ppm). The asphaltene portion of certain oil fields in the Absheron peninsula contains a high concentration of microelements including Co, Br, Ag, Au, La, Sb, and Sc (Table 1). Neutron activation research revealed that the crude oil from the Balakhani oil field contains 1.2 × 10–3% Rb, 6.3% × 10–5% Cs, and 2.1% × 10–5% Eu

[17]. Typically, the asphaltene-tar fractions of oil contain the majority of heavy metals. Additionally, they can be found in the makeup of oil cuttings and drilling fluids.

Table 1. Metal content of the Absheron oil fields.

Oil fields	Metals (%)														
	Fe	Ni	Cr	Zn	Ba	V	Co	Hg	Ag	Au	La	Sb	Sc	Hf	Se
		× $^{4-}$10				× $^{5-}$10				× $^{6-}$10				× $^{7-}$10	
Asphaltenes															
1.	600	57	96	750	53	151	110	nd*	289	140	87	300	1500	28	47
2.	138	58	174	270	265	260	64	160	nd	8	28	19	20	26	32
3.	400	63	68	110	540	142	35	570	nd	6	nd	4	13	14	nd
Tars															
4.	170	8	20	80	5	37	2	14	10	150	4	2	10	23	75
5.	34	23	12	3	nd	62	10	nd	2.4	6	nd	nd	2	52	60
6.	110	19	9	8	99	47	4	2250	14	4	nd	9	11	127	46
7.	180	65	27	14	8	53	6	nd	nd	1	nd	1	nd	18	Nd

*nd, compound not detected or below instrumental direction limits.

According to the researches of specialists, oil and gas complex creates pedochemical factors of having an impact due to the existing natural technogenic hydrogeological conditions resulting in over-moistening and extent of contamination. There is a potential hazard of washing out of oil and oil products into rivers, bays, ground, and surface waters. Abnormal redistribution of chemical elements forming the secondary lithochemical areas has been recorded in several oil fields of the Absheron peninsula. Thus, some mobile elements including Pb, Zn, Cu, Ni, and Cr are accumulated in soils jointly and form paragenic association. They are able to migrate and accumulate in the appropriate barriers. Their concentrations exceed the normal values typical of the natural landscapes in the Absheron peninsula [5]. There is a tendency of the growth of heavy metals concentration both in the bottom sediments and in some hydrobionts in the Absheron territory. As a result of contamination by wastewater from various anthropogenic sources including offshore oil fields, heavy metals penetrate into the coastal line, natural lakes, and other reservoirs.

Usually, the most commonly encountered metals in polluted areas are lead (Pb), chromium (Cr), mercury (Hg), arsenic (As), zinc (Zn), cadmium (Cd), copper (Cu), and nickel (Ni). It is well known that heavy metals are nondegradable and persistent pollutants and accumulate in the environment for a long time. In soil, they could be adsorbed and gathered in different parts of plants through root systems. It has been revealed that heavy metal pollution causes adverse effects on the quality and yield of agricultural plants, resulting in the changes of the number, composition, and activity of microorganisms. Practice shows that the geochemical cycle of heavy metals under anthropogenic impacts represents serious, at times unpredictable, environmental effects. As mentioned earlier, due to their migration and accumulation in the environment, most heavy metals can easily enter the food chain and create serious threat to human health. Many heavy metals have a strong affinity for sulfur and disturb the enzyme function forming bonds with sulfur groups. They can get into a food chain and accumulate in an organism reaching toxic levels. These may cause massive mortality of various animal species [18].

In order to find a proper solution to the problem, we studied the level of heavy metal pollution in ecosystem while evaluating the contribution of anthropogenic factors and the potential ecological risk posed by individual elements.

The main purpose of this work was to investigate the heavy metal pollution of

- Soils from two types of land uses in the study site, including the areas of oil wells and transport road;
- Ground and groundwaters in a zone of influence of the oil-industrial waste;
- Waters and sediments of two natural water basins—the Boyuk shore and Bulbula Lakes—that had undergone a long-term impact of anthropogenic sources.
- The studies have revealed increased levels of some heavy metals in soils, ground, and surface water basins in the territories subjected to the impact of various anthropogenic sources.

2. THE OBJECTS AND METHODS OF RESEARCH

The objects of the present research were the territories of oil production enterprises, transport roads, as well as ground and groundwaters around an aeration station and surface water basins in industrial territories (Fig. 2). The main sources of ecosystem pollution in the study site are the industrial discharges formed during oil fields development and production activities. Along with the produced waters, these discharges also contain drilling solutions and various chemicals used in the drilling and production operations. Thus, the territory is subject to the impact of oil, oil products, heavy metals, and other technogenic waste. The main pollution sources of ground and groundwaters in the territory are the substances characterized with a high migration ability such as crude oil and heavy metals. It should be noted that the atmosphere emissions of oil-gas refineries and petrochemical industries are among the factors contributing to the ecological situation in the study area. In order to evaluate the impact of transport emissions on the heavy metal pollution, soil samples were collected at a 10-m distance of Baku-Mardakan highway.

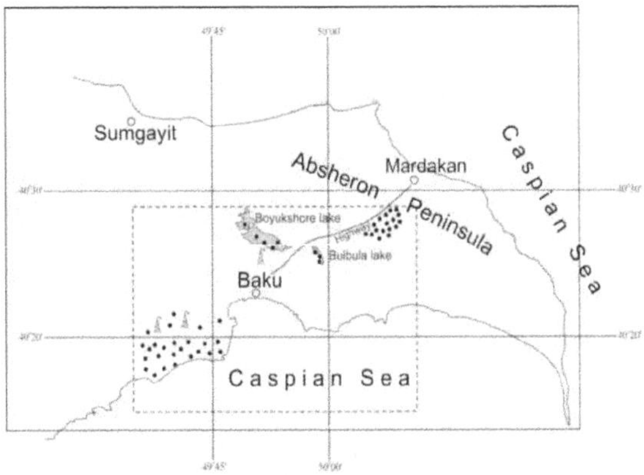

Fig. 2. The map of the study area.

One of the major objects of the research—Boyuk-shore lake—is located at the center of the Absheron peninsula at an absolute height of 12 m. It is the largest and most polluted lake of the peninsula with a 45-million-m3 volume, a 15-km^2 surface area, and a 4.2-m maximum depth. Being an enclosed water body, the lake feeds underground and surface waters of water catchment. Starting from 1930s, a huge amount of oil-industrial wastewater has been discharged to this basin (Picture 1). According to the information of Ministry of Ecology and Natural Resources, the total daily volume of wastewater released from industrial and communal-domestic objects to the Boyuk-shore lake amounted to 18,000 m^3 [19].

Picture 1. Oil-polluted area in the northern part of Boyuk-shore Lake (taken in 2017).

The Bulbula Lake is situated at the center of the Absheron peninsula in the northeastern part of Baku city at 8 m below the ocean level. It is an enclosed saline basin with a 2-km^2 surface area and a 2.05 million m^3 volume. Its maximal depth is 3–4 m. Till the end of the nineteenth century, the water regime of this lake was dependent on seasonal fallings and inflows from adjacent areas. Starting from that time, a large volume of oil-industrial wastewater have been discharged into this basin that resulted in dynamics of its level and flooding the residential buildings, social objects, and infrastructures in the territory (Picture 2). In order to lower the lake level, a pumping station with a 60,000-m^3 daily capacity was built for pumping about 30,000-m^3 water to Hovsan treatment plant. The lake is surrounded with five population settlements occupying an area of 122.5 ha with 3690-m length and 320–430 m width and is therefore permanently subject to the impact of domestic waste.

Picture 2. Wastewater flows to Bulbula Lake from its northern side (taken in 2015).

Analysis of water, sediment, and soil samples was carried out according to standard methods [20, 21]. Inductive Coupled Plasma-Optical Emission Spectrometry (ICP-OES) device was used in the analysis of heavy metals (As, Ba, Cd, Cr, Cu, Fe, Mn, Pb, and Zn), except Hg that was analyzed by Cold Vapor Atomic Fluorescence (CVAF) apparatus.

For the first time, the overall pollution status of the study area was evaluated with the calculation of pollution indices of heavy metals.

Various methods have been used to evaluate the heavy metal pollution. The calculation of single pollution index (PIi), enrichment factor (EF), geoaccumulation index (Igeo), Nemerow pollution index (PIN), and potential ecological risk index (RI) is widely applied for the assessment of heavy metals' level in soils and sediments [19, 22].

The pollution indexes used during our researches to assess the heavy metal contamination in the study sites are as follows:

2.1. Pollution index

The pollution level of each heavy metal was evaluated using pollution index (PIi), calculated as the ratio between the metal concentration (Ci) in the sample and its reference value—the national criteria of metal (Si):

$$PI_i = \frac{C_i}{S_i} \quad (1)$$

The maximum permissible concentrations (MPCs) of pollutants established by Azerbaijan legislation were taken as Si values.

2.2. Enrichment factor

Enrichment factor (EF) was initially calculated to identify the origin of elements in atmosphere, precipitation, or seawater, and it was further extended to the study of soils, aquatic sediments, and other components of ecosystem [23]. In this study, the enrichment factors of heavy metals in soils and sediments are calculated to assess the contribution of anthropogenic sources to the natural levels of heavy metals in the Absheron soils and lake sediments. The following formula was used to calculate EF:

$$EF = \frac{C_i/C_r}{B_i/B_r} \quad (2)$$

where Ci and Cr are the concentrations of the target metal and the reference metal in the sample, while Bi and Br are the background concentrations (BCs) of the target metal and the reference metal for the natural soils of the Absheron peninsula. Immobile elements such as Al, Fe, Ti, or Mn have been used as the reference metals for EF calculation. EFs for all the heavy metals were calculated using Mn as the reference metal in our study.

2.3. Index of geoaccumulation (Igeo)

A geoaccumulation index was originally defined by Muller [24] for the evaluation of metal contamination in aquatic sediments, but it was also applied in assessing the metal contamination of soils. The formula used for the calculation of geoaccumulation index soils and sediments is.

$$I_{geo} = Log_2\left(\frac{C_n}{1.5\,B_n}\right) \quad (3)$$

where Cn is the measured content of element (n), Bn is the BC of element n, and 1.5 is the background matrix correction factor due to lithogenic effects (the constant 1.5 is introduced to minimize the effect of possible variations in the background values which may be attributed to lithologic variations in soils).

2.4. Ecological risk factor

The potential ecological risk of heavy metal pollution in the soils and sediments of water basins in the study area was assessed using the ecological risk index (RI) [25]. RI is the comprehensive potential ecological index, which is the sum of individual heavy metals—Ei. It represents the sensitivity of the biological community to the toxic substance and illustrates the potential ecological risk caused by the overall contamination. The RI was calculated as the sum of risk factors of heavy metals:

$$RI = \sum E_i \tag{4}$$

where Ei is the single risk factor for heavy metal i and is defined as.

$$E_i = T_{i.}f_i = T_i \frac{C_i}{B_i} \tag{5}$$

where Ti is the toxic-response factor for heavy metal i. The ratio fi is the metal pollution factor calculated from the measured concentration Ci and the background concentration Bi of metal in soils. MPC and BC, and the Ti values defined by Hakanson for the measured heavy metals that were used during the calculation of pollution indices are given in Table 2.

Table 2. MPC, BC, and toxic-response factors of heavy metals.

Metals	Hg	Cd	As	Pb	Cu	Cr	Ni	Zn
MPC (mg/kg)	2.1*	1.0	2.0	32.0	3.0	6.0	4.0	23.0
BC (mg/kg)	0.4	3.0	15	20	100	40	20	70
T_i	40	30	10	5	5	2	5	1

*The value of 2.1 is acceptable only for Hg pollution of soils in nonresidential areas.

The calculated values of PIi, Ei, EF, and Igeo given in subsequent tables are the means of these indices from at least 10 samples.

3. RESULTS AND DISCUSSION

The initial investigations were primarily concerned with evaluating the extent of heavy metal contamination in the areas surrounding oil and gas production facilities. The study was conducted several years ago. Analyzed were 10 soil samples for the presence of metals, namely potentially harmful elements such as Arsenic (As), Cadmium (Cd), Chromium (Cr), Mercury (Hg), Lead (Pb), Copper (Cu), Zinc (Zn), and Barium (Ba). The obtained values from the analyses were compared to both the background concentrations and the maximum allowable amounts of these components set by the Cabinet of Ministers of the Republic of Azerbaijan. The findings are displayed in Table 3.

Table 3. The content of heavy metals in oil-contaminated soils.

Samples	Heavy metals (mg/kg)									
	As	Ba	Cd	Cr	Cu	Fe	Mn	Hg	Pb	Zn
1	5.8	730	0.28	20.4	67.1	16,500	380	0.04	27.2	60.7
2	7.6	214	0.10	27.1	30.7	30,600	494	0.03	25.3	32.5
3	5.4	1290	0.41	21.6	44.4	28,600	691	0.11	37.3	105
4	8.2	335	0.09	27.7	26.6	15,500	541	0.07	27.2	54.9

Samples	Heavy metals (mg/kg)									
5	7.6	330	0.12	14.4	20.8	16,200	294	0.02	26.2	17.1
6	3.4	360	0.15	18.9	57.2	7900	188	0.04	37.1	37.5
7	3.9	45	0.02	6.5	5.0	4500	130	0.01	25.1	11.5
8	4.4	760	0.23	17.7	38.4	13,900	435	0.19	29.8	32.4
9	7.5	570	0.37	16.9	60.0	13,500	474	0.07	42.2	76.2
10	7.8	354	0.08	28.4	29.9	15,600	485	0.03	14.3	52.1
MPC (mg/kg)	2.0	200–350	1.0	6.0	3.0		1500	2.1	6.0	23
BC (mg/kg)	15		3	40	100		250	0.4	20	70

The values in Table 3 show that the concentrations of the majority of heavy metals, including As, Ba, Cr, Cu, Pb, and Zn, exceed their permissible levels established in the republic. The contents of most toxic metals like As, Pb, and Zn are found to be from 3.4 to 8.2, 14.3 to 42.2 and 11.5 to 105 mg/kg, respectively. Their concentrations were higher than the background levels in some sampling areas. Meanwhile, no exceeding of maximum permissible and background levels on Hg, Cd, and Mn was observed in this area during our researches.

According to their impact on the ecosystem, many of these metals belong to the following categories, with exception of Fe, which is low toxic:

- As, Hg, Pb, Cd, and Zn—super dangerous
- Cr and Cu—dangerous
- Ba and Mn—moderately dangerous

It is necessary to have knowledge about sediment pollution situation to evaluate the level of heavy metals' impact on water biological system. With this purpose, sediment samples from 12 distinct water bodies (including natural and artificial basins) in the study site were analyzed for heavy metal contamination. It should be noted that along with oil-gas production enterprises, there are some other industrial objects in this area. The results of analyses are presented in Table 4. The table indicates that significantly high quantities of heavy metals are accumulated in the bottom sediments of surface waters.

Table 4. The content of heavy metals in bottom sediments of surface waters within the Absheron industrial region.

Samples	Heavy metals (mg/kg)								
	Ni	Co	Pb	Mn	Cr	Zn	Cu	Cd	V
1	44.7	18.4	18.9	136.4	11.9	20.4	14.9	33.0	3.3
2	3.0	9.6	7.0	155.0	16.1	20.1	12.9	ND	3.0
3	36.4	7.6	6.9	213.4	11.3	16.9	58.9	17.8	1.8
4	15.2	10.9	6.9	162.0	14.9	21.0	14.8	ND	3.2
5	13.2	9.8	6.3	159.1	18.4	20.3	13.6	ND	2.1
6	50.1	19.9	12.1	636.0	23.9	30.0	13.1	29.8	6.9
7	44.6	9.8	9.6	351.0	16.9	86.8	83.1	28.4	4.0
8	52.1	25.0	10.9	546.0	20.6	32.3	11.0	29.4	8.9

Samples	Heavy metals (mg/kg)								
	Ni	Co	Pb	Mn	Cr	Zn	Cu	Cd	V
9	49.6	24.6	11.3	513.2	23.9	29.6	10.6	1.0	9.3
10	46.8	21.0	11.6	596.0	25.1	28.3	12.9	31.0	5.7
11	15.8	12.1	7.0	169.3	15.9	21.9	16.1	ND	3.4
12	13.4	11.3	6.8	156.8	13.8	19.8	15.4	ND	2.4

As can be seen from Table 4, in many samples, the concentrations of toxic heavy metals, especially Cd, Cr, Cu, and Zn, were found to be greater than their permissible levels accepted in the republic—1, 6, 3, and 23 mg/kg, respectively. In several stations, the contents of most toxic metals like Cd and Zn ranged from 17.8 to 33.0 mg/kg and from 28.3 to 86.8 mg/kg, respectively. The highest concentrations of Cd were recorded in the study sites located closer to cement manufacture and nonferrous metal-processing plants. In a large majority of samples, the values of Cd were higher than its background level (3 mg/kg) to a considerable extent. The content of Cr was significantly higher than its MPC at most of sampling stations. However, no exceeding of the MPC on Pb (32 mg/kg) was detected in the samples during the study period.

Further studies were carried out to assess the potential ecological risk caused by the heavy metal contamination in the Absheron industrial zone through the calculation of single pollution index (PIi), enrichment factor (EF), geoaccumulation index (Igeo), and potential ecological risk index (RI). The studies focused on the evaluation of the contamination levels of various ecosystem components with toxic metals. At first, the contamination of ground and groundwaters was studied in one of the polluted areas of the Absheron peninsula, in the territory of Hovsan Aerator Station. For the evaluation of the pollution level and the potential ecological risk caused by heavy metal pollution in soils, enrichment factor and ecological risk index were calculated for the most important heavy metals with regard to potential hazard such as Cd, Cr, Pb, Zn, and Ni, with contamination values exceeding MPC.

Table 5. The content of heavy metals in groundwaters (mg/l).

Metals	MPC, mg/l	Sampling stations								
		1	2	3	4	5	6	7	8	9
Cd	0.001	0.010	0.008	0.009	0.009	0.005	0.008	0.014	0.012	0.006
Zn	0.5	0.09	0.09	0.012	0.015	0.014	0.116	0.248	0.190	0.120
Pb	0.03	0.009	0.013	0.006	0.014	0.007	0.012	0.019	0.015	0.008
Cr	0.1	0.007	0.008	0.005	0.012	0.005	0.025	0.040	0.011	0.007
Ni	0.1	0.005	0.018	0.014	0.012	0.006	0.008	0.021	0.010	0.005
Mn	0.05	0.013	0.014	0.016	0.008	0.015	0.027	0.014	0.010	0.008

The data derived from heavy metal analysis of ground and groundwaters of the study site are presented in Tables 5 and 6. The results indicated that in some groundwater samples, the concentrations of Cd and Zn were found to be higher than their MPC established by the national legislation. No exceeding of MPC was recorded in the values of Pb, Cr, Ni, and Mn during our observations. Overall, the concentrations of the majority of the analyzed heavy metals in groundwaters were at the level of national and international standards. According to the international water-quality standards, the minimal risk concentration of Zn, Cu, Ni, and Pb should be 20, 10, 2, and 10 μq/l, respectively [26, 27].

Table 6. The content of heavy metals in groundwaters (mg/l).

Metals	Sampling stations								
	1	2	3	4	5	6	7	8	9
Cd	3.9	1.4	4.4	6.4	2.2	5.6	3.5	3.3	2.5
Zn	99	140	130	120	110	77	110	80	130
Pb	7	9	8	14	16	9	5	12	14
Cr	30	92	68	55	70	46	130	57	59
Ni	20	30	40	36	25	21	35	53	18
Mn	360	230	270	320	220	320	400	280	260

Analysis of the data presented in Table 6 shows unequal distribution of heavy metals in the polluted grounds of the territory. The variability in the concentrations of metals may be associated with the origins of their pollution sources. The maximum heavy metal pollution levels observed were 6.4 mg/kg for Cd, 140 mg/kg for Zn, 130 mg/kg for Cr, and 53 mg/kg for Ni. With exception of Mn and Pb, the average concentrations of Cd, Zn, Cr, and Ni in the ground samples were found to be greater than their MPC about 3, 5, 10, and 9 times, respectively.

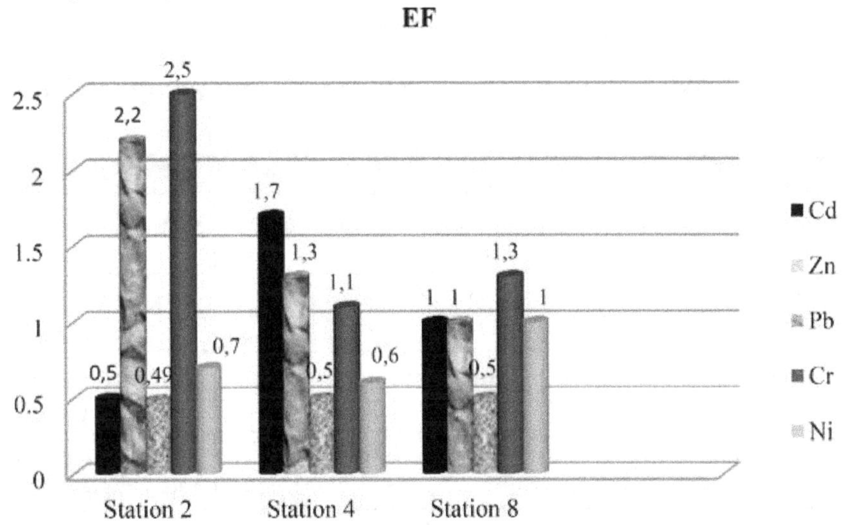

Fig. 3. Change of the values of enrichment factor of heavy metals in the grounds of three study stations.

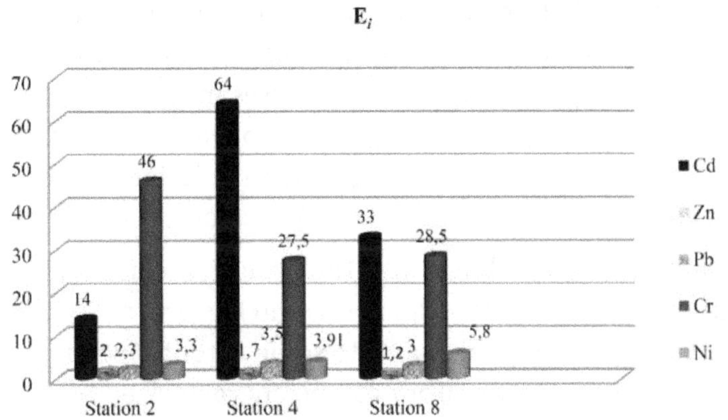

Fig. 4. Change of the values of potential ecological risk index of heavy metals in the grounds of three study stations.

Fig.s 3 and 4 show the calculated EF and Ei values of Cd, Zn, Pb, Cr, and Ni. The results of calculation showed that the values of EF of Zn and Cr are 2.18 and 2.50 in the ground from station 2, indicating moderate enrichment by these elements, whereas the EF values of Cd, Pb and Ni were 0.51, 0.49, and 0.71, respectively, exhibiting depletion to mineral enrichment (<2). EF values of all the studied metals in both stations 4 and 8 recorded the values lower than 2, indicating that the grounds of these areas are in the state of depletion to mineral enrichment with respect to these metals. The data presented in Fig. 1 show the same EF values for Pb at all three stations, which can be associated with the composition of the parent materials of soils. A relatively high EF value of Cd (1.67) was observed at sampling station 4.

Comparative analysis of the calculated values of Ei index with the data given in Table 7 shows that among other elements, Cr and Cd demonstrated a higher ecological risk during the observations, indicating that the site was impacted by the anthropogenic sources of these metals. The calculated Ei values were 46 for Cr at station 2, and 64 for Cd at station 4 exhibiting a moderate potential ecological risk. According to the results of calculations, the area of station 8 can be characterized by a relatively lower enrichment and a potential ecological risk with regard to heavy metal pollution. As can be seen from Fig. 4, the potential environmental risk of metal in a greater extent depends on its toxicity to the ecosystem, rather than the content of the element in the soil. At station 8, the concentration of Cd exceeded MPC by 3.3 times, and the concentration of Zn exceeded MPC by 3.5 times, whereas according to the calculation results, the values of their ecological risk index Ei were 33 and 1.2, respectively.

Table 7. Classification of heavy metal pollution levels in soils and sediments based on Igeo, EF, and Ei values.

Class	I_{geo}		EF		E_i	
1	<0	Unpolluted/slightly polluted	<2	Depletion to mineral enrichment	<40	Low potential ecological risk
2	0–1	Moderately polluted	2–5	Moderate enrichment	40–80	Moderate potential ecological risk
3	1–3	From moderately to strongly polluted	5–20	Significant enrichment	80–160	Considerable potential ecological risk
4	3–5	Strongly polluted	20–40	Very high enrichment	160–320	High potential ecological risk
5	>5	Extremely polluted	>40	Extremely high enrichment	320	Very high potential ecological risk

Part of the research was carried out to assess the current level of heavy metal pollution of soils and surface water systems in the oil-industrial zone of the Absheron peninsula. The researches focused on the heavy metal pollution of soils from two types of land uses including two natural lakes—Boyuk shore and Bulbula, which have been subject to a long-term impact of various anthropogenic sources. The heavy metal pollution of the study site was assessed through the calculation of the single pollution index (PIi), geoaccumulation index (Igeo), enrichment factor (EF), and ecological risk factor (Ei).

Comparison of the results given in Table 8 with the pollution classes in Table 7 shows that oil field area can be categorized as unpolluted by Pb, Cd, Zn, Cu, and Hg, and moderately to strongly polluted by As and Cr, which mean Igeo values are higher than 1 to some extent. The mean values of EF are 0.0367 for Cd and 0.8900 for Pb, indicating that the territory can be characterized by the first classification level—depletion to mineral enrichment. According to the mean Ei values of individual metals within the range of 0.68 (Zn)–7.28 (Pb), the soils of oil field area demonstrate a low potential ecological risk regarding heavy metal contamination.

Table 8. The content and pollution indices of heavy metals in soils in oil field area.

Elements	Concentration (mg/kg)			PI_i	E_i	EF	$I_{geo.}$
	Min.	Max	Mean	Mean	Mean	Mean	Mean
As	3.4	8.2	4.16	2.8	4.16	0.2683	1.4623
Cd	0.02	0.41	0.18	0.18	1.85	0.0367	-4.5658
Cr	6.5	28.4	19.9	3.36	0.99	0.3000	1.7297
Cu	5.0	67.1	37.1	12.7	1.89	0.2250	-1.9808
Hg	0.01	0.19	0.06	0.03	6.10	0.0912	-3.3219
Pb	14.3	42.2	29.2	0.9	7.28	0.8900	-0.0389
Zn	11.5	105	47.9	2.3	0.68	0.4107	-1.1292
Mn	130	691	410				

Fig. 5 presents the percentage of individual heavy metals in potential ecological risk index (RI) for the soils of oil field area. It is seen from this Fig. that the soils from this area had the highest Ei percentage for Pb, Hg, and As compared to other heavy metals.

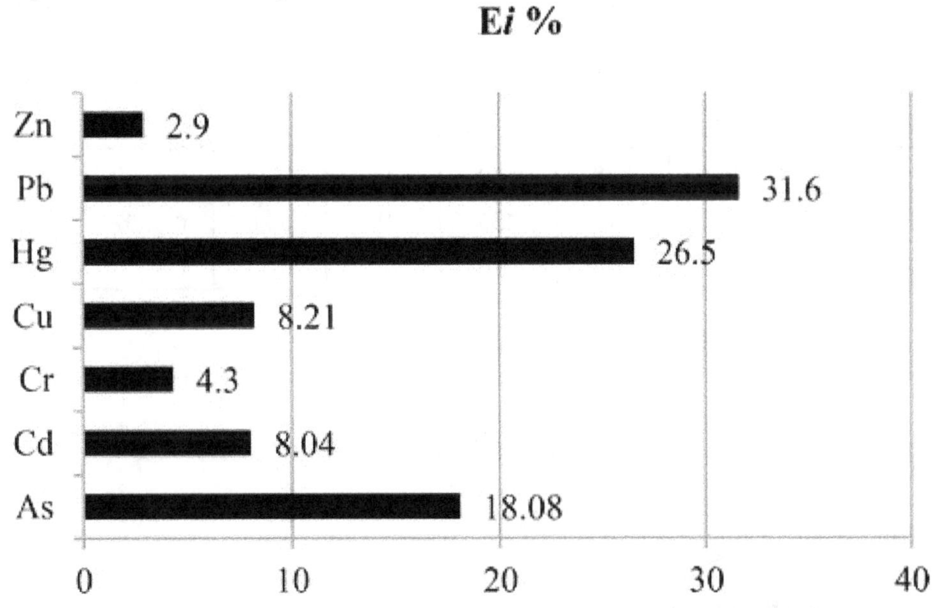

Fig. 5. Percentage of heavy metals in potential ecological risk (RI) for the soils of oil fields area.

In order to study the impact of transport emissions on heavy metal contamination of soils, the content of heavy metals was measured in soil samples collected in the vicinity of two main roads. The results are given in Table 9.

Table 9. The content of metals along transport roads.

Roads	Concentration of metals, mg/kg									
	Pb	Cd	Ni	Zn	Mn	Hg	Cr	As	Co	Cu
Baku-Sumqait	470	4.5	375	430	1950	0.019	860	4.08	52	115
Baku-Mardakan	565	5.8	440	515	2175	0.023	909	5.29	57	150

The calculated values of pollution indices for the soils near to Baku-Mardakan highway that was polluted predominantly by transport emissions (Table 10) varied from their values for oil-polluted site. The highest pollution indices in this area were recorded for Pb. The mean values of Igeo, EF, and Ei of this metal were 4.1429, 2.1250, and 132.5, respectively, demonstrating strongly polluted, moderate enrichment, and considerable ecological risk levels. The calculated Igeo values indicate that the site can be categorized as strongly polluted with respect to Pb and Cr and from moderately to strongly polluted area with respect to Zn and Cu. Relatively low pollution levels were detected for Hg and As in the soils of this area.

Fig. 6 indicates that an increasingly high Ei percentage was recorded for Pb, Cd, and Cr in the soils from the area of transport road. Combustion of petroleum products by transport facilities is a major source of environment pollution by these metals.

Table 10. Concentration and mean values of pollution indices of heavy metals in soils in the area of Baku-Mardakan highway.

Elements	Concentration (mg/kg)			PI_i	E_i	EF	$I_{geo.}$
	Min.	Max	Mean	Mean	Mean	Mean	Mean
As	1.25	6.17	5.30	2.65	3.5	0.0266	-2.0858
Cd	1.30	5.80	4.50	4.5	45.0	0.1416	0
Cr	215	926	860	143	43.0	1.5000	3.8413
Cu	150	740	575	191	28.7	0.5500	1.9386
Hg	0.007	0.034	0.028	0.013	2.80	0.0219	-7.7433
Pb	75	588	530	16	132.5	2.1250	4.1429
Zn	126	517	435	19	6.20	0.5357	2.0506
Mn	1755	3390	2170				

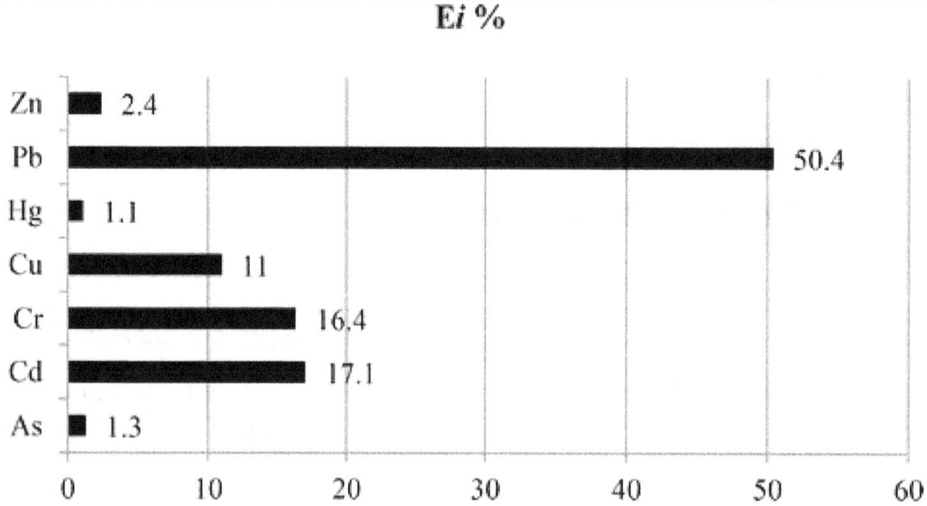

Fig. 6. Percentage of heavy metals in potential ecological risk (RI) for the soils of the area close to transport road.

Preliminary investigations were carried out to assess the extent of pollution in the lakes described above, namely in terms of oil products and heavy metals present in the lake waters. The samples obtained from the industrial wastewater flow to Boyuk beach Lake during the study showed the greatest levels of heavy metal and oil contamination, as indicated in Table 11. Zinc (Zn), lead (Pb), and cadmium (Cd) showed the highest quantities of harmful heavy metals, surpassing the maximum permissible concentration (MPC) by 70, 19, and 44 times, respectively. Significantly elevated levels of petroleum compounds were detected in the waters of both lakes, above the Maximum Permissible Concentration (MPC) by a factor of 10. Copper pollution was detected at high levels in all sample stations.

Table 11. The content of petroleum compounds and heavy metals in two natural lakes of the Absheron Peninsula.

Pollutants (mg/l)	MPC (mg/l)	Boyuk shore Lake	Bulbula Lake
Petroleum compounds	0.05	2.6	2.05
Zn	0.5	35.06	ND
Cu	0.01	1.72	0.37
Ni	0.1	6.07	0.006
Pb	0.03	0.56	0.01
Cd	0.001	0.044	<0.001

The contents and calculated results of pollution indices of heavy metals in the sediments of Boyuk shore and Bulbula Lakes are presented in Tables 12 and 13. Minor concentrations of As were detected in the sediments of both lakes during the studies. Consequently, the pollution indices were calculated for six metals including Cd, Cr, Cu, Hg, Pb, and Zn.

Table 12. The content and mean values of pollution indices of heavy metals in the sediments of Boyuk shore Lake.

Elements	Concentration (mg/kg)			PI_i	E_i	EF	$I_{geo.}$
	Min.	Max	Mean	Mean	Mean	Mean	Mean
Cd	1.0	5,0	1.7	1.7	17.0	0.3916	-1.4043
Cr	15.6	59.0	25.1	4.2	1.21	0.4562	-1.2572
Cu	10.6	24.4	14.7	4.9	0.74	0.1125	-3.3510
Hg	0.004	0.063	0.0085	0.04	0.85	0.1500	-6.1413
Pb	16.7	47.2	28.5	0.9	7.12	1.0040	-0.0740
Zn	20.3	510	86.8	3.8	1.24	0.8714	-0.2746
Mn	213.4	546	355				

Table 13. The content and mean values of pollution indices of heavy metals in the sediments of Bulbula Lake.

Elements	Concentration (mg/kg)			PI_i	E_i	EF	$I_{geo.}$
	Min.	Max	Mean	Mean	Mean	Mean	Mean
Cd	<0.5	1.4	0.6	0.7	6.00	0.2150	-2.9068
Cr	11.3	28.9	17.7	2.9	0.88	0.2937	-1.7612
Cu	13.1	35.0	22.6	7.5	1.13	0.1475	-2.7305
Hg	0.002	0.029	0.006	0.003	0.60	0.0993	-6.6438
Pb	6.9	30.2	18.4	0.6	4.60	0.6101	-0.7052
Zn	6.6	36.8	20.3	0.9	0.29	0.1928	-2.3708
Mn	179.4	636.0	376.2				

It is seen from Table 12 that the average PIi values of Cd, Cr, Cu, and Zn in the Boyuk shore Lake sediments showed the mean concentrations to be significantly higher than the MPC of these metals. The maximum concentrations of Cd, Pb, and Zn each were 5, 47.2, and 510 mg/kg, respectively. Laboratory analyses showed that the concentrations of the majority of heavy metals in the sediments of Bulbula Lake were lower than those in Boyuk shore Lake, exclusively Cu which had the highest mean PIi value of 7.5. The variations in the heavy metals pollution levels can be explained with the origins of their sources.

The results of calculations suggest that in both lakes, the mean values of Igeo. and EF for the studied metals represent unpolluted condition, while there are some elevated levels of Cd, Pb, and Zn, especially in the sediments of Boyuk shore Lake. The results showed that the maximum EF value in the sediments of Boyuk shore Lake observed in Zn is greater than 3, indicating that the lake was severely impacted by anthropogenic sources of this metal.

Table 13 demonstrates that the EF values computed for the metals analyzed in the sediments of Bulbula Lake are below 1, indicating a decline in mineral enrichment. The Igeo values for all metals in the Bulbula Lake sediments were found to be negative, indicating that the lake falls into the "unpolluted/slightly polluted" category. The computed values of Ei suggest that in both lakes, Pb and Cd pose comparatively higher potential dangers compared to other metals (Fig.s 7 and 8). The sediments of Boyuk beach Lake showed the greatest risk for Cd contamination.

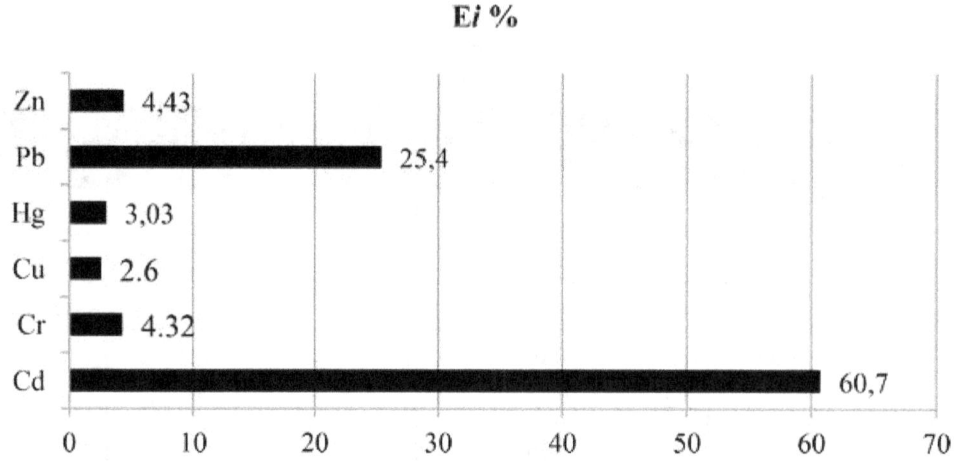

Fig. 7. Percentage of heavy metals in potential ecological risk (RI) for the sediments of Boyuk shore Lake.

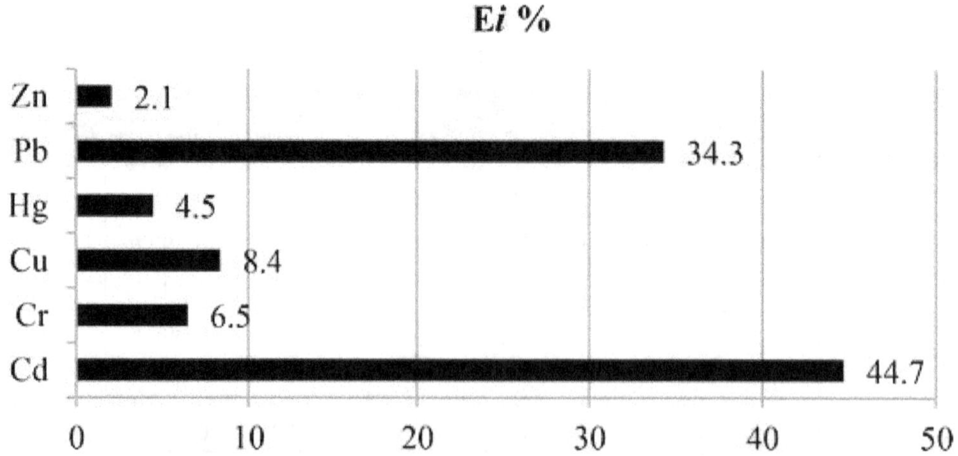

Fig. 8. Percentage of heavy metals in potential ecological risk (RI) for the sediments of Bulbula Lake.

4. CONCISE SUMMARY

In summary, the levels of heavy metal contamination of the main ecosystem components including soils, grounds, surface, and groundwaters were assessed in the oil-industrial region of the Absheron peninsula. In order to evaluate the contribution of anthropogenic sources to heavy metal contamination, pollution indexes were calculated for most toxic metals—As, Cd, Cr, Hg, Pb, Cu, Zn, and Ni.

Based on the results of researches, it can be concluded that the long-term impact of anthropogenic discharges has led to the pollution of all ecosystem components in the study site. The majority of the studied metals exceeded their permissible levels. The laboratory analysis revealed high levels of toxic heavy metals in grounds and groundwaters in the territory of aerator station. The studies showed that the concentrations of Cr, Cd, Zn, and Ni in the grounds of this site exceeded their MAC by several times. In order to evaluate the metal pollution status, the EF and Ei indexes of metals were calculated for three areas. Despite the fact that in the majority of samples, the concentrations of heavy metals were higher than their permissible levels, the calculated EF and Ei values suggest that the grounds in the site can be categorized as depletion to mineral enrichment area representing a moderate potential ecological risk with respect to heavy metals. Thus, the grounds in the territory of the aerator station cannot be characterized by a serious potential risk from heavy metals.

Based on the results of the comparative studies carried out in the oil-industrial area and the Baku-Mardakan transport road, we can say that the long-term impact of anthropogenic sources has led to the pollution of ecosystem by heavy metals in this area. The researches have revealed increased levels of heavy metal contamination. High concentrations of Pb, Cr, Zn, and Cu were detected in soils exceeding their permissible limits by several magnitudes that pose serious threat. Both oil industry and transport are the main sources of soil pollution in this site, where transport emission is the dominant factor.

Calculations showed that there is some correlation between the values of EF and Igeo. for Boyuk shore and Bulbula Lakes, respectively. The accumulation of metals in the sediments of these lakes was not insignificant, indicating the condition of depletion to mineral enrichment. Analysis of the calculated EF, Igeo, and Ei values confirmed that in both lakes, the highest and lowest values of the mentioned indexes were recorded almost in the same metals that can be explained by their lithogenic sources. Higher quantities of heavy metals were detected in the samples from Boyuk shore Lake. This is the evidence of the fact that anthropogenic factors strongly influence the contamination of the lake sediments by heavy metals.

In recent years, according to the Presidential Decree, comprehensive works were implemented on the rehabilitation of nine natural lakes of the Absheron peninsula [28]. The works implemented on the rehabilitation and protection of the Boyuk shore Lake were divided into two phases. The first phase covered a 283-ha lake area and an area of 500 m from the lake. At this stage, the division of the lake into two parts was intended for the eastern part of the lake, adjacent to the Baku Olympic Stadium. The first phase was completed in 2015. However, it is important to have a detailed knowledge on the ecological situation of these lakes and surrounding areas to evaluate the extent of the works to be undertaken for further rehabilitation. Overall, the potential ecological risk indices indicated a low potential risk (<40) from individual heavy metals, with exception of Pb posing a considerable potential risk level in the territory of Baku-Mardakan highway.

The calculated values suggest that the study area cannot be characterized by a high potential risk from heavy metals but increasing industrial development and urban growth can lead to a higher potential risk in the future years. In order to prevent pollution and protect the ecosystem from the future disasters, it is possible to recommend

- The creation and execution of intricate strategies, rooted in scientifically validated methods, to safeguard the ecosystem from the detrimental effects of harmful substances, such as heavy metals.
- Implementation of ICT systems to create an electronic database of ecosystem contamination, including maps that show the distribution of heavy metals and identify the human-caused sources of each element.
- The creation of a comprehensive program to study the effects of heavy metals on soil and water ecosystems, considering their geological changes, movement, and accumulation conditions, such as soil acidity and alkalinity, oxidation-reduction processes, humus content, and the presence of easily soluble salts, among other factors.
- The focus will be on creating and using environmentally-friendly technologies to reduce industrial discharges and heavy metal pollution throughout all stages of operation.
- A comprehensive plan will be developed and put into action to reduce transport emissions in the Absheron region, with a particular emphasis on the city of Baku.

REFERENCES

1. Aliyev F, Khalilova H. The anthropogenic impact on surface water resources in Azerbaijan. Energy and Environment. 2014;25(2):343-356
2. Second National Communication to the United Nations Framework Convention on Climate Change. Report of the Ministry of Ecology and Natural Sources of the Republic of Azerbaijan, Baku. 2010. 83 p
3. Alekperov A. Absheron: Problems of Hydrogeology and Geoecology. Baku: State Book Chamber of Azerbaijan; 2000. 484 p. (in Rus.)
4. Talibov A. Cartographic Analysis of Landscape-Ecological Conditions of the Absheron Peninsula. Baku: Chashioglu Publishing; 2004. 191 p. (in Rus.)
5. Isayev S, Babayev F, Ragimzadeh A, Sultanov R. Ecologic-Geochemical Assessment of Changes in Biosphere of the Absheron Peninsula. Baku: MBM Publishing House; 2007. 470 p. (in Rus.)
6. Khalilova H, Mamedov V. Assessing heavy metal pollution of sediments of the Boyuk-Shore lake in the Absheron industrial region of Azerbaijan. In: Proceedings of the 18th Conference "Engineering geology and geoecology. Fundamental problems and applied tasks" dedicated to 25th anniversary of foundation of Institute of Geoecology of Russian Academy of Sciences named after E. M. Sergeyev; 25-26 March 2006; Moscow. pp. 429-433 (in Rus.)
7. Aliyev M. Ecological Monitoring of the Caspian Sea Water Area. Baku: Tahsil Publishing; 2009. 290 p. (in Azerbaijani)
8. Reymers N. Ecology (Theory, Rules, Laws, Principles and Hypotheses). Moscow: Nauka; 1994. 367 p. (in Rus.)
9. Puschenreiter M, Horak F, Hartt W. Low cost agricultural measures to reduce heavy metal transfer into food chain, a review. Plant, Soil and Environment. 2005;51:1-11
10. Teplaya G. Heavy metals as a factor environmental pollution, a review. Bulletin of Environmental Education. 2013;1(23):182-192 (in Rus.)
11. Smith C, Hopmans P, Cook F. Accumulation of Cr, Pb, cu, Ni, Zn and cd in soil following irrigation with treated urban effluent in Australia. Environmental Pollution. 1996;94:317-323

12. Mammadov V. Lakes of Absheron Peninsula. National Atlas of the Republic of Azerbaijan. Cartography Factory: Baku; 2014. 187 p

13. Dobrovolskiy V. Textbook for the Students of Higher Education Institutions. Moscow: "Academy" Publishing House; 2003. 400 p. (in Rus.)

14. Osunkiyesi A, Taiwo A, Olawunmi O, et al. Environmental impact of heavy metal contaminants and micro nutrients in soil samples of metal dumpsites in Abeokuta, Ogunstate, Nigeria. IOSR Journal of Applied Chemistry. 2008;7(5):52-55

15. Samedov P, Bababekova L, Aliyeva V, Mamedzade V. Biological Characteristics of Technogenic Contaminated Soils. Baku: Elm Publishing; 2011. 104 p. (in Rus.)

16. Urushadze T, Ghambashidze G, Blum W, Mentler A. Soil contamination with heavy metals in Imereti region (Georgia). Bulletin of the Georgian National Academy of Sciences. 2007;175(1):122-130

17. Mirbabayev M. Microelement composition of oils according to neutron activation analysis. Chemistry and Technology of Fuels and Oils. 1997;5:46-47 (in Rus)

18. Qiu H. Migration mechanism of organic pollutants in national water-body sediments. Geography and Geology. 2011;1:239-246

19. Khalilova H, Mammadov V. Assessing the anthropogenic impact on heavy metal pollution in soils and sediments in urban areas of Azerbaijan's oil industrial region. Polish Journal of Environmental Studies. 2016;25(1):159-166. DOI: 10.152.44/pjoes/60723

20. Standard Methods for Examination of Water and Wastewater. 19th ed. Washington, DC: American Public Health Association/American Water Works Association/Water Environment Federation; 1995

21. Test Methods for Evaluating Solid Wastes. USEPA; 1986

22. Hu Y, Liu X, Bai J, et al. Assessing heavy metal pollution in the surface soils of a region that had undergone three decades of intense industrialization and urbanization. Environmental Science and Pollution Research. 2013;20:6150-6159. DOI: 10.1007/s11356-013-1668-z

23. Liu J, Zhuo Z, Sun S, et al. Concentrations of heavy metals in six municipal sludges in Guangzhou and their potential ecological risk assessment for agricultural land use. Polish Journal of Environmental Studies. 2015;24(1):165-174. DOI: 10.15244/pjoes/28348

24. Muller G. Schwermetallein den sediments des Rheins-VeraderungenSeitt. Umschan. 1979;79:778-783

25. Hakanson L. An ecological risk index for aquatic pollution control. A sedimentological approach. Water. Research. 1980;14(8):975-1001

26. Uzoekwe S, Achudume A. Pollution status and effect of crude oil spillage in Ughoton stream ecosystem in Niger Delta. Journal of Ecology and the Natural Environment. 2011;3(15):469-473. DOI: 10.5897/JENE11.071

27. Water Quality Criteria (WQC). A Report of the Committee on Water Quality Criteria. Washington, DC: NAS; 1972. 593 p

28. Report of Ministry of Emergency Situations on the Management of Lakes, Reservoirs, Surface and Underground Waters in the Absheron Peninsula. Baku; 2015. 263 s

21 Chapter

HEAVY METAL POLLUTION POSES A SIGNIFICANT THREAT TO BIODIVERSITY

1. INTRODUCTION: AN OVERVIEW

Natural sources and human activities can release heavy metals into the environment, with human activities being the primary contributors to these emissions. Out of all the activities mentioned, mining operations pose the most significant danger to the integrity of ecosystems. This is because heavy metals remain in the environment for hundreds of years even after mining operations have stopped [1]. These toxic substances have several ways of causing harm in environmental exposures. The impacts found in wild animal populations and communities, which have ramifications at the ecosystem level, are primarily caused by chronic exposures to low doses of complex combinations of metals [2]. Hence, this form of exposure poses a significant risk to biodiversity. Exposed individuals assimilate the presence of pollutants in their surroundings and exhibit discernible and anticipated reactions, which can be observed at several biological levels [3]. Therefore, in order to gain a deeper comprehension of the ecological impacts of metal exposure, it is imperative to utilize biological indicators or biomarkers. Biomarkers are instruments that facilitate the assessment of the degree of exposure and the impacts of environmental chemical contamination [4]. These measurements are excellent indicators of environmentally significant impacts. Nevertheless, in the field of ecotoxicology, which focuses on studying the impact of toxic chemicals on populations, communities, and ecosystems, utilizing biomarkers is a challenging endeavor. This is due to the fact that the physiological responses to chemical stress become less specific and are influenced by numerous variables. Within this particular framework, Bickham and colleagues elucidate that while xenobiotic exposure causes harm at the cellular or genetic levels, its impacts can also be recognized at higher levels of biological organization, known as emergent effects [5]. It is crucial to recognize that the biomarker response must be closely and consistently linked to reactions at higher levels, especially if the biomarkers are to be utilized as indications of effects [6]. In addition, it is necessary to utilize biomarkers of exposure such as external dose, internal dosage, and bioaccumulation levels, as well as biomarkers of biological effective dose like DNA adducts, and biomarkers of effect such as DNA breaks, in order to examine the correlation between cellular and genetic effects and ecological responses. When studying the emergent impacts, it is important to consider certain factors at the population level. Several effects of this phenomenon include changes in sex ratios, modifications in age structure, decreased reproductive success, inbreeding, alterations in genetic structure and diversity, reduced fitness, and population decreases [7]. Nevertheless, these impacts are not exclusive to metal exposures in the environment. Therefore, variations in the way biomarkers respond within different populations of a species can be thoroughly examined by considering geographical factors, environmental impacts, population susceptibility (unique to the particular population), and exposure history [6]. Over the past ten years, researchers have examined the impact of environmental exposure on populations and identified changes in their genetic makeup. These changes, referred to as permanent biomarkers by Mussali-Galante and colleagues, have been the focus of evaluation.

At the community level, the emergent consequences may include shifts in variety and species richness, changes in dominating species, alterations in species composition, and loss of biodiversity. Nevertheless, because of the intricate nature of species interactions, it is not possible to reliably forecast these effects based just on observations at the population level, as acknowledged several years ago [8, 9, 10]. Research examining the reactions of communities at the community level to metal stress in the environment is primarily carried out in aquatic environments, focusing on invertebrate and fish communities. Within the limited number of research undertaken in terrestrial environments, the focus has been on insect communities [11, 12]. These studies assess the levels of bioaccumulation in several groups of invertebrates and examine the relationship between bioaccumulation levels and community consequences, primarily species richness and composition. At the ecosystem level, the process of biomagnification, which refers to the accumulation of certain metals within successive trophic levels, has been extensively studied and documented. Observations have shown that individuals at lower trophic levels tend to be exposed to lower quantities of metals in the food chain. In such research, primary producers (plants) play a crucial role in the transfer of metals as they form the basis of the food chain. Therefore, specific metals have the ability to be carried from plants to higher levels of the food chain, posing a risk to biodiversity and the overall health of the ecosystem [13].

2. THE IMPACT OF HEAVY METAL POLLUTION ON POPULATIONS AND COMMUNITIES OF TERRESTRIAL WILD ANIMALS

Heavy metal (HM) exposure affect the health and survival of the individuals, resulting in negative impacts in the subsequent levels of biological organization, like populations and communities. The first step in ecotoxicological studies is to determine HM concentrations in soil and in the organism in question and then, to analyze the effects at the population level using emergent properties like male/female ratio, age class, reproductive success, inbreeding, genetic diversity and fitness. At the community level, species richness and diversity, dominant species, and species composition [7] are some of the characteristics that have been analyzed. In general, the Shannon-Wiener diversity index is one of the most used parameters in ecotoxicological studies. For a better understanding of the effects of HM on terrestrial ecosystems, the study of the functioning of detritivore soil communities [14, 15, 16] has been incorporated, where parameters such as biomass, soil organic matter content, microbial respiration, microbial biomass carbon, and the phosphatase activity have been analyzed.

2.1. Accumulation of heavy metals in invertebrates living on land

Heavy metals may enter the trophic chain trough primary producers and invertebrates that live in soils [17]. Invertebrate are widely used in ecotoxicological studies due to their easy capture, wide distribution, high abundance, their key ecological roles such as soil decomposers, constitute the first step in trophic chains, low mobility, and are in close contact with soils [18, 19, 20]. For example, Gramigni et al. detected a relationship between HM (Zn, Ni, Mn, Cd and Pb) in soils and their bioaccumulation in ants Crematogaster scutellaris [21].

In invertebrate communities HM bioaccumulation has been observed in target organs. For example: Wilczek and Babczyńska found that spiders (P. amentata, L. triangularis, M. segmentata, A. diadematus, A. marmoreus) had higher bioaccumulation levels of Cu, Zn and Cd in their hepatopancreas than in the gonads [22]. Also, Gramigni et al. documented that ants (Crematogaster scutellaris) bioaccumulate Mn and Zn in their intestines, being Zn accumulated specifically in Malpighi tubules and low Zn concentrations were find in fat tissue [21]. In contrast, Ni, Pb, and Cd did not bioaccumulate specifically in target organs.

Some of the most invertebrate Phyla used in ecotoxicological studies are Arthropoda and Annelida [18]. Bioaccumulation patterns depend on the species or taxonomic group in question (Class, Order, Family, Genera). For example, Wilczek and Babczyńska studied different spider species (Pardosa amentata, Linyphia triangularis, Metellina segmentata, Araneus diadematus, Araneus marmoreus) [22]. They found that the spider *P. amentata* bioaccumulates more Zn in the hepatopancreas than other spider species that inhabit in the same polluted site. Also, HM bioaccumulation differences have been observed in different taxonomic classes. It was evidenced that bioaccumulation of Cr, Cu, Ni, Pb and Zn in the order Araneida (Class Arachnida) were higher than those detected in for the order Coleoptera (Class Insecta) [17]. Moreover, they found an effect of HM bioaccumulation depending on the analyzed order, being the order Araneida who presented a positive correlation between Zn bioaccumulation in spiders and Zn concentrations in soils. In contrast, in the order Coleoptera a positive correlation was register between Cd and Pb bioaccumulation and soil concentrations, these differences may be attributed to feeding preferences between both orders (predators and herbivores).

2.2. The impact of heavy metal exposure on populations and communities of land-dwelling invertebrates

It has been documented that HM bioaccumulation in organisms may modify their body size. For example: Jones and Hopkin studied woodlice populations (Porcellio scaber) in polluted environments (Zn, Cd, Pb and Cu) [23]. The author reported that HM bioaccumulation had an effect on head size independently of the gender (male vs. female), being bigger in individuals from contaminated sites. They propose that environmental stressors generate costs in individuals because of the detoxification process, a fact that results in negative effects on their health and survival rates. Hence, HM exposure may have consequences in higher levels of biological organization such as populations and communities.

In ecotoxicology, gradient studies are necessary. They offer the visualization of gradual changes in HM soil concentrations in polluted sites and their relationship with population distribution and abundance and with some community structure parameters. Spurgeon and Hopkin found a negative correlation between the distance from the pollution source and HM concentrations (Pb, Cd, Zn and Cu) [24]. Also, a negative correlation was registered between HM concentration and the absolute abundance of six earthworm species (*Phyllum Annelida: L. rubellus, L. cataneus, L. terrestris, A. rosea, A. caliginosa, A. chlorotica*). Particularly, the species *L. rubellus, L. castaneus and L. terrestris* were present in most sites, whereas the species *A. rosea, A. caliginosa and A. chlorotica* were absent at the nearest sites to the pollution source. These differences were attributed to differences in calcium metabolism between earthworm species. Also, earthworm communities (*L. terrestres, L. rubellus, L. castaneus, A. chlorotica, A. rosea, A. caliginosa, A. longa, E. tetraedra, M. minúscula, O. cyaneum*) in a polluted site (Pb, Cd, Zn and Cu) were studied [25]. They found that the relative abundance and the Shannon-Weiner diversity index were lower on the nearest sites to the pollution source and higher in the farthest sites from the pollution source. However, the dominance index (Berger-Parker) was higher in sites near the pollution source because *L. rubellus, L. castaneus and L. terrestres* were the dominant species in those sites. The study by Jung et al. demonstrated that although the Shannon-Wiener diversity index values were similar between contaminated and less contaminated sites (Pb, Cd) for six different spider families, the abundance of the Linyphiidae family was correlated with metal concentrations in soils and enabled the discrimination between contaminated and less contaminated sites [12]. In contrast, others analyzed the community of the Phyla Annelida and Arthopoda, they found that absolute and relative abundances of all the organisms was lower in the most polluted sites [26]. Specifically, the earthworm species A. caliginosa had higher densities in non-polluted sites and was absent in polluted sites, on the contrary, the larvae of the coleopteran Hoplinae predominated in highly Zn polluted sites.

Also, there are studies that have shown that HM exposure does not affect some taxonomic groups. For example, Zaitsev and van Straalen studied the mite community (Phyllum Arthropoda) from contaminated soils (Pb, Zn, Cu, Fe, Cd) [19]. They evidenced that although a metal contamination gradient was found in the soils, this gradient was not detected in bioaccumulation of HM in mites, and no effects were found in the community structure. Likewise, Migliorini et al. did not find differences between the abundance of some arthropod groups (Collembola, Protura and Diplura), in contaminated sites by Pb and Sb, while other groups were absent (Symplyla) [20]. Hence, HM exposure affects differently the community structure of different invertebrate groups. Through the study of the functioning of detritivore soil communities [14, 15, 16] some parameters like biomass, soil organic matter content, microbial respiration, microbial biomass carbon, and the phosphatase activity are analyzed as biomarkers for HM effects at the community level. Hobbelen and colleagues studied millipedes, isopods, and earthworms in contaminated zones (Zn, Cu, Cd), where no correlation was found between community structure (richness and density) and soil metal content [14]. On the contrary, Zn concentration correlated positively with biomass of the earthworm *Lumbricus rubellus*. On the other hand, the soil organic matter content explained the variation in species density, showing that HM concentration in soils is not the only variable that influences the community structure.

The aforementioned studies evidenced that community structure and function of terrestrial invertebrates, facilitates the evaluation of HM impact on the first trophic chain levels, as well as their incorporation and biomagnification patterns. Therefore, studies assessing HM bioaccumulation in other trophic levels like terrestrial vertebrates complete the knowledge of the effects of HM in the ecosystem health.

2.3. The impact of heavy metal bioaccumulation on the health of land-dwelling vertebrates.

In wild vertebrates, information regarding HM bioaccumulation and their effects on target organs is vast. Some examples of wild vertebrate species used in ecotoxicological studies are: Brown bears (Ursus arctos), Gray wolfs (Canis lupus), Eurasian badgers (Meles meles) and Pine martens (Martes martes) [27], bank vole (Clethrionomys glareolus), yellow-necked mouse (Apodemus flavicollis) [28, 29, 30], wood mice (Apodemus sylvaticus) [31, 32], tuco-tuco (Ctenomys torquatus) [33], greater white-toothed (Crosidura russula) [34], Peromyscus melanophrys, pygmy mice (Baiomys musculus) [7, 35].

In general, studies on HM bioaccumulation on wild life, have detected an effect of the study species and the target organ on bioaccumulation patterns. For example, Bilandžić and collaborators analyzed HM bioaccumulation on wild carnivores, the authors report that the highest Cd concentration was present in kidney and liver of the Eurasian badger (M. meles) [27]. While Cu concentration in liver decreased among studied species showing the next pattern: Eurasian badger > Brown bear > Pine marten > Eurasian lynx > Gray wolf. The Eurasian badger registered the highest concentrations in muscles (As, Cu, Pb), liver (As, Cd, Cu, Pb) and kidneys (Cd, Pb) and the Pine marten accumulated the highest concentrations in kidneys (As, Cu, Hg). However, in the scientific literature, contrasting results about the effects of HM exposure on the health of the individuals are present. In this context, Levengood and Heske showed that white-footed mice that inhabit in a Cd and Zn polluted site, registered the highest Cd, Cu and Zn concentrations in liver, in comparison to unexposed individuals [36]. In spite of the bioaccumulation levels observed, they did not detect changes in the health of the individuals (reproductive and fitness parameters).

In contrast, other studies have detected histological changes in exposed individuals to HM. Damek-Poprawa and Sawicka-Kapusta found that yellow-necked mouse individuals that live in a polluted site, bioaccumulate more Pb and Cd that unexposed individuals [28]. In particular, individuals bioaccumulate Pb in their femur and Cd in kidneys. Also, histopathological studies showed that exposed individuals presented multiple organ alterations in liver, kidneys and testicles. Similar results were found for wood mice and the greater white-toothed living in landfill zones [31].

Studies assessing the effects of HM bioaccumulation on population and community parameters are scarce, a fact that may be attributed to sampling technique, which is influenced by the size and mobility of the individuals and trapping success (e.g. site perturbance, water availability, predator activity, migration index, etc.), among others. Therefore, in order to infer the population health status, some studies have considered the gender, age, (age class), reproductive condition, litter size (number of embryos, placental scars; embryos/scars per female) embryos weight, trap success, and condition index. In this regard, Santolo documented that male and female individuals of the deer mice exposed to Se from a contaminated site, had a lower condition Index than those from the reference site [37]. In addition, the ratio of males to females age class and reproductive condition were similar between individuals from both sites. Except from individuals from the polluted site that their reproductive condition was lower. This last result suggest that Se exposure affects negatively rodent populations, among other factors.

In addition, in terrestrial vertebrate communities, changes or alterations in community parameters may be due to the competitive selection of the most tolerant species. Moreover, some species may be opportunistic and HM residues may serve a protection mechanism against their predators [38].

Other parameters used as population level biomarkers are: residual index "RI" (linear regression between body weight and body length without tail) and kidney size proportion. RI is used as "energy reserve" measure. Individuals with positive RI values are considered as better fitted individuals, and the increase in kidney relative weight suggest the presence of a stressor [38].

2.4. The impact of heavy metal exposure on the genetic diversity of affected populations

Although HM exposure has immediate effects at the molecular and cellular levels, they may extend to higher levels of biological organization, like the genetic structure and diversity of the exposed populations [3]. Chronic exposures at low doses is one of the factor implicated in changes in the genetic pool of the populations, especially if chemical agents are capable of inducing DNA damage, such as HM. In general, there are four mechanisms by which HM exert their effects on the genetic diversity of exposed populations: (1) Some HM are genotoxic, mutagenic and alter DNA repair processes, increasing the mutational load of the individuals; (2) HM exposure favors the presence of tolerant genotypes and the elimination of intolerant ones, changing the genetic composition of the exposed population; (3) HM may cause bottlenecks and (4) alter migration patterns, increasing or decreasing genetic flow between populations [39, 40, 41].

Exposed population to HM pollution may have two types of response on genetic diversity levels: (a) increase in genetic diversity levels as a consequence of induced mutations by genotoxins or (b) decrease in genetic diversity levels as a result of bottlenecks [7]. In both cases, these responses are the consequence of the adaptation of the population to polluted environments [3, 41, 42, 43, 44].

In general, studies where 11 mammal species were analyzed for the effects of HM exposure (being Cd, Zn, Cu and Pb the most common) on genetic diversity, the pattern found was that 45.4% of the analyzed species displayed a decrease in genetic diversity levels in comparison to non-exposed populations. Some examples are: Peromyscus melanophrys: As, Pb, Cd, Cu, Zn [7], Cognettia sphagnetorum: Cu [45]; Talitrus saltator: Cd, Hg, Cu, [46]; Pachygrapsus marmoratus: As, Pb, Cd, Co [47]; Ficedula hypoleuca: Cd, Zn, Pb, Cu, Ni, Al, As, Cr, Se [48]. In contrast, an increase in genetic diversity levels in 36.4% of the analyzed species was found, like *Lumbricus rubellus*: Cd, Zn, Cu, Pb [49]; Cepaea nemoralis: Cd, Cr, Cu, Ni, Pb, Zn [50]; Parus major: Cd, Zn, Cu, Pb, Ni, Al, As, Cr, Sn [48] and Larus argentatus (steel), [51]. Meanwhile, the remaining 18.2% of the studied species did not register changes in genetic diversity levels, such as: Apodemos sylvaticus: Cd, Zn, Cu, Pb, Ni, Al, Ag, As, Co, Mn, Fe [42] and Succinea putris: Cd, Cr, Cu, Ni, Pb, Zn [50].

Changes in genetic diversity levels as a consequence of metal pollution may serve as a biomarker of permanent effects. Mussali-Galante et al. defined permanent biomarkers as changes in genetic structure and diversity due to metal pollution that cannot be the same as they were before the exposure [2].

Finally, polluted environments may be considered as unique systems because their different origin, pollution type and degree are specific for each site. Additionally, edaphic characteristics, weather and vegetation type differ between sites. Hence, it is expected that there will be differences among exposed sites that may or may not alter populations and communities; therefore, the use of bioindicator o sentinel species becomes important. Species that represent biological diversity in terms of feeding preferences, life cycles, trophic chain position, etc. are the point of interest. This last initiative permits to identify susceptible species to environmental stressors such as HM. Basu et al. suggested that sentinel species should have: wide geographical distribution, high abundance, capacity to bioaccumulate HM, easy capture and sampling, low mobility and well known biology [52]. In terrestrial environments, small mammals are commonly used because of their similar physiological systems to humans.

3. THE IMPACT OF HEAVY METAL POLLUTION ON ECOSYSTEMS

Ecosystems are open thermodynamic systems of matter and energy effluxes, which maintain stable from the balance of their biotic and abiotic components [53]. Ecosystem stability may be altered because of the incorporation of HM, derived from mining activities [54]. HM incorporation in the ecosystem depends mainly on their bioavailability and thereafter, through their incorporation into the trophic chain, reaching their highest concentrations in the last levels of the chain, a process known as "biomagnification" [55, 56]. Under this situation, ecosystems may be or may not be affected by HM, depending on the magnitude and exposure time, or if one of the functions that maintain the ecosystem integrity is compromised (e.g. nutrient cycles, energy efflux) due to biodiversity loss [57].

Many ecotoxicological studies that assess HM effects on terrestrial ecosystems have focused on the analysis of HM concentrations in soils, their bioavailability and bioaccumulation, but few have analyzed their biomagnification trough the trophic chain and their effects on ecosystem integrity [54].

3.1. Heavy metal incorporation into the ecosystems

The first step of HM incorporation into the ecosystems is because their bioavailability potential and soil mobility, where metallic cations adhere to negative charged particles like clay and organic matter, when metals separate from these soil particles, they enter the soluble soil fraction, being bioavailable and having the potential to bioaccumulate in different organisms [56]. Microorganisms, plants and invertebrate species have mechanisms to incorporate trace metals for their development and survival (e.g. Cu, Ni, Fe, Co, Mn and Mg), however, these can be toxic in higher concentrations. Also, these same mechanisms facilitate the entrance of non-trace metals (As, Cd, Hg and Pb) in the organisms, which are highly toxic at low concentrations [58].

Microorganism are vital elements of soils, they participate in nutrient and inorganic element recycling, like minerals and trace metals, for plants that constitute the first trophic level in terrestrial ecosystems [59]. However, HM pollution may affect microorganism communities, generating changes in their structure and biodiversity, which in turn, has consequences on the soil processes in which they participate [60, 61]. For example, development alterations and in biochemical processes of microorganism have been reported [61, 62, 63]. Such alterations affect organic matter decomposition process, reducing nutrient accumulation and availability for plants and compromising matter and energy fluxes at the base of the trophic chains [61]. On the other hand, soil invertebrates can bioaccumulate HM because of their feeding preferences, like crustaceans, snails, and earthworms that inhabit leaf litter and feed on organic matter with high HM concentrations [64, 65]. In fact, it has been proven that these invertebrate groups had the highest HM concentrations in comparison to beetles or butterflies, or even higher concentrations than some vertebrate groups [18, 64].

3.2. Heavy metal transfer along food chain

In particular, it has been suggested that HM hyperaccumulation by plants as a defense mechanism against herbivores, may transcend to higher trophic levels. For example, the plant Streptanthus polygaloides (Brassicaceae) is a Ni hyperaccumulator species, that is consumed by the herbivore Melanotrichus boydi (Hemiptera) who bioaccumulates Ni in its body as a defense mechanism against the predator species Misumena vatia (Araneae) [66].

HM transfer along the trophic chains varies depending on the type of metal, the trophic level in question and the number and type of species that integrate it. For example, [18] report that Cd, Cu, Pb and Zn concentrations among invertebrate groups registered the next pattern: Isopoda > Lumbricidae > Coleoptera which is attributed to their different feeding patterns [57]. Another interesting example is that Cd is more mobile towards herbivores and their predators, while Zn is less efficient in its transfer to higher trophic levels [67, 68].

HM transfer along the trophic chain has been reported, for example for the Ni hyperaccumulator plant Alyssum pintodalsilvae (Brassicaceae), that transfers Ni to grasshoppers (herbivore) and spiders (carnivorous insect), having the spiders higher Ni concentrations [69]. Similar results were reported by Boyd and Wall [66]. These studies demonstrate that HM can be transferred among invertebrate species, mobilizing metals from one trophic level to another, reaching animals such as small mammals [70, 71, 72]. These last studies evaluated HM concentration in small mammals, reporting higher HM levels in carnivorous or omnivorous mammals in comparison to those that feed only by plants. So, HM transfer along the trophic chain not only depends on the magnitude of exposure, but on the species type, season, gender, age and metal type [73].

Additionally, HM bioaccumulation in plants may also affect interactions with their pollinators, since HM can transfer to nectar, a fact that alters pollinators feeding patterns, suggesting that metals and metalloids such as Se found in pollen and nectar affect negatively the pollinators, which results in changes in plant communities due to the nonappearance of pollinators on such plants [74, 75, 76, 77].

Most of the studies about metal transfer along trophic chains in terrestrial ecosystems focus on at least three trophic chains levels. In contrast, a study by Hsu and collaborators in more than three trophic levels, they report high levels of Cd, Hg, Pb and Sn and their biomagnification in all analyzed levels, being the snails and the earthworms the groups who registered the highest metal concentrations [64].

All the aforementioned studies have evaluated HM transfer in small trophic chains in terrestrial ecosystems, the majority of them analyze three trophic levels. These studies are of great importance because the information generated helps to understand the general patterns of HM transfer along trophic chains, especially for the most common metals like Cd, Cu, Pb and Zn. Also, these studies highlight that HM transfer, assimilation and excretion in organisms along the trophic chain, can have extended effects, at the individual level (altering their health and physiology) at the population level (modifying population dynamics, abundance, distribution and their genetic pool) at the community level (altering species richness and diversity), affecting then, the ecosystem dynamics. Therefore, is very important to conduct studies where more than three trophic chain levels are analyzed and to integrate new biomarkers (e.g. stable isotope techniques; which enable to follow HM transfer along trophic chains by knowing the extent of the pollutant flux in the chain [78]).

4. CONCISE SUMMARY

Prolonged exposure to environmental metals has detrimental impacts on individuals' health, with implications for the well-being of populations and communities, and posing a threat to the integrity of ecosystems. Nevertheless, the identification and application of biomarkers in the field of ecotoxicology has proven to be a challenging endeavor, mostly because of the non-specific reactions and numerous factors that influence physiological responses to toxic stress. Consequently, it is imperative for ecotoxicological studies to incorporate the following elements: measurements of heavy metal concentrations in soils, assessment of bioaccumulation parameters in both vertebrate and invertebrate species, and examination of the correlation between these biomarkers and any morphological, anatomical, or physiological changes that could potentially impact population dynamics. Specifically, the utilization of bioindicator or sentinel species is essential to demonstrate the effects of heavy metal exposure in natural populations. Terrestrial invertebrates have been utilized as a suitable paradigm for assessing community reactions to environmental chemical stress. This is because they are easily captured, widely distributed, very abundant, have limited mobility, and are in close proximity to heavy metals found in soils. Earthworms and arthropods are the organisms that have been extensively investigated. However, there is a limited number of research that assess the impact of HM on the organization of vertebrate communities. This scarcity is likely attributed to the challenges posed by the size, mobility, and sampling methods of these organisms. Nevertheless, when conducting research on vertebrates, a highly effective alternative has been the examination of small mammal species that function as reliable bioindicators, allowing for convenient comparisons to be made with humans. Furthermore, a commonly employed approach in numerous research involves utilizing pollution gradients to visually represent subtle variations in concentrations of heavy metals (HMs) across different soil levels. These variations are then correlated with other characteristics related to community structure. Currently, we can deduce that HM have varying effects on the community structure and functioning of the many animal groups that have been investigated thus far. At the local level, the ongoing pursuit of novel biomarkers persists. In this particular situation, variations in the abundance of distinct groups within the community might also function as a biomarker. This is because changes in abundance or the disappearance of some groups in communities that have been exposed to chemicals can indicate an ecological response to the chemical stress. Ecotoxicological research are significantly lacking at the ecosystem level. Reports have indicated changes in trophic chains, biomagnification, and adjustments in food and energy cycles. Studies typically evaluate the transfer of heavy metals (HM) across three levels of a food chain. These studies have determined that the movement of metals is influenced by the biological characteristics of the species involved, their location in the food chain, and the specific type or combination of metals being studied. The primary focus of research has been on the transfer of harmful substances from plants to herbivorous invertebrates (insects), and from insects to other invertebrates (spiders) or predatory vertebrates (small mammals). The data from these research has garnered significant attention, particularly due to the fact that human people occupy the highest position in the trophic hierarchy, as seen in agroecosystems. Using stable carbon and nitrogen isotopes as biomarkers is preferable for assessing the transport of heavy metals along terrestrial trophic chains in ecosystem research. Future initiatives must incorporate diverse biological and ecological responses at all levels of biological organization through the use of biomarker techniques. Furthermore, it is imperative to enhance the rigor of study designs by incorporating several species and biomarkers. This will enable a more accurate assessment of heavy metal (HM) exposure, leading to improved prediction, comprehension, and resolution of global HM pollution issues.

REFERENCES

1. Duruibe JO, Ogwuegbu MOC, Egwurugwu JN. Heavy metal pollution and human biotoxic effects. International Journal of Physical Sciences. 2007;2:112-118
2. Mussali-Galante P, Tovar-Sánchez E, Valverde M, Rojas E. Genetic structure and diversity of animal populations exposed to metal pollution. Reviews of Environmental Contamination and Toxicology. 2014;227:79-106. DOI: 10.1007/978-3-319-01327-5_3
3. Bickham J, Sandhu S, Hebert P, Chikhi L, Athwal R. Effects of chemical contaminants on genetic diversity in natural populations: Implications for biomonitoring and ecotoxicology. Mutation Research. 2000;463:33-51. DOI: 10.1016/S1383-5742(00)00004-1
4. Handy R, Galloway T, Depledege M. A proposal for the use of biomarkers for the assessment of chronic pollution and in regulatory toxicology. Ecotoxicology. 2003;12:331-343. DOI: 10.1023/A:1022527432252
5. Bickham J, Smolen M. Somatic and heritable effects of environmental genotoxins and the emergence of evolutionary toxicology. Environmental Health Perspectives. 1994;102:25-28
6. Forbes VE, Annemette P, Lis B. The use and misuse of biomarkers in ecotoxicology. Environmental Toxicology and Chemistry. 2006;25:272-280. DOI: 10.1897/05-257R.1
7. Mussali-Galante P, Tovar-Sánchez E, Valverde M, Rojas E. Biomarkers of exposure for assessing environmental metal pollution: From molecules to ecosystems. Revista Internacional de Contaminación Ambiental. 2013;29:117-140
8. Forbes T, Forbes V. A critique of the use of distribution-based extrapolation models in ecotoxicology. Functional Ecology. 1993;7:249-254. DOI: 10.2307/2390202
9. Hopkin S. Ecological implications of 95% protection levels for metals in soil. Oikos. 1993;66:137-141. DOI: 10.2307/3545206
10. Lagadic L, Caquet T, Ramade F. The role of biomarkers in environmental assessment. Invertebrate populations and communities. Ecotoxicology. 1994;3:193-208. DOI: 10.1007/BF00117084
11. Nahmani J, Lavelle P. Effects of heavy metal pollution on soil macrofauna in a grassland of northern France. Ecotoxicology and Environmental Safety. 2012;86:188-197. DOI: 10.1016/j.ecoenv.2012.09.013
12. Jung MP, Kim ST, Kim H, Lee J. Species diversity and community structure of ground-dwelling spiders in unpolluted and moderately heavy metal-polluted habitats. Water, Air, and Soil Pollution. 2008;195:15-22. DOI: 10.1007/s11270-008-9723-y
13. Peralta-Videa J, Laura Lopez M, Narayan M, Saupe G, Gardea-Torresdey J. The biochemistry of

environmental heavy metal uptake by plants: Implications for the food chain. International Journal of Biochemistry & Cell Biology. 2009;41:1665-1677. DOI: 10.1016/j.biocel.2009.03.005

14. Hobbelen HF, van de Brinkb PJ, Hobbelena JF, van Gestela CA. Effects of heavy metals on the structure and functioning of detritivore communities in a contaminated floodplain area. Soil Biology and Biochemistry. 2006;38:1596-1607. DOI: 10.1016/j.soilbio.2005.11.010

15. van Gestel CA, Koolhaas JE, Hamers T, van Hoppe M, van Roovert M, Korsman C, Reinecke SA. Effects of metal pollution on earthworm communities in a contaminated floodplain area: Linking biomarker, community and functional responses. Environmental Pollution. 2009;157:895-903. DOI: 10.1016/j.envpol.2008.11.002

16. Niemeyer JC, Nogueira MA, Carvalho GM, Cohin-De-Pinho SJ, Outeiro US, Rodrigues GG, da Silva EM, Sousa JP. Functional and structural parameters to assess the ecological status of a metal contaminated area in the tropics. Ecotoxicology and Environmental Safety. 2012;86:188-197. DOI: 10.1016/j.ecoenv.2012.09.013

17. Schipper AM, Wijnhoven S, Leuven RS, Ragas AM, Hendriks AJ. Spatial distribution and internal metal concentrations of terrestrial arthropods in a moderately contaminated lowland floodplain along the Rhine River. Environmental Pollution. 2008;151:17-26. DOI: 10.1016/j.envpol.2007.03.007

18. Heikens A, Peijnenburg WJ, Hendriks AJ. Bioaccumulation of heavy metals in terrestrial invertebrates. Environmental Pollution. 2001;113:385-393. DOI: 10.1016/S0269-7491(00)00179-2

19. Zaitsev AS, van Straalen NM. Species diversity and metal accumulation in oribatid mites (Acari, Oribatida) of forests affected by a metallurgical plant. Pedobiologia. 2001;45:467-479. DOI: 10.1078/0031-4056-00100

20. 20.Migliorini M, Pigino G, Bianchi N, Bernini F, Leonzio C. The effects of heavy metal contamination on the soil arthropod community of a shooting range. Environmental Pollution. 2004;129:331-340. DOI: 10.1016/j.envpol.2003.09.025

21. Gramigni E, Calusi N, Gelli L, Giuntini M, Massi G, Delfino G, Chelazzi D, Barachi D, Frizzi F, Santini G. Ants as bioaccumulators of metals from soils: Body content and tissue-specific distribution of metals in the ant Crematogaster scutellaris. European Journal of Soil Biology. 2013;58:24-31. DOI: 10.1016/j.ejsobi.2013.05.006

22. Wilczek G, Babczyńska A. Heavy metals in the gonads and hepatopancreas of spiders (Araneae) from variously polluted areas. Ekológia (Bratislava). 2000;19:283-292

23. Jones DT, Hopkin SP. Reduced survival and body size in the terrestrial isopod Porcellio scaber from a metal-polluted environment. Environmental Pollution. 1998;99:215-223. DOI: 10.1016/S0269-7491(97)00188-7

24. Spurgeon DJ, Hopkin SS. The effects of metal contamination on earthworm populations around a smelting works: Quantifying species effects. Applied Soil Ecology. 1996;4:147-160. DOI: 10.1016/0929-1393(96)00109-6

25. Spurgeon DJ, Hopkin SP. Seasonal variation in the abundance, biomass and biodiversity of earthworms in soils contaminated with metal emissions from a primary smelting works. Journal of Applied Ecology. 1999;36:173-183. DOI: 10.1046/j.1365-2664.1999.00389.x

26. Nahmani J, Lavelle P. Effects of heavy metal pollution on soil macrofauna in a grassland of northern France. European Journal of Soil Biology. 2002;38:297-300. DOI: 10.1016/S1164-5563(02)01169-X

27. Bilandžić N, Dežđek D, Sedak M, Đokić M, Simić B, Rudan N, Brstilo M, Lisicin T. Trace elements in

tissues of wild carnivores and omnivores in Croatia. Bulletin of Environmental Contamination and Toxicology. 2012;88:94-99. DOI: 10.1007/s00128-011-0449-y

28. Damek-Poprawa M, Sawicka-Kapusta K. Damage to the liver, kidney, and testis with reference to burden of heavy metals in yellow-necked mice from areas around steelworks and zinc smelters in Poland. Toxicology. 2003;186:1-10. DOI: 10.1016/S0300-483X(02)00595-4

29. Metcheva R, Teodorova S, Topashka-Ancheva M. A comparative analysis of the heavy metal loading of small mammals in different regions of Bulgaria I: Monitoring points and bioaccumulation features. Ecotoxicology and Environmental Safety. 2003;54:176-187. DOI: 10.1016/S0147-6513(02)00052-0

30. Al Sayegh PS, Kopušar N, Kryštufek B. Small mammals as biomonitors of metal pollution: A case study in Slovenia. Environmental Monitoring and Assessment. 2014;186:4261-4274. DOI: 10.1007/s10661-014-3696-7

31. 31.Sánchez-Chardi A, Peñarroja-Matutano C, Borrás M, Nadal J. Bioaccumulation of metals and effects of a landfill in small mammals. Part III: Structural alterations. Environmental Research. 2009;109:960-967. DOI: 10.1016/j.envres.2009.08.004

32. Tête N, Durfort M, Rieffel D, Scheifler R, Sánchez-Chardi A. Histopathology related to cadmium and lead bioaccumulation in chronically exposed wood mice, Apodemus sylvaticus, around a former smelter. Science of the Total Environment. 2014;481:167-177. DOI: 10.1016/j.scitotenv.2014.02.029

33. da Silva J, de Freitas TR, Heuser V, Marinho JR, Bittencourt F, Cerski CT, Kliemann LM, Erdtmann B. Effects of chronic exposure to coal in wild rodents (Ctenomys torquatus) evaluated by multiple methods and tissues. Mutation Research. 2000;470:39-51. DOI: 10.1016/S1383-5718(00)00094-2

34. Sánchez-Chardi A, Nadal J. Bioaccumulation of metals and effects of landfill pollution in small mammals. Part I. The greater white-toothed shrew, Crocidura russula. Chemosphere. 2007;68:703-711. DOI: 10.1016/j.chemosphere.2007.01.042

35. Tovar-Sánchez E, Cervantes T, Martínez C, Rojas E, Valverde M, Ortiz-Hernández L, Mussali-Galante P. Comparison of two wild rodent species as sentinels of environmental contamination by mine tailings. Environmental Science and Pollution Research. 2012;19:1677-1686. DOI: 10.1007/s11356-011-0680-4

36. Levengood JM, Heske EJ. Heavy metal exposure, reproductive activity, and demographic patterns in white-footed mice (Peromyscus leucopus) inhabiting a contaminated floodplain wetland. Science of the Total Environment. 2008;389:320-328. DOI: 10.1016/j.scitotenv.2007.08.050

37. Santolo GM. Small mammals collected from a site with elevated selenium concentrations and three reference sites. Archives of Environmental Contamination and Toxicology. 2009;57:741-754. DOI: 10.1007/s00244-009-9322-y

38. Sánchez-Chardi A, Peñarroja-Matutano C, Ribeiro CA, Nadal J. Bioaccumulation of metals and effects of a landfill in small mammals. Part II. The wood mouse, Apodemus sylvaticus. Chemosphere. 2007;70:101-109. DOI: 10.1016/j.chemosphere.2007a.06.047

39. Van Straalen N. Genetic biodiversity in toxicant-stressed populations. Progress in Environmental Science. 1999;1:195-201

40. Van Straalen N, Timmermans M. Genetic variation in toxicant stressed populations: An evaluation of the "genetic erosion" hypothesis. Human and Ecological Risk Assessment. 2002;8:983-1002. DOI: 10.1080/1080-700291905783

41. Maes GE, Raeymaekers JAM, Pampoulie C, Seynaeve A, Goemans G, Belpaire C, Volckaert FAM. The catadromous European eel Anguilla anguilla (L.) as a model for freshwater evolutionary ecotoxicology: Relationship between heavy metal bioaccumulation, condition and genetic variability. Aquatic Toxicology. 2005;73:99-114. DOI: 10.1016/j.aquatox.2005.01.010

42. Berckmoes V, Scheirs J, Jordaens K, Blust R, Backeljau T, Verhagen R. Effects of environmental pollution on microsatellite DNA diversity in wood mouse (Apodemus Sylvaticus) populations. Environmental Toxicology and Chemistry. 2005;24:2898-2907. DOI: 10.1897/04-483R.1

43. Gardeström J, Dahl U, Kotsalainen O, Maxson A, Elfwing T, Grahn M, Bengtsson B, Breitholtz M. Evidence of population genetic effects of long-term exposure to contaminated sediments: A multiendpoint study with copepods. Aquatic Toxicology. 2008;86:426-436. DOI: 10.1016/j.aquatox.2007.12.003. Get rights and content

44. Durrant J, Stevens JR, Hogstrand C, Burya NR. The effect of metal pollution on the population genetic structure of brown trout (*Salmo trutta* L.) residing in the river Hayle, Cornwall, UK. Environmental Pollution. 2011;59:3595-3603. DOI: 10.1016/j.envpol.2011.08.005

45. Haimi J, Knott K, Selonen S, Laurikainen M. Has long-term metal exposure induced changes in life history traits and genetic diversity of the enchytraeid worm Cognettia sphagnetorum? (Vejd). Environmental Pollution. 2006;140:463-470. DOI: 10.1016/j.envpol.2005.08.009

46. Ungherese G, Mengoni A, Somigli S, Baroni D, Focardi S, Ugolini A. Relationship between heavy metals pollution and genetic diversity in Mediterranean population of the sandhopper Talitrus saltator (Montagu) (Crustaceae, Amphipoda). Environmental Pollution. 2010;158:1638-1643

47. Fratini S, Zane L, Ragionieri L, Vannini M, Cannicc S. Relationship between heavy metal accumulation and genetic variability decrease in the intertidial crab Pachygrapsus marmoratus (Decapoda; Grapsidae). Estuarine, Coastal and Shelf Science. 2008;79:679-686. DOI: 10.1016/j.ecss.2008.06.009

48. Eeva T, Belskii E, Kuranov B. Environmental pollution affects genetic diversity in wild bird populations. Mutation Research. 2006;608:8-15. DOI: 10.1016/j.mrgentox.2006.04.021

49. Peles JD, Towler WI, Guttman SI. Population genetic structure of earthworms (Lumbricus rubelluz) in soils contaminated by heavy metals. Ecotoxicology. 2003;12:379-386. DOI: 10.1023/A:1026269804938

50. Jordaens K, De Wolf H, Van Houtte N, Vandecasteele B, Backeljau T. Genetic variation in two land snails Cepaea nemoralis and Succinea putris (Gastropoda), from sites differing in heavy metal content. Genetica. 2006;128:227-239. DOI: 10.1007/s10709-005-5705-9

51. 51.Yauk C, Fox G, McCarry B, Quinn J. Induced minisatellite germline mutations in herring gulls (Larus argentatus) living near steel mills. Mutation Research. 2000;452:211-218. DOI: 10.1016/S0027-5107(00)00093-2

52. Basu N, Scheuhammer AM, Bursian SJ, Elliott J, Rouvinen-Watt K, Chan HM. Mink as a sentinel species in environmental health. Environmental Research. 2007;103:130-144. DOI: 10.1016/j.envres.2006.04.005

53. Maintaining NR. Ecological integrity of landscape and eco-region. In: Noss RF, editor. Ecological Integrity: Integrating Environmental, Conservation and Health. Washington, DC: Island Press; 2001. pp. 191-208

54. Gall JE, Boyd RS, Rajakaruna N. Transfer of heavy metals through terrestrial food webs: A review. Environmental Monitoring and Assessment. 2015;187:201

55. Hossain MA, Piyatida P, Teixeria da Silva JA, Fujita M. Molecular mechanism of heavy metal toxicity and tolerance in plants: central role of glutathione in detoxification of reactive oxygen species and methylglyoxal and in heavy metal chelation. Journal of Botany. 2012;2012:37. DOI: 10.1155/2012/872875

56. Kim RY, Yoon JK, Kim TS, Yang JE, Owens G, Kim KR. Bioavailability of heavy metals in soils: Definitions and practical implementation—A critical review. Environmental Geochemistry and Health. 2015;37:1041. DOI: 10.1007/s10653-015-9695-y

57. Mann RM, Vijver MG, Peijnenburg WJGM. Metals and metalloids in terrestrial systems: Bioaccumulation, biomagnification and subsequent adverse effects. In: Sánchez-Bayo PJ, van den Brink PJ, Mann RM, editors. Ecological Impacts of Toxic Chemicals. Sharjah: Bentham; 2011. pp. 43-62

58. Gadd GM. Metals, minerals and microbes: Geomicrobiology and bioremediation. Microbiology. 2010;156:609-643. DOI: 10.1099/mic.0.0371430

59. Harris J. Soil microbial communities and restoration ecology: Facilitators or followers? Science. 2009;325:573-574. DOI: 10.1126/science.1172975

60. Giller KE, Witter E, McGrath SP. Heavy metals and soil microbes. Soil Biology and Biochemistry. 2009;41:2031-2037. DOI: 10.1016/j.soilbio.2009.04.026

61. Boshoff M, De Jonge M, Dardenne F, Blust R, Bervoets L. The impact of metal pollution on soil faunal and microbial activity in two grassland ecosystems. Environmental Research. 2014;134:169-180. DOI: 10.1016/j.envres.2014.06.024

62. Smejkalova M, Mikanova O, Boruvka L. Effects of heavy metal concentrations on biological activity of soil micro-organisms. Plant, Soil and Environment. 2003;49:321-326. DOI: 10.1.1.557.375

63. Abdousalam A. Effect of heavy metals on soil microbial processes and population. Egyptian Academic Journal of Biological Sciences. 2010;2:9-14

64. Hsu MJ, Selvaraj K, Agoramoorthy G. Taiwan's industrial heavy metal pollution threatens terrestrial biota. Environmental Pollution. 2006;43:327-334. DOI: 10.1016/j.envpol.2005.11.023

65. Rogival D, Scheirs J, Blust R. Transfer and accumulation of metals in a soil–diet–wood mouse food chain along a metal pollution gradient. Environmental Pollution. 2007;145:516-528. DOI: 10.1016/j.envpol.2006.04.019

66. Boyd RS, Wall MA. Responses of generalist predators fed high-Ni Melanotrichus boydi (Heteroptera: Miridae): Elemental defense against the third trophic level. The American Midland Naturalist. 2001;146:186-198. DOI: 10.1674/0003-0031(2001)146[0186:ROGPFH]2.0.CO;2

67. Green ID, Diaz A, Tibbett M. Factors affecting the concentration in seven-spotted ladybirds (*Coccinella septempunctata* L.) of Cd and Zn transferred through the food chain. Environmental Pollution. 2010;158:135-141. DOI: 10.1016/j.envpol.2009.07.032

68. Dar MI, Khan FA, Green ID, Naikoo MI. The transfer and fate of Pb from sewage sludge amended soil in a multi-trophic food chain: A comparison with the labile elements Cd and Zn. Environmental Science and Pollution Research. 2015;22:16133-16142. DOI: 10.1007/s11356-015-4836-5

69. Peterson LR, Trivett V, Baker AJM, Aguiar C, Pollard AJ. Spread of metals through an invertebrate food chain as influenced by a plant that hyperaccumulates nickel. Chemoecology. 2003;13:103-108. DOI: 10.1007/s00049-003-0234-4

70. Hamers T, van den Berg JHJ, van Gestel CAM, van Schooten FJ, Murk AJ. Risk assessment of metals and organic pollutants for herbivorous and carnivorous small mammal food chains in a polluted floodplain (Biesbosch, The Netherlands). Environmental Pollution. 2006;144:581-595. DOI: 10.1016/j.envpol.2006.01.020

71. Veltman K, Huijbregts MAJ, Hamers T, Wijnhoven S, Hendriks AJ. Cadmium accumulation in herbivorous and carnivorous small mammals: Meta-analysis of field data and validation of the bioaccumulation model optimal modeling for ecotoxicological applications. Environmental Toxicology and Chemistry. 2007;26:1488-1496. DOI: 10.1897/06-518R.1

72. Wijnhoven S, Leuven RSEW, Van der Velde G, Jungheim G, Koelemij EI, de Vries FT, Eijsackers HJP, Smits AJM. Heavy-metal concentrations in small mammals from a diffusely polluted floodplain: Importance of species- and location-specific characteristics. Archives of Environmental Contamination and Toxicology. 2007;52:603-613

73. Van den Brink N, Lammertsma D, Dimmers W, Boerwinkel MC, Van der Hout A. Effects of soil properties on food web accumulation of heavy metals to the wood mouse (Apodemus sylvaticus). Environmental Pollution. 2010;158:245-251. DOI: 10.1016/j.envpol.2009.07.013

74. Hladun KR, Parker DR, Trumble JT. Selenium accumulation in the floral tissues of two *Brassicaceae* species and its impact on floral traits and plant performance. Environmental and Experimental Botany. 2011;74:90-97. DOI: 10.1016/j.envexpbot.2011.05.003

75. Quinn CF, Prins CN, Freeman JL, Gross AM, Hantzis LJ, Reynolds RJB, Yang S, Covey PA, Bañuelos GS, Pickering IJ, Fakra SC, Marcus MA, Arathi HS, Pilon-Smits EAH. Selenium accumulation in flowers and its effects on pollination. The New Phytologist. 2011;192:727-737. DOI: 10.1111/j.1469-8137.2011.03832.x

76. Meindl GA, Ashman TL. The effects of aluminum and nickel in nectar on the foraging behavior of bumblebees. Environmental Pollution. 2013;177:78-81. DOI: 10.1016/j.envpol.2013.02.017

77. Meindl GA, Bain DJ, Ashman TL. Nickel accumulation in leaves, floral organs and rewards varies by serpentine soil affinity. AoB Plants. 2014;6:1-9. DOI: 10.1093/aobpla/plu036

78. Croteau MN, Luoma SN, Stewart AR. Trophic transfer of metals along freshwater food webs: Evidence of cadmium biomagnification in nature. Limnology and Oceanography. 2005;50:1511-1519. DOI: 10.4319/lo.2005.50.5.151

INDEX

www.ingramcontent.com/pod-product-compliance
Lightning Source LLC
LaVergne TN
LVHW062145200926
848484LV00007B/1862
* 9 7 8 1 7 8 7 1 5 3 2 1 9 *